Lecture Notes in Mathematics

Edited by A. Dold and B. Eckm

T0215488

561

Function Theoretic Methods for Partial Differential Equations

Proceedings of the International Symposium Held at Darmstadt, Germany, April 12–15, 1976

Edited by
V. E. Meister, N. Weck and W. L. Wendland

Springer-Verlag
Berlin · Heidelberg · New York 1976

Editors
V. Erhard Meister
Wolfgang L. Wendland
Fachbereich Mathematik
Technische Hochschule Darmstadt
Schloßgartenstraße 7
6100 Darmstadt/BRD

Norbert Weck
Fachbereich Mathematik
Universität Essen
Gesamthochschule
Universitätsstraße 2
4300 Essen/BRD

AMS Subject Classifications (1970): 30A24, 30A92, 30A94, 30A96, 30A97, 32D99, 35A20, 35C15, 35E05, 35Q05, 35Q15, 35J05, 35J15, 35J45, 35J65, 45M05, 35R25, 35R30, 45E05, 45E99, G05, 78A45

ISBN 3-540-08054-6 Springer-Verlag Berlin · Heidelberg · New York
ISBN 0-387-08054-6 Springer-Verlag New York · Heidelberg · Berlin

Printing and binding: Beltz Offsetdruck, Hemsbach/Bergstr.

P R E F A C E

These Proceedings form a record of the lectures delivered to the International Symposium on Function-Theoretic Methods for Partial Differential Equations held at the Technische Hochschule Darmstadt, Germany, April 12-15, 1976. The volume includes also a few papers from authors who were unable to attend the conference.

The conference was attended by about loo mathematicians from the following countries: Austria, Belgium, Canada, Czechoslovakia, Finland, France, Germany, Israel, Italy, Libya, The Netherlands, Poland, Rumania, Union of Soviet Socialist Republics, United Kingdom, United States of America.

Due to the large number of non German speaking participants the editors decided to publish the Proceedings with an English title and to include this English preface. One of the main objects of this conference was to bring together mathematicians from different places working on this special field. The organizers were very happy to see that participants from so many different countries, among them some of the original inventors of function-theoretic methods, took part. Besides the more classical topics of analytic and generalized analytic functions in elliptic partial differential equations, the conference showed also the strong influence of singular integral equations on the function-theoretic methods, and moreover extensions of these methods to non-elliptic problems as well as to higher dimensional problems. Many neighboring areas, including fields of applications, will probably be influenced by these methods.

The organizers take this opportunity to thank all mathematicians who took part in the work of the conference for their contributions. They thank the "Deutsche Forschungsgemeinschaft" (German Research Council) for the main financial support provided by admitting the conference to its special program of "Internationale Fachtagungen" (International Specialist Symposia). Thanks are also due to the German Academic Exchange Service (DAAD) which provided financial support for the travel expenses of colleagues from Eastern European Countries. Moreover, the organizers would like to express their warmest thanks to the Society of Applied Mathematics and Mechanics (GAMM), especially to its president, Professor E. Becker, for his enduring encouragement and for sponsoring this kind of activity for the GAMM-committee on Applied Analysis and Mathematical Physics.

The organizers thank the Technische Hochschule Darmstadt (Technical University Darmstadt) and the Hessian Ministry of Cultural Affairs for providing facilities and for covering the material expenses. They express their thanks to many members of the University for freely offering their help and advice, and to colleagues in the Department of Mathematics, especially to the members of the Organizing Committee. Finally thanks are due to Mrs. Abou El-Seoud, Mrs. Beltzig and Mrs. Karl, Secretaries in the Department of Mathematics, for considerable assistance in the preparation of the papers for the conference and many of the manuscripts of these Proceedings.

E. Meister, N. Weck, W. Wendland

Organizers and Editors

C O N T E N T S

I. 45 minutes Lectures

45 minutes Lectures

30 minutes Lectures

30 minutes Lectures

30 minutes Lectures

30 minutes Lectures

Papers submitted to the editors which could not be read by the authors

List of speakers and those who submitted manuscripts

"Symposium on Function Theoretic Methods
for Partial Equations"

K. W. Bauer, Prof. Dr.,	Technische Hochschule Graz, I. Lehrkanzel und Inst. für Mathem., Kopernikusgasse 24 A-8010 Graz, Austria
H. Begehr, Prof. Dr.,	FU Berlin, Fachbereich Mathematik, Hüttenweg 9, 1000 Berlin 33, Germany
St. Bergman, Prof. Dr.,	Stanford University, Department of Mathematics, Stanford, Cal. 94305, USA
A.V. Bitsadze, Prof. Dr.,	Steklov Institute of Mathematics, Acad. Sc. USSR, 117333, Vavilova 42, Moscow, USSR
B. Bojarski, Prof. Dr.,	Instytut Matematyki, Uniwersytet Warszawski, P.K.iN.9p., Warsaw, Poland
R. Böhme, Prof. Dr.,	Fachbereich Mathematik, Universität Erlangen Bismarckstr. 1 1/2, 8520 Erlangen, Germany
F. F. Brackx, Dr.,	Seminarie voor Wiskundige Analyse, Rijksuniversiteit Gent J. Plateaustraat 22, B-9000 Gent , Belgium

D. L. Colton, Prof. Dr., Department of Mathematics,
University of Strathclyde,
26, Richmond Street,
Glasgow G1 1XH, Great Britain

Mme. Coroi-Nedelcu, Prof. Dr., Institutul Politehnic Bucuresti,
Catedra Matematica,
Splaiul Independentei 313
Bucuresti, Rumania

I. Daniljuk, Prof. Dr., Akademia Nauk SSSR, Institute for
Applied Mathematics and Mechanics
Universitetskaja 77
Donezk 48, USSR

R. Delanghe, Prof. Dr., Fakulteit van de Wetenschappen,
Rijksuniversiteit,
Seminarie voor Hogere Analyse
Krijgslaan 271, B-9000 Gent, Belgium

J. Donig, Dr., Fachbereich Mathematik der
Technischen Hochschule Darmstadt
Schlossgartenstrasse 7
6100 Darmstadt, Germany

A. Džuraev, Prof. Dr., Academician of the Tajik. Academy
of Sc. Math. Inst.,
Šosse Ordžonikidzeabadskoc, km. 8
Dužanbe 30, USSR

J. Edenhofer, Dr., Mathematisches Institut der
TU München
Arcisstr. 21, 8000 München, Germany

G. Fichera, Prof. Dr., Universitá di Roma
Via Pietro Mascagni 7
00199 Roma, Italy

N. Friedrich, Dr.,	Fachbereich Mathematik der Universität des Saarlandes Im Stadtwald 6600 Saarbrücken, Germany
R. P. Gilbert, Prof. Dr.,	Department of Mathematics, University of Delaware 223 Sharp Lab., Newark, Delaware 19711, USA
I. Z. Gohberg, Prof. Dr.,	Department of Mathematics, University of Tel Aviv Tel Aviv, Israel
D. Gronau, Dr.,	II. Mathematisches Institut der Universität Graz, 3. Lehrkanzel Steyrergasse 17, 8010 Graz, Austria
K. Habetha, Prof. Dr.,	Lehrstuhl II für Mathematik der Technischen Hochschule Aachen Templergraben 55 5100 Aachen, Germany
R. Heersink, Dr.,	II. Mathematisches Institut der Universität Graz, 3. Lehrkanzel Steyrergasse 17, 8010 Graz, Austria
G. N. Hile, Prof. Dr.	Department of Mathematics, University of Hawaii 2565 The Mall, Honolulu 96822, USA
G. C. Hsiao, Prof. Dr.,	Dep. of Mathem., Univ. of Delaware 223 Sharp Lab., Newark, Delaware 19711, USA 1975/76 TH Darmstadt Fachbereich Mathematik
T. Iwaniec, Prof. Dr.,	Instytut Matematyki, Uniwersytet Warszawski P. K. iN. 9p., Warsaw, Poland

G. Jank, Dr.,	Institut für Mathematik der Technischen Hochschule Graz, 1. Lehrkanzel Kopernikusgasse 24, A-8010 Graz, Austria
J. Kisýnski, Prof. Dr.	Wydzial Mathematyki i. Mechaniki, Inst. Matem., Uniwersytet Warszawski Palac Kultury i Nauki IX p. 00-901 Warszawa, Poland
R. Kleinman, Prof. Dr.,	Department of Mathematics, University of Delaware 223 Sharp Lab., Newark, Del. 19711, USA
P. Kopp, Dipl.-Math.,	Fachbereich Mathematik der TH Darmstadt, Schlossgartenstrasse 7 6100 Darmstadt, Germany
T. Kori, Prof. Dr.,	Equipe d'Analyse, Université Paris IV 4, place Jassieu, 75005 Paris, France, and University of Waseda, Japan
M. Kracht, Dr.,	Mathem. Inst. der Univ. Düsseldorf Universitätsstrasse 1 4000 Düsseldorf, Germany
M. Kremer, Dr.,	Kornblumenstrasse 2 - 4 6115 Münster b. Dieburg, Germany
V. D. Kupradze, Prof. Dr.,	Mathem. Inst. , Z. Ruchadze Str. 1 380093 Tbilissi, Georg. SSR, USSR
N. Latz, Prof. Dr.,	Fachbereich Mathematik, TU Berlin Strasse des 17. Juni 135 1000 Berlin 12, Germany
H. Löffler, Dipl.-Math.,	Fachbereich Mathematik der TH Darmstadt, Schlossgartenstrasse 7 6100 Darmstadt, Germany

L. G. Mikhailov, Prof. Dr. Akademitscheskaja 3, kw 16
Dušanbe 13, 734013, USSR

R. F. Millar, Prof. Dr., Department of Mathematics,
University of Alberta
Edmonton, Canada T6G 2G1

J. Mitchell, Prof. Dr., State University of New York at
Buffalo, 4246 Ridge Lea Road,
Amherst, N. Y., 14226, USA

E. Pehkonen, Dr., Mathem. Institut, Univ. Jyväskylä
Sammonkatu 6, Jyväskylä 10, Finland

A. Piskorek, Prof. Dr., Przasnyska 10/50,
01-756 Warszawa, Poland

R. Rautmann, Prof. Dr., Fachbereich 17,
Gesamthochschule Paderborn
Pohlweg 55, 4790 Paderborn, Germany

L. Reich, Prof. Dr., II. Mathematisches Institut der
Universität Graz
Steyrergasse 17, A-8010 Graz, Austria

P. E. Ricci, Prof. Dr., Università di Roma
Via Albertazzi n. 92
00137 Roma, Italy

G. Roach, Prof. Dr., Dep. of Mathem., Univ. of Strathclyde
26, Richmond Street,
Glasgow, G1 1XH, Great Britain

St. Rolewicz, Prof. Dr. Instytut Matematyczny, Polskiej
Akademii Nauk, Sniadeckich 8
00950 Warszawa, skr. poczt. 137
Poland

S. Ruscheweyh, Prof. Dr., Abt. Mathematik der Univ. Dortmund
Postfach 500500
4600 Dortmund 50, Germany

B. Schüppel, Dr.,	Fachbereich Mathematik der TH Darmstadt, Schlossgartenstrasse 7 6100 Darmstadt, Germany
W. Schuster, Dr.,	Heideweg 8 5307 Wachtberg-Niederbachem, Germany
B. Sleeman, Prof. Dr.,	Department of Mathematics, University of Dundee Dundee, DD1 4HN, Great Britain
F.-O. Speck, Dr.,	Fachbereich Mathematik der TH Darmstadt, Schlossgartenstrasse 7 6100 Darmstadt, Germany
I. N. Vekua, Prof. Dr.,	Institute of Mathematics, University of Tbilissi 380093 Tbilissi, Georg. SSR, USSR
Chung-Ling Yu, Dr., Ass. Prof.,	Faculty of Engineering, University of Benghazi Benghazi, Libya

ON THE MATHEMATICAL THEORY OF FLOW PATTERNS OF COMPRESSIBLE FLUIDS

A SURVEY

STEFAN BERGMAN

1. The Derivation of the Equations for Potential and Stream Functions of a Compressible Fluid

The mathematical theory of two-dimensional, irrotational, steady flow patterns of an incompressible fluid is closely related to the theory of analytic functions of a complex variable. Generalizing this approach, one can investigate flow patterns of a compressible fluid. A two-dimensional steady flow of a perfect fluid can be described either by its potential ϕ or by the stream function ψ. In the incompressible fluids, ϕ and ψ are connected by Cauchy-Riemann equations, so that $f = \phi + i\psi$ is an analytic function of a complex variable $z = x + iy$. Taking the real and imaginary part of f, we obtain ϕ and ψ, respectively. This process can obviously be interpreted as an operation transforming analytic functions of a complex variable into solutions of equations arising in the theory of incompressible fluid, that is, of Laplace's equation. A much more complicated situation occurs in the case of compressible fluids. A flow of a fluid is initially defined in the *physical plane*, that is, in the plane where the motion occurs. At every point x,y of this plane the velocity vector (u,v) is determined, see Fig. 1. The pair of

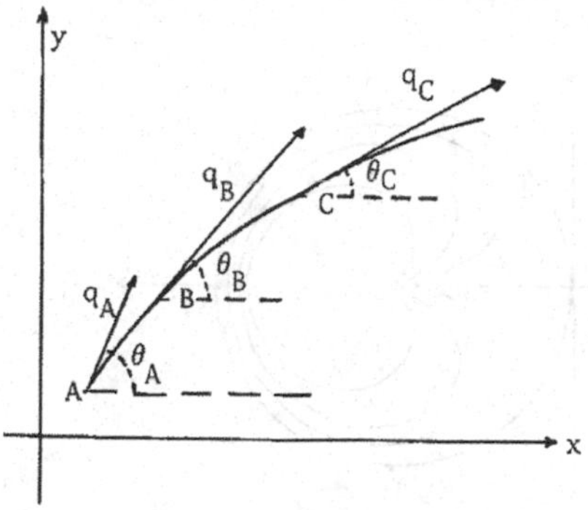

Fig. 1. A streamline in the physical plane.

functions $[u(x,y), -v(x,y)]$, where u and v are cartesian components of the velocity vector q at (x,y), determines a mapping of the domain of the x,y-plane in which the motion takes place into a (not necessarily schlicht) domain of the $(u,-v)$-plane, the so-called *hodograph* of the flow. See Figs. 2, 3, 4.

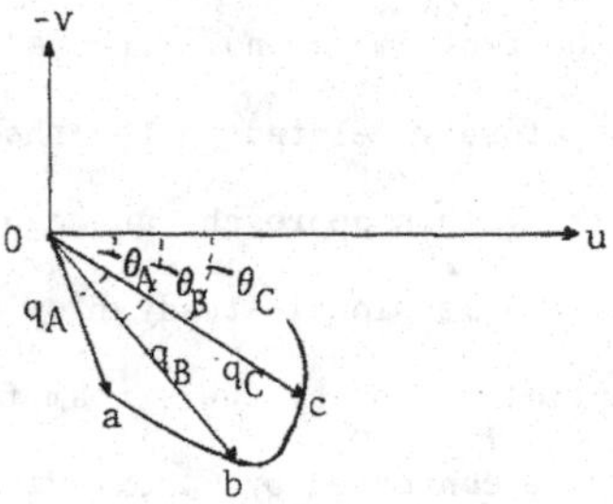

Fig. 2. The image in the hodograph plane of the streamline indicated in Fig. 1.

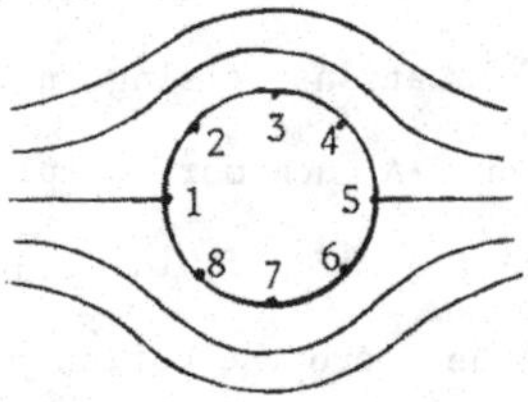

Fig. 3. A flow (in the physical plane) around a circle.

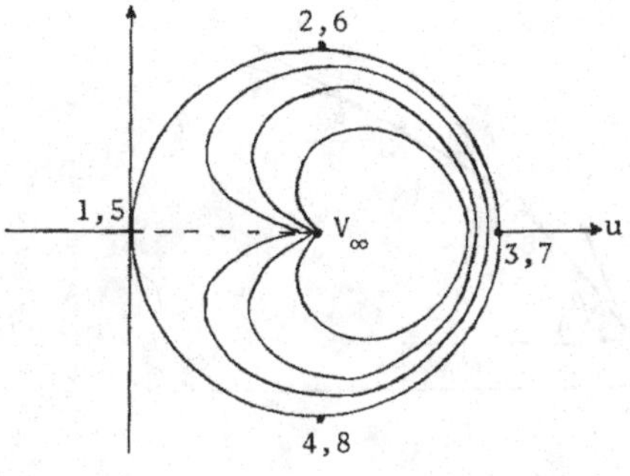

Fig. 4. The image in the hodograph plane of a flow around a circle.

The complex potential can be investigated in either of these planes, that is, one can investigate directly $\phi(x,y)$, $\psi(x,y)$, or primarily $\phi^{(1)}(u,v)=\phi[x(u,v),y(u,v)]$, $\psi^{(1)}(u,v) = \psi[x(u,v),y(u,v)]$ (the hodograph method). In the case of an incompressible fluid, $\phi(x,y)$ and $\psi(x,y)$, as well as $\phi^{(1)}(u,v)$, $\psi^{(1)}(u,v)$, satisfy Laplace's equation, and the flow patterns (except for some special problems) are investigated directly in the physical plane. In the case of a compressible fluid, $\phi(x,y)$ and $\psi(x,y)$ satisfy a system of complicated *nonlinear* partial differential equations, while, as Chaplygin and Molenbroek showed, $\phi^{(1)}(u,v)$ and $\psi^{(1)}(u,v)$ satisfy a system of *linear* equations.

Chaplygin [6], who introduced the hodograph method into the theory of compressible fluids, used the method of separation of variables in order to obtain solutions of the compressibility equation.

Another procedure (see [1] and [2]) for generating solutions of linear partial differential equations of the form

$$(1) \qquad \approx{u}_{xx} + \approx{u}_{yy} + a_1\approx{u}_x + a_2\approx{u}_y + a_3\approx{u} = 0$$

has been developed in the case where $a_p \equiv a_p(z,z^*)$, $z = x+iy$, $z^* = x-iy$, $p=1,2,3$, are analytic functions of two complex variables, which are regular in a sufficiently large domain. By the integral operator

$$(2) \qquad P_1(f) \equiv \int_{-1}^{1} E(z,z^*,t)f[\tfrac{1}{2}z(1-t^2)]\frac{dt}{\sqrt{1-t^2}},$$

analytic functions $f(z)$ are transformed into solutions of the differential equation (1). Here

$$(3) \qquad E(z,z^*,t) = \{\exp[-\int_0^{z^*} a_1(z,z^*)dz^* + n(z)]\}\tilde{E}(z,z^*,t),$$

where $\tilde{E}$ satisfies

$$(4) \qquad B(\tilde{E}) = (1-t^2)\tilde{E}_{z^*t} - \frac{1}{t}\tilde{E}_{z^*} + 2tz[\tilde{E}^*_{zz^*} + D\tilde{E}_{z^*} + F\tilde{E}] = 0,$$

is denoted as a generating function of the operator, $n(z)$ is an arbitrary function of z, which is regular at $z=0$. E has further to satisfy certain conditions, see for details [2], p. 10, and [8], p. 362 ff.

Using the integral operator P_1 of the first kind, it is possible to generalize various theorems in the theory of analytic functions of one complex variable to the case of functions satisfying (1). Florian, Gilbert, Kreyszig, Kracht, Jank, Lanckau, Rosenthal, Watzlawek and others generalized and extended this approach. I am referring to their lectures and publications in which they present their important results. To apply the same approach to the theory of compressible fluids, one has still to make an additional step. In this case the continuity equation has the form

$$\frac{\partial(\rho u)}{\partial x} + \frac{\partial(\rho v)}{\partial y} = 0\,, \tag{5}$$

where ρ is the density and $q = (u,v)$. The assumption that the flow is irrotational is expressed by the equation

$$\frac{\partial u}{\partial y} - \frac{\partial v}{\partial x} = 0\,. \tag{6}$$

It follows from (5) and (6) that for every flow we can introduce a potential ϕ and a stream function ψ, such that

$$u = \frac{\partial\phi}{\partial x} = \frac{1}{\rho}\frac{\partial\psi}{\partial y}\,, \qquad v = \frac{\partial\phi}{\partial y} = \frac{1}{\rho}\frac{\partial\psi}{\partial x}\,. \tag{7}$$

We shall consider only adiabatic flows for which the thermodynamical equation of state may be expressed in the form

$$p = \sigma\rho^{\gamma} + \beta\,, \tag{8}$$

where σ, γ, β are constants and p is the pressure. By combining the Bernoulli equation $\frac{q^2}{2} + \int_{p_0}^{p} \frac{dp}{\rho(p)} = 0$ (where $q^2 = u^2 + v^2$ and p_0 is the pressure at a stagnation point) with the equation of state, we can eliminate the pressure, and we obtain

$$\rho = \left(1 - \frac{\gamma-1}{2}\, q^2\right)^{1/(\gamma-1)}. \tag{9}$$

Here the units of mass and velocity are so chosen that at a stagnation point $p = 1$ and $(dp/d\rho) = 1$. Since $q^2 = \phi_x^2 + \phi_y^2$, $\phi_x \equiv \frac{\partial\phi}{\partial x}$, $\phi_y \equiv \frac{\partial\phi}{\partial y}$, the equations (5, 6, 9) represent a system of three (nonlinear) partial differential equations for ϕ, ψ and ρ.

An important simplification in the study of the motion of a compressible fluid has been achieved by Molenbroek [9] and Chaplygin [6]. They showed that if ϕ and ψ are considered as functions of the speed q and θ, $qe^{i\theta} = u + iv$, instead of x and y, ϕ and ψ satisfy the system

$$(10) \qquad \frac{\partial\phi}{\partial\theta} = \frac{q}{\rho}\frac{\partial\psi}{\partial q}, \qquad \frac{\partial\phi}{\partial q} = -\frac{1-M^2}{\rho q}\frac{\partial\psi}{\partial\theta},$$

where $M = q/[1-(\gamma-1)q^2/2]^{\frac{1}{2}}$ is the Mach number, and θ is the angle which the velocity vector forms with the positive direction of the x-axis of the physical plane.

Since ρ is a known function of q, see (9), equations (10) represent a system of two *linear* partial differential equations. In order to simplify the form of equations (10), it is convenient to introduce in the subsonic case instead of q the variable

$$(11) \qquad \lambda = \frac{1}{2}\,\mathrm{lg}\left[\frac{1-(1-M^2)^{\frac{1}{2}}}{1+(1-M^2)^{\frac{1}{2}}}\left(\frac{1+h(1-M^2)^{\frac{1}{2}}}{1-h(1-M^2)^{\frac{1}{2}}}\right)^{1/h}\right], \qquad h = \left(\frac{\gamma-1}{\gamma+1}\right)^{\frac{1}{2}}, \qquad \gamma > 1.$$

The plane whose cartesian coordinates are θ and λ will be denoted as the *pseudo-logarithmic plane.*

In the pseudo-logarithmic plane, equations (10) assume the form

$$(12) \qquad \phi_\theta - \ell^{\frac{1}{2}}\psi_\lambda = 0, \qquad \phi_\lambda + \ell^{\frac{1}{2}}\psi_\theta = 0, \qquad \phi_\theta \equiv \frac{\partial\phi}{\partial\theta}, \ldots,$$

where

$$(13) \qquad \ell = \ell(\lambda) = \frac{1-M^2}{\rho^2}.$$

Sometimes it is convenient to use the complex notation. Let

$$(14) \qquad \zeta = \theta + i\lambda, \qquad \zeta^* = \theta - i\lambda.$$

The equations (12) can now be written

$$(15) \qquad \phi_\zeta - i\ell^{\frac{1}{2}}\psi_\zeta = 0, \qquad \phi_{\zeta^*} + i\ell^{\frac{1}{2}}\psi_{\zeta^*} = 0.$$

Eliminating ψ and ϕ, respectively, we obtain

$$(15a) \quad \phi_{\zeta\zeta^*} - iN(\phi_\zeta - \phi_{\zeta^*}) = 0, \qquad (15b) \quad \psi_{\zeta\zeta^*} + iN(\psi_\zeta - \psi_{\zeta^*}) = 0, \qquad N = -\frac{\gamma+1}{8}\frac{M^4}{(1-M^2)^{3/2}}.$$

2. Solutions of Compressibility Equations Generated by Operator P_2 and some of Their Properties

As shown in the preceding section, the mathematical theory of compressible fluid flows can be reduced to the study of solutions of equation

$$L(\psi) \equiv \psi_{ZZ^*} + N(\lambda)(\psi_Z + \psi_{Z^*}) = 0, \tag{16}$$

$$N = \frac{1}{8}\ell^{-3/2}\ell_H, \qquad Z = \lambda + i\theta, \qquad Z^* = \lambda - i\theta, \qquad \ell(x) = \sum_{n=1}^{\infty} a_n(-x)^n, \quad a_1 > 0,$$

$$-\lambda(-H) = \int_{t=0}^{-H} [\ell(t)]^{-\frac{1}{2}} dt,$$

where $N(\lambda)$ is singular at $\lambda = 0$. Let us assume that N is an analytic function of the variable λ, $\lambda \neq 0$. It is convenient to continue N to complex values of the argument. If we set $\tau = \lambda + i\Lambda$, the domains $[-\infty < \theta < \infty, H < 0]$ and $[-\infty < \theta < \infty, H > 0]$ correspond respectively to $[-\infty < \theta < \infty, \operatorname{Re}\tau < 0, \operatorname{Im}\tau = 0]$. We assume further that

1° in the neighborhood of the point $H = 0$, $N(\lambda)$ has an expansion of the form

$$N(\lambda) = (-\lambda)^{-1}\left[-\frac{1}{12} + \sum_{\nu=1}^{\infty} \beta_\nu (-\lambda)^{\frac{2\nu}{3}}\right] \tag{17}$$

valid for $-\lambda_o < \lambda < 0$, $\lambda_o > 0$;

2° $N(\lambda)$ is an analytic function for $-\infty < \lambda < 0$ and is real for $\lambda < 0$;

3° the expression $\exp[\int_{-\infty}^{\lambda} 2N(t)dt]$ exists for all $\lambda < 0$.

Under these assumptions $S_o = \lim_{\lambda \to 0^-} (-\lambda)^{1/6} N(\lambda)$ exists. Also, we shall assume that $S_o > 0$. Let H and θ be cartesian coordinates of the plane. We investigate at first the special case

$$\ell(H) = -a_1 H, \qquad a_1 > 0, \tag{18}$$

which we refer to as the *Tricomi case*. In this case $N(\lambda) = (12\lambda)^{-1}$.

In the Tricomi case the generating function E^* of the integral operator P_2 is the hypergeometric function of the variable $u = Zt^2/(Z+Z^*) = t^2Z/2\lambda$

$$E^{*\dagger} = A_1 F(\tfrac{1}{6}, \tfrac{5}{6}, \tfrac{1}{2}, u) + B_1 u^{\frac{1}{2}} F(\tfrac{2}{3}, \tfrac{4}{3}, \tfrac{3}{2}, u) , \tag{19}$$

A_1, B_1 are constants, F is the hypergeometric function. It follows from the theory of hypergeometric functions that $E^{*\dagger}$ admits two series representations: One development of F is valid for $|Z/2\lambda| > 1$, and the other is valid for $|Z/2\lambda| < 1$. See (3.8) - (3.11), p. 452 of [5].

The above mentioned results can be extended to the general case. For the Tricomi case these series developments have been given in (4.1) and (4.2), respectively, p. 453 ff. of [5]. In (4.1) and (4.2) of [3] the corresponding developments for the generating function E are given in the general case. We denote the integral operator which we obtain in this way as the *integral operator* P_2 *of the second kind*. The use of P_2 permits us to generate solutions of the differential equations

$$L(\psi) \equiv \psi_{ZZ^*} + N(\lambda)(\psi_Z + \psi_{Z^*}) = 0 , \qquad N(\lambda) = \frac{1}{8}\, \ell^{-3/2}\, \ell_H , \tag{20}$$

see (1.1a), p. 445 of [5], which one obtains considering equations

$$M(\psi) \equiv \psi_{HH} + \ell(H)\psi_{\theta\theta} = 0 , \tag{21}$$

$$\ell(H) = \sum_{n=1}^{\infty} a_n(-H)^n , \qquad a_1 > 0, \quad a_2 < 0 , \tag{21a}$$

of mixed type. Here $\ell(H) > 0$ for $H < 0$, and $\ell(H) < 0$ for $H > 0$.

The integral operator P_1 of the first kind permits us to translate various results of the theory of functions of one complex variable into theorems about solutions of equation (1). Integral operator P_2 permits us at first to generate solutions of equation (15a). Further, one sees that for certain partial differential equations with singular coefficients results are valid which have a similarity with results which follow from Fuchs' theory in the case of ordinary differential equations (see also p. 884 of [4]). However, the problem of concluding the properties of generated solutions from the properties of the associate function f(z) is in the case of P_2 much more difficult than that in the case of P_1. In many instances, when studying solutions of equation (15a), it is useful to limit the considerations to the

Tricomi case.

One of the important questions is generating solutions of differential equations possessing at a prescribed point singularities of certain kind. Rosenthal solved this problem for the Tricomi case of operator P_2.

Theorem (Rosenthal). Let $W = \{(x,y) \,|\, 3^{\frac{1}{2}}|x| < y,\ x \leq 0,\ y > 0\}$, and let

$$(22)\qquad f(q) = (a_m q^m + \cdots + a_0)/(b_p q^p + \cdots + b_0), \qquad p > m,\ \ p > 1,$$

whose poles $x_j + iy_j = A_j \in W$, $|A_j| = M$, $|2x_j| > \max|x_j|$, $|x_j| > 0$, $1 \leq j \leq n$. Let $R = \{(x,y) \,|\, (x,y) \in W,\ \min \frac{2}{3}|x_j| \leq |x| < M,\ 0 < y < \min 2y_j,\ 1 \leq j \leq n\}$, and $S = \{(x,y) \,|\, (x,y) \in W,\ \min \frac{2}{3}|x_j| > x > 0,\ 1 \leq j \leq n\}$, $D = (R \cup S) \subset W$. Then, (1) for all $(x,y) \in D$, $P_2(f)$ is analytic, and (2) $(2x_j, 2y_j)$, $(\frac{2}{3}x_j, 2y_j)$ are singular points of $P_2(f)$, see [10].

3. Remarks about Supersonic and Mixed Flows

To obtain supersonic flow patterns, we proceed analogously as before. The pseudo-logarithmic plane in this case is the plane whose cartesian coordinates are

$$(23)\qquad \Lambda = \frac{1}{h}\arctan[h(M^2-1)^{\frac{1}{2}}] - \arctan[(M^2-1)^{\frac{1}{2}}]$$

and θ. Here M is the local Mach number, i.e.,

$$(24)\qquad M = \frac{q}{a} = \frac{q}{[a_0^2 - (k-1)q^2]^{\frac{1}{2}}}, \qquad h = \sqrt{(k-1)/(k+1)},$$

a is the speed of sound, a_0 is the speed of sound at a stagnation point. For the air $k = 1.4$ and $1/h = 2.45$. The stream function ψ satisfies in this case

$$(25)\qquad \psi_{\Lambda\Lambda} - \psi_{\theta\theta} - 4N_1\psi_\Lambda = 0$$

where

$$(26)\qquad N_1 = \frac{k+1}{8}\,\frac{M^4}{(M^2-1)^{\frac{3}{2}}}\,.$$

Remark. $a = \sqrt{dp/d\rho}$, $M = q/a$, see v. Mises [8], (3), p. 49.

If we write

(27) $$\psi^* = \psi[-\int_a^{\xi+\eta} N_1(\tau)d\tau] ,$$

where a is an abitrary constant, ψ^* satisfies

(28) $$\psi^*_{\xi\eta} + F_1\psi^* = 0$$

(29) $$F_1 = \frac{k+1}{64}\left[\frac{5(k+1)}{B^6} + \frac{12k}{B^4} + \frac{6k-14}{B^2} + (4k+8)-(8k-1)B^2\right], \quad B^2 = M^2 - 1 ,$$

see (2.14) - (2.17) of [4].

Bibliography

[1] Bergman, S.: Zur Theorie der Funktionen, die eine lineare partielle Differentialgleichung befriedigen, Mat. Sb. (2) 44 (1937), 1169-1198.

[2] Bergman, S.: Integral Operators in the Theory of Partial Differential Equations, Third Printing, Vol. 23, Ergebnisse der Mathematik und ihrer Grenzgebiete, Springer-Verlag Berlin, Heidelberg, New York, 1971.

[3] Bergman, S.: Two-dimensional subsonic flows of a compressible fluid and their singularities, Trans. Amer. Math. Soc. 62 (1947), 452-498.

[4] Bergman, S.: Two-dimensional transonic flow patterns, Amer. J. Math. 70 (1948), 856-891.

[5] Bergman, S.: On solutions of linear partial differential equations of mixed type, Amer. J. Math. 74 (1952), 444-474.

[6] Chaplygin, C.A.: On gas jets, Scient. Memoirs, Moscow Univ., Phys.-Math. Sect., vol. 21 (1904), 1-127 (also NACA Techn. Memorandum 1063 (1944)).

[7] Gilbert, R.: Function Theoretic Methods in Partial Differential Equations, Math. in Science and Engineering. vol. 54, Academic Press, New York, 1969.

[8] Mises, R.v.: Mathematical Theory of Compressible Fluid Flow, Academic Press, New York, 1958.

[9] Molenbroek, P.: Über einige Bewegungen eines Gases mit Annahme eines Geschwindigkeitspotentials, Arch. Math. Phys. (2) 9 (1890), 157-195.

[10] Rosenthal, P.: On the location of the singularities of the function generated by the Bergman operator of the second kind, Proc. Amer. Math. Soc. 44, (1974), 157-162.

[11] Rosenthal, P.: On the singularities of functions generated by the Bergman operator of the second kind, Pacific J. Math. (to appear).

ON A CLASS OF NONLINEAR PARTIAL DIFFERENTIAL EQUATIONS

A. V. Bitsadze

(Moscow)

1°. We consider a class of second order partial differential equations

$$\sum_{j,k=1}^{n} a_{jk}(x)\left[U_{x_j x_k} - f(U)U_{x_j}U_{x_k}\right] + \sum_{j=1}^{n} b_j(x)U_{x_j} + C(x,U) = 0, \qquad (1)$$

where $x = (x_1, \dots, x_n)$, $a_{jk}(x)$, $b_j(x)$, $C(x,U)$ and $f(U)$ are given sufficiently smooth functions of their arguments.

Transforming the dependent variable by

$$U = \phi(V) \qquad (2)$$

we find

$$\phi'\left(\sum_{j,k=1}^{n} a_{jk}(x)V_{x_j x_k} + \sum_{j=1}^{n} b_j(x)V_{x_j}\right) + C(x,\phi)$$

$$= -\left[\phi'' - f(\phi)\phi'^2\right]\sum_{j,k=1}^{n} a_{jk}(x)V_{x_j}V_{x_k}\,. \qquad (3)$$

If we choose the function ϕ in such a way as to satisfy

$$\phi'' - f(\phi)\phi'^2 = 0 \qquad (4)$$

under the assumption $\phi' \neq 0$ the equation (1) assumes the form

$$\sum_{j,k=1}^{n} a_{jk}(x)V_{x_j x_k} + \sum_{j=1}^{n} b_j(x)V_{x_j} + C(x,\phi)/_{\phi'} = 0\,. \qquad (5)$$

The general solution of the ordinary differential equation (4) is

$$v = \alpha \int_0^{\phi} \exp\left(-\int_0^{\tau} f(t)dt\right)d\tau + \beta\,, \qquad (6)$$

where α and β are arbitrary constants.

If

$$C(x,\phi)/_{\phi'} = C_0(x)V + C_1(x)$$

the equation (5) is linear

$$\sum_{j,k=1}^{n} a_{jk}(x)V_{x_j x_k} + \sum_{j=1}^{n} b_j(x)V_{x_j} + C_o(x)V + C_1(x) = 0 \qquad (7)$$

and when $C = C(U)$ the equation (5) can be written in the form

$$\sum_{j,k=1}^{n} a_{jk}(x)V_{x_j x_k} + \sum_{j=1}^{n} b_j(x)V_{x_j} + w(V) = 0 , \qquad (8)$$

where

$$w(V) = C[\phi(V)]/\phi'(V) .$$

Provided that $\phi(V)$ and $V(x)$ satisfy the equations (4) and (7), respectively, the solution of the given nonlinear partial differential equation (1) can be obtained by inserting $V(x)$ into (2). Moreover, any problem (boundary value, initial value, and others) well-posed for the equation (7), is also well-posed for the equation (1) if the conditions on V impose conditions on U by formula (2).

It should be noted that in case of complex-valued coefficients and complex-valued variables x, U the equations (1), (4), (7), (8) are equivalent to systems of real equations.

We now pass to a more detailed consideration of some important equations of class (1).

2°. For the case

$$f(U) = \frac{\delta}{U} ,$$

where δ is any constant, it follows from (6) that

$$V = \beta + \frac{\alpha}{1-\delta}\,\phi^{1-\delta}, \quad \delta \neq 1, \quad \phi = \beta\, e^{\alpha V}, \quad \delta = 1. \qquad (9)$$

After the choice of ϕ by (9) the equation (5) obtained for $V(x)$ is sometimes linear, the general solution of which being well known.

Consider two simple examples

Example 1.

$$U_{x_1 x_1} + U_{x_2 x_2} - \frac{1}{U}(U_{x_1}^2 + U_{x_2}^2) = 0.$$

By virtue of (9) and (5) we have

$$U = \phi(V) = \beta e^{\alpha V}, \quad V = \operatorname{Re} \Phi(z),$$

where α, β are arbitrary constants and $\Phi(z)$ is an arbitrary analytic function of the complex variable $z = x_1 + ix_2$.

Example 2.

$$x_1 x_2 U_{x_1 x_2} - x_2^2 U_{x_2 x_2} - 2x_1 U_{x_1} + 2x_2 U_{x_2} + C(x_1, x_2, U) - \delta U^{-1}(x_1 x_2 U_{x_1} U_{x_2} - x_2^2 U_{x_2}^2) = 0, \quad (10)$$

where

$$C = -\frac{2}{1-\delta} U, \quad \delta \neq 1 \quad (11)$$

and

$$C = -2U \log U, \quad \delta = 1. \quad (12)$$

By (10), (11) and (12) from (6) and (5) we obtain

$$\phi = V^{\frac{1}{1-\delta}}, \quad \delta \neq 1, \quad \phi = e^V, \quad \delta = 1$$

and

$$x_1 x_2 V_{x_1 x_2} - x_2^2 V_{x_2 x_2} - 2x_1 V_{x_1} + 2x_2 V_{x_2} - 2V = 0. \quad (13)$$

It is known that the general solution of equation (13) is represented by the formula

$$V = x_2^2 F_1(x_1) + x_2 F_2(x_1 x_2),$$

where F_1 and F_2 are arbitrary functions of their arguments [1].

Hence

$$U = \left[x_2^2 F_1(x_1) + x_2 F_2(x_1 x_2)\right]^{\frac{1}{1-\delta}}$$

in the case $\delta \neq 1$ and

$$U = \exp\left[x_2^2 F_1(x_1) + x_2 F_2(x_1 x_2)\right]$$

in the case $\delta = 1$.

3°. The complex potential $U(x_1, x_2)$ of axial symmetric gravitational fields satisfies the equation

$$\left[(x_1^2-1)U_{x_1}\right]_{x_1} + \left[(1-x_2^2)U_{x_2}\right]_{x_2} - \frac{2\bar{U}}{U\bar{U}-1}\left[(x_1^2-1)U_{x_1}^2 + (1-x_2^2)U_{x_2}^2\right] = 0, \quad (14)$$

where x_1, x_2 are the prolate spheroidal coordinates [2].

In this case according to (4) and (5) the functions ϕ and V must satisfy the equations

$$\phi'' - \frac{2\overline{\phi}}{\phi\overline{\phi} - 1}\phi'^2 = 0 \qquad (4')$$

and

$$\left[(x_1^2-1)V_{x_1}\right]_{x_1} + \left[(1-x_2^2)V_{x_2}\right]_{x_2} = 0 \,, \qquad (5')$$

respectively.

The equation (5') is elliptic for $(x_1^2-1)(1-x_2^2) > 0$, hyperbolic for $(x_1^2-1)(1-x_2^2) < 0$, with parabolic degeneracy along the curves $x_1 = \pm 1$, $x_2 = \pm 1$.

After transforming the independent variables by

$$z = \sqrt{(x_1^2-1)(1-x_2^2)} + ix_1x_2, \quad \overline{z} = \sqrt{(x_1^2-1)(1-x_2^2)} - ix_1x_2$$

equation (5') in its domain of ellipticity reduces to the form

$$V_{z\overline{z}} + \frac{1}{2(z+\overline{z})}(V_z+V_{\overline{z}}) = 0. \qquad (5'')$$

The solutions of the equation (4') are given by

$$\phi(V) = e^{i\alpha}\frac{1+\beta e^{\gamma V}}{1-\overline{\beta}e^{\gamma V}} \,, \qquad (15)$$

where α, γ are arbitrary real constants and β is an arbitrary complex constant.

A class of analytic solutions of the equation (5") is given by the formula

$$V^{(1)}(\xi, \eta) = \mathrm{Re}\int_0^1 f(\eta+i\xi-2i\xi t)\frac{dt}{\sqrt{t(1-t)}} \,, \quad \xi+i\eta = z \,, \qquad (16)$$

where $f(\tau)$ is an arbitrary analytic function of the complex variable τ.

Another class of solutions of the equation (5") analytic in the half-plane $\xi > 0$ with a logarithmic singularity at $\xi = 0$ is given by

$$V^{(2)}(\xi,\eta) = \mathrm{Re}\int_0^1 f(\eta+i\xi-2i\xi t)\log[\xi t(1-t)]\,\frac{dt}{\sqrt{t(1-t)}}\ . \qquad (17)$$

Substituting the expressions (15), (16), (17) for ϕ and V into (2) we get the solutions of equation (14) [3].

The application of the non-singular transformation

$$\xi = 2y^{1/2}\ ,\ \eta = x$$

reduces equation (5") in its domain of ellipticity $\{(x_1^2-1)(1-x_2^2) > 0\}$ to the form

$$V_{xx} + y\cdot V_{yy} + V_y = 0\ . \qquad (18)$$

Let $\mathcal{D}$ be a simply-connected domain, contained in the region $\mathcal{D}_o := \{y > 0\}$, the boundary of which consists of two parts $\Gamma = \sigma \cup AB$, where σ is a smooth Jordan curve with its endpoints $A(0,0)$, $B(1,0)$, lying inside of $\mathcal{D}_o$ and AB is a segment of the line $y = 0$. In this case the following problem has a unique solution: find the solution $V(x,y)$ of equation (18) which is regular in the domain $\mathcal{D}$, remains bounded for $y \to 0$, and takes given continuous values only on σ [4], [5].

In the case of $(x_1^2-1)(x_2^2-1) > 0$ after transforming by

$$x = x_1x_2\ ,\quad 4y = (1-x_1^2)(1-x_2^2)$$

equation (5') takes the form

$$V_{xx} - y\cdot V_{yy} - V_y = 0\ . \qquad (19)$$

The general solution of equation (19) in the region $\{y > 0\}$ is

$$V = \int_0^1 f_1[x+2y^{1/2}(1-2t)]\,\frac{dt}{\sqrt{t(1-t)}}$$

$$+ \int_0^1 f_2[x+2y^{1/2}(1-2t)]\log[yt(1-t)]\,\frac{dt}{\sqrt{t(1-t)}}\ ,$$

where f_1 and f_2 are arbitrary twice differentiable functions [6].

4°. We now consider a special case of Eq. (1) which is of particular interest to hydromechanics:

$$x_2^{2m} U_{x_1x_1} + x_2 U_{x_2x_2} + k\,U_{x_2} - U^{-1}(x_2^{2m} U_{x_1}^2 + x_2 U_{x_2}^2) = 0 \,, \tag{20}$$

where m is a positive integer, k is a real constant, $1/2-m \leq k < 1$.

From (5) and (6) we see that

$$x_2^{2m} V_{x_1x_1} + x_2 V_{x_2x_2} + k\,V_{x_2} = 0 \tag{21}$$

and

$$U = \phi(V) = \alpha e^{\beta V} \,.$$

In particular when $m = 1$, $k = 0$ equation (21) coincides with Tricomi's equation

$$x_2 V_{x_1x_1} + V_{x_2x_2} = 0 \,.$$

In the case of $k = 1/2-m$, after transforming the independent variables by

$$x = x_1 \,, \quad y = \frac{2}{2m+1} \left|x_2\right|^{\frac{2m+1}{2}} \operatorname{Sgn} x_2 \qquad (x_2 \neq 0) \,,$$

we find that equation (21) can be rewritten in the form

$$V_{xx} + \operatorname{sgn} y \cdot V_{yy} = 0 \,.$$

R e f e r e n c e s

[1] J. L. Reid and P. B. Burt,
Solution of nonlinear partial differential equations from base equations
J. Math. Anal. Appl., 47, no. 3 (1974), 520 - 530.

[2] F. J. Ernst,
Complex potential formulation of the axially symmetric gravitational field problem
J. Math. Phys., 15, no. 9 (1974), 1409 - 1412.

[3] A. V. Bitsadze,
On a gravitational field equation
Dokl. Akad. Nauk SSSR, 222, no. 4 (1975), 765 - 768.

[4] M. V. Keldysh,
On certain classes of elliptic equations with singularity on the boundary of their domain
Dokl. Akad. Nauk SSSR, 77, no. 2 (1951), 181 - 183.

[5] A. V. Bitsadze,
Equations of the mixed type
Pergamon Press (1964), 65.

[6] G. Darboux,
Leçons sur la theorie generale des surfaces,
II, Paris (1915), 66 - 69.

Integral Operators and Inverse Problems in Scattering Theory

David Colton*

I. Introduction.

In this talk we shall survey some recent results we have obtained on the use of integral operators in the investigation of certain inverse problems connected with the scattering of acoustic waves by a bounded obstacle. In particular suppose an incoming plane acoustic wave of frequency ω moving in the direction of the z axis is scattered by a "soft" bounded obstacle D which may be surrounded by a pocket of rarefied or condensed air (contained in a ball of radius a) in which the local speed of sound is given by the continuously differentiable function c(r) where $r = |\underset{\sim}{x}|$ for $\underset{\sim}{x} \in \mathbb{R}^3$. Let the speed of sound in the undisturbed medium be given by c_0(a constant)and let $u_s(\underset{\sim}{x})e^{i\omega t}$ be the velocity potential of the scattered wave. Then, setting $\lambda = \frac{\omega}{c_0}$, $B(r) = \left(\frac{c_0}{c(r)}\right)^2 - 1$, $u_s(\underset{\sim}{x}) = v(\underset{\sim}{x}) + u(\underset{\sim}{x})$, we have that $u(\underset{\sim}{x})$ satisfies

$$\Delta_3 u + \lambda^2(1+B(r))u = 0 \qquad \text{in } \mathbb{R}^3 \setminus D \tag{1.1}$$

$$u(\underset{\sim}{x}) = -(e^{i\lambda z} + v(\underset{\sim}{x})) \qquad \text{on } \partial D \tag{1.2}$$

$$\lim_{r\to\infty} r\left(\frac{\partial u}{\partial r} - i\lambda u\right) = 0 \tag{1.3}$$

where $B(r)=0$ for $r \geqslant a$, and $v(\underset{\sim}{x})$ is such that $e^{i\lambda z} + v(\underset{\sim}{x})$ is a

* This research was supported in part by AFOSR Grant 74-2592.

solution of (1.1) in $\mathbb{R}^3 \setminus D$ where $v(\underset{\sim}{x}) = 0$ for $r \geq a$. If $u(\underset{\sim}{x})$ is a solution of (1.1) - (1.3) then at infinity we have

$$f(\theta,\phi;\lambda) = \lim_{r\to\infty} re^{-i\lambda r}u(\underset{\sim}{x}) \tag{1.4}$$

where r,θ,ϕ are spherical coordinates. The function $f(\theta,\phi;\lambda)$ is known as the <u>far field</u> (or <u>radiation</u>) <u>pattern</u> and, if $B(r)$ is known, uniquely determines $u(\underset{\sim}{x})$ (c.f. [17]). Our aim is to investigate certain <u>inverse problems</u> associated with (1.1) - (1.3), in particular, given the far field pattern, to determine either the function $B(r)$ or the shape of the obstacle D. We shall confine ourselves to the simplest problem in each of these cases, i.e.

<u>Problem I</u>: Let D be the sphere of radius one centered at the origin and assume $a > 1$. Then given $f(\theta,\phi;\lambda)$ for $0 \leq \theta \leq \pi$. $0 \leq \phi \leq 2\pi$, $0 < \lambda_o < \lambda < \lambda_1$, where λ_o and λ_1 are constants, determine $B(r)$ for $1 \leq r \leq a$.

<u>Problem II</u>: Let $B(r) \equiv 0$ (i.e. $c(r) = c_o$). Then given $f(\theta,\phi;\lambda)$ for $0 \leq \theta \leq \pi$, $0 \leq \phi \leq 2\pi$, $0 < \lambda_o < \lambda < \lambda_1$, determine the shape of D.

In both of these problems we shall not discuss the problem of continuous dependence on the data, i.e. we shall assume that $f(\theta,\phi;\lambda)$ is known exactly.

<u>Open Question I</u>: Is it possible to approximate the solutions to Problems I and II if $f(\theta,\phi;\lambda)$ is not known exactly, but satisfies certain other a priori assumptions?

As will be seen from the discussion which follows, both Problems I and II are in general unstable and thus "improperly posed". For a discussion of methods used in the study of improperly posed problems the reader is referred to [13] and [18].

II. Problem I.

We shall now investigate Problem I through the use of an integral operator recently constructed by Colton and Wendland in [8]. This operator maps solutions of the Helmholtz equation

$$\Delta_3 h + \lambda^2 h = 0 \tag{2.1}$$

defined in the exterior of a bounded starlike domain D onto solutions of (1.1) defined in the exterior of D by the relation

$$\begin{aligned} u(r,\theta,\phi) &= (\underset{\sim}{I}+\underset{\sim}{K}_0)h \\ &= h(r,\theta,\phi) + \int_r^\infty K(r,s;\lambda)h(s,\theta,\phi)ds, \end{aligned} \tag{2.2}$$

where $K(r,s;\lambda)$ is the twice continuously differentiable solution of

$$r^2\left[K_{rr} + \frac{2}{r}K_r + \lambda^2(1+B(r))K\right] = s^2\left[K_{ss} + \frac{2}{s}K_s + \lambda^2 K\right] \tag{2.3}$$

for $s > r$ satisfying the rather unusual boundary conditions

$$K(r,s;\lambda) = 0 \qquad \text{for } r\cdot s \geqslant a^2 \tag{2.4}$$

$$K(r,r;\lambda) = -\frac{\lambda^2}{2r}\int_r^\infty sB(s)ds \tag{2.5}$$

$$K(r,s;\lambda) = 0 \qquad \text{for } r < s. \tag{2.6}$$

The solution of (2.3) - (2.6) can be obtained in the form

$$K(r,s;\lambda) = (rs)^{-\frac{1}{2}} \sum_{j=0}^{\infty} \lambda^{2j+2} N_j(\log r, \log s) \tag{2.7}$$

with

$$N_0(\log r, \log s) = -\frac{1}{2}\int_{(rs)^{\frac{1}{2}}}^{a} \xi B(\xi)d\xi \tag{2.8}$$

and the functions $N_j(\log r, \log s)$, $j = 0,1,\ldots$, being determined recursively. Due to the fact that $B(r) = 0$ for $r \geqslant a$ it can be shown ([4], [8]) that the series (2.7) is uniformly convergent for

$1 \leqslant r \leqslant s < \infty$ and is an entire function of λ.

In the case of Problem I we note that $u(r,\theta,\phi) = u(r,\theta)$ and $h(r,\theta,\phi) = h(r,\theta)$ are independent of ϕ, and the solution of (1.1) - (1.3) can be obtained by separation of variables in the form ([4], [6])

$$u(r,\theta) = -\sqrt{\frac{\pi}{2}} \sum_{n=0}^{\infty} \frac{(2n+1)i^n\, j_{n+\frac{1}{2}}(1)}{h_{n+\frac{1}{2}}(1)} h_{n+\frac{1}{2}}(r) P_n(\cos\theta) \qquad (2.9)$$

where

$$\begin{aligned} j_{n+\frac{1}{2}}(r) &= (\underset{\sim}{I}+\underset{\sim}{K}_0)\left[(\lambda r)^{-\frac{1}{2}} J_{n+\frac{1}{2}}(\lambda r)\right] \\ h_{n+\frac{1}{2}}(r) &= (\underset{\sim}{I}+\underset{\sim}{K}_0)\left[(\lambda r)^{-\frac{1}{2}} H^{(1)}_{n+\frac{1}{2}}(\lambda r)\right] \end{aligned} \qquad (2.10)$$

and we have used standard notation for Bessel, Hankel, and Legendre functions. From (2.9) we have that the far field pattern $f(\theta,\phi;\lambda) = f(\theta;\lambda)$ is given by

$$\begin{aligned} f(\theta;\lambda) &= \sum_{n=0}^{\infty} a_n(\lambda)\, P_n(\cos\theta) \\ &= \sum_{n=0}^{\infty} \frac{i(2n+1) j_{n+\frac{1}{2}}(1)}{\lambda\, h_{n+\frac{1}{2}}(1)} P_n(\cos\theta) \end{aligned} \qquad (2.11)$$

where for Problem I the $a_n(\lambda)$, $n = 0,1,\ldots$, are assumed known. Note however that the functions $j_{n+\frac{1}{2}}(r)$ and $h_{n+\frac{1}{2}}(r)$ are unknown in this case since $B(r)$ is of yet unknown. We now equate like powers of λ for each term in the series (2.11) and use (2.8) to arrive at the identity

$$\mu_n = \int_1^a B(s)\left[s^{2n+2} + s^{-2n} - 2s\right] ds \quad ; \quad n = 0,1,\ldots \qquad (2.12)$$

where the μ_n are constants obtained from the Taylor coefficients of $a_n(\lambda)$ ([4], [6]). The following theorem shows that (2.12) uniquely determines B(r):

Theorem ([4], [6]): The functions $P_n(r) = r^{2n+2} + r^{-2n} - 2r$, $n = 0,1,2,\ldots$, are complete in $L^2[1,a]$.

Proof: Let $f(r) \in C^0[1,a]$ and for $r \in [\frac{1}{a},1]$ define f(r) by $f(r) = r^{-4} f(\frac{1}{r})$. Then if

$$\int_1^a f(s)P_n(s)ds = 0 \tag{2.13}$$

for n=0,1,2,... we have

$$\begin{aligned} 0 &= \int_1^a f(s)\left[P_n(s) - P_{n+1}(s)\right]ds \\ &= \frac{1}{2}\int_{1/a^2}^{a^2} f(s^{1/2})\left[s^{1/2}-s^{3/2}\right]s^n ds , \end{aligned} \tag{2.14}$$

and since the set $\{r^n\}_{n=0}^{\infty}$ is complete in $L^2[\frac{1}{a^2}, a^2]$ we have $f(r^{1/2})[r^{1/2}-r^{3/2}] = 0$ for $r \in [\frac{1}{a^2}, a^2]$. This implies that $f(r) = 0$ for $r \in [1,a]$ and the Theorem follows.

We can now approximate B(r) in $L^2[1,a]$ by orthonormalizing the set $\{P_n(r)\}_{n=0}^{\infty}$ over the interval $[1,a]$ to obtain the orthonormal set $\{\phi_n(r)\}_{n=0}^{\infty}$ and then approximating B(r) by the function

$$B_N(r) = \sum_{n=0}^{N} b_n \phi_n(r) \tag{2.15}$$

where

$$\begin{aligned} b_n &= \int_1^a \phi_n(s)B(s)ds \\ &= \text{linear combination of the } \mu_j,\ 0 \leqslant j \leqslant n. \end{aligned} \tag{2.16}$$

For an analysis of a similar problem using the theory of integral equations see [19].

Open Question II: Can similar results be obtained in the case when D is not a sphere? What changes must be made when B(r) no longer has compact support?

III. Problem II.

We now consider Problem II, i.e. given the far field pattern $f(\theta,\phi;\lambda)$ to determine the shape of the scattering body D in a homogeneous medium. Under the assumption that D is bounded we can conclude (c.f. [17]) that $u(r,\theta,\phi)$ is known in the exterior of the smallest ball S containing D in its interior, where S can be determined from a knowledge of $f(\theta,\phi;\lambda)$. In particular if the radius of S is a, we can write (c.f. [17])

$$u(r,\theta,\phi) = \sum_{n=0}^{\infty} \sum_{m=-n}^{n} a_{mn}(\lambda) h_n^{(1)}(\lambda r) Y_{nm}(\theta,\phi) \quad ; \quad r \geq a \tag{3.1}$$

where the coefficients $a_{nm}(\lambda)$ are determined from the far field pattern, $h_n^{(1)}$ denotesa spherical Hankel function, Y_{nm} a spherical harmonic, and the series (3.1) is uniformly convergent for $r \geq a$, $0 \leq \theta \leq \pi$, $0 \leq \phi \leq 2\pi$, $\lambda_o \leq \lambda \leq \lambda_1$. Hence to find D we must analytically continue $u(r,\theta,\phi)$(as given by (3.1)) across the boundary of S and look for the locus ∂D of points where $u(r,\theta,\phi)+\exp(i\lambda r\cos\theta)=0$. (We note that it follows from the results of [14], pp.173-174, that ∂D is unique). From a practical point of view we would first determine the domain of regularity of $u(r,\theta,\phi)$ by the methods we are about to describe, establish an a priori bound on $u(r,\theta,\phi)$ from physical considerations, and then use a stabilized numerical analytical continuation approach (c.f. [13]) to determine the locus ∂D.

We note that the problem of the analytic continuation of solutions to the Helmholtz equation and its connection with Problem II has been investigated by many mathematicians, in particular Colton ([3],[5]), Hartman and Wilcox ([11]), Karp ([12]), Millar ([15],[16]), Müller ([17]), Sleeman ([20]), and Weston, Bowman and Ar ([22]). We shall describe a new approach to this problem based on the theory of integral operators for partial differential equations.

We shall need two operators which map solutions of Laplace's equation

$$\Delta_3 h = 0 \tag{3.2}$$

onto solutions of the Helmholtz equation

$$\Delta_3 u + \lambda^2 u = 0 \ . \tag{3.3}$$

These operators are related to a class of operators constructed by I.N. Vekua ([21], pp.57-61) and to R.P. Gilbert's "method of ascent" ([10]). Let D* be a bounded starlike domain containing the ball S and let $h(r,\theta,\phi) \in C^2(D^* \setminus \bar{S}) \cap C^0(D^* \setminus S)$ be a solution of (3.2) such that

$$h(a,\theta,\phi) = 0 \ . \tag{3.4}$$

Then we define the operator $\underset{\sim}{I}+\underset{\sim}{K}_1$ by ([7])

$$\begin{aligned} u(r,\theta,\phi) &= (\underset{\sim}{I}+\underset{\sim}{K}_1)h \\ &= h(r,\theta,\phi) + \int_a^r K(r,s;\lambda)h(s,\theta,\phi)ds \end{aligned} \tag{3.5}$$

where $K(r,s;\lambda)$ is the twice continuously differentiable solution of

$$r^2\left[K_{rr} + \frac{2}{r}K_r + \lambda^2 K\right] = s^2\left[K_{ss} + \frac{2}{s}K_s\right] \tag{3.6}$$

in the cone $\{(r,s) : 0 < r \leq s \leq a \text{ or } r \geq s \geq a > 0\}$ satisfying the initial data

$$K(r,r;\lambda) = -\frac{\lambda^2}{4r}(r^2-a^2) \tag{3.7}$$

$$K(r,a;\lambda) = 0. \tag{3.8}$$

$K(r,s;\lambda)$ can be constructed by standard methods ([7]) and it can be easily verified that the operator $\underset{\sim}{I}+\underset{\sim}{K}_1$ maps solutions of (3.2) defined in $D^*\setminus S$ and satisfying (3.4) onto solutions of (3.3) defined in the same domain and satisfying the same boundary data on ∂S. We now define a second operator $\underset{\sim}{I}+\underset{\sim}{K}_2$ which maps solutions $h(r,\theta,\phi)$ of (3.2), $h(r,\theta,\phi) \in C^2(D^*\setminus\bar{S})\cap C^1(D^*\setminus S)$ defined in $D^*\setminus\bar{S}$ and satisfying

$$h_r(a,\theta,\phi) + \frac{1}{2a}h(a,\theta,\phi) = 0, \tag{3.9}$$

onto solutions $u(r,\theta,\phi)$ of (3.3) in $D^*\setminus\bar{S}$ satisfying the same boundary data on ∂S. This operator is defined by

$$\begin{aligned} u(r,\theta,\phi) &= (\underset{\sim}{I}+\underset{\sim}{K}_2)h \\ &= h(r,\theta,\phi) + \int_a^r K(r,s;\lambda)h(s,\theta,\phi)ds \end{aligned} \tag{3.10}$$

where $K(r,s;\lambda)$ is the twice continuously differentiable solution of (3.6) in the cone $\{(r,s) : 0 < r \leqslant s \leqslant a \text{ or } r \geqslant s \geqslant a > 0\}$ satisfying the initial data

$$K(r,r;\lambda) = -\frac{\lambda^2}{4r}(r^2-a^2) \tag{3.11}$$

$$K_s(r,a;\lambda) + \frac{1}{2a}K(r,a;\lambda) = 0. \tag{3.12}$$

$K(r,s;\lambda)$ can again be constructed by standard methods ([7]).

By using the operator $\underset{\sim}{I}+\underset{\sim}{K}_1$, the Schwarz reflection principle for harmonic functions, and the fact that $\underset{\sim}{I}+\underset{\sim}{K}_1$ is invertible, we immediately have the following theorem ([7]):

Theorem (Reflection Principle): Let $u(r,\theta,\phi) \in C^2(D^*\setminus\bar{S})\cap C^0(D^*\setminus S)$ be a solution of (3.3) such that $u(a,\theta,\phi) = 0$ and let $\bar{S}\setminus D$ denote the set obtained by inverting $D^*\setminus S$ across ∂S, i.e. $(r,\theta,\phi) \in \bar{S}\setminus D$ if and only if $(\frac{a^2}{r},\theta,\phi) \in D^*\setminus S$. Then $u(r,\theta,\phi)$ is a twice continuously differentiable (and hence analytic) solution of (3.3) in $D^*\setminus S\cup\bar{S}\setminus D$.

The above Theorem is not strong enough for the purpose of studying Problem II since if $u(r,\theta,\phi)$ is a solution of (3.3) in $\mathbb{R}^3\setminus D$ satisfying the radiation condition (1.3) and vanishing on ∂S then $u(r,\theta,\phi)$ is identically zero by Rellich's uniqueness theorem. Hence we now use the operator $\underset{\sim}{I}+\underset{\sim}{K}_2$ to obtain a stronger version of the above reflection principle ([7]):

Theorem (Generalized Reflection Principle): Let $h(r,\theta,\phi)$ be the (unique) harmonic function defined in the exterior of the ball S such that $h(a,\theta,\phi) = u(a,\theta,\phi)$ on ∂S where $u(r,\theta,\phi)$ is a solution of (3.3) in the exterior of S. If $h(r,\theta,\phi)$ can be continued to a harmonic function defined in the exterior of a starlike domain $D \subset S$, then $u(r,\theta,\phi)$ can be continued as a solution of (3.3) into the exterior of D.

Proof: Let $\tilde{h}(r,\theta,\phi)$ be the harmonic function defined by

$$\tilde{h}(r,\theta,\phi) = \frac{1}{2}\left[h(r,\theta,\phi) + \left(\frac{a}{r}\right)h\left(\frac{a^2}{r},\theta,\phi\right)\right]. \tag{3.13}$$

Then $\tilde{h}_r(a,\theta,\phi) + \frac{1}{2a}\tilde{h}(a,\theta,\phi) = 0$ and $\tilde{u} = (\underset{\sim}{I}+\underset{\sim}{K}_2)\tilde{h}$ is a solution of (3.3) in $D^* \cup S\setminus D$ such that $\tilde{u}(a,\theta,\phi) = u(a,\theta,\phi)$ (D^* denotes the inversion of $\bar{S}\setminus D$ across ∂S). Hence $w(r,\theta,\phi) = u(r,\theta,\phi) - \tilde{u}(r,\theta,\phi)$ is a solution of (3.3) in D^* such that $w(a,\theta,\phi) = 0$ and hence by the Reflection Principle $w(r,\theta,\phi)$ is a solution of (3.3) in $D^*\cup S\setminus D$. We can now conclude that $u(r,\theta,\phi)$ is a solution of (3.3) in the exterior of D.

In order to apply the Generalized Reflection Principle to Problem II it is necessary to have a method for determining the location of the singularities of the harmonic function

$$h(r,\theta,\phi) = \sum_{n=0}^{\infty} \sum_{m=-n}^{n} a_{nm} h_n^{(1)}(\lambda a)\left(\frac{r}{a}\right)^{-n-1} Y_{nm}(\theta,\phi). \qquad (3.14)$$

But this theory has been extensively developed by Bergman ([1]) and Gilbert ([9],[10]). In particular by using Gilbert's envelope method and the Bergman-Whittaker operator we have that the singular points of $h(r,\theta,\phi)$ in $\mathbb{R}^3$ can be determined from a knowledge of the singular points of the analytic function

$$g(z_1,z_2) = \sum_{n=0}^{\infty} \sum_{m=-n}^{n} a_{nm} h_n^{(1)}(\lambda a) z_1^n z_2^m \qquad (3.15)$$

in $\mathbb{C}^2$, the space of two complex variables. Methods for determining the singular points of $g(z_1,z_2)$ can be found in [9]. As an example of the type of result which can be obtained, consider the case when $u(r,\theta,\phi) = u(r,\theta)$ is axially symmetric (i.e. independent of ϕ). Then $g(z_1,z_2) = g(z_1)$ is an analytic function of a single complex variable and the far field pattern $f(\theta,\phi) = h(\cos\theta)$ is a function only of $\xi = \cos\theta$. In this case we have the following result:

<u>Theorem ([3],[5]):</u> Let

$$F(z) = \int_{-1}^{1} h(\xi) \frac{(1+4z^2)d\xi}{(1-4iz\xi-4z^2)^{3/2}} ; \quad |z| < \frac{1}{2} .$$

Then $F(z)$ can be continued to an entire function of exponential type. If I is the indicator diagram of $F(z)$ (c.f. [2]) then $u(r,\theta)$ is regular in the exterior of $I \cup \bar{I}$ (where the bar denotes conjugation).

<u>Open Question III:</u> Can the analysis of this section be extended to the case when D may be unbounded, but is contained in some paraboloid of revolution?

References

1. S.Bergman, Integral Operators in the Theory of Linear Partial Differential Equations, Springer-Verlag, Berlin, 1969.

2. R.P. Boas, Entire Functions, Academic Press, New York, 1954.

3. D.Colton, Partial Differential Equations in the Complex Domain, Pitman Press, London, 1976.

4. D.Colton, The Solution of Boundary Value Problems by the Method of Integral Operators, Pitman Press, London, to appear.

5. D.Colton, On the inverse scattering problem for axially symmetric solutions of the Helmholtz equation, Quart.J.Math.22 (1971), 125-130.

6. D.Colton, The inverse scattering problem for acoustic waves in a spherically stratified medium, Proc.Edin.Math.Soc., to appear.

7. D.Colton, A reflection principle for solutions to the Helmholtz equation and an application to the inverse scattering problem, Glasgow Math.J. to appear.

8. D.Colton and W.Wendland, Constructive methods for solving the exterior Neumann problem for the reduced wave equation in a spherically symmetric medium, Proc.Roy.Soc.Edin., to appear.

9. R.P.Gilbert, Function Theoretic Methods in Partial Differential Equations, Academic Press, New York, 1969.

10. R.P.Gilbert, Constructive Methods for Elliptic Equations, Springer-Verlag Lecture Note Series Vol.365, Berlin, 1974.

11. P.Hartman and C.Wilcox, On solutions of the Helmholtz equation in exterior domains, Math.Zeit. 75 (1961), 228-255.

12. S.N.Karp, Far field amplitudes and inverse diffraction theory, in Electromagnetic Waves, R.E.Langer, editor, University of Wisconsin Press, Madison, 1962, 291-300.

13. M.M.Lavrentiev, Some Improperly Posed Problems of Mathematical Physics, Springer-Verlag, Berlin, 1967.

14. P.D.Lax and R.S.Phillips, Scattering Theory, Academic Press, New York, 1967.

15. R.F.Millar, Singularities of two-dimensional exterior solutions of the Helmholtz equation, Proc.Camb.Phil.Soc. 69 (1971), 175-188.

16. R.F.Millar, Singularities of solutions to exterior analytic boundary value problems for the Helmholtz equation in three independent variables I. The plane boundary, SIAM J.Math.Anal., to appear.

17. C.Müller, Radiation patterns and radiation fields, J.Rat.Mech.Anal. 4 (1955), 235-246.

18. L.E.Payne, Improperly Posed Problems in Partial Differential Equations, SIAM Publications, Philadelphia, 1975.

19. C.Rorres, Low energy scattering by an inhomogeneous medium and by a potential, Arch.Rat.Mech.Anal.39 (1970), 340-357.

20. B.D.Sleeman, The three-dimensional inverse scattering problem for the Helmholtz equation, Proc.Camb.Phil.Soc.73 (1973), 477-488.

21. I.N.Vekua, New Methods for Solving Elliptic Equations, John Wiley, New York, 1967.

22. V.H.Weston, J.J.Bowman and E.Ar, On the inverse electromagnetic scattering problem, Arch.Rat.Mech.Anal. 31 (1968),199-213.

Department of Mathematics,
University of Strathclyde,
Glasgow, Scotland.

STUDY of PARTIAL DIFFERENTIAL EQUATIONS by the MEANS of GENERALIZED ANALYTICAL FUNCTIONS

by

A. DŽURAEV

Introduction

It is well known that solutions of partial differential equations of elliptic type are smooth if the coefficients are sufficiently smooth functions. For a subclass of elliptic equations the structure of the solutions is described by some integral operators mapping analytic or generalized analytic functions on the solutions. This representation allows to investigate the properties of solutions of these equations and also of boundary value problems for them based on the properties of analytic and generalized analytic functions.

Let us recall that a complex valued function $w(z) = u(x,y) + iv(x,y) \in W_p^1(G)$ is called a generalized analytic function of the class $\mathfrak{U}_p$ in the domain $G \subset R^2$ if w satisfies an elliptic equation

$$\frac{\partial w}{\partial \bar{z}} + A(z)w + B(z)\bar{w} = 0 \quad , \qquad (1.1)$$

where $A(z)$, $B(z)$ - are given functions in $L_p(\bar{G})$, $p > 2$ and $\frac{\partial w}{\partial \bar{z}} = \frac{1}{2}(\frac{\partial}{\partial x} + i\frac{\partial}{\partial y})w$ [1]. A more general class of elliptic equations is given by

$$\frac{\partial w}{\partial \bar{z}} - q(z)\frac{\partial w}{\partial z} + A(z)w + B(z)\bar{w} = f(z) , \qquad (1.2)$$

where $q(z)$ is a sufficiently smooth function fulfilling

$$|q(z)| < 1$$

for all z and where

$$\frac{\partial w}{\partial z} = \frac{1}{2}(\frac{\partial w}{\partial x} - i\frac{\partial w}{\partial y}) . \qquad (1.3)$$

Functions satisfying the homogeneous system (1.2) with $f(z) \equiv 0$ have all the properties of generalized analytic functions, too. Moreover, for every right hand side $f \in L_p(\bar{G})$, $p > 2$, the inhomogeneous equation (1.2) has always solutions in $\mathfrak{U}_p$. In the case $G \equiv R^2$, the equation (1.2) can also be solved in the class $\mathfrak{U}_{p,2}$ provided that $A(z)$, $B(z)$, $f(z) \in L_{p,2}(R^2)$. The space $L_{p,2}(R^2)$ consists of all functions $f(z)$ for which f and $f^*(z) \equiv |z|^{-2}f(\frac{1}{z})$ belong to $L_p(|z| \leq 1)$. In this case, Liouville's theorem holds for the solutions w of (1.2) if $f \equiv 0$:

Every solution w of the homogeneous equation (1.2) vanishing at infinity must vanish in the whole plane.

But both properties do not hold anymore if the coefficients and the right hand side are not in $L_{p,2}(R)$ for some $p > 2$ (see [2]). An example for such an equation is

$$\frac{\partial w}{\partial \bar{z}} - e^{i\phi}\,\frac{k(|z|)}{2}\,w = f(z), \quad \phi = \arg z \tag{1.3'}$$

in R^2 where $k(\tau)$ is an increasing real function with $\omega(\tau) = \int_o^\tau k(\rho)d\rho \to \infty$ for $\tau \to \infty$.

It's easy to see that the homogeneous equation

$$\frac{\partial v}{\partial \bar{z}} + e^{i\phi}\,\frac{k(|z|)}{2}\,v = o \tag{1.3''}$$

adjoint to (I.3') has infinitely many linearly independent solutions

$$v_n(z) = z^n \cdot e^{-\omega(|z|)}, \quad n = o,1,2,\ldots \tag{1.4}$$

which are continuous on R^2 and vanish at infinity. Moreover, generally speaking, the inhomogeneous equation (1.3) is unsolvable in the class of continuous functions in R^2.

If $f(z)$ has compact support then the solution of the inhomogeneous equation may be represented formally by

$$w(z) = -\frac{e^{\omega(|z|)}}{\pi} \iint \frac{e^{-\omega(|\zeta|)}}{\zeta - z}\, f(\zeta)d\sigma_\zeta \tag{1.5}$$

Choosing R sufficiently large, (I.5) has an expansion

$$w(z) = e^{\omega(|z|)} \sum_{k=o}^{\infty} \frac{a_k}{z^{k+1}}, \quad a_k = \frac{1}{\pi}\iint \zeta^k \cdot e^{-\omega(|\zeta|)} \cdot f(\zeta)d\sigma_\zeta \tag{1.6}$$

for $|z| > R$. Hence, w is continuous in R^2 if and only if the conditions

$$\pi a_k = \iint v_n(z)f(z)d\sigma = o, \quad n = o,1,2,\ldots \tag{1.7}$$

are satisfied

where the $v_n(z) = z^n e^{-\omega(|z|)}$ are the linear independent, continuous solutions of the adjoint equation (1.3")in R^2 vanishing at infinity.

2. Systems of equations of composite type and their boundary value problems

The above elliptic equations are special cases of the following real, first order system of partial differential equations in R^2,

$$U_x - AU_y - BU = F,$$

where A, B are given real quadratic $n \times n$ matrix valued functions and F is a given real vector valued function. $U = (u_1, u_2, \dots, u_n)$ denotes the unknown vector function.

The ellipticity of the system (2.1) in some domain $G \subset R^2$ means that its symbol

$$\sigma(x,y,\xi) = \det (I\xi_1 - A(x,y)\xi_2) \tag{2.2}$$

never vanishes for all $(x,y) \in G$ and any real vector $\xi = (\xi_1, \xi_2) \in R^2 \setminus 0$. Such symbols are called elliptic in G. Obviously, for elliptic systems, n has to be even, necessarily.

The simple system

$$\frac{\partial u_1}{\partial x} = 0, \quad \frac{\partial u_2}{\partial x} - \frac{\partial u_3}{\partial y} = 0, \quad \frac{\partial u_3}{\partial x} + \frac{\partial u_2}{\partial y} = 0 \tag{2.3}$$

has the symbol $\sigma = \xi_1(\xi_1^2 + \xi_2^2)$ and is evidently not elliptic in R^2.
The system (2.3) belongs to the class of systems of composite type. The simplest scalar equation of composite type is

$$\frac{\partial^3 u}{\partial x^3} + \frac{\partial^3 u}{\partial x \partial y^2} = 0$$

which has the symbol $\sigma = \xi_1(\xi_1^2 + \xi_1^2)$. Boundary value problems for this equation have been investigated by J. Hadamard [3]. Further investigations on scalar equations of composite type are contained in [4] - [7]. Note also the work [8] where the boundary value problems for the equation

$$y\frac{\partial^3 u}{\partial x^3} + \frac{\partial^3 u}{\partial y^3} = 0$$

have been investigated. This equation has the symbol $\sigma(y,\xi) = \xi_1(y\xi_1^2 + \xi_2^2)$ which behaves essentially differently for $y > 0$ and for $y < 0$. For the system (2.1) of three equations of composite type the author formulated 1964 the boundary value problem [9]. Then these investigations were continued by the author and summed in his monography [10]. They were continued by some other authors, too, among which we may note, for example the work [11].

Here we shall consider only systems (2.1) of composite type consisting of three first order equations in a bounded domain G. It can be shown that these systems can be transformed by means of a linear transformation of the unknowns and a homeomorphic transformation of (x,y) into a new system of the form

$$\begin{aligned} \frac{\partial u}{\partial y} &= A_1(z)u + \operatorname{Re}\,[B_1(z)w], \\ \frac{\partial w}{\partial \bar z} - q(z)\frac{\partial w}{\partial z} &= A_0(z)u + B_0(z)w + C_0(z)\bar w, \end{aligned} \tag{2.4}$$

where $A_1(z)$ is a given real valued function $|q(z)| < 1$, where A_0, B_j, C_0

complex valued functions in $L_p(\bar{G})$, $p > 2$ and where $u(z)$ and $w(z)$ are the new real valued and complex valued unknowns, respectively. The characteristics of the system (2.4) are the straight lines $x = \text{const}$. Now we assume that every characteristic passing through the domain G intersects the boundary Γ in exactly two points, the "incoming" and "outcoming" points of G. Further we assume that two of the characteristics are tangent to Γ limiting the domain G from the left and the right side, respectively. These two characteristics devide the boundary Γ into two parts γ and $\Gamma - \gamma$.

Now one of the main boundary value problems for (2.4) can be formulated as follows.

Problem A: Find all regular solutions of the system (2.4) satisfying the boundary conditions,

$$\begin{aligned} a_o(t)u(t) + \operatorname{Re}\,[a^o(t)w(t)] &= h_o(t) \quad \text{for } t \in \Gamma, \\ a_1(t)u(t) + \operatorname{Re}\,[a^1(t)w(t)] &= h_1(t) \quad \text{for } t \in \gamma \;. \end{aligned} \tag{2.5}$$

Let us assume that the coeffitions and the right hand sides in (2.5) are Hölder continuous and that, moreover, the following conditions are fulfilled:

$$\begin{aligned} &\Delta(t) = a^o(t)a_1(t) - a^1(t)\, a_o(t) \neq o \;, \quad t \in \bar{\gamma}, \\ &a^o(t) \neq o \;, \quad t \in \overline{\Gamma - \gamma} \end{aligned} \tag{2.6}$$

If $h_o = h_1 \equiv o$ then the homogeneous problem (2.5) is denoted by A_o.

Besides the problem A we consider also the homogeneous

Adjoint problem A*: Find all regular solutions of the adjoint system

$$\frac{\partial u^*}{\partial y} = - A_1(z)u^* + \operatorname{Re}[A_o(z)w^*] \;,$$

$$\frac{\partial w^*}{\partial \bar{z}} - \frac{\partial}{\partial z}(q(z)w^*) = B_1(z)u^* - B_o(z)w^* - \overline{C_o(z)w^*} \;,$$

satisfying the boundary conditions,

$$- \operatorname{Im} \frac{d(t)}{\Delta(t)} \cdot x'(s)u^*(t) - 1/2 \operatorname{Re}\, [\frac{\theta(t)}{\Delta(t)} w^*(t)] = o \qquad \text{on } \gamma$$

$$x'(s)u^*(t) + \frac{a_o(t)}{2} (\operatorname{Re}\,[\frac{i\theta(t)}{a^o(t)} w^*(t)] = o \;, \tag{2.7}$$

$$\operatorname{Re}\,[\frac{\theta(t)}{a^o(t)} w^*(t)] = o \qquad \text{on } \Gamma - \gamma \;,$$

$$\text{where } \; d(t) = i\,(a^o(t)\operatorname{Re}\frac{ia^1(t)}{\Delta(t)} - a^1(t)\operatorname{Re}\frac{ia^o(t)}{\Delta(t)}), \; \theta(t) = \frac{dt}{ds} + q(t)\,\overline{\frac{dt}{ds}}$$

and where $t = t(\zeta)$ denotes the parametric equation of the boundary curve Γ.

If the coefficients satisfy the conditions $a_o = o$ and $a_1 = 1$ at the endpoints of γ then the function a^* defined by

$$a^*(t) = \begin{cases} \Delta(t) & \text{on } \gamma , \\ a^o(t) & \text{on } \Gamma - \gamma \end{cases}$$

becomes continuous on Γ. In this case, the index of the vector field a^* on Γ,

$$\kappa \equiv (2\pi)^{-1} \{\arg \overline{a^*(t)}\}_\Gamma , \tag{2.8}$$

is well defined in the usual way.

Now we are in the position to formulate the main results.

Theorem 1: The homogeneous problems A_o and A_o^* have only finitely many linearly independent solutions. The index of the problem is given by

$$\ell - \ell^* = 2\kappa + 1 \tag{2.9}$$

where ℓ and ℓ^* denote the numbers of linearly independent solutions of A_o and A_o^*, respectively. Here, the linear independence is understood over the field of the real numbers.

Theorem 2: The conditions for h^o, h^1,

$$\int_\gamma h^o(t)\left[\frac{\theta(t)}{2i\Delta(t)} \overset{*}{w}(t) + \frac{d(t)}{\Delta(t)} x'(s)\, u^*(t)\right] ds + \\ + \int_\gamma h^1(t)u^*(t)dx + \int_{\Gamma-\gamma} h_o(t) \frac{\theta(t)}{2ia^o(t)} w^*(t)ds = o \tag{2.10}$$

with all solutions ($u^*(t)$, $w^*(t)$) of the adjoint problem A_o^* are necessary and sufficient for the solvability of the boundary value problem A .

The proofs of both theorems are based on a reduction of A and A_o^* to certain boundary value problems for generalized analytic functions in G . For this purpose, a special homeomorphism $\alpha(t)$ of Γ onto itself is introduced which yields the boundary condition with displacement on Γ, for the generalized analytic function ϕ

$$\text{Re}\, [a^*(t)\, \phi(t) + b^*(t)\, \phi(\alpha(t)) + K(\phi)] = h^*(t), \tag{2.11}$$

where K is some smooth integral operator and the function $b^*(t)$ satisfies on Γ the relations

$$b^*(t)b^*(\alpha(t)) \equiv o \quad \text{but} \quad b^*(t) \not\equiv o \ . \tag{2.12}$$

Both the reduced problems "with displacement" corresponding to A and A_o^* lead to singular integrofunctional equations which are adjoint mutually. The solvability theory for the latter hinges on condition (2.12).

This method can be extended to more general boundary value problems for (2.4) where the boundary operators (2.5) are replaced by arbitrary linear differential operators. Finally, this method allows also the investigation of boundary value problems for higher order systems of composite type in two independent variables.

Another interesting question is the extension of these results to higher dimensions than two. In three dimensions we investigated the special first order system of type A,

$$\frac{\partial u_1}{\partial x_1} - \frac{\partial u_2}{\partial x_3} = 0 \ , \quad \frac{\partial u_2}{\partial x_1} + \frac{\partial u_1}{\partial x_3} - \frac{\partial u_3}{\partial x_2} = 0, \quad \frac{\partial u_3}{\partial x_1} + \frac{\partial u_2}{\partial x_2} = 0, \tag{2.13}$$

in the three dimensional unit ball. The symbol of (2.13) is $\sigma = \xi_1(\xi_1^2 + \xi_2^2 + \xi_3^2)$ and, hence, (2.13) is a direct generalization of the composite type system (2.3) to three dimensions. Analogously, in four dimensions with coordinates (t,x_1,x_2,x_3), the corresponding system of composite type is given by

$$\frac{\partial u}{\partial t} + \text{rot} \quad u = 0 \ , \quad u = (u_1,u_2,u_3) \tag{2.14}$$

with the symbol $\sigma = \tau(\tau^2 + \xi_1^2 + \xi_2^2 + \xi_3^2)$. For (2.14) and for systems

$$\frac{\partial u}{\partial t} + \text{rot} \quad u + ku = f \ ,$$

some boundary value problems were recently investigated in [12] by the use of functional analytic methods. It is still an interesting problem to find the solvability conditions and the index in terms of the coefficients in analogy to the two dimensional problem A .

3. Other problems

As we have seen above, for the equation (1.3'') the Liouville theorem is no longer valid and the equation (1.3') is in general not solvable in R^2. Both difficulties arise from the behavior of the coefficients at infinity when they do not belong to any $L_{p,2}(R^2)$ with $p > 2$. In the following let us consider some degenerating elliptic equations in the plane and, moreover, a class of equations in R^3 including the famous equation of Hans Levy for which take place analogous phenomena.
To this end let us consider the infinite system of functions

$$w_n(z) = \begin{cases} z^n & \text{for} \quad |z| < 1 \ , \\ 1/\bar{z}^n & \text{for} \quad |z| < 1 \ , \quad n = 1,2,3,\dots. \end{cases} \tag{3.1}$$

Obviously, these functions are linearly independent, they are continuous in the whole plane vanishing at infinity and they satisfy the equation

$$\text{sign}\,(1-|z|)\,\frac{\partial w}{\partial x} + i\,\frac{\partial w}{\partial y} = 0 \ . \tag{3.2}$$

On the other hand, the infinite system of functions

$$u_k(z) = Z_n^k(z)/W_n^{k+2}(z) , \quad k = 0,1,2,\ldots, \quad \text{where}$$

$$Z_n(z) = 2(n+1)x + i(y^{2(n+1)}-2(n+1)), \quad W_n(z) = 2(n+1)x + i(y^{2(n+1)}+2(n+1)) \tag{3.3}$$

form a system of solutions of the equation

$$y^{2n+1} \frac{\partial u}{\partial x} + i \frac{\partial u}{\partial y} = 0 \tag{3.4}$$

where $n \geq 0$ is any integer.

Using these functions it can be shown that the inhomogeneous equation

$$\operatorname{sign}(1-|z|) \frac{\partial w}{\partial x} + i \frac{\partial w}{\partial y} = f(z)$$

is solvable in $C^1(R^2)$ if and only if its right hand side $f(z)$ satisfies the conditions

$$\iint\limits_{|z|<1} f(z)z^n \, dxdy + \iint\limits_{|z|>1} \frac{f(z)}{\bar{z}^{n+2}} \, dxdy = 0, \quad n = 0,1,2,\ldots \tag{3.5}$$

Correspondingly, the equation

$$y^{2n+1} \frac{\partial u}{\partial x} + i \frac{\partial u}{\partial y} = f(z)$$

is solvable in $C^1(R^2)$ if and only if the conditions

$$\iint\limits_{R^2} f(z) \, Z_h^k(z)/W_n^{k+2}(z) \, dxdy = 0, \quad k = 0,1,2,\ldots \tag{3.6}$$

are fulfilled. Using generalized analytic functions we can show, for example, the following

<u>Theorem 3. If $b(z) \neq 0$ for $|z| = 1$ then the equation</u>

$$\operatorname{sign}(1-|z|) \frac{\partial w}{\partial x} + i \frac{\partial w}{\partial y} + a(z)w + b(z)\bar{w} = f(z) \tag{3.7}$$

<u>is solvable with $w \in C^1(R^2)$ and w vanishing at infinity if and only if the finitely many conditions</u>

$$\operatorname{Re} \iint\limits_{R^2} f(z)v_j(z)dxdy = 0, \quad j = 1,2,\ldots,\ell^* , \tag{3.8}$$

<u>are fulfilled, where the $\{v_j(z)\}$ denote a certain system of linearly independent functions in R^2. Moreover, the index of equation (3.7) is equal to</u>

$$(2/\pi) \; \{\arg b(z)\}_{|z|=1} - 3 \; .$$

In three dimensions, let us consider the equation

$$\frac{\partial u}{\partial \bar{z}} - ie^{i\phi} \frac{k(|z|)}{2} \frac{\partial u}{\partial t} = f(z,t), \; z = x+iy, \; \phi = \arg z \tag{3.9}$$

where $k(|z|)$ is a function as in (1.3'). For $k(|z|) \equiv 2|z|$ (3.9) this is the Levy equation. For the right hand side $f(z,t)$ in (3.9) let us assume that f is a bounded function in z for every fixed t and absolutely integrable with respect to t for every fixed z. The unknown $u(z,t)$ is supposed to be absolutely integrable in t for every fixed z. The Fourier transform of (3.9) with respect to t yields the equation

$$\frac{\partial \tilde{u}}{\partial \bar{z}} + \tau e^{i\phi} \frac{k(|z|)}{2} \tilde{u} = \tilde{f}. \tag{3.10}$$

This is an equation of the form (1.3) depending on the parameter τ. According to chapter 1, for the solvability of equation (3.10) in $C^1(R^2)$ it is necessary and sufficient that the solvability conditions,

$$\iint e^{\tau\omega(|z|)} z^n \tilde{f}(z,\tau)d\sigma = o , \quad n = o,1,2,\ldots,$$

are fulfilled for every $\tau < o$. Moreover, it is easy to see that an infinite system of the functions

$$u_n(z,t) = z^n/ (z + i(\omega(|z|) + 1))^{n+1} , \quad n = o,1,2,\ldots$$

itself presents a system of linearly independent solutions in $C^1(R^3)$ of the homogeneous equation (3.9) vanishing at infinity.

All solutions of the homogeneous equation (3.9) in some three dimensional domain G can be represented by an analytic function of the two variables

$$z = x + iy \quad \text{and} \quad \zeta = t + i\,\omega(|z|) .$$

In particular, if the domain G is the exterior of the cylinder $|z| < 1, -\infty<t<\infty$ then every solution of the homogeneous equation (3.9) in G is of the form

$$u(z,t) = \sum_{k=o}^{\infty} \left[\frac{\phi_k(\zeta)}{z^{k+1}} + \psi_k(\zeta) z^k \right] , \tag{3.11}$$

where $\phi_k(\zeta)$, $\psi_k(\zeta)$ are holomorphic functions of ζ in the domain $\mathrm{Im}\,\zeta > \omega(1) > o$, $\zeta = t + i\,\omega(|z|)$. The first part of the right hand side of (3.11),

$$u(z,t) = \sum_{k=o}^{\infty} \phi_k(\zeta)/z^{k+1} , \tag{3.11'}$$

is a holomorphic function of z.

Now we formulate the following boundary value problem:

Find all C^1-solutions of the homogeneous equation (3.9) which are holomorphic with respect to z in the domain $G = \{(z,t) |\ |z| > 1 , -\infty < t < \infty\}$ satisfying the boundary condition

$$\mathrm{Re}\ [z^m \lambda(t)\, u(z,t)] = h(z,t) \tag{3.12}$$

on the cylindrical boundary surface $\{(z,t) |\ |z| = 1, -\infty < t < \infty\}$ where $\lambda(t)$ and $h(z,t)$ are given Hölder continuous functions and moreover,

$\lambda(t) \neq o , \quad -\infty \leq t \leq \infty.$

The representation (3.11) implies that the problem can be separated into a Riemann-Hilbert problem in the exterior of the unit circle and a family of linear adjoint boundary value problems in the half-plane. If $\kappa = (2\pi)^{-1} \{\arg \lambda(t)\}_{-\infty}^{\infty}$ then for $m \geq o$, and $\kappa \geq o$, the homogeneous problem $(h \equiv o)$ has exactly $(2m+1)(2\kappa+1)$ linearly independent solutions, whereas the inhomogeneous problem is always solvable. For $m < o$, the inhomogeneous problem is solvable if $h(z,t)$ satisfies finitely many solvability conditions. Moreover, these conditions can be formulated explicitely.

The Mathematical Institute with a Computation Centre of the Academy of Science of the Tajik SSR, Dushanbe, USSR.

BIBLIOGRAPHY.

I. И.Н. Векуа. Обобщенные аналитические функции. Физматгиз. М., I959.

2. I.N. Vekua. On one class of the elliptic systems with singularities. Proc. Int. Conf. on Functional Analysis and Related Topics, Tokyo, 1969.

3. J. Hadamard. Properties d'une equation lineair aux derivees partielles du quatrieme ordre; The Tohoku Math. J., v. 37, (1933), 133-150.

4. Sjöstrand O. Sur une equation aux derivees partielles du type composite, Arkiv f. Math. Astr. och Fys. Bd. 25A, N 21 (1936), 1-11.

5. Sjöstrand O. Sur une equation aux derivees partielles du type composite. Arkiv f. Math. Astr. och Fys. Bd. 26A, N 1 (1937), 1-10.

6. Davis R.B. A boundary value problem for third-order linear partial differential equation of composite type, Proc. Amer.Math.Soc. v. 3, N 5 (1952), 751-756.

7. Эскин Г.И., Краевая задача для уравнения $\frac{\partial}{\partial t} P\left(\frac{\partial}{\partial t}, \frac{\partial}{\partial x}\right) u = f$, где $P\left(\frac{\partial}{\partial t}, \frac{\partial}{\partial x}\right)$ - эллиптический оператор, Сиб. мат. ж., т. 3, № 6 (1962), 882-911.

8. Бицадзе А.В., Салахитдинов М.С. К теории уравнений смешанно-составного типа, Сиб. мат. ж., т. 2, № 1 (1961), 7-19.

9. А. Джураев. Граничные задачи для системы уравнений первого порядка составного типа, Доклады АН Таджикской ССР т. УП, № 10 (1964).

10. А. Джураев. Системы уравнений составного типа, "Наука", М., 1972.

11. Wolfersdorf L. von Sjöstrandsche Probleme der Richtungableitung bei einer Gleichung vom zusammengesetzten Typ. Math. Nachr., Bd.45, N 1-6 (1970), 263-277.

12. Берхин П.Е. О краевых задачах для некоторых дифференциальных уравнений и систем составного типа. Автореф.канд. диссерт. Новосибирск, 1976.

13. А. Джураев. О свойствах некоторых вырождающихся эллиптических систем на плоскости, Доклады АН СССР, т.223, № 3 (1975).

14. А. Джураев. Об одном способе исследования систем уравнений первого порядка в трехмерном пространстве, Доклады АН СССР, т. 223, № 5 (1975).

THE SINGLE LAYER POTENTIAL APPROACH IN THE THEORY OF BOUNDARY VALUE PROBLEMS FOR ELLIPTIC EQUATIONS.

Gaetano Fichera - Paolo Emilio Ricci (Rome).

Everybody knows that the Dirichlet boundary value problem for the Laplace equation in two independent variables

(1) $$\frac{\partial^2 u}{\partial x^2} + \frac{\partial^2 u}{\partial y^2} = 0$$

in a bounded domain A with a smooth boundary ∂A can be solved by representing the unknown function $u(z)$ $(z = x + iy)$ either by a double layer potential

(2) $$u(z) = \int_{\partial A} \psi(\zeta) \frac{\partial}{\partial \nu_\zeta} \log |z - \zeta| \, ds_\zeta$$

($\frac{\partial}{\partial \nu_\zeta}$ denotes differentiation along the inward normal at the point ζ of ∂A ; s_ζ is the curvilinear abscissa of ζ) or by a simple layer potential

(3) $$u(z) = \int_{\partial A} \varphi(\zeta) \log |z - \zeta| \, ds_\zeta .$$

By using representation (2), assuming that ∂A is a closed Jordan curve of class $C^{1+\lambda}$ [1], and imposing on u to assume on ∂A prescribed boundary values $f(z)$, one is led to a classical, well known Fredholm equation.

The use of the simple layer potential (3) requires that the boundary values of u on ∂A are expressed by a function $f(z)$ continuously differentiable on ∂A, then, by imposing that the derivative $\frac{\partial u}{\partial s}$ of u, with respect to the arc length, coincides with $\frac{\partial f}{\partial s}$, one gets

---

[1] We say that a closed Jordan curve is of class $C^{n+\lambda}$ (n positive integer) if it can be parametrically represented by a complex function $\zeta = \zeta(s)$ $(0 \le s \le \Lambda)$ which has continuous derivatives up to the order n in the closed interval $[0, \Lambda]$, satisfying the conditions: $\zeta^{(k)}(0) = \zeta^{(k)}(\Lambda)$ $(k = 0, 1, \ldots, n)$ and the Hölder condition

(*) $$|\zeta^{(n)}(s_1) - \zeta^{(n)}(s_2)| \le H |s_1 - s_2|^\lambda , \quad \forall s_1, s_2 \in [0, \Lambda]$$

where H is a positive constant and λ is some positive number ≤ 1.

a singular integral equation of the regular type (see [1] , p. 184).

By using (2) one gets that the solution $u(z)$ of the Dirichlet problem belongs to $C^0(\bar{A})$ provided $f \in C^0(\partial A)$.

The simple layer potential approach requires more restrictive conditions on f ,i.e. $f \in C^{1+\lambda}(\partial A)$ (2) , however this method gives a more regular solution for the Dirichlet problem, i.e. a solution belonging to the class $C^{1+\lambda}(\bar{A})$ (3) .

In 1957 S.Agmon [3] considered the equation

$$\sum_{k=0}^{2l} a_k \frac{\partial^{2l} u}{\partial x^{2l-k} \partial y^k} = 0 \tag{4}$$

with constant real coefficients. Supposing the equation (4) elliptic, Agmon considered the Dirichlet problem consisting in prescribing on ∂A the boundary values of u and its partial derivatives up to the order $l-1$. He succeeded in obtaining the solution of this problem by a potential-theoretic method which extends to Eq.(4) the double layer potential approach for Eq.(1).

Later, in 1960, G.Fichera considered the extension to higher order elliptic equations in two independent variables of the simple layer potential method [4]. Fichera considered the following elliptic equation with variable coefficients:

$$\sum_{k=0}^{2l} a_k(z) \frac{\partial^{2l} u}{\partial x^{2l-k} \partial y^k} + \sum_{r=0}^{2l-1} \sum_{j=0}^{2l-1-r} a_j^{(2l-1-r)}(z) \frac{\partial^{2l-1-r} u}{\partial x^{2l-1-r-j} \partial y^j} = 0 \; ; \tag{5}$$

under suitable hypotheses on the coefficients of Eq.(5), Fichera con-

---

(2) This means that $\frac{\partial f}{\partial s}$ is uniformly Hölder continuous in $[0, \Lambda]$ with some Hölder exponent λ.

(3) The class $C^{n+\lambda}(\bar{A})$ is formed by the functions which are continuously differentiable up to the order n with respect to the real variables x and y , moreover each partial derivative of order n is uniformly Hölder continuous in $\bar{A}$ with some exponent λ . In this paper, when we consider either functions or curves belonging to the class $C^{n+\lambda}$,we do not mean that the Hölder exponent λ must be the same in any case. For instance by saying that the solution of the Dirichlet problem obtained by (3) belongs to $C^{1+\lambda}(\bar{A})$ we do not mean that the Hölder exponent λ of u_x and u_y is the same λ which enters in the Hölder conditions satisfyed by $\zeta'(s)$ and $\frac{\partial f}{\partial s}$. In fact it is known that in general the Hölder exponent λ of u_x and u_y is such that $\lambda < \lambda'$,where λ' is the smallest of the two Hölder exponents pertaining to $\zeta'(s)$ and to $\frac{\partial f}{\partial s}$ (see [1], Chapt.6 ; [2], p.99).

structs a principal fundamental solution, in the sense of Giraud, $F(z,\zeta)$ for Eq. (5). In the particular case that Eq. (5) reduces to Eq. (4), one can assume as $F(z,\zeta)$ the same fundamental solution considered by Agmon.

Fichera defines the single layer potential for Eq. (5) as follows:

$$(6) \qquad u(z) = \sum_{k=0}^{l-1} \int_{\partial A} \varphi_k(\zeta) \frac{\partial^{l-1}}{\partial \xi^{l-1-k} \partial \eta^k} F(z,\zeta) ds_\zeta .$$

Supposing that on the boundary ∂A the following Dirichlet conditions are prescribed:

$$\frac{\partial^{l-1} u}{\partial x^{l-1-h} \partial y^h} = f_h(s) \qquad (h = 0, 1, \ldots, l-1),$$

by imposing on the function u represented by (6) the boundary conditions

$$\frac{\partial}{\partial s} \frac{\partial^{l-1} u}{\partial x^{l-1-h} \partial y^h} = \frac{\partial}{\partial s} f_h(s) \qquad (h = 0, 1, \ldots, l-1),$$

one gets a system of singular integral equations of the regular type. The analysis of this system leads to the existence of a solution of the Dirichlet problem which belongs to the class $C^{\phi \cdot + \lambda}(\bar{A})$, provided $f_h \in C^{1+\lambda}(\partial A)$.

More recently Fichera's method has been adroitly used, with some modifications, by R.C. MacCamy [5], G.Hsiao & R.C. MacCamy [6]; G.Hsiao & W.L.Wendland [7] for solving concrete problems of applications and getting numerical approximations.

P.E.Ricci [8] has extended Fichera's simple layer method to more general boundary conditions. Actually Ricci considers Eq. (4) and general boundary conditions expressed by non-homogeneous differential operators of order l defined on ∂A.

In the present paper the simple layer potential approach is further generalized. The elliptic equation (4) is considered in the case that the "coefficients" a_k are $q \times q$ matrices with constant complex entries and u is a q-vector with complex components. For this equation quite general boundary conditions, expressed by non-homogeneous $q \times q$ matrix differential operators of order $l+n$ (n non-negative integer) on ∂A, are considered. Solutions of class $C^{l+n+\lambda}(\bar{A})$ are sought. In order to investigate these problems an extension of the concept of potential of simple layer, as formerly given by Fichera, is introduced. To this end a <u>fundamental solution of order</u> n is introduced for the elliptic system (4) and, accordingly, a <u>single layer potential</u>

of order n defined.

In this paper we shall survey the main results connected with this new extensions of the two-dimensional simple layer potential theory. Proofs and details will appear in two forthcoming papers [9],[10] by P.E.Ricci.

It is very likely that results of this paper may be extended to more general elliptic systems, for instance to systems with variable coefficients containing lower order terms like the Eq.(5) considered by Fichera in his first approach to the simple layer potential method.

It must be remarked that function-theoretic methods and the theory of one-dimensional singular integral equations for boundary value problems, connected with elliptic equations in two variables,have been, since long time, considered by I.N.Vekua [11]in his pioneering work in the theory of boundary value problems for higher order elliptic equations, which has inspired many researchers in this area.The procedures used by I.N.Vekua are quite different from the methods founded on the simple layer potential approach.

Nowadays a complete theory of boundary value problems for general elliptic systems in any number of variables has been established. See, for instance, [12],[13],[14],[15],[16],[17],[18]. The general theory is developed along lines which, except for boundary value problems in half spaces, generally do not use explicit representation of the solution, but results of functional analysis combined with some " a priori " estimates.

Although our theory refers only to the two-dimensional case and to particular elliptic systems, it seems to us that it is worth to be considered, first, because the global representation which it furnishes for the solutions of the boundary value problems is very suitable for the concrete construction of these solutions and for their numerical approximations as MacCamy, Hsiao and Wendland have shown. Second, because the main results concerning the boundary value problems are obtained by using very simple tools of Analysis, namely the Plemelj formula, the Muskhelishvili theorem on one-dimensional singular integral systems on a contour, and the residue theorem for analytic functions of one complex variable.

1. An extension of the Plemelj formula.

Let A be a bounded domain (open set) of the plane. The boundary ∂A is supposed to be a closed Jordan curve of class $C^{1+\lambda}$.

From now on we shall suppose that $z = x+iy \in A$, $\zeta = \xi + i\eta \in \mathscr{C}A$ [4]. Let us denote by Γ a rectifiable closed Jordan curve of the w-plane contained in the half plane $\Im m\, w < 0$. Let $w = w(\sigma)$ $(0 \le \sigma \le \Lambda)$ be a parametric representation of Γ. Let Γ^* be the symmetric of Γ with respect to the real axis.

Let us consider, for fixed $z \in A$ and $\zeta \in \mathscr{C}A$, the closed curve of the complex τ-plane

$$\tau = \tau(\sigma) = (x-\xi)w(\sigma) + (y-\eta).$$

Let us choose a determination of $\log \tau(0)$. This determination depends on z and ζ. In the sequel we shall suppose that the chosen determination for $\log \tau(0) = \log[(x-\xi)w(0) + (y-\eta)]$ depends continuously on z and ζ. For any σ such that $0 \le \sigma < \Lambda$ a determination of $\log \tau(\sigma)$ is determined by analytic continuation along the curve $\tau = \tau(\sigma)$. If we choose $\log \overline{\tau(0)} = \overline{\log \tau(0)}$ [5] we have, by analytic continuation along the curve $\tau = \overline{\tau(\sigma)}$, a determination of $\log \overline{\tau(\sigma)}$ which is such that $\log \overline{\tau(\sigma)} = \overline{\log \tau(\sigma)}$.

Let $L(w)$ be a complex valued function which is continuous on Γ and Γ^* and never vanishing on these two curves. Denote by α a non-negative integer. We set

$$(1.1)\quad \begin{aligned} P_\alpha(z-\zeta) = &-\frac{1}{4\pi^2\alpha!}\int_{+\Gamma} \frac{[(x-\xi)w+(y-\eta)]^\alpha \log[(x-\xi)w+(y-\eta)]}{L(w)}\,dw \\ &+ \frac{1}{4\pi^2\alpha!}\int_{+\Gamma^*} \frac{[(x-\xi)w+(y-\eta)]^\alpha \log[(x-\xi)w+(y-\eta)]}{L(w)}\,dw, \end{aligned}$$

where the determinations of $\log[(x-\xi)w+(y-\eta)]$ on the curves Γ and Γ^* are chosen as above specified.

---

[4] By $\mathscr{C}A$ we mean the complement set of A; i.e. the set of all the points of the plane which do not belong to A.

[5] If a is a complex number, by $\bar{a}$ we denote its complex conjugate.

The function $P_\alpha(z-\zeta)$ is well defined for $z\in A$, $\zeta\in\mathcal{C}A$ and is an analytic function of the four real variables x, y, ξ, η in the cartesian product $A\times\mathcal{C}A$. Moreover each partial derivative with respect to these variables can be obtained by differentiating under the integral signs on the right hand side of (1.1).

Let $\varphi(\zeta)$ be a complex function belonging to $C^\lambda(\partial A)$. Set for $z\in A$: [(6)]

$$v_{\alpha k}(z)=\int_{\partial A}\varphi(\zeta)\frac{\partial^{\alpha+1}}{\partial x^{\alpha+1-k}\partial y^{k}}P_\alpha(z-\zeta)ds_\zeta \qquad (k=0,1,\dots,\alpha+1)\ ;$$

by using the same technique employed in [4], essentially founded on the well known Plemelj formula, we get, for any $z_0\in\partial A$, the following limit relation (uniform with respect to z_0):

$$\lim_{z\to z_0} v_{\alpha k}(z)=\frac{i\varphi(z_0)}{4\pi}\left\{\int_{+\Gamma}\frac{w^{\alpha+1-k}}{L(w)(\dot x_0 w+\dot y_0)}dw+\int_{+\Gamma^*}\frac{w^{\alpha+1-k}}{L(w)(\dot x_0 w+\dot y_0)}dw\right\}$$
$$(1.2)\qquad -\frac{1}{4\pi^2}\left\{\int_{\partial A}\varphi(\zeta)ds_\zeta\left[\int_{+\Gamma}\frac{w^{\alpha+1-k}}{L(w)[(x_0-\xi)w+(y_0-\eta)]}dw-\int_{+\Gamma^*}\frac{w^{\alpha+1-k}}{L(w)[(x_0-\xi)w+(y_0-\eta)]}dw\right]\right\}.$$

By $\dot x$, $\dot y$ we denote derivatives with respect to the arc length (increasing counter-clockwise). By $\dot x_0$, $\dot y_0$ we mean that these derivatives have been computed in the point z_0 of ∂A. The integral over ∂A on the right hand side of (1.2) must be understood as a Cauchy singular integral.

Formula (1.2) may be regarded as a formal extension of the Plemelj formula which is included in (1.2) by assuming $k=\alpha+1$, $L(w)=w+i$ and as Γ a contour enclosing the point $w=-i$.

2. Simple layer potential of order n.

Let us consider the matrix differential operator (4) where a_k ($k=0,1,\dots,2l$; $l>0$) is a $q\times q$ matrix with constant complex entries.

We assume the following hypotheses:

1) If we set $$L(w)=\det\left\{\sum_{k=0}^{2l}a_k w^{2l-k}\right\},$$

---

(6) If we denote by $\zeta=\zeta(s)$ $(0\le s\le\Lambda_1)$ a parametric representation of ∂A, for defining the function $v_{\alpha k}(z)$ we must integrate on the rectangle $(0\le s\le\Lambda_1\ ;\ 0\le\sigma\le\Lambda)$ two functions which contain as a factor, the first $\log\tau(s,\sigma)\equiv\log[(x-\xi(s))w(\sigma)+(y-\eta(s))]$ and the second $\log\overline{\tau(s,\sigma)}\equiv\log[(x-\xi(s))\overline{w(\sigma)}+(y-\eta(s))]$. According to the explainations we have given above, that is done without any ambiguity by chosing a determination of $\log\tau(0,0)$ (continuously depending on z) and defining in the above rectangle $\log\tau(s,\sigma)$ by analytic continuation. For $\log\overline{\tau(s,\sigma)}$ we take $\overline{\log\tau(s,\sigma)}$.

we have $L(w) \neq 0$ for any real w and $\det a_0 \neq 0$, i.e. the operator

$$E\left(\frac{\partial}{\partial x}, \frac{\partial}{\partial y}\right) \equiv \sum_{K=0}^{2l} a_K \frac{\partial^{2l}}{\partial x^{2l-K} \partial y^K} \tag{2.1}$$

is elliptic in the sense of Petrowski.

2) The polynomial $L(w)$ of degree $2m = 2lq$ has m zeroes with positive imaginary part and m zeroes with negative imaginary part.

Let us denote by $a_K^{\mu\nu}$ $(\mu,\nu=1,\dots,q)$ the entries of the matrix a_K. Set

$$E(\xi,\eta) = \sum_{K=0}^{2l} a_K \xi^{2l-K} \eta^K$$

$$E^{\mu\nu}\left(\frac{\partial}{\partial x}, \frac{\partial}{\partial y}\right) \equiv \sum_{K=0}^{2l} a_K^{\mu\nu} \frac{\partial^{2l}}{\partial x^{2l-K} \partial y^K} \quad ; \qquad E^{\mu\nu}(\xi,\eta) = \sum_{K=0}^{2l} a_K^{\mu\nu} \xi^{2l-K} \eta^K ;$$

let us denote by $E_{\mu\nu}(\xi,\eta)$ the co-factor of $E^{\mu\nu}(\xi,\eta)$ in the $q\times q$ matrix $((E^{\mu\nu}(\xi,\eta)))$ $(\mu,\nu=1,\dots,q)$. If we set

$$L^{\mu\nu}(w) = E^{\mu\nu}(w,1) \quad ; \qquad L_{\mu\nu}(w) = E_{\mu\nu}(w,1) ,$$

we have

$$\sum_{\rho=1}^{q} L^{\mu\rho}(w) L_{\nu\rho}(w) = \delta_{\mu\nu} L(w) = \delta_{\mu\nu} E(w,1) .$$

Let us consider the closed curves Γ and Γ^* introduced in the preceeding Section. Suppose that Γ is such that the bounded domain which has Γ as contour contains all the zeroes of $L(w)$ which belong to the half plane $\Im m\, w < 0$ and moreover the bounded domain which has Γ^* as contour contains all the zeroes of $L(w)$ which belong to the half plane $\Im m\, w > 0$.

Let n be a non-negative integer. The following $q\times q$ matrix will be denoted as the <u>fundamental solution</u> <u>(matrix)</u> <u>of order</u> n for the operator (2.1):

$$\begin{aligned} F^{(n)}(z-\zeta) &\equiv ((F_{\mu\nu}^{(n)}(z-\zeta))) \\ &\equiv \left(\left(E_{\nu\mu}\left(\frac{\partial}{\partial x}, \frac{\partial}{\partial y}\right) P_{2m+n-2}(z-\zeta)\right)\right) \qquad (\mu,\nu=1,\dots,q) , \end{aligned} \tag{2.2}$$

(7)

where $P_{2m+n-2}(z-\zeta)$ is given by (1.1) with the definitions of $L(w)$ and Γ assumed in this Section.

---

(7) For simplicity we do not indicate explicitly the dependence of the $q\times q$ matrix on m, i.e. on l.

Let $\varphi_{\sigma j}(\zeta)(\sigma = 1,\dots,q;\ j = 0,\dots,l-1)$ be complex functions belonging to $C^{\lambda}(\partial A)$. Set

(2.3) $$u_\nu(z) = \sum_{\sigma=1}^{q} \sum_{j=0}^{l-1} \int_{\partial A} \varphi_{\sigma j}(\zeta) \frac{\partial^{l-1}}{\partial \xi^{l-1-j} \partial \eta^{j}} F^{(n)}_{\cdot \nu \sigma}(z-\zeta)\, ds_\zeta \qquad (\nu = 1,\dots,q).$$

The vector valued function defined by (2.3) will be denoted as single layer potential of order n for the operator (2.1). The following theorem holds:

I. If $\varphi_{\sigma j}(\zeta) \in C^{\lambda}(\partial A)$, the single layer potential of order n defined by (2.3) is a solution of the system

(2.4) $$E\left(\frac{\partial}{\partial x}, \frac{\partial}{\partial y}\right) u = 0$$

and belongs to the class $C^{l+m+\lambda}(\bar{A})$.

3. Representation theorem for the solutions of Eq. (2.4).

In addition to hypotheses 1) and 2) of Sect. 2, concerning the operator E, we shall assume also the following one:

3) Let us consider on ∂A the following boundary conditions for the q-vector u:

(3.1) $$\frac{\partial^{\alpha} u}{\partial x^{\alpha-k} \partial y^{k}} = 0 \qquad (k = 0,\dots,\alpha;\ \alpha = 0,\dots,l-1)$$

i) Only the vector-valued function $u \equiv 0$ belongs to $C^{l}(\bar{A})$, is a solution of $E^{*}u = 0$ in A and satisfies on ∂A the boundary conditions (3.1); E^{*} is obtained from E by substituting for a_k the transposed matrix a_k^{*}.

ii) Only the vector-valued function $u \equiv 0$ belongs to $C^{l}(\mathscr{C}A)$, is a solution of $E^{*}u = 0$ in $\mathscr{C}A - \partial A$, satisfies on ∂A the boundary conditions (3.1) and satisfies, for $|z| \to \infty$, the estimates

$$\frac{\partial^{\beta} u}{\partial x^{\beta-h} \partial y^{h}} = O\left(|z|^{l-2-\beta} \log |z|\right) \qquad (h = 0,\dots,\beta;\ \beta = 0,1,2,\dots).$$

II. If the operator $E\left(\frac{\partial}{\partial x}, \frac{\partial}{\partial y}\right)$ is strongly elliptic, i.e. if

$$\mathrm{Re}\left(\sum_{\mu,\nu}^{1,q} \sum_{k=0}^{2l} a_k^{\mu\nu} w^{2l-k} \eta_\nu \bar{\eta}_\mu\right) > 0$$

for any real w and any non-zero complex q-vector $\eta = (\eta_1,\dots,\eta_q)$, then hypotheses 1), 2), 3) are satisfied.

It must be remarked that the strong ellipticity of the operator $E\left(\frac{\partial}{\partial x}, \frac{\partial}{\partial y}\right)$ is a sufficient condition but not necessary for 1), 2), 3) to hold.

The following theorem substantially inverts Theor.I.

III. Suppose that hypotheses 1),2),3) hold and that ∂A is of class $C^{n+1+\lambda}$. If $u(z) \equiv (u_1(z), \ldots, u_q(z))$ is any solution in A of Eq.(2.4) belonging to $C^{l+n+\lambda}(\bar{A})$, then for $z \in A$ the following representation holds:

$$u_\nu(z) = \sum_{\sigma=1}^{q} \sum_{j=0}^{l-1} \int_{\partial A} \varphi_{\sigma j}(\zeta) \frac{\partial^{l-1}}{\partial \xi^{l-1-j} \partial \eta^j} F^{(n)}_{\nu\sigma}(z-\zeta)\, ds_\zeta$$

(3.2)

$$+ \sum_{k=0}^{l-1} \sum_{i=0}^{k} c_\nu^{ki} x^i y^{k-i} \qquad (\nu = 1, \ldots, q),$$

where $\varphi_{\sigma j}(\zeta) \in C^\lambda(\partial A)$ and the c_ν^{ki} are constants.

4. General boundary value problem for Eq.(2.4).

From now on we shall assume that hypotheses 1),2),3) hold.

Let us introduce the following boundary operators:

$$B^{\rho\nu}\left(z; \frac{\partial}{\partial x}, \frac{\partial}{\partial y}\right) = B_0^{\rho\nu}\left(z; \frac{\partial}{\partial x}, \frac{\partial}{\partial y}\right) + B_1^{\rho\nu}\left(z; \frac{\partial}{\partial x}, \frac{\partial}{\partial y}\right)$$

$$(\rho = 1, \ldots, m; \quad \nu = 1, \ldots, q)$$

where

$$B_0^{\rho\nu}\left(z; \frac{\partial}{\partial x}, \frac{\partial}{\partial y}\right) = \sum_{h=0}^{l+n} b_h^{\rho\nu}(z) \frac{\partial^{l+n}}{\partial x^{l+n-h} \partial y^h}$$

$$B_1^{\rho\nu}\left(z; \frac{\partial}{\partial x}, \frac{\partial}{\partial y}\right) = \sum_{s=0}^{l+n-1} \sum_{i=0}^{l+n-1-s} b_{is}^{\rho\nu}(z) \frac{\partial^{l+n-1-s}}{\partial x^{l+n-1-s-i} \partial y^i}.$$

The functions $b_h^{\rho\nu}(z)$, $b_{is}^{\rho\nu}(z)$ are defined for $z \in \partial A$ and belong to $C^\lambda(\partial A)$.

We shall consider the following boundary value problem:

B.V.P. Given the complex valued functions $\psi_\rho(z) \in C^\lambda(\partial A)$ $(\rho = 1, \ldots, m)$, find a solution u of (2.4) in A which belongs to $C^{l+n+\lambda}(\bar{A})$ and satisfies on ∂A the boundary conditions:

$$\sum_{\nu=1}^{q} B^{\rho\nu}\left(z; \frac{\partial}{\partial x}, \frac{\partial}{\partial y}\right) u_\nu(z) = \psi_\rho(z) \qquad (\rho = 1, \ldots, m).$$

From (1.2), (2.2) and from Theorem III the following "Equivalence Theorem" follows:

IV. Let us consider the following system of integral equations:

$$(-1)^{l-1} 2\pi \psi_\rho(z) = \sum_{j=0}^{l-1} \sum_{\sigma=1}^{q} \varphi_{\sigma j}(z)\, \alpha_{\rho,\sigma j}(z)$$

(4.1)

$$+ \frac{1}{\pi i} \sum_{j=0}^{l-1} \sum_{\sigma=1}^{q} \beta_{\rho,\sigma j}(z) \int_{+\partial A} \frac{\varphi_{\sigma j}(\zeta)}{\zeta - z}\, d\zeta$$

$$+ \sum_{j=0}^{l-1} \sum_{\sigma=1}^{q} \int_{\partial A} \varphi_{\sigma j}(\zeta)\, Q_{\rho,\sigma j}(z, \zeta)\, ds_\zeta$$

where

$$\alpha_{\rho,\sigma j} = \frac{i}{2}\int_{+\Gamma} \frac{\sum_{\nu=1}^{q} B_0^{\rho\nu}(z;w,1)L_{\nu\sigma}(w)\,w^{l-1-j}}{L(w)(\dot{x}w+\dot{y})}\,dw + \frac{i}{2}\int_{+\Gamma^*} \frac{\sum_{\nu=1}^{q} B_0^{\rho\nu}(z;w,1)L_{\nu\sigma}(w)\,w^{l-1-j}}{L(w)(\dot{x}w+\dot{y})}\,dw$$

$$\beta_{\rho,\sigma j} = \frac{i}{2}\int_{+\Gamma} \frac{\sum_{\nu=1}^{q} B_0^{\rho\nu}(z;w,1)L_{\nu\sigma}(w)\,w^{l-1-j}}{L(w)(\dot{x}w+\dot{y})}\,dw - \frac{i}{2}\int_{+\Gamma^*} \frac{\sum_{\nu=1}^{q} B_0^{\rho\nu}(z;w,1)L_{\nu\sigma}(w)\,w^{l-1-j}}{L(w)(\dot{x}w+\dot{y})}\,dw$$

$$Q_{\rho,\sigma j} = \frac{1}{2\pi}\Big[\Big(\int_{+\Gamma} \frac{\sum_{\nu=1}^{q} B_0^{\rho\nu}(z;w,1)M_{\nu\sigma}(w,z,\zeta)}{\zeta-z}\,w^{l-1-j}dw - \int_{+\Gamma^*} \frac{\sum_{\nu=1}^{q} B_0^{\rho\nu}(z;w,1)M_{\nu\sigma}(w,z,\zeta)}{\zeta-z}\,w^{l-1-j}dw$$

$$+\,2\pi\,\frac{\partial^{l-1}}{\partial x^{l-1-j}\partial y^{j}}\sum_{\nu=1}^{q} B_1^{\rho\nu}\Big(z;\frac{\partial}{\partial x},\frac{\partial}{\partial y}\Big)F_{\nu\sigma}^{(n)}(z-\zeta)$$

$$M_{\nu\sigma}(w,z,\zeta) = \frac{L_{\nu\sigma}(w)\,[K(w,z,\zeta)-1]}{L(w)(\dot{x}w+\dot{y})}\,\dot{\zeta}$$

$$K(w,z,\zeta)\begin{cases} = \dfrac{\dot{\zeta}(\dot{x}w+\dot{y})}{\dfrac{x-\xi}{z-\zeta}w+\dfrac{y-\eta}{z-\zeta}}, & z\neq\zeta \\ = 1, & z=\zeta \end{cases}$$

$$(\rho=1,\dots,m\ ;\ \sigma=1,\dots,q\ ,\ j=0,\dots,l-1).$$

If u is a solution of the above stated B.V.P., the functions $\varphi_{\sigma j}$ considered in the right hand side of (3.2) are solutions of the integral system (4.1).

Viceversa, if the $\varphi_{\sigma j}$ are solutions of the system (4.1), the vector-function defined by (3.2) is a solution of B.V.P.

Set

$$H_{\rho\sigma}(z,w) = \sum_{\nu=1}^{q} B_0^{\rho\nu}(z;w,1)L_{\nu\sigma}(w)$$

$$(\rho=1,\dots,m\ ;\ \sigma=1,\dots,q)$$

and denote by $H_\rho(z,w)$ the q-vector $(H_{\rho 1}(z,w),\dots,H_{\rho q}(z,w))$. Denote by $L^+(w)$ [by $L^-(w)$] a polynomial of degree m which has as zeroes all the zeroes of $L(w)$ such that $\Im m\, w>0$ $(\Im m\, w<0)$.

V. Necessary and sufficient condition for the system (4.1) to be of the regular type, i.e. $\det((\alpha_{\rho,\sigma j}+\beta_{\rho,\sigma j}))\neq 0$, $\det((\alpha_{\rho,\sigma j}-\beta_{\rho,\sigma j}))\neq 0$ $(\rho=1,\dots,m\ ;\ \sigma=1,\dots,q\ ,\ j=0,\dots,l-1)$ is the following:

"For every $z\in\partial A$ the vectors

(L) $$H_1(z,w),\dots,H_m(z,w)$$

are linearly independent mod $L^+(w)$ and mod $L^-(w)$."

The algebraic conditions expressed by (L) are equivalent to the so called Lopatinskii conditions for the B.V.P. we have considered.

The conditions (L) mean that, if for some $z \in \partial A$ we have $\sum_{\varrho=1}^{m} c_\varrho H_\varrho(z,w) = L^+(w) R(w)$, where $R(w)$ is a q-vector and $c_1,\dots,c_m$ are constant, then $c_1 = \dots = c_m = 0$ and that the same must hold if we replace $L^+(w)$ by $L^-(w)$.

From our theory it turns out that the B.V.P. is an "index problem" if and only if the algebraic conditions (L) are satisfied.

From the Muskhelishvili theory for singular integral systems it follows that the index of the problem B.V.P. (i.e. the difference between the dimension of the space of eigensolutions of the problem and the maximum number of linearly independent compatibility conditions) is given by

$$\varkappa = \frac{1}{2\pi i}\left[\log \frac{\det((\alpha_{\varrho,\sigma j} + \beta_{\varrho,\sigma j}))}{\det((\alpha_{\varrho,\sigma j} - \beta_{\varrho,\sigma j}))}\right]_{\partial A} .$$

References

[1] N.I.MUSKHELISHVILI, Singular integral equations, 2nd ed. (Moscow 1946); transl.from the Russian by J.R.M.Radok, P.Noordhoff,1953.

[2] N.M.GÜNTER, Die Potentialtheorie und ihre Anwendung auf Grundaufgaben der mathematischen Physik, B.G.Teubner Verlagsgesellschaft, 1957.

[3] S.AGMON, Multiple Layer Potentials and the Dirichlet Problem for Higher Order Elliptic Equations in the Plane I, Comm. on Pure and Appl.Math.Vol.X,N.2,pp.179-239,1957.

[4] G.FICHERA, Linear elliptic equations of higher order in two independent variables and singular integral equations , with applications to anisotropic inhomogeneous elasticity, Proceed. of the Symp. "Partial Differential Equations and Continuum Mechanics" (Madison Wisc.1960) edited by R.E.Langer, The Univ. of Wisconsin Press,1961.

[5] R.C.MacCAMY, On a class of two-dimensional Stokes flows, Arch.Rat. Mech.& Anal.,21,pp.256-258,1966.

[6] G.HSIAO-R.C.MacCAMY, Solution of boundary value problems by integral equations of the first kind, SIAM Review 15,pp.687-705,1973.

[7] G.HSIAO-W.L.WENDLAND, A finite element method for some integral equations of the first kind, Journal Math. Anal. Appl., to appear.

[8] P.E.RICCI, Sui potenziali di semplice strato per le equazioni ellittiche di ordine superiore in due variabili, Rend.di Matem. (1),Vol.7,Serie VI,pp.1-39,1974.

[9] P.E.RICCI, Un teorema di rappresentazione per le soluzioni di un particolare sistema lineare ellittico in due variabili, Rend. di Matem. (to appear).

[10] P.E.RICCI, Studio dei problemi al contorno per sistemi ellittici in due variabili mediante potenziali di semplice strato, Atti Accad.Naz.Lincei (to appear).

[11] I.N.VEKUA, New methods for solving elliptic equations, transl. from the Russian by D.E.Brown, North-Holland Publ.Co., 1967.

[12] S.AGMON-A.DOUGLIS-L.NIRENBERG, Estimates near the boundary for solutions of elliptic partial differential equations satisfying general boundary conditions, Comm.Pure Appl.Math.I,Vol. XII,N.1,pp.623-727,1959 ; II,Vol.XVII,N.4,pp.35-92,1964.

[13] L.HÖRMANDER, Linear partial differential operators, 2nd ed., Springer Verlag, 1964.

[14] L.BERS-F.JOHN-M.SCHECHTER, Partial differential equations, Interscience Pub.-J.Wiley & Sons, 1964.

[15] S.AGMON, Lectures on elliptic boundary value problems, D. Van Nostrand Co.Inc.,1965.

[16] G.FICHERA, Linear elliptic differential systems and eigenvalue problems, Lecture Notes in Mathem.Vol.8, Springer Verlag, 1965.

[17] A.FRIEDMAN, Partial differential equations, Holt, Rinehart and Winston, Inc., 1969.

[18] J.L.LIONS-E.MAGENES, Non homogeneous boundary value problems and applications, Vol.I, Springer Verlag, 1972.

CONSTRUCTIVE FUNCTION THEORETIC METHODS FOR HIGHER ORDER PSEUDOPARABOLIC EQUATIONS‡

R.P. Gilbert and G.C. Hsiao

0. INTRODUCTION

In this work we will develop a constructive method for solving pseudoparabolic equations of order $2n$ in the plane. More precisely, we investigate equations of the form

$$\mathcal{L}[u] := \underset{\sim}{M}[u_t] + \underset{\sim}{L}[u] \quad , \tag{0.1}$$

where $\underset{\sim}{M}$ and $\underset{\sim}{L}$ are the respective elliptic operators

$$\underset{\sim}{M}[u] := \Delta^n u + \sum_{k=1}^{n} \underset{\sim}{M}_k(\Delta^{n-k}u) \quad , \tag{0.2}$$

$$\underset{\sim}{M}_k[\phi] := \sum_{p,q=o}^{p+q\le k} a_k^{pq}(x,y)\,\frac{\partial^{p+q}\phi}{\partial x^p \partial y^q} \quad ,$$

and

$$\underset{\sim}{L}[u] := \Delta^m u + \sum_{k=1}^{m} \underset{\sim}{L}_k(\Delta^{m-k}u) \quad , \quad m < n \ , \tag{0.3}$$

$$\underset{\sim}{L}_k[\psi] := \sum_{p,q=o}^{p+q\le k} b_k^{pq}(x,y)\,\frac{\partial^{p+q}\psi}{\partial x^p \partial y^q} \quad .$$

The coefficients of $\underset{\sim}{M}_k$, $\underset{\sim}{L}_k$ are taken, furthermore, to be analytic functions of x and y for $(x,y) \in D \subset C^1$.

Recently, the integral operator methods of BERGMAN [2] and VEKUA [22], which have been very successful for developing representations for solutions of elliptic equations in the plane, have been extended by COLTON to treat the cases of parabolic [9] and second order pseudoparabolic equations [10], [11] in the plane. BROWN, GILBERT and HSIAO [8] and BROWN and GILBERT [7] developed analogous techniques for fourth-order pseudoparabolic equations using

---

‡ This research was supported in part by the U.S. Air Force Office of Scientific Research through AF-AFOSR Grant No. 76-2879, and in part by the Alexander von Humboldt Foundation.

respectively the methods of VEKUA and BERGMAN to treat elliptic operators. BROWN [6], on the other hand, completed the study of fourth-order, analytic, parabolic equations in two space variables.

Investigations concerning integral operators which generate solutions to parabolic and pseudoparabolic equations in three and four space dimensions have been made by RUNDELL [12], RUNDELL and STECHER [13], and by BHATNAGAR and GILBERT [3], [4], [5].

Pseudoparabolic equations arise in a variety of physical problems, such as the velocity of a non-steady flow of a viscous fluid [21], the theory of seepage of homogeneous fluids through fissured rock [1], hydrostatic excess pressure during the consolidation of clay [19], and the stability of liquid filled shells [18], [23], [24]. GILBERT and ROACH are presently investigating the last mentioned problem as an application of some of the ideas presented in this current work.

I. THE FUNDAMENTAL SOLUTION

If the coefficients of $\underset{\sim}{M}_k$ as analytic functions of x,y have an analytic extension to $(z,z^*) \in D \times D^*$, $D^* := \{z : \bar{z} \in D\}$, where $z = x + iy$, $z^* = x - iy$, then $\underset{\sim}{M}[u]$ has a representation

$$(1.1)\qquad \underset{\sim}{M}[u] := \sum_{k,j=0}^{n} A_{kj}(z,z^*) \frac{\partial^{k+j}U}{\partial z^k \partial z^{*j}} , \quad U(z,z^*) := u\left(\frac{z+z^*}{2}, \frac{z-z^*}{2i}\right) ,$$

with $A_{nn} \equiv 1$. We assume also that the operator $\underset{\sim}{L}$ has a complex form

$$(1.2)\qquad \underset{\sim}{L}[U] := \sum_{k,j=0}^{n-1} B_{kj}(z,z^*) \frac{\partial^{k+j}U}{\partial z^k \partial z^{*j}} .$$

The adjoint operator to $\underset{\sim}{\mathcal{L}}$ is given by $\underset{\sim}{\mathcal{L}}^* = \underset{\sim}{\mathcal{M}}$,

$$(1.3)\qquad \underset{\sim}{\mathcal{M}}[U] := \underset{\sim}{M}^*[U_t] - \underset{\sim}{L}^*[U] ,$$

with

$$(1.4)\qquad \underset{\sim}{M}^*[U] := \sum_{k,j=0}^{n} (-1)^{k+j} \frac{\partial^{k+j}}{\partial z^k \partial z^{*j}} (A_{kj}U) ,$$

$$(1.5)\qquad \underset{\sim}{L}^*[U] := \sum_{k,j=0}^{n-1} (-1)^{k+j} \frac{\partial^{k+j}}{\partial z^k \partial z^{*j}} (B_{kj}U) .$$

A function S of the form

(1.6) $$S(x,y,t;\xi,\eta,\tau) := A(x,y,t;\xi,\eta,\tau).\ \ln\frac{1}{r} + B(x,y,t;\xi,\eta,\tau)$$

where $r = [(x-\xi)^2 + (y-\eta)^2]^{\frac{1}{2}}$ will be called a <u>fundamental solution</u> of (1.1) if it satisfies the following conditions

(c-1) As a function of (x,y,t), S is a solution of the adjoint equation $\underset{\sim}{\mathcal{M}}[S] = 0$ and is an analytic function of its argument except at $r = 0$, where $\Delta^{n-1}S$ has a logarithmic singularity.

(c-2) At the parameter point $x = \xi$, $y = \eta$ we have

$$\frac{\partial^{p+q}A_t}{\partial x^p \partial y^q} = 0 \quad \text{for} \quad p+q \leq 2n-3\ , \quad \text{and} \quad \Delta^{n-1}A_t = 4^{n-1}.$$

(c-3) A and B are analytic functions of (x,y,t) at $r = 0$ and vanish at $t = \tau$.

<u>Remark</u>: The above implies that A may be written as $r^{2n-2}\hat{A}(x,y,t;\xi,\eta,\tau)$ with $\hat{A}$ regular at the parameter point. The above notation is computationally easy to work with.

We intend to show that it is possible to develop the coefficients A and B as analytic functions with the expansions

(1.7) $$A(z,z^*,t;\zeta,\zeta^*,\tau) = \sum_{j=1}^{\infty} A_j(z,z^*;\zeta,\zeta^*)\frac{(t-\tau)^j}{j!}\ ,$$

(1.8) $$B(z,z^*,t;\zeta,\zeta^*,\tau) = \sum_{j=1}^{\infty} B_j(z,z^*;\zeta,\zeta^*)\frac{(t-\tau)^j}{j!}\ .$$

Furthermore, we shall identify $A_1(z,z^*;\zeta,\zeta^*)$ as the Riemann function corresponding to the operator $\underset{\sim}{M}$. The other coefficients will be seen to satisfy homogeneous Goursat conditions on $z = \zeta$ and $z^* = \zeta^*$.

Inserting (1.6) into $\underset{\sim}{\mathcal{M}}[U] := \underset{\sim}{M}^*[U_t] - \underset{\sim}{L}^*[U] = 0$ we obtain, after some manipulation of terms

(1.9) $$\underset{\sim}{\mathcal{M}}[S] \equiv \underset{\sim}{\mathcal{M}}[A]\ \ln\frac{1}{r} + \underset{\sim}{\mathcal{M}}[B] + I_n + I_n^* = 0$$

with

$$I_n := \left\{ \frac{1}{2}(n-1)! \sum_{j=0}^{n} (-1)^j \frac{\partial^j}{\partial z^{*j}} (A_{nj}A_t) \right\} (z-\zeta)^{-n}$$

(1.10)
$$+ \frac{1}{2} \sum_{p=1}^{n-1} \left\{ \binom{n}{n-p} \sum_{j=0}^{n} (-1)^{n+j} \frac{\partial^{n+j-p}}{\partial z^{n-p} \partial z^{*j}} (A_{nj}A_t) \right.$$
$$+ \sum_{k=p}^{n-1} \binom{k}{k-p} \left[(-1)^{k+n} \frac{\partial^{k+n-p}}{\partial z^{k-p} \partial z^{*n}} (A_{kn}A_t) \right.$$
$$\left.\left. + \sum_{j=0}^{n-1} (-1)^{k+j} \frac{\partial^{k+j-p}}{\partial z^{k-p} \partial z^{*j}} (A_{kj}A_t - B_{kj}A) \right] \right\} \frac{(-1)^p (p-1)!}{(z-\zeta)^p}$$

and

$$I_n^* := \left\{ \frac{1}{2}(n-1)! \sum_{k=0}^{n} (-1)^k \frac{\partial^k}{\partial z^k} (A_{kn}A_t) \right\} (z^*-\zeta^*)^{-n}$$

(1.11)
$$+ \frac{1}{2} \sum_{q=1}^{n-1} \binom{n}{n-q} \sum_{k=0}^{n} (-1)^{k+n} \frac{\partial^{k+n-q}}{\partial z^k \partial z^{*n-q}} (A_{kn}A_t)$$
$$+ \sum_{j=q}^{n-1} \left\{ \binom{j}{j-q} \left[(-1)^{n+j} \frac{\partial^{n+j-q}}{\partial z^n \partial z^{*j-q}} (A_{nj}A_t) \right. \right.$$
$$\left.\left. + \sum_{k=0}^{n-1} (-1)^{k+j} \frac{\partial^{k+j-q}}{\partial z^k \partial z^{*j-q}} (A_{kj}A_t - B_{kj}A) \right] \right\} \frac{(-1)^q (q-1)!}{(z^*-\zeta^*)^q} .$$

Because of the multivaluedness of the logarithmic singularity it is necessary to set $\underset{\sim}{\mathcal{M}}[A] = 0$. To cancel the poles at $z = \zeta$ and $z^* = \zeta^*$ we ask that the coefficients of $(z-\zeta)^p$, $(z^*-\zeta^*)^p$, $(p=1,\ldots,n)$ vanish. These latter conditions provide us with so-called Goursat data for the A_j, B_j coefficients in the representations (1.9). Setting (1.7) into (1.10) and (1.11) we obtain the following conditions which $A_1(z,z^*;\zeta,\zeta^*)$ must satisfy

(1.12)
$$\sum_{k=p}^{n} \binom{k}{p} \sum_{j=0}^{n} (-1)^{k+j} \frac{\partial^{k+j-p}}{\partial z^{k-p} \partial z^{*j}} \left[A_{kj}(\zeta,z^*) A_1(\zeta,z^*;\zeta,\zeta^*) \right] = 0$$

$(p=1,2,\ldots,n)$, $z = \zeta$,

and

$$(1.13)\qquad \sum_{j=q}^{n} \binom{j}{q} \sum_{k=0}^{n} (-1)^{k+j} \frac{\partial^{k+j-q}}{\partial z^k \partial z^{*j-q}} \left[A_{kj}(z,\zeta^*)\ A_1(z,\zeta^*;\zeta,\zeta^*)\right] = 0$$

$$(q=1,2,\ldots,n),\quad z^* = \zeta^* \ .$$

We recall that condition (c-2) implies that

$$(1.14)\qquad \frac{\partial^{p+q}}{\partial z^p \partial z^{*q}} A_j(\zeta,\zeta^*;\zeta,\zeta^*) = 0 \ ,\quad \text{for } j=1,2,3,\ldots \ ,$$

$$(p,q=0,1,2,\ldots,2n-3\ ,\ \text{with } p+q \le 2n-3) \ .$$

In order to show that A_1 is actually the Riemann function, it is sufficient for us to show that the above conditions are equivalent to the characteristic conditions which uniquely determine it. We recall from VEKUA [22], Chapter V, the following conditions imposed on the Riemann function $R(z,z^*;\zeta,\zeta^*)$ and for convenience we label these conditions using his equation numbers :

$$(37.28)\qquad \underset{\sim}{M}^*[R] = 0$$

$$(37.29)\qquad \frac{\partial^k R}{\partial z^k}(z,z^*;\zeta,\zeta^*)\bigg|_{z=\zeta} = 0 \ ,\quad \frac{\partial^k R}{\partial \zeta^k}(z,z^*;\zeta,\zeta^*)\bigg|_{z^*=\zeta^*} = 0 \ ,$$

$$(k=0,1,2,\ldots,n-2)$$

$$(37.30)\qquad \begin{aligned} &\frac{\partial^{n-1} R}{\partial z^{n-1}}(z,z^*;\zeta,\zeta^*)\bigg|_{z=\zeta} = X(z^*,\zeta^*,\zeta) \ ,\\ &\frac{\partial^{n-1} R}{\partial z^{*n-1}}(z,z^*;\zeta,\zeta^*)\bigg|_{z^*=\zeta^*} = X^*(z,\zeta,\zeta^*) \ . \end{aligned}$$

Here X, and X* are solutions respectively of the ordinary differential equations

$$(37.31)\qquad \begin{aligned} &\sum_{k=0}^{n} (-1)^k \frac{\partial^k}{\partial z^{*k}} \left[A_{nk}(\zeta,z^*)\ X\right] = 0 \ , \quad \text{and}\\ &\sum_{k=0}^{n} (-1)^k \frac{\partial^k}{\partial z^k} \left[A_{kn}(z,\zeta^*)\ X^*\right] = 0 \ . \end{aligned}$$

Furthermore, X and X* are seen to satisfy the intial conditions

$$(37.32)\qquad \frac{\partial^k X}{\partial z^{*k}}(z^*,\zeta^*,\zeta)\Big|_{z^*=\zeta^*} = 0\ ,\quad \frac{\partial^{n-1} X}{\partial z^{*n-1}}(z^*,\zeta^*,\zeta)\Big|_{z^*=\zeta^*} = 1\ ,$$

$$(k=0,1,\ldots n-2)$$

$$\frac{\partial^k X^*}{\partial z^k}(z,\zeta,\zeta^*)\Big|_{z=\zeta} = 0\ ,\quad \frac{\partial^{n-1} X^*}{\partial z^{n-1}}(z,\zeta,\zeta^*)\Big|_{z=\zeta} = 1\ .$$

$$(k=o,1,\ldots n-2)$$

We note that $A_1(z,z^*;\zeta,\zeta^*)$ automatically satisfies (37.28), (37.29), (37.32) by virtue of our conditions (c-1) and (c-2). This suggests that we check to see if the remaining conditions (37.30), (37.31) are compatible with ours. Using (37.29) in combination with condition (1.12) we obtain for $p > 1$

$$(1.15)\qquad \sum_{j=o}^{n}(-1)^{n+j}\frac{\partial^{n+j-p}}{\partial z^{n-p}\partial z^{*j}}\Big[A_{nj}(z,z^*)\,A_1(z,z^*;\zeta,\zeta^*)\Big]\Big|_{z=\zeta} \equiv 0\ .$$

In the case where $p = 1$, this becomes

$$(1.16)\qquad \sum_{j=o}^{n}(-1)^{n+j}\sum_{\ell=o}^{n-1}\binom{n-1}{\ell}\frac{\partial^j}{\partial z^{*j}}\left[\frac{\partial^{n-1-\ell}}{\partial z^{n-1-\ell}}A_{nj}(z,z^*)\,\frac{\partial^\ell A_1}{\partial z^\ell}(z,z^*;\zeta,\zeta^*)\right]\Big|_{z=\zeta}$$
$$= \sum_{j=o}^{n}(-1)^{n+j}\frac{\partial^j}{\partial z^{*j}}\left[A_{nj}(\zeta,z^*)\,\frac{\partial^{n-1}A_1}{\partial z^{n-1}}(z,z^*;\zeta,\zeta^*)\Big|_{z=\zeta}\right] = 0\ .$$

Identifying temporarily A_1 with R and identifying the $(n-1)^{st}$ derivative with respect to z as X as given in (37.30), (1.16) becomes the first of equations (37.31). Repeating this analysis with (1.13) and identifying X^* as the $(n-1)^{st}$ derivative with respect to z^*, (37.30) yields likewise the second equation of (37.31). This exhausts our conditions and leaves us free to impose the additional initial conditions on X and X^* prescribed by (37.32). We conclude that we are permitted to identify $A_1 \equiv R$.

<u>Lemma:</u> <u>The first coefficient</u> $A_1(z,z^*;\zeta,\zeta^*)$ <u>for</u> A <u>in the representation</u> (1.7) <u>may be taken as the Riemann function associated</u>

with $\underset{\sim}{M}[u] = 0$. *The coefficients* $A_p(z,z^*;\zeta,\zeta^*)$, $p \geq 2$, *are determined recursively by the nonhomogeneous equations*

$$\underset{\sim}{M}^*[A_p] = \underset{\sim}{L}^*[A_{p-1}] \quad , \tag{1.17}$$

and the homogeneous Goursat data,

$$\begin{aligned} \frac{\partial^\ell}{\partial z^\ell} A_p(z,z^*;\zeta,\zeta^*)\Big|_{z=\zeta} &= 0 \,, \quad \ell=0,1,\ldots,n-1 \,, \\ \frac{\partial^\ell}{\partial z^{*\ell}} A_p(z,z^*;\zeta,\zeta^*)\Big|_{z^*=\zeta^*} &= 0 \,, \quad \ell=0,1,\ldots,n-1 \,. \end{aligned} \tag{1.18}$$

Proof: Since $S = A \log \frac{1}{r} + B$ satisfies the adjoint equation, it follows that $\underset{\sim}{\mathcal{M}}[A] = 0$. This in turn implies $\underset{\sim}{M}^*[A_1] = 0$, and $\underset{\sim}{M}^*[A_{j+1}] = \underset{\sim}{L}^*[A_j]$, $(j=0,1,\ldots)$. A moment's reflection concerning our condition (c-1) indicates that it can be satisfied using the above conditions (1.18), when A_1 is taken to be the Riemann function for $\underset{\sim}{M}[U] = 0$.

II. DETERMINATION OF THE COEFFICIENTS $A_p(z,z^*;\zeta,\zeta^*)$

The series representations (1.9) for the coefficients A and B of the singular solution

$$S(z,z^*,t;\zeta,\zeta^*,\tau) := A(z,z^*,t;\zeta,\zeta^*,\tau)\ \ell n\ \frac{1}{r} + B(z,z^*,t;\zeta,\zeta^*,\tau)$$

suggest that we try to determine A_p and B_p successively. To this end we develop a Green's formula based on the formal identity

$$V\ \underset{\sim}{M}[U] - U\ \underset{\sim}{M}^*[v] \equiv \frac{\partial P}{\partial z}(z,z^*) + \frac{\partial P^*}{\partial z^*}(z,z^*) \,, \tag{2.1}$$

where $\underset{\sim}{M}$ and $\underset{\sim}{M}^*$ are given by (1.4) and (1.7 and

$$P(z,z^*) := \sum_{\substack{j=o\\k=1}}^{n} \sum_{p=0}^{k-1} (-1)^p \frac{\partial^p}{\partial z^p} (VA_{kj}) \frac{\partial^{k+j-p-1}U}{\partial z^{k-p-1}\partial z^{*j}} \,, \tag{2.2}$$

and

$$P^*(z,z^*) := \sum_{\substack{k=o\\j=1}}^{n} \sum_{q=0}^{j-1} (-1)^{k+q} \frac{\partial^{q+k}}{\partial z^{*q}\partial z^k} (VA_{kj}) \frac{\partial^{j-q-1}U}{\partial z^{*j-q-1}} \,.$$

If $\underset{\sim}{M}_s$ is meant as the $\underset{\sim}{M}$ operator with s, s^* replacing z, z^*, then, setting $U(s,s^*) := A_1(z,z^*;s,s^*)$ and noting that A_1 is also a solution of $\underset{\sim}{M}_s[U] = 0$, the identity (2.1) yields

$$(2.4) \qquad - U(s,s^*)\ \underset{\sim}{M}_s^*[V] \equiv \frac{\partial P}{\partial s}(s,s^*) + \frac{\partial P^*}{\partial s^*}(s,s^*) \quad .$$

Lemma 2: The coefficients $A_{p+1}(z,z^*;\zeta,\zeta^*)$ may be formally determined by the recursive scheme

$$A_{p+1}(z,z^*;\zeta,\zeta^*) = \int_{\zeta^*}^{z^*} ds^* \int_{\zeta}^{z} ds\ A_1(z,z^*;s,s^*)\ \underset{\sim}{L}^*[A_p(s,s^*;\zeta,\zeta^*)] \ ,$$
$$(2.5) \qquad\qquad p=1,2,\ldots \quad .$$

Proof: Integrating (2.4) yields

$$- \int_{\zeta^*}^{z^*} ds^* \int_{\zeta}^{z} ds\ U(s,s^*)\ \underset{\sim}{M}^*[V] = \int_{\zeta^*}^{z} P(z,s^*)ds^* + \int_{\zeta}^{z} P^*(s,z^*)ds - Q(z,z^*;\zeta,\zeta^*) \ ,$$

with

$$Q(z,z^*;\zeta,\zeta^*) = \int_{\zeta^*}^{z^*} P(\zeta,s^*)ds^* - \int_{\zeta}^{z} P^*(s,\zeta^*)ds \quad .$$

Using the conditions (37.29) and recalling $U := A_1$, we can simplify the integral of P above as

$$(2.6) \qquad \int_{\zeta^*}^{z^*} P(z,s^*)ds^* = \sum_{j=o}^{n} \int_{\zeta^*}^{z^*} V(z,s^*)\ A_{nj}(z,s^*)\ \frac{\partial^{n-1+j}U(z,s^*)}{\partial z^{n-1}\partial s^{*j}}\ ds^* \ .$$

According to VEKUA [22] we may identify the function

$$(37.10) \qquad g(s^*;z,z^*) := \frac{\partial^{n-1}}{\partial z^{n-1}} A_1(z,z^*;z,s^*) \quad ,$$

which, furthermore, satisfies an ordinary differential equation

$$(37.9) \qquad \frac{d^n g}{d\zeta^n} + \sum_{m=o}^{n-1} A_{nm}(z,\zeta)\ \frac{d^m g}{d\zeta^m} = 0 \quad .$$

Using (37.9), the right-hand side of (2.6) is seen to vanish identically.

We next turn our attention to the P^*-integral, and note that (37.29) implies that $\frac{\partial^{j-q-1} U}{\partial z^{*\,j-q-1}}(s,z^*) = 0$, except for $q = 0,\ j = n$. Hence

$$\int_\zeta^z P^*(s,z^*)ds = \sum_{k=o}^{n} (-1)^k \int_\zeta^z \left\{ \frac{\partial^k}{\partial s^k} \left[V(s,z^*)\ A_{kn}(s,z^*)\right] \cdot \frac{\partial^{n-1} U(s,z^*)}{\partial z^{*\,n-1}} \right\} ds \ . \tag{2.7}$$

The associated functions

$$g^*(s;z,z^*) := \frac{\partial^{n-1}}{\partial s^{*\,n-1}} A_1(z,z^*;s,s^*) \Big|_{s^*=z^*} \tag{37.10}$$

are known [24] to satisfy the ordinary differential equation

$$\frac{d^n g^*}{ds^n} + \sum_{m=o}^{n-1} A_{mn}(s,z^*) \frac{d^m g^*}{ds^m} = 0 \ , \tag{37.9}$$

and the initial conditions

$$\begin{aligned} \frac{\partial^\ell g^*}{\partial s^\ell}(s;z,z^*)\Big|_{s=z} &= 0 \ , \qquad (\ell=0,1,\ldots,n-2) \\ \frac{\partial^{n-1} g^*}{\partial s^{n-1}}(s;z,z^*)\Big|_{s=z} &= 1 \ . \end{aligned} \tag{37.11}$$

Consequently, after some regrouping, it may be seen that

$$\begin{aligned} \int_\zeta^z P^*(s,z^*)ds &= \sum_{k=1}^{n} \int_\zeta^z \frac{\partial}{\partial s} \left\{ \sum_{\ell=o}^{k-1} (-1)^{\ell+k} \frac{\partial^{k-1-\ell}}{\partial s^{k-1-\ell}} \left[V(s,z^*)\ A_{kn}(s,z^*)\right] \cdot \frac{\partial^\ell g^*(s;z,z^*)}{\partial s^\ell} \right\} ds \\ &= (-1)^{2n-1} V(z,z^*)\ A_{nn}(z,z^*) \frac{\partial^{n-1}}{\partial z^{n-1}} g^*(z;z,z^*) + Q^*(\zeta;z,z^*) \\ &= -\ V(z,z^*) + Q^*(\zeta;z,z^*) \ , \end{aligned} \tag{2.8}$$

where

$$Q^*(\zeta;z,z^*) := -\sum_{k=1}^{n} \sum_{\ell=o}^{k-1} (-1)^{\ell+k} \frac{\partial^{k-1-\ell}}{\partial s^{k-1-\ell}} \left[V(\zeta,z^*)\ A_{kn}(\zeta,z^*)\right] \cdot \frac{\partial^\ell g^*(\zeta;z,z^*)}{\partial \zeta^\ell} \ . \tag{2.9}$$

Putting together the above terms we note that the following representation for $V(z,z^*)$ has been obtained

$$(2.10)\quad \begin{aligned} V(z,z^*) &= Q^*(\zeta;z,z^*) - Q(z,z^*;\zeta,\zeta^*) \\ &\quad + \int_{\zeta^*}^{z^*} ds^* \int_{\zeta}^{z} ds\; A_1(z,z^*;s,s^*)\; \underset{\sim}{M}^*[V(s,s^*)] \ . \end{aligned}$$

Setting $V(z,z^*) := A_{p+1}(z,z^*;\zeta,\zeta^*)$ and recalling the initial conditions (1.18), we conclude $Q^*(\zeta;z,z^*) \equiv 0$. The expression for $Q(z,z^*;\zeta,\zeta^*)$ may also be seen to vanish identically by virtue of the identities

$$\int_{\zeta^*}^{z^*} P(\zeta,s^*)ds^* = \sum_{\substack{j=o\\k=1}}^{n} \sum_{\ell=o}^{k-1} (-1)^{\ell} \int_{\zeta^*}^{z^*} ds^* \frac{\partial^{\ell}}{\partial s^{\ell}} \left[A_{p+1}(\zeta,s^*;\zeta,\zeta^*)\, A_{kj}(\zeta,s^*)\right] \cdot \frac{\partial^{k+j-\ell-1}}{\partial z^{k-\ell-1}\partial s^{*j}} U(\zeta,s^*) \equiv 0 \ ,$$

and

$$\int_{\zeta}^{z} P^*(s,\zeta^*)ds = \sum_{\substack{k=o\\j=1}}^{n} \sum_{q=o}^{j-1} (-1)^{k+q} \int_{\zeta}^{z} \frac{\partial^{q+k}}{\partial z^{*q}\partial s^{k}} \left(A_{p+1}(s,\zeta^*;\zeta,\zeta^*) \cdot A_{kj}(s,\zeta^*)\right) \frac{\partial^{j-q-1}}{\partial z^{*j-q-1}} U(s,\zeta^*)ds \equiv 0 \ .$$

Our Lemma is proved at this point by recognizing that $\underset{\sim}{M}^*[A_{p+1}) = \underset{\sim}{L}^*(A_p)$.

<u>Lemma 3</u>: <u>The coefficients</u> $A_{p+1}(z,z^*;\zeta,\zeta^*)$ <u>may be computed recursively by</u>

$$(2.11)\quad A_{p+1}(z,z^*;\zeta,\zeta^*) = \int_{\zeta^*}^{z^*} ds^* \int_{\zeta}^{z} ds\; A_p(z,z^*;s,s^*)\; \underset{\sim}{L}^*[A_1(s,s^*;\zeta,\zeta^*)] \ ,$$

$p=1,2,\ldots$. <u>Furthermore, the series representation for</u> $A(z,z^*;t;\zeta,\zeta^*,\tau)$ <u>converges uniformly in the domain</u> $(G \times G^* \times T)^2$ <u>where</u> G <u>is the domain of regularity of the coefficients in the</u> x-y <u>coordinates and</u> T <u>is a disk in the complex</u> t-<u>plane</u>.

<u>Proof</u>: Recalling the form of $\underset{\sim}{L}$ and $\underset{\sim}{L}^*$, we note that we may write

$$A_1(z,z^*;s,s^*)\ \underset{\sim}{L}^*[A_p(s,s^*;\zeta,\zeta^*)] - A_p(s,s^*;\zeta,\zeta^*)\ \underset{\sim}{L}[A_1(z,z^*;s,s^*)]$$

$$= -\sum_{\substack{j=o\\k=1}}^{n-1} \frac{\partial}{\partial s}\left\{\sum_{\ell=0}^{k-1}(-1)^{\ell}\frac{\partial^{\ell}}{\partial s^{\ell}}\Big[A_p(s,s^*;\zeta,\zeta^*)\ B_{kj}(s,s^*)\Big]\frac{\partial^{k+j-\ell-1}}{\partial s^{k-\ell-1}\partial s^{*j}}\ A_1(z,z^*;s,s^*)\right\}$$

$$-\sum_{\substack{k=o\\j=1}}^{n-1}(-1)^k\frac{\partial}{\partial s^*}\left\{\sum_{m=0}^{j-1}(-1)^m\frac{\partial^{m+k}}{\partial s^{*m}\partial s^k}\Big[A_p(s,s^*;\zeta,\zeta^*)B_{kj}(s,s^*)\Big]\frac{\partial^{j-m-1}}{\partial s^{*j-m-1}}A_1(z,z^*;s,s^*)\right\}.$$

Integration gives the following identity

$$\int_{\zeta^*}^{z^*} ds^* \int_{\zeta}^{z} ds\ A_1(z,z^*;s,s^*)\ \underset{\sim}{L}^*[A_p(s,s^*;\zeta,\zeta^*)]$$
$$= \int_{\zeta^*}^{z^*} ds^* \int_{\zeta}^{z} ds\ A_p(s,s^*;\zeta,\zeta^*)\ \underset{\sim}{L}[A_1(z,z^*;s,s^*)] + H + H^* ,$$

where

$$H := -\sum_{\substack{j=o\\k=1}}^{n-1}\sum_{\ell=0}^{k-1}(-1)^{\ell}\int_{\zeta^*}^{z^*} ds^*\left\{\frac{\partial^{\ell}}{\partial s^{\ell}}\Big[A_p(s,s^*;\zeta,\zeta^*)\ B_{kj}(s,s^*)\Big]\ \frac{\partial^{k+j-\ell-1}}{\partial s^{k-\ell-1}\partial s^{*j}} \cdot A(z,z^*;s,s^*)\right\}_{s=\zeta}^{s=z} \tag{2.12}$$

and

$$H^* := \sum_{\substack{k=o\\j=1}}^{n-1}(-1)^k\sum_{m=0}^{j-1}(-1)^m\int_{\zeta}^{z} ds\left\{\frac{\partial^{m+k}}{\partial s^{*m}\partial s^k}\Big[A_p(s,s^*;\zeta,\zeta^*)\ B_{kj}(s,s^*)\Big] \cdot \frac{\partial^{j-m-1}}{\partial s^{*j-m-1}}\ A_1(z,z^*;s,s^*)\right\}_{s^*=\zeta^*}^{s^*=z^*} . \tag{2.13}$$

Using the conditions (1.18) for A_{p+1} and (37.29) for A_1, the terms H and H^* are seen to identically vanish. This establishes the recursive definition (2.11). To show that the series for A converges we note that according to VEKUA [22], p. 186 $A_1(z,z^*;\zeta,\zeta^*)$ is dominated by

$$\frac{|z-\zeta|^{n-1}|z^*- \ ^*|^{n-1}}{[(n-1)!]^2} \quad \text{in} \quad [G\times G^*]^2 .$$

Consequently, using (2.11) it follows by induction that

$$(2.14)\qquad |A_p(z,z^*;\zeta,\zeta^*)| \le C \frac{|z-\zeta|^{n+p-2}|z^*-\zeta^*|^{n+p-2}}{[(n+p-2)!]^2} \| \underset{\sim}{L}[A_1] \|^{p-1}$$

where

$$\| \underset{\sim}{L}[A_1] \| := \sup_{(G \times G^*)^2} |\underset{\sim}{L}[A_1]|$$

and

$$C := \sup_{(G \times G^*)^2} |A_1| \quad .$$

From (2.14) it is clear that the series for A converges in the stated domain.

III. <u>DETERMINATION OF</u> $B(z,z^*,t;\zeta,\zeta^*,\tau)$

From (1.10) we obtain as the partial differential equation for B,

$$(3.1)\qquad \underset{\sim}{\mathcal{M}}[B] = - I_n - I_n^* \;,$$

where I_n and I_n^* are given by (1.11) and (1.12) respectively. Putting the series expansion (1.8) for B into the left-hand side of (3.1) yields then a recursive scheme for the coefficients B_j, namely

$$(3.2)\qquad \underset{\sim}{M}^*[B_1] = f_o(z,z^*;\zeta,\zeta^*) := -\frac{1}{2}\sum_{p=1}^{n} \frac{(-1)^p(p-1)!}{(z-\zeta)^p} \left\{ \sum_{k=p}^{n} \binom{k}{p} \cdot \sum_{j=o}^{n} (-1)^{k+j} \frac{\partial^{k+j-p}}{\partial z^{k-p}\partial z^{*j}} (A_{kj}A_1) \right\}$$

$$-\frac{1}{2}\sum_{q=1}^{n} \frac{(-1)^q(q-1)!}{(z^*-\zeta^*)^q} \left\{ \sum_{j=q}^{n} \binom{j}{q} \sum_{k=o}^{n} (-1)^{k+j} \frac{\partial^{k+j-q}}{\partial z^{k}\partial z^{*j-q}} (A_{kj}A_1) \right\} ,$$

$$(3.3)\qquad M^*[B_{\ell+1}] = f_\ell(z,z^*;\zeta,\zeta^*) := L^*[B_\ell] - \frac{1}{2}\frac{(n-1)!}{(z-\zeta)^n} \sum_{j=o}^{n} (-1)^j \frac{\partial^j}{\partial z^{*j}} (A_{nj}A_{\ell+1})$$

$$-\frac{1}{2}\sum_{p=1}^{n-1} \frac{(-1)^p(p-1)!}{(z-\zeta)^p} \left\{ \sum_{k=p}^{n} \binom{k}{p} \sum_{j=0}^{n} (-1)^{k+j} \frac{\partial^{k+j-p}}{\partial z^{k-p}\partial z^{*j}} (A_{kj}A_{\ell+1}) - \sum_{k=p}^{n-1} \binom{k}{p} \sum_{j=o}^{n-1} (-1)^{k+j} \frac{\partial^{k+j-p}}{\partial z^{k-p}\partial z^{*j}} (B_{kj}A_\ell) \right\} +$$

$$- \frac{1}{2} \frac{(n-1)!}{(z^*-\zeta^*)^n} \sum_{k=o}^{n} (-1)^k \frac{\partial^k}{\partial z^k} (A_{kn} A_{\ell+1})$$

$$- \frac{1}{2} \sum_{q=1}^{n-1} \frac{(-1)^q (q-1)!}{(z^*-\zeta^*)^q} \left\{ \sum_{j=q}^{n} \binom{j}{q} \sum_{k=o}^{n} (-1)^{k+j} \frac{\partial^{k+j-q}}{\partial z^k \partial z^{*j-q}} (A_{kj} A_{\ell+1}) \right.$$

$$\left. - \sum_{j=q}^{n-1} \binom{j}{q} \sum_{k=o}^{n-1} (-1)^{k+j} \frac{\partial^{k+j-q}}{\partial z^k \partial z^{*j-q}} (B_{kj} A_\ell) \right\} ,$$

$\ell=1,2,\ldots$.

If we specify that the B_ℓ satisfy the homogeneous Goursat data

$$\left. \frac{\partial^k B_\ell(z,z^*;\zeta,\zeta^*)}{\partial z^{*k}} \right|_{z^*=\zeta^*} = 0 \ ,$$

(3.4) $$\left. \frac{\partial^k B_\ell(z,z^*;\zeta,\zeta^*)}{\partial z^k} \right|_{z=\zeta} = 0 \ , \qquad k=0,1,2,\ldots,n-1, \quad \ell=1,2,\ldots \ ,$$

then

$$B_{\ell+1}(z,z^*;\zeta,\zeta^*) = \int_{\zeta^*}^{z^*} ds^* \int_{\zeta}^{z} ds \, f_\ell(s,s^*;\zeta,\zeta^*) \, A_1(z,z^*;s,s^*) \ ,$$

$\ell=0,1,\ldots$.

The majoration for the $B_{\ell+1}$, while technically involved, proceeds by the usual methods [10], [11], [8].

We summarize the discussion of Sections II and III in the following theorem.

Theorem 1: Assume that the coefficients A_{jk}, B_{jk} are analytic functions of two complex variables z, z^* in the bicylinder $D \times D^*$. Then $A(z,z^*,t;\zeta,\zeta^*,\tau)$ and $B(z,z^*,t;\zeta,\zeta^*,\tau)$ are analytic functions of their six independent variables for all (complex) t, τ, and $z,\zeta \in D$, $z^*,\zeta^* \in D^*$. Moreover, both A and B can be represented by a uniformly convergent series expansion in the form of (1.7) and (1.8) respectively.

IV. BOUNDARY VALUE PROBLEMS

We now proceed to use the fundamental solution (1.8) to develop representations for boundary value problems. The solutions to boundary value problems we will seek in the class

$$(4.1) \quad \mathcal{G} := \{u(x,y,t) : u \in C^{2n-2}(\bar{D}\times T);\ u_t \in C^{2n}(D\times T)\cap C^{2n-1}(\bar{D}\times T)\},$$

where $T := \{t : 0 \leq t < t_o\}$ and where t_o is a fixed constant. We begin by considering the identity

$$(4.2) \quad \begin{aligned} v_t \mathfrak{L}[u] - u_t \mathfrak{M}[v] &\equiv \{v_t\, \underset{\sim}{M}[u_t] - u_t\, \underset{\sim}{M}^*[v_t]\} \\ &- \{v\, \underset{\sim}{L}[u_t] - u_t\, \underset{\sim}{L}^*[v]\} + \frac{\partial}{\partial t}\{v\, \underset{\sim}{L}[u]\} \quad . \end{aligned}$$

If we replace v by the fundamental solution

$$S(x,y,t;\xi,\eta,\tau) = A \log \frac{1}{r} + B \quad ,$$

then (4.2) leads to an integral representation for solutions which satisfy the homogeneous initial data $u(x,y,o) = 0$ in D, namely

$$(4.3) \quad u(\xi,\eta,\tau) = \frac{1}{2\pi[(n-1)!]^2}\int_0^\tau dt \int_{\partial D} H[u(x,y,t),\ S(x,y,t;\xi,\eta,\tau)] \quad .$$

Here H is a bilinear form in $\dfrac{\partial^{k+j}u_t}{\partial x^k \partial y^j}$, $\dfrac{\partial^{k+j}S_t}{\partial x^k \partial y^j}$, $(k+j \leq 2n-1)$ as well as $\dfrac{\partial^{k+j}u}{\partial x^k \partial y^j}$, $\dfrac{\partial^{k+j}S}{\partial x^k \partial y^j}$, $(k+j \leq 2n-3)$. The introduction of a special, Green-type, fundamental solution, namely one which satisfies on ∂D

$$(4.4) \quad S_t = \frac{\partial S_t}{\partial \nu} = \ldots = \frac{\partial^{n-1}S_t}{\partial \nu^{n-1}} = 0 \qquad (\nu = \text{inward normal})$$

permits us to reduce the bilinear form $H[\ ,\ ]$ to the case where only the boundary data of the first kind

$$(4.5) \quad u_t^+ = f_o\ ,\ \frac{\partial u_t^+}{\partial \nu} = f_1\ ,\ \ldots\ ,\ \frac{\partial^{n-1}u_t^+}{\partial \nu^{n-1}} = f_n \quad \text{on } \partial D$$

appears.

This representation is more easily computed using the complex notation. To this end we recall the elementary identities,

$$(4.6)\qquad \frac{d}{ds}\left(\frac{\partial^{k+m}u}{\partial z^k \partial z^{*m}}\right)^+ = \left(\frac{\partial^{k+m+1}u}{\partial z^{k+1}\partial z^{*m}}\right)^+ \frac{dz}{ds} + \left(\frac{\partial^{k+m+1}u}{\partial z^k \partial z^{*m+1}}\right)^+ \frac{dz^*}{ds} ,$$

$$\left(\frac{d^k u}{d\nu^k}\right)^+ = i^k \sum_{\ell=0}^{k} (-1)^{k-\ell}\binom{k}{\ell}\left(\frac{\partial^k u}{\partial z^{k-\ell}\partial z^{*\ell}}\right)^+ \left(\frac{dz}{ds}\right)^{k-2\ell} ,$$

which hold on ∂D. The first two brackets of (4.2) we compute directly as

$$(4.7)\qquad \begin{aligned} &\{V_t\, M[U_t] - U_t\, M^*[V_t]\} - \{V\, L[U_t] - U_t\, L^*[V]\} \\ &= \frac{\partial}{\partial z}\tilde{P} + \frac{\partial}{\partial z^*}\tilde{P}^* , \end{aligned}$$

where

$$(4.8)\qquad \begin{aligned} \tilde{P} := &\sum_{\substack{k=1\\ j=0}}^{n} \sum_{\ell=0}^{k-1} (-1)^{\ell} \frac{\partial^{\ell}}{\partial z^{\ell}} (V_t A_{kj}) \frac{\partial^{k+j-\ell-1} U_t}{\partial z^{k-\ell-1}\partial z^{*j}} \\ &- \sum_{\substack{k=1\\ j=0}}^{n-1} \sum_{\ell=0}^{k-1} (-1)^{\ell} \frac{\partial^{\ell}}{\partial z^{\ell}} (VB_{kj}) \frac{\partial^{k+j-\ell-1} U_t}{\partial z^{k-\ell-1}\partial z^{*j}} , \end{aligned}$$

$$(4.9)\qquad \begin{aligned} \tilde{P}^* := &\sum_{\substack{k=0\\ j=1}}^{n} (-1)^k \sum_{m=0}^{j-1} (-1)^m \frac{\partial^{m+k}}{\partial z^{*m}\partial z^k} (V_t A_{kj}) \frac{\partial^{j-m-1} U_t}{\partial z^m} \\ &- \sum_{\substack{k=0\\ j=1}}^{n-1} (-1)^k \sum_{m=0}^{j-1} (-1)^m \frac{\partial^{m+k}}{\partial z^{*m}\partial z^k} (VB_{kj}) \frac{\partial^{j-m-1} U_t}{\partial z^m} . \end{aligned}$$

Then (4.3) takes on the form

$$(4.10)\qquad U(z,z^*,\tau) = \int_0^{\tau} dt \int_{\partial D} N_o(U,S)\, ds ,$$

where

$$(4.11)\qquad N_o(u,v) := i\left(\tilde{P}\frac{dz^*}{ds} - \tilde{P}^*\frac{dz}{ds}\right) .$$

This in turn may be expressed in terms of tangential and normal derivatives of the data using (4.6). When S is a Green's function then (4.10) is directly evaluated in terms of (4.5).

REFERENCES

[1] Barenblat, G., Zeltov, I., and Kochiva, I. "Basic concepts in the theory of seepage of homogeneous fluids in fissured rock", J. Appl. Math. Mech. 24(1960), 1286-1303.

[2] Bergman, S. "Integral Operators in the Theory of Linear Partial Differential Equations", Springer-Verlag, Berlin, 1961.

[3] Bhatnagar, S., and Gilbert, R.P. "Bergman type operators for pseudoparabolic equations in several space variables", Bull. Mat. Soc. Mat. Rep. Soc. Rom. 18 (66) (1974).

[4] Bhatnagar, S., and Gilbert, R.P. "A function theoretic approach to higher order pseudoparabolic equations", Math. Nachrichten (to appear).

[5] Bhatnagar, S., and Gilbert, R.P. "Constructive methods for solving $\Delta_4^2 u + A\Delta_4 u_t + Bu_t + C\Delta_4 u + Du = 0$", Math. Nachrichten (to appear).

[6] Brown, P.M. "Integral operators for fourth order linear parabolic equations", SIAM J. Math. Anal. (to appear).

[7] Brown, P.M., and Gilbert, R.P. "Constructive methods for higher order, analytic, Sobolev-Galpern equations", Bulletin Mathématique (to appear).

[8] Brown, P.M., Gilbert, R.P., and Hsiao, G.C. "Constructive function theoretic methods for fourth order pseudoparabolic equations in two space variables", Rendiconti di Matematica (to appear).

[9] Colton, D.L. "Bergman operators for parabolic equations in two space variables", Proc. Amer. Math. Soc. 38(1970), 119-126.

[10] Colton, D.L. "Integral operators and the first initial boundary value problem for pseudoparabolic equations with analytic coefficients", J. Diff. Equats. 13(1973), 506-522.

[11] Colton, D.L. "On the analytic theory of pseudoparabolic equations", Quart. J. Math. 23(1972), 179-192.

[12] Rundell, W. "The solution of the initial-boundary value problems for pseudoparabolic partial differential equations", Proc. Royal Soc. Edinburgh (to appear).

[13] Rundell, W., and Stecher, M. "A method of ascent for parabolic and pseudoparabolic partial differential equations", SIAM J. Math. Anal. (to appear).

[14] Showalter, R.E. "Partial differential equations of Sobolev-Galpern type", Pac. J. Math. 31(1969, 789-794.

[15] Showalter, R.E.
"Well-posed problems for a partial differential equation of order 2m+1", SIAM J. Math. Anal. 1(1970), 214-231.

[16] Showalter, R.E., and Ting, T.W.
"Pseudoparabolic partial differential equations", SIAM J. Math. Anal. 1(1970), 1-26.

[17] Sobolev, S.
"Some new problems in mathematical physics", Izv. Akad. Nauk. SSSR Ser. Mat. 18(1954), 3-50 (Russian).

[18] Stewartson, K.
"On the stability of a spinning top containing liquid", J. Fluid Mech. 5(1959), 577-592.

[19] Taylor, D.W.
"Research on Consolidation of Clays", MIT Press, Cambridge, 1942.

[20] Ting, T.W.
"Parabolic and pseudoparabolic partial differential equations", J. Math. Soc. Japan 21(1969), 440-453.

[21] Ting, T.W.
"Certain non-steady flows of second order fluids", Arch. Rat. Mech. Anal. 14(1963), 1-26.

[22] Vekua, I.N.
"New Methods for Solving Elliptic Equations", John Wiley, New York, 1967.

[23] Wedemeyer, E.H.
"The unsteady flow within a spinning cylinder", J. Fluid Mech. 20(1964), 383-391.

[24] Wedemeyer, E.H.
"Dynamics of liquid-filled shell ; theory of viscous corrections to Stewartson's stability problem", BRL Report 1287 (1965).

Über die Lösung einiger nichtklassischer Probleme der Elastizitätstheorie

V. D. Kupradze
(Tbilissi)

Für das erste und zweite Randwertproblem der Elastizitäts- oder Thermoelastizitätstheorie haben wir gegenwärtig eine allgemeine Lösungstheorie und die verschiedenen Methoden zur angenäherten Lösung.

Es gibt jedoch wichtige Probleme, die in der klassischen Theorie nicht genügend untersucht sind, sowohl im Bezug auf die Existenz als auch auf effektive Lösungsverfahren. Solche Probleme nennen wir hier (konventionell) "nichtklassisch".

Eines solcher Probleme ist das folgende:
Im endlichen, durch S_o begrenzten dreidimensionalen Gebiet $\mathcal{D}_o$, das mit den isotropen, elastischen Medien der Konstanten λ_o, μ_o, angefüllt ist, seien m isolierte elastische Einschließungen $\mathcal{D}_k$ $k=1,\ldots,m$, aus verschiedenen elastischen Materialien mit Konstanten λ_k, μ_k, $k=1,\ldots,m$, gegeben.

Es soll das Gleichgewicht des derart zusammengesetzten Körpers bestimmt werden, wenn die Sprünge der Verschiebungen (u) und Spannungen (Tu) längs S_k, $k=1,\ldots,m$, sowie die Werte einer dieser Größen auf S_o gegeben sind.

Dieses Problem läßt sich in verschiedenen Richtungen verallgemeinern; man kann z.B. einige Einschließungen für "Löcher" halten, oder statt der statischen Probleme dynamische oder statt der elastischen Gleichungen thermoelastische betrachten usw.

Natürlich kann ich mich jetzt mit diesen Fragen nicht aufhalten, ich werde auch die Fragen der Existenz und Eindeutigkeit nicht berühren; bemerke nur, daß sie zur Zeit gut untersucht sind durch zwei Methoden; einerseits mit Hilfe der singulären Integralgleichungen und andererseits mit Hilfe der funktionentheoretischen Methoden.

Mein Hauptziel ist zu zeigen, daß die Lösung des obengenannten Problems sich explizit durch Quadraturen ausdrücken läßt.

Anfangs betrachten wir unsere Aufgabe im Sonderfall, daß es keine Einschließungen gibt, d.h., daß $\mathcal{D}_o$ homogen ist und auf S_o etwa die Verschiebungen vorgegeben sind. Dann liegt ein gewöhnliches erstes Randwertproblem vor, dessen Greenschen Tensor man als Limes des Ausdrucks

$$G_p^o(x,y;N) = \Gamma_p^o(x-y) - \sum_{i=1}^{N} \mathfrak{f}_{pi}^o(x) W_i(y), \qquad x,y \in \mathcal{D}_o, \tag{1}$$

für $N \to \infty$ betrachten kann. Hierbei sind $\Gamma_p^o(x-y)$; $p=1,2,3$; die Spalten der bekannten Somigliana-Matrix

$$\Gamma^o(x-y) = \| \Gamma_1^o(x-y),\ \Gamma_2^o(x-y),\ \Gamma_3^o(x-y) \|_{3x3},$$

und $W_i(y)$ sind ganz bestimmte Linearkombinationen, die sukzessive aus Elementen der Menge

$$\{\Gamma_p^o(x_k-y)\}_{k=1}^{i}, \qquad p=1,2,3;$$

durch Orthonormierung konstruiert werden.

Die x_k; $k=1,..,\infty$, sind dabei beliebige Punkte außer $\mathcal{D}_o$ und

$$\mathfrak{f}_{pi}^o(x) = \int_S \Gamma_p^o(x-y)\, W_i(y) d_y S.$$

Die (3x3) - Matrix mit Spalten $G_1^o(x,y;N), G_2^o(x,y;N), G_3^o(x,y;N)$

bezeichnen wir durch $G^\sigma(x,y;N)$.

Führen wir noch einige Bezeichnungen ein; der Rand von $\mathcal{D}_k$;

$k=1,\ldots,m$; sei S_k, das Gebiet zwischen den Flächen S_o und $\bigcup_{k=1}^{m} S_k$ sei $\mathcal{D}'_o$.

Für den Vektor $\Gamma_p(x-y)$ in $\mathcal{D}_k$; $k=0,1,\ldots,m$; schreiben wir

$$\Gamma_p^k(x-y),$$

Die Grenzwerte von u bzw. (Tu) in Punkten auf S_k werden durch u^+ bzw. $(Tu)^+$ bezeichnet, wenn die Annäherung an S_k aus $\mathcal{D}_k$ geschieht und durch u^-, $(Tu)^-$, wenn sie aus $\mathcal{D}'_o$ erfolgt.

Wenn wir berücksichtigen, daß die Differenzen $(u^+ - u^-)$, $(Tu)^+ - (Tu)^-$

auf S_k ; $k=1,\ldots m$; bekannt sind, so liefern die Greenschen Formeln mit Kernen $\Gamma^k(x-y)$, $G^o(x,y)$; $k=1,\ldots,m$; folgendes

$$2u(x) \int_{S_k} \Gamma^k(x-y) \cdot (Tu)^- dS - \int_{S_k} (T\Gamma^k(y-x))' u^-(y) dS + \mathfrak{f}_k(x), \tag{2}$$

$\forall x \in \mathcal{D}_k$:

$$(3)\qquad 0=\int_{S_k}\Gamma^k(x-y)(Tu)^- dS-\int_{S_k}\left((T\Gamma^k(y-x)\right)' u^-(y)\,dS+F_k(x),\quad \forall x\in\mathcal{D}_0'$$

$$(4)\qquad 2u(x)=-\int_{\overset{m}{\underset{i=1}{U}}S_i} G^0(x,y;N)(Tu)^- dS+\int_{\overset{m}{\underset{i=1}{U}}S_i}\left(TG^0(y,x;N)\right)' u^-(y)\,dS-F_{m+1}(x)\,\forall x\in\mathcal{D}_0'$$

$$(5)\qquad 0=-\int_{\overset{m}{\underset{i=1}{U}}S_i} G^0(x,y;N)(Tu)^- dS+\int_{\overset{m}{\underset{i=1}{U}}S_i}\left(TG^0(y,x;N)\right)' u^-(y)\,dS-F_{m+1}(x)\,\forall x\in\overset{m}{\underset{i=1}{U}}\mathcal{D}_i\,,$$

wobei $F_{m+1}(x)$, $F_k(x); k=1,\ldots,m$ bekannte Funktionen sind; das Zeichen bei Matrizen bedeutet Transponierung.

Die Idee der Methode besteht nun darin, u^- und $(Tu)^-$ aus Gleichungen (3) und (5) zu bestimmen. Danach ergeben (2) und (4) die endgültige Lösung.

Führen wir noch einige Bezeichnungen ein:

$M^k(x,y)$ - eine Matrix der Dimension (3x6)-

$$(6)\qquad M^k(x,y)=\begin{cases}\left\|(T\Gamma^k(y-x))',-\Gamma^k(x-y)\right\|_{3x6}, & y\in S_k, k=1,\ldots,m\\ 0 & ,y\in S_j\,, j\neq k\,, x\in E_3;\end{cases}$$

und die Matrix

$$(7)\qquad M^0(x,y)=\left\|\left(TG^0(y,x;N)\right)',-G^0(y,x;N)\right\|_{3x6},\ y\in\overset{m}{\underset{i=1}{U}}S_i\,,\ x\in E_3$$

$X(y)$ - sei der Vektor mit sechs Komponenten gegeben durch

$$X(y)=\left(u^-(y),(Tu(y))^-\right)\ .$$

Nun erhalten (3) und (5) folgende Gestalt

$$(2_1)\qquad \int_{\overset{m}{\underset{i=1}{U}}S_i} M^k(x,y)X(y)\,d_yS=F_k(x)\ \forall x\in\mathcal{D}_k;\ k=1,\ldots m\ ;$$

$$(4_1)\qquad \int_{\overset{m}{\underset{i=1}{U}}S_i} M^0(x,y)X(y)\,d_yS=F_{m+1}(x)\qquad \forall\, x\in\ \overset{m}{\underset{k=1}{U}}\mathcal{D}_k$$

Seien $\{x_{kr}\}_{r=1}^{\infty}$; $k=0,\ldots m$; abzählbare Mengen von Punkten, die beliebig auf $\overline{S_k}$; $k=o,\ldots m$; fixiert seien, wobei $\overline{S}_o'$ S_o umschließt, ohne gemeinsame Punkte zu haben: $\overline{S_k}$; $k=1,\ldots m$; aber seien in $\mathcal{D}_k$ enthalten ohne gemeinsame Punkte mit S_k.

Es ist klar, daß in (2_1) wir $x=x_{or}\in\overline{S}_o$ und in (4_1) $x = x_{kr}\in -S_k, k=1,\ldots m$, setzen dürfen. Daraus folgt

$$(8) \qquad \int_{\bigcup_{i=1}^{m} S^i} M^k(x_{or},y)X(y)dS=F(x_{or}); \quad k=1,\ldots,m; r=1,2\ldots ;$$

$$(9) \qquad \int_{\bigcup_{i=1}^{m} S_i} M^o(x_{1r},y)X(y)dS=F_{m+1}(x_{1r}),\ldots\ldots \int_{\bigcup_{i=1}^{m} S_i} M^o(x_{mr},y)X(y)dS=F_{m+1}(x_{mr})$$

$r=1,2\ldots.$

Links in diesen Gleichungen haben wir Skalarprodukte des Vektors $X(y)$ mit 6-komponentigen Vektoren aus der abzählbaren Menge

$$\left(T\Gamma_p^1(y-x_{or}),-\Gamma_p^1(y-x_{or})\right)_{p=1,2,3},\ldots,\left(T\Gamma_p^m(y-x_{or}),\right.$$
$$\left.-\Gamma_p^m(y-x_{or})\right)_{p=1,2,3}$$

$$(10) \qquad \left(TG_p^o(y,x_{1r};N),-G_p^o(y,x_{1r};N)\right)_{p=1,2,3}\ldots,\left(TG_p^o(y,x_{mr};N),\right.$$
$$\left.-G_p^o(y,x_{mr};N)\right)_{p=1,2,3}$$

rechts aber stehen die bekannten Größen.

Numerieren wir die Elemente von (10) so durch, daß jedes eine bestimmte Nummer erhält; diese neue Menge bezeichnen wir mit $\{\varphi^r(y)\}_{r=1}^{\infty}$. Es läßt sich zeigen, daß die Menge $\{\varphi^r(y)\}_{r=1}^{\infty}$ linear unabhängig und vollständig im Raume $\mathcal{L}_2\left(\bigcup_{i=1}^{m} S_i\right)$ ist.

Definieren wir die Konstanten a_k; $k=1,\ldots,n$; durch die Bedingung

$$(11) \qquad \| X(y) - \sum_{k=1}^{n} a_k \varphi^k(y) \|_{\mathcal{L}_2 (\bigcup_{i=1}^{m} S_i)} = \min. \quad ,$$

so erhalten wir das lösbare System

$$(12) \qquad \sum_{k=1}^{n} a_k (\varphi^k, \varphi^l) = (X, \varphi^l), \quad l=1,\ldots,n, \qquad \text{mit}$$

$$(13) \qquad (\varphi^k, \varphi^l) = \int_{\bigcup_{i=1}^{m} S_i} \varphi^k(y) \varphi^l(y) dS, \quad (X, \varphi^l) = \int_{\bigcup_{i=1}^{m} S_i} X(y) \varphi^l(y) dS \quad ,$$

wobei n eine beliebig fixierte natürliche Zahl sei.

Wir bilden den sechskomponentigen Vektor

$$(14) \qquad X(y;n) = \sum_{k=1}^{n} a_k \varphi^k(y),$$

und nehmen seine ersten drei Komponenten für u^-, die drei letzten aber für $(Tu)^-$. Wir tragen diese Werte von u^-, $(Tu)^-$ in (2) und (4) ein und bezeichnen das Resultat mit $u(x;n)$.

Nun ist leicht zu beweisen, daß es für jedes $\varepsilon > 0$, eine natürliche Zahl $n_0(\varepsilon)$ gibt, so daß für $n > n_0(\varepsilon)$ und jedes $x \in \mathcal{D}_0' \cup (\bigcup_{i=1}^{m} \mathcal{D})$ gilt:

$$|u(x;n) - u(x)| < \varepsilon,$$

wobei $u(x)$ die exakte Lösung der Aufgabe sei.

$u(x;n)$ ist mithin eine gesuchte angenäherte Lösung unseres Problems: sie ist in Quadraturen ausgedrückt.

Die Rechnungen werden vereinfacht, wenn es keine äußere Fläche S_0 gibt und $\mathcal{D}_0'$ unendlich ist; in diesem Falle ist statt des Greenschen Tensors $G^0(x,y)$ der Somiglianische $\Gamma^0(x-y)$ zu nehmen.

THE SINGULARITIES OF SOLUTIONS TO ANALYTIC ELLIPTIC BOUNDARY VALUE PROBLEMS

R.F. MILLAR

Department of Mathematics, University of Alberta, Edmonton, Canada T6G 2G1

1. Introduction. The mathematical description of many problems in science and engineering leads to analytic boundary value problems for second order, elliptic partial differential equations. This paper is concerned with locating the singularities of solutions to these problems. Such information is useful in many practical contexts. As one example, we mention inverse scattering problems like those considered by Colton [3,4], Sleeman [19], and Weston, Bowman, and Ar [23]. Here a wave is scattered by an unknown object. From knowledge of the solution at large distances from the scatterer, one wishes to deduce its physical properties; in fact, one is led to the singularities (sources) of the solution. Then there are inverse potential problems, like those arising in geophysics where one wishes to determine the sources of a gravitational anomaly from the field at the Earth's surface. Here, too, we are led to the singularities. For a discussion and extensive bibliography on these, and other improperly posed problems, see [17,18].

For certain boundary value problems, the location of the singularities plays a decisive role in determining domains of validity of representations for the solution. It was in this context that the procedure to be described was developed.

There are many methods for solving analytic boundary value problems; see, for example [1,6,7,20]. However, to use them to locate the singularities, one would have to solve the problem first. Unfortunately, in many practical situations this information is required at the outset.

The procedure to be described has been developed for just this reason - to locate *a priori* the singularities of solutions to analytic elliptic boundary value problems; we make no attempt to construct the solution. Thus it complements the other methods. We assume that the solution u to the boundary value problem is holomorphic on one side of, and up to, an analytic boundary Σ on which a linear, analytic boundary condition is prescribed. At first sight, it seems that we must continue u analytically across Σ to find its singularities. Fortunately we can locate these without actually performing the continuation; for they are related to singularities in the continuation of the boundary data (u and the normal derivative $\partial u/\partial \nu$ ($\equiv v$) on Σ) into the complex domain of their arguments.

Thus we are faced with two major problems. The first is to locate the singularities of the boundary data. This is not a trivial matter, since the data are not prescribed completely. But if an appropriate fundamental solution is known, an integral equation for the unknown data may be obtained. By extending this equation into the complex domain, we can locate the singularities of the unknown data; moreover, this is accomplished without first solving the integral equation. We shall describe how this goal has been achieved quite satisfactorily in two and, to a lesser extent at present, in three dimensions.

The second problem - that of relating the singularities in the data to real singularities in the solution - has not yet been dealt with in quite so satisfactory a manner. It is known that, in a neighbourhood of an initial, non-characteristic, analytic manifold in which is embedded an analytic, singularity manifold S, the singularities are borne by complex characteristic manifolds emanating from S [22]. To our knowledge, global results are lacking and may be difficult to obtain, especially in certain cases of symmetry where coefficients in the differential equation are singular and focussing may result. Nevertheless, in order to proceed, we shall assume that for a boundary value problem the singularities are borne on some of these characteristics, and that real singularities arise where the characteristics pierce the real domain. Plausibility arguments will be used to decide which characteristics are significant.

In the next section, we shall discuss these ideas in their simplest context, namely that of differential equations in two independent variables. Then, in section 3, we describe some recent work on axially symmetric problems for the Helmholtz equation, In section 4, strictly three-dimensional problems are examined. We consider specifically only exterior problems, although interior problems could be treated in the same way. [The results given in section 3 and parts of sections 2 and 4 have not appeared elsewhere.]

2. Boundary Value Problems in Two Independent Variables. To illustrate the methods, we shall consider the simplest possible differential equation, namely Laplace's equation. The problem for the general, homogeneous, analytic elliptic equation of the second order has been examined elsewhere [14]. Let D denote the exterior domain in the x,y-plane that contains a full neighbourhood of infinity, and is bounded internally by a simple, closed analytic curve Σ. We consider the following boundary value problem:

$$u_{xx} + u_{yy} = 0, \quad (x,y) \in D, \tag{2.1}$$

$$\alpha u + \beta \partial u/\partial \nu = \gamma \quad \text{on} \quad \Sigma, \tag{2.2}$$

$$u \to 0, \quad |(x^2+y^2)\nabla u| < \infty, \quad \text{as} \quad x^2 + y^2 \to \infty. \tag{2.3}$$

Here α, β, and γ are holomorphic functions of s, the arclength parameter on Σ, and $\partial/\partial\nu$ denotes differentiation along the unit normal $\hat{\nu}$ to Σ, drawn into D.

By means of Green's theorem, we find

$$2\pi\, u(x,y) = \int_{-\ell}^{\ell} [u(s) \frac{\partial}{\partial\nu} \log(1/R) - v(s)\log(1/R)]ds, \quad (x,y) \in D,$$

where $v(s) \equiv (\partial u/\partial\nu)(s)$, R is distance between (x,y) and $(\xi(s),\eta(s))$ on Σ, and 2ℓ is the length of Σ. We let (x,y) tend to a point on Σ specified by arclength t, and obtain the integral equation

$$\pi\, u(t) = \int_{-\ell}^{\ell} v(s)\log R\, ds - \int_{-\ell}^{\ell} u(s)R^{-1}\partial R/\partial\nu\, ds, \quad t \in I, \tag{2.4}$$

where I is the real interval $[-\ell,\ell]$, and

$$R = \{[\xi(s)-\xi(t)]^2 + [\eta(s)-\eta(t)]^2\}^{\frac{1}{2}}, \quad s,t \in I. \tag{2.5}$$

We now wish to extend (2.4) into the complex domain of t. On account of periodicity of ξ,η and, hence, of u and v, we need consider only the case for which $\mathrm{Re}\,t \in I$. Since

$$(2.6)\qquad R^{-1}\partial R/\partial\nu = \frac{[\xi(s)-\xi(t)]\eta'(s) - [\eta(s)-\eta(t)]\xi'(s)}{[\xi(s) - \xi(t)]^2 + [\eta(s) - \eta(t)]^2} \equiv K(s,t)$$

is, for $s \in I$, an analytic function of t in a complex neighbourhood N of the interval I, we may continue the second integral in (2.4) into N merely by replacing the real parameter t by $t_1 + it_2$. This simplicity is not characteristic of the situation that is usually encountered.

More typical is the continuation of the first integral in (2.4). Let us determine an analytic function in N that reduces to this integral when $\mathrm{Im}\,t \to 0$. We do so in a manner reminiscent of that used to continue integrals of the form $\int_L P(z,\zeta)\log|z-\zeta|dz$, $\zeta \in L$, away from L [5, §53]. According to (2.5), we have $R \geq 0$ for $s,t \in I$. We define r as an analytic function of t, for $s \in I$, by

$$(2.7)\qquad r = r(s,t) = \{[\xi(s)-\xi(t)]^2 + [\eta(s)-\eta(t)]^2\}^{1/2} \equiv |r|e^{i\psi},$$

with ψ fixed in the following manner: $\psi \to 0$ if $t \to t_0$ and $-\ell < t_0 < s < \ell$. In particular,

$$(2.8)\qquad \psi \to \mp\pi \quad \text{as} \quad t \to t_0 \pm i0, \quad -\ell < s < t_0 < \ell.$$

Let $N = N^+ \cup N^-$, where $\mathrm{Im}\,t \gtrless 0$ in $N^\pm$, respectively. Then $r(s,t) \neq 0$ for $t \in N^+ \cup N^-$, $s \in I$, and $\log r \equiv \log|r| + i\psi$ is analytic for $t \in N^+ \cup N^-$, $s \in I$.

From (2.8), we find that

$$(2.9)\qquad \int_{-\ell}^{\ell} v(s)\log r\,ds \pm \pi i\int_{-\ell}^{t} v(s)ds \to \int_{-\ell}^{\ell} v(s)\log R\,ds, \quad \text{as} \quad \mathrm{Im}\,t \to \pm 0;$$

here each integral on the left-hand side is analytic in t for $t \in N^+ \cup N^-$.

In $N^+ \cup N^-$, (2.4) and (2.9) give

$$(2.10)\qquad \pi[u(t) \mp i\int_{-\ell}^{t} v(s)ds] = \int_{-\ell}^{\ell} v(s)\log r\,ds - \int_{-\ell}^{\ell} u(s)K(s,t)ds, \quad \mathrm{Im}\,t \gtrless 0,$$

and these reduce to (2.4) when $\mathrm{Im}\,t \to \pm 0$. It is assumed that the integration path from $-\ell$ to t avoids singularities of v. We conclude that (2.10) provides an analytic continuation of (2.4) into $N^+ \cup I \cup N^-$.

For our purposes, an important property of (2.10) is that the right-hand side, in its dependence on u and v, involves only values in I.

The result (2.10) (and its limit as $\mathrm{Im}\,t \to 0$) is valid in N. It may be continued further into $-\ell \leq \mathrm{Re}\,t \leq \ell$ where singularities of $\xi(t)$ or $\eta(t)$ may be encountered. Also it is here that curves called root loci play a role. These are analytic curves in the t-plane, defined by $r(s,t) = 0$ and described as s runs from $-\ell$ to ℓ. The first integral on the right in (2.10) is continued across a root locus in the same manner that (2.10) was derived from (2.4). A pole of $K(s,t)$ will usually cross I when t crosses a root locus, and the continuation of the second integral on the right in (2.10) will involve the addition of a residue term. This procedure is described more fully in [12,14]. By proceeding in this manner, eventually we either are prevented from further continuation by a continuum of singularities, or we may proceed indefinitely. For simplicity, we shall assume that (2.10) provides the continuation throughout the domain of analyticity in $-\ell \leq \mathrm{Re}\,t \leq \ell$.

A second useful relation is obtained by differentiating (2.10):

$$\text{(2.11)} \qquad \pi[u'(t) \mp i\,v(t)] = \int_{-\ell}^{\ell} v(s)Q(s,t)ds - \int_{-\ell}^{\ell} u(s)K_t(s,t)ds, \quad \mathrm{Im}\,t \gtrless 0.$$

Here

$$Q(s,t) \equiv \frac{[\xi(t)-\xi(s)]\xi'(t) + [\eta(t)-\eta(s)]\eta'(t)}{[\xi(t)-\xi(s)]^2 + [\eta(t)-\eta(s)]^2}.$$

For a Neumann boundary value problem, $\alpha = 0$ and $\beta = 1$ in (2.2), and v is known. Thus (2.10) expresses $u(t)$ for $\mathrm{Im}\,t \gtrless 0$ in terms of known functions and values of $u(s)$ on I. If we consider a Dirichlet problem, then $\alpha = 1$, $\beta = 0$, u is known, and (2.11) may be used to examine $v(t)$. Equation (2.10) also may be employed for the general linear boundary condition (2.2), in which case a simple integral equation for $u(t)$ is obtained.

We assume that the singularities in the given data have been found. Then, by using (2.10) or (2.11) as the case may be, we can locate the singularities in the unknown data. It is necessary to find the singularities in the right-hand side of (2.10) or (2.11), and this may be done in the manner described above.

Although the existence of root loci will increase the complexity of the analysis, it is never necessary to know the values of the unknown data on I.

As a simple example, we consider a Neumann problem for the circle Σ:

(2.12) $\xi(s) = a \cos s/a, \quad \eta(s) = a \sin s/a,$

for $-\pi a \leq s \leq \pi a$. Suppose that we have a point singularity or source at $(x_0,0)$, with $x_0 > a$. Its potential is $U(x,y) = \log r_0$, where $r_0 = [(x-x_0)^2 + y^2]^{\frac{1}{2}}$. We write the potential in the presence of the circular boundary as $U(x,y)+u(x,y)$. Then u is harmonic outside the circle and satisfies (2.3). If we assume that the normal derivative of the potential vanishes on Σ, we find that $v(s) = (x_0 \cos s/a - a)/(a^2 + x_0^2 - 2ax_0 \cos s/a)$. Moreover, from (2.7) and (2.12),

(2.13) $r = 2a \sin\left(\frac{s-t}{2a}\right),$

and (2.6), (2.12) and (2.13) give $K(s,t) = 1/(2a)$. Consequently, for sufficiently small values of $\mathrm{Im} t$, (2.10) yields

(2.14) $$u(t) = \pm i \int_{-\pi a}^{t} \frac{x_0 \cos s/a - a}{a^2 + x_0^2 - 2ax_0 \cos s/a}\, ds - \frac{1}{2\pi a}\int_{-\pi a}^{\pi a} u(s)ds$$

$$+ \frac{1}{\pi}\int_{-\pi a}^{\pi a} \frac{x_0 \cos s/a - a}{a^2 + x_0^2 - 2ax_0 \cos s/a} \cdot \log[2a \sin\left(\frac{s-t}{2a}\right)]ds, \quad \mathrm{Im} t \gtrless 0.$$

Since the solutions to $\sin\left(\frac{s-t}{2a}\right) = 0$ are $t = s + 2n\pi a$ $(n = 0, \pm 1, \pm 2, \cdots)$, it follows that the last integral in (2.14) is analytic throughout $\mathrm{Im} t > 0$ and $\mathrm{Im} t < 0$. Thus, in the finite t-plane, the only singularities of $u(t)$ occur where $\cos t/a = (a^2+x_0^2)/(2ax_0)$. If we solve this we find that $t = \pm t'$, where $e^{it'/a} = a/x_0$.

The characteristics through a point specified by the parameter value t are

(2.15) $$x \pm iy = \xi(t) \pm i\eta(t)$$
$$= a \exp[\pm it/a].$$

If we insert $t = \pm t'$ in turn into (2.15), we obtain the two real solutions $x = a^2/x_0$, $y = 0$ and $x = x_0$, $y = 0$. The first lies inside the circle at the point $(a^2/x_0, 0)$; it is the image of the source in the circle. The second lies outside the circle where the solution is analytic. Thus no singularity is borne by those characteristics through $t = \pm t'$ that meet the real domain in $(x_0, 0)$. The explanation is simple: we are considering a boundary value problem so the data u and v are not independent; they must be related in precisely the manner needed so that no singularity appears at $(x_0, 0)$.

We can also obtain the image singularity in an ellipse of an exterior point singularity. If the ellipse is defined by $x = a\cos\theta$, $y = b\sin\theta$, $0 \leq \theta \leq 2\pi$, $a > b$, and if the singularity is at $x = a_0\cos\theta_0$, $y = b_0\sin\theta_0$, with $a^2 - b^2 = a_0^2 - b_0^2$, then the image is at

$$(2.16)\qquad x = A\cos\theta_0, \quad y = B\sin\theta_0,$$

where

$$(2.17)\qquad A = \frac{1}{2}\left[\frac{(a+b)(a_0-b_0)^2 + (a-b)^3}{(a-b)(a_0-b_0)}\right], \quad B = \frac{1}{2}\left[\frac{(a+b)(a_0-b_0)^2 - (a-b)^3}{(a-b)(a_0-b_0)}\right].$$

The foregoing analysis illustrates the essentials for elliptic equations in canonical form with analytic coefficients in two independent variables. In the more general cases, we still obtain relations corresponding to (2.10) and (2.11). These lead to Volterra integral equations for the unknown data in the complex domain; for details see [14]. The procedure can be extended readily to the consideration of simple boundary curves that are composed of analytic arcs [15].

3. Axially Symmetric Boundary Value Problems. We now consider axially symmetric solutions to the Helmholtz equation in three independent variables. In terms of cylindrical coordinates (ρ,ϕ,z), the solution to the boundary value problem is independent of ϕ; thus $u = u(\rho,z)$ and, in the region exterior to the axially symmetric analytic boundary Σ,

$$(3.1)\qquad u_{\rho\rho} + u_{zz} + \rho^{-1}u_\rho + k^2 u = 0, \qquad k = \text{constant} \geq 0.$$

Consequently we again have a problem in two independent variables. We shall consider u in a meridian plane Π in which a point is specified by Cartesian coordinates (ρ,z). The origin is taken to lie inside Σ. Usually we shall assume that $\rho \geq 0$; however, a solution that is analytic on a portion of the z-axis may be extended into $\rho < 0$, and $u(-\rho,z) = u(\rho,z)$; see [8]. If Π cuts Σ in a closed analytic curve γ, if σ denotes that portion of γ in $\rho \geq 0$, and if u satisfies an appropriate radiation condition at infinity, then at an exterior point (ρ,z),

$$4\pi\, u(\rho,z) = \int_\sigma [u(s)\partial G/\partial\nu - v(s)G]\rho(s)ds.$$

Here s denotes arclength on σ, measured from the point where the negative z-axis intersects γ, $u(s)$ and $v(s)$ are the values of u and $\partial u/\partial\nu$ on σ, $G \equiv 2\int_0^\pi e^{ikR}/R\, d\phi$,

$$\text{(3.2)} \qquad R \equiv \{\rho(s)^2 + \rho^2 + [z(s)-z]^2 - 2\rho(s)\rho\cos\phi\}^{\frac{1}{2}},$$

and $(\rho(s),z(s))$ is a point of σ. If we let (ρ,z) tend to $(\rho(t),z(t))$ on σ, and set $\partial G/\partial\nu = H$, we find the integral equation

$$\text{(3.3)} \qquad 2\pi\, u(t) = \int_0^\ell [u(s)H(s,t)-v(s)G(s,t)]\rho(s)ds, \qquad 0 \leq t \leq \ell,$$

where ℓ is the length of σ.

Once $G(s,t)$ and $H(s,t)$ have been extended properly into the complex domain of s and t, we may analytically continue (3.3) in essentially the same manner as is described in section 2. This will be discussed fully elsewhere, and we give here only the results. In a neighbourhood of $0 < t < \ell$, we find

$$\text{(3.4)} \qquad 2\pi\, u(t) \pm \int_\ell^t u(s)P(s,t)\rho(s)ds$$

$$= \pm\int_\ell^t v(s)M(s,t)\rho(s)ds + \int_0^\ell [u(s)N(s,t)-v(s)L(s,t)]\rho(s)ds, \quad \operatorname{Im} t \gtrless 0,$$

which is useful for a Neumann or general linear boundary condition. The analytic functions L, M, N, and P are defined by $L(s,t) \equiv 2\int_\Gamma e^{ikr}/r\, d\phi$, $M(s,t) \equiv \int_\Lambda e^{ikr}/r\, d\phi$, $N(s,t) \equiv 2\int_\Gamma \frac{\partial}{\partial\nu}(e^{ikr}/r)d\phi$, $P(s,t) \equiv \int_\Lambda \frac{\partial}{\partial\nu}(e^{ikr}/r)d\phi$. Here r is the analytic function of s and t that corresponds to R in the same sense that $\log r$ and $\log R$ are related in the previous section:

$r \equiv (\beta - \alpha \cos\phi)^{\frac{1}{2}}$, with $\alpha \equiv 2\rho(s)\rho(t)$, $\beta \equiv \rho(s)^2 + \rho(t)^2 + [z(s)-z(t)]^2$, and $\arg r = 0$ for $0 < s < t < \ell$, $0 \leq \phi \leq \pi$. The contour Λ in the complex ϕ-plane encloses a cut joining the two branch points of r that coalesce on $\phi = 0$ when $\alpha = \beta$. The arc Γ joins $\phi = 0$ to $\phi = \pi$; we may take it to be the real interval $0 \leq \phi \leq \pi$ if $0 < s < t < \ell$ but, as s and t vary, it is necessary to deform Γ away from this real segment to avoid the branch point singularities of the integrand.

An equation analogous to (2.11) follows by differentiation of (3.4). In a complex neighbourhood of $0 < t < \ell$, we have

(3.5)
$$\mp 2\pi i\, v(t) \mp \int_\ell^t v(s)M_t(s,t)\rho(s)ds$$
$$= -2\pi\, u'(t) \pm \pi i z'(t)u(t)/\rho(t) \mp \int_\ell^t u(s)P_t(s,t)\rho(s)ds$$
$$+ \int_0^\ell [u(s)N_t(s,t) - v(s)L_t(s,t)]\rho(s)ds, \qquad \operatorname{Im} t \gtrless 0,$$

which is useful when $u(s)$ is prescribed. Here we have used the fact that $M(t,t)$ and $P(t,t)$ may be calculated explicitly: $M(t,t) = 2\pi i/\rho(t)$, $P(t,t) = -\pi i\, z'(t)/\rho(t)^2$.

In order to continue (3.4) and (3.5) further, knowledge of the singularities of L, M, N, and P, is needed. Singularities will generally occur where any of $\rho(s)$, $z(s)$, $\rho(t)$, or $z(t)$ are singular. Other possible singularities arise where $\zeta(\equiv \beta/\alpha) = \pm 1$, or when $\alpha = 0$; $\zeta = +1$ does not give a singularity of M or P. The solutions to $\zeta = \pm 1$ are analogous to the root loci, referred to in section 2.

As examples, we have considered Neumann boundary value problems for a sphere, and for oblate, and prolate, spheroids. For a boundary condition like that in the earlier example, that corresponds to an axially symmetric ring singularity at (ρ_0, z_0) exterior to the boundary, we have

(3.6)
$$v(s) = 2\int_0^\pi e^{ikR}(ikR-1)R^{-3}\{[z(s)-z_0]\rho'(s) - [\rho(s)-\rho_0\cos\phi_0]z'(s)\}d\phi_0 ,$$

where R is given by (3.2) with ρ, ϕ, z replaced by ρ_0, ϕ_0, z_0.

By locating the singularities of $v(s)$ from (3.6) and of $u(t)$ from (3.4), and by using the theory of characteristics, we find the classical result for the image of the ring singularity in the sphere.

For an oblate spheroid, arclength is not the most convenient parameter, and we denote a point on σ by $(\rho(\theta),z(\theta))$, where

(3.7) $$\rho(\theta) = a\cos\theta, \quad z(\theta) = b\sin\theta,$$

with $-\pi/2 \leq \theta \leq \pi/2$, and $a > b$. If $\rho_0 = a_0\cos\theta_0$, $z_0 = b_0\sin\theta_0$, where $a_0^2 - b_0^2 = a^2 - b^2$, we find an image singularity in the spheroid at

(3.8) $$\rho = A\cos\theta_0, \quad z = B\sin\theta_0,$$

where A and B are determined by (2.17).

The result for a prolate spheroid is found by interchanging a with b, and a_0 with b_0 in the above equations.

These results for spheroids are believed to be new. We note that the result (3.8) for the spheroid (3.7) is the same as the result (2.16) for the corresponding ellipse.

4. Strictly Three-Dimensional Problems. We consider exterior solutions to the Helmholtz equation in three independent variables:

(4.1) $$u_{xx} + u_{yy} + u_{zz} + k^2u = 0,$$

of which axially symmetric solutions are a special case. Although the general principles used in previous sections still apply, the addition of a third independent variable introduces significant differences and difficulties. In particular, the integral equations that are used to continue the data are two-dimensional, and the characteristics that carry the singularities are four-dimensional manifolds in $\mathbb{C}^3$.

If the analytic boundary surface is again denoted by Σ, if P is an exterior point and Q' a point on Σ, then Helmholtz's formula gives

(4.2) $$4\pi\, u(P) = \int_\Sigma [u(Q')\partial/\partial\nu' - v(Q')]e^{ikr}/r\, d\Sigma\ ,$$

where $\hat{\nu}'$ is the unit normal to Σ at Q', directed into the exterior domain, and r is distance between P and Q'. If $P \to Q \in \Sigma$ we obtain a relationship between the analytic boundary data u and v:

$$(4.3) \qquad 2\pi\, u(Q) - \int_\Sigma u(Q') \frac{\partial}{\partial \nu'} (e^{ikr}/r) d\Sigma = - \int_\Sigma v(Q')(e^{ikr}/r) d\Sigma,$$

with $r = r(Q,Q')$.

If we differentiate (4.2) along the normal $\hat{\nu}$ at Q, we find

$$(4.4) \qquad 2\pi\, v(Q) + \int_\Sigma v(Q') \frac{\partial}{\partial \nu} (e^{ikr}/r) d\Sigma = \frac{\partial}{\partial \nu} \int_\Sigma u(Q') \frac{\partial}{\partial \nu'} (e^{ikr}/r) d\Sigma,$$

which may be manipulated into

$$(4.5) \qquad 2\pi\, v(Q) + \int_\Sigma v(Q') \frac{\partial}{\partial \nu} (e^{ikr}/r) d\Sigma = \int_\Sigma^* w(Q,Q') \frac{\partial^2}{\partial \nu \partial \nu'} (e^{ikr}/r) d\Sigma$$
$$+ u(Q) \int_\Sigma \frac{\partial^2}{\partial \nu \partial \nu'} [(e^{ikr}-1)/r]\, d\Sigma;$$

here the asterisk (*) denotes a singular integral, and $w(Q,Q') \equiv u(Q') - u(Q)$. All integrals in (4.5) may be expressed as sums of convergent integrals of the form $\int_\Sigma r^{-1}\, p(Q,Q') d\Sigma$, where $p(Q,Q')$ is bounded as $Q' \to Q$. The result is complicated and we shall not give it here.

If Q is determined by real parameters α and β, then local analyticity in α and β of convergent integrals of the above type can be demonstrated by a method of E.E. Levi [10]. In essence, such an integral is defined for complex values of α and β by deforming the integration manifold M through the complex domain so that α and β remain on it and manifolds on which $r = 0$ meet M only in the point (α,β). Then it may be shown that this integral is analytic and reduces properly when α and β become real; see, for example, [9] or [2, chapter II, §6]. Global results should follow by using Levi's method step-by-step, together with the Cauchy-Poincaré theorem [21, chapter IV, §22], and knowledge of the singularity manifolds of the integrands in $\mathbb{C}^2$.

The case in which Σ is a general analytic surface has not yet been considered. However, the problem for which Σ is the plane $z = 0$ has been examined [16]. Although in most ways this is more simple than the general case, the unboundedness of Σ introduces minor difficulty, and in particular the reduction of the right-hand-side of (4.4) proceeds somewhat differently from that above. We content ourselves with a brief description.

In the planar case, (4.3) becomes

$$(4.6) \qquad 2\pi\, u(x,y) = -\int_\Sigma v(\xi,\eta)(e^{ikr}/r)\,d\xi d\eta, \quad (x,y) \in \Sigma,$$

and if v is prescribed, with suitable behaviour at infinity to ensure convergence, we have an integral representation for $u(x,y)$. Its singularities may be examined by studying the integral. We have the following theorem:

Let $u(x,y)$ *be determined for real* x *and* y *by* (4.6), *in which* v *is holomorphic for real* ξ *and* η *and has suitable behaviour at infinity. Suppose that the singularity manifolds of* v *may be represented in the form* $F^j(\xi,\eta) = 0$, *for* $j = 1,2,\cdots,p$, *where the* F^j *are analytic and where* $F_\xi{}^j$, $F_\eta{}^j$ *do not vanish simultaneously on* $F^j = 0$. *If none of the* p *sets of simultaneous equations* $F^j(\xi,\eta) = 0$, $r(\xi,\eta;x,y) = 0$, *and* $F_\xi{}^j/(r^2)_\xi = F_\eta{}^j/(r^2)_\eta$ *has a solution* $(\xi,\eta) \neq (x,y)$, *then* u *can be continued analytically into the complex* x,y*-domain as far as a singularity manifold of* v. *If any of these sets of equations has a solution* $(\xi,\eta) \neq (x,y)$, *then the corresponding point* (x,y) *may lie on a singularity manifold of* u.

[Remark: The equations referred to in the theorem give necessary conditions for the complex integration manifold to be pinched between the singularity manifolds of v and of e^{ikr}/r.]

A corresponding result for the Dirichlet problem follows from (4.4). For a general linear boundary condition of the form $v(\xi,\eta) = a(\xi,\eta)u(\xi,\eta) + b(\xi,\eta)$, with a and b holomorphic for real ξ and η, (4.6) yields an integral equation for $u(x,y)$. The solution to this may be continued analytically into the complex domain of x and y, in many cases up to singularities of a and b [16].

Possible real singularities occur where characteristics issuing from the singularity manifolds of the data meet the real domain. If S: $z = 0$, $F(x,y) = 0$, is a singularity manifold, and if $\phi(x,y,z) = 0$ is a characteristic emanating from S, we have $\phi_x{}^2 + \phi_y{}^2 + \phi_z{}^2 = 0$, $\phi(x,y,0) = F(x,y)$. Point singularities have been found in this way [16], but completely satisfactory results have not been obtained. A case in point is the known solution to the potential problem for a prolate spheroid situated in $z < 0$ and with the z-axis as axis of symmetry. In $z > 0$, the solution may be represented in terms of data on $z = 0$. We have been unable to predict the continuum of logarithmic singularities on the interfocal segment by this naive use of characteristics [16, §6]. Moreover, the method of section 3 also fails in this respect. These singularities are geometric in origin. We believe that real singularities arising from singularities in the data that are not of this nature can be located in the above way.

5. Concluding Remarks. We have outlined a method for locating singularities in the solution to an analytic boundary value problem in either two or three independent variables. Attention has been confined to locating singularities because this is important in numerous practical contexts; in particular, a simple recipe for determining the singularities *a priori* would be useful. But it is also possible to use the integral equations to study the qualitative behaviour of the data near a singularity. This idea is not new; see [11] (Math. Reviews 25 (1963), #1413). However its application to problems of the present type seems to be novel.

We have noted an apparent limitation of the procedure concerning the method of characteristics for associating singularities in the data with real singularities in the solution. Certainly there are other ways to relate these singularities. For example, if the data are on a plane and their singularities are known, we may employ Gilbert's procedure [7, pp. 198-209] to locate the real singularities. Nevertheless, the simplicity of the calculations for the method of characteristics would appear to justify further study of the procedure described here.

REFERENCES

[1] S. Bergman, Integral Operators in the Theory of Linear Partial Differential Equations (Ergeb. Math. Grenzgeb., new series, vol. 23), Springer-Verlag, Berlin, 1961.

[2] S. Bochner and W.T. Martin, Several Complex Variables, Princeton University Press, Princeton, 1948.

[3] D. Colton, On the inverse scattering problem for axially symmetric solutions of the Helmholtz equation, Quart. J. Math. Oxford (2), 22 (1971), 125-130.

[4] D. Colton, Integral operators and inverse problems in scattering theory, this Conference.

[5] F.D. Gakhov, Boundary Value Problems, Pergamon Press, Oxford, 1966.

[6] R.P. Gilbert, Function Theoretic Methods in Partial Differential Equations, Academic Press, New York, 1969.

[7] R.P. Gilbert, Constructive Methods for Elliptic Equations, Lecture Notes in Mathematics, No. 365, Springer-Verlag, Berlin, Heidelberg, New York, 1974.

[8] P. Henrici, Zur Funktionentheorie der Wellengleichung, Comm. Math. Helv. 27 (1953), 235-293.

[9] E. Hopf, Über den funktionalen, insbesondere den analytischen Charakter der Lösungen elliptischer Differentialgleichungen zweiter Ordnung, Math. Z. 34 (1932), 194-233.

[10] E.E. Levi, Sulle equazioni lineari totalmente ellittiche alle derivate parziali, Rend. Circ. Mat. Palermo, 24 (1907), 275-317.

[11] G.S. Litvinčuk, Integral equations with analytic kernels, Izv. Vysš. Učebn. Zaved. Matematika, 3 (1958), no. 2, 197-209.

[12] R.F. Millar, The location of singularities of two-dimensional harmonic functions, I, SIAM J. Math. Anal. 1 (1970), 333-344.

[13] R.F. Millar, Singularities of two-dimensional exterior solutions of the Helmholtz equation, Proc. Camb. Phil. Soc. 69 (1971), 175-188.

[14] R.F. Millar, Singularities of solutions to linear, second order analytic elliptic equations in two independent variables. I. Applicable Analysis 1 (1971), 101-121.

[15] R.F. Millar, Singularities of solutions to linear, second order, analytic elliptic equations in two independent variables. II. Applicable Analysis 2 (1973), 301-320.

[16] R.F. Millar, Singularities of solutions to exterior analytic boundary value problems for the Helmholtz equation in three independent variables. I. SIAM J. Math. Anal. 7 (1976), 131-156.

[17] L.E. Payne, Some general remarks on improperly posed problems for partial differential equations, Symposium on Non-Well-Posed Problems and Logarithmic Convexity, Lecture Notes in Mathematics, No. 316, Springer-Verlag, Berlin, Heidelberg, New York, 1973, 1-30.

[18] L.E. Payne, Improperly Posed Problems in Partial Differential Equations, Regional Conference Series in Applied Mathematics, No. 22, SIAM Publications, Philadelphia, 1975.

[19] B.D. Sleeman, The three-dimensional inverse scattering problem for the Helmholtz equation, Proc. Camb. Phil. Soc. 73 (1973), 477-488.

[20] I.N. Vekua, New Methods for Solving Elliptic Equations, John Wiley and Sons, Inc., New York, 1967.

[21] V.S. Vladimirov, Methods of the Theory of Functions of Many Complex Variables, M.I.T. Press, Cambridge, Mass., 1966.

[22] C. Wagschal, Sur le problème de Cauchy ramifié, J. Math. pures et appl. 53 (1974), 147-163.

[23] V.H. Weston, J.J. Bowman, and E. Ar, On the electromagnetic inverse scattering problem, Arch. Rational Mech. Anal. 31 (1968/69), 199-213.

Über einige neuere Anwendungen der verallgemeinerten Cauchy-Riemannschen Gleichungen in der Schalentheorie

I. N. VEKUA

(Tbilissi)

1. Im folgenden werden wir die Grundtatsachen der Flächentheorie und der Tensoranalysis benutzen, insbesondere die Summationskonvention, wie sie in dem Buch [1] (siehe auch [4]) verwendet wird. Die lateinischen bzw. griechischen Indizes durchlaufen die Werte 1,2,3 bzw. 1,2.

Wir bezeichnen mit Ω die Schale, aber auch den durch sie bestimmten räumlichen Bereich. Wir parametrisieren den Bereich Ω mit Hilfe eines Koordinatensystems, dessen eine Schar von Koordinatenlinien parallel ist zu den Normalen einer regulären Fläche S. Diese Fläche S nennt man dann Basis der Parametrisierung. Als eine solche Basis kann man auch eine beliebige aber zu S äquidistante Fläche nehmen. Den Radiusvektor des Punktes mit den Koordinaten x^1, x^2, x^3 wird man durch die Formel

$$\underline{\hat{r}}(x^1, x^2, x^3) = \underline{r}(x^1, x^2) + x^3 \underline{n}(x^1, x^2) \tag{1}$$

darstellen, wobei x^1, x^2 die Gauß'schen Parameter der Punkte der Fläche S sind, $\underline{r}$ ist der Radiusvektor und $\underline{n}$ ist der Einheitsvektor der Normalen im Punkte $(x^1, x^2) \in S$. Die skalare Koordinate x^3 ist die Entfernung eines Punktes des Bereichs Ω von der Fläche S.

Die Vektoren $\underline{\hat{r}}_i = \partial_i \underline{\hat{r}} \equiv \frac{\partial \underline{\hat{r}}}{\partial x^i}$ bilden die Basis der S-Parametrisierung des Bereiches. Offensichtlich ist $\underline{\hat{r}}_\alpha = \underline{r}_\alpha + x^3 \underline{n}_\alpha$, $\underline{\hat{r}}_3 = n$. Es sei $\hat{a}_{ij} = \hat{a}_{ji} = \underline{\hat{r}}_i \cdot \underline{\hat{r}}_j$ und $a_{\alpha\beta} = a_{\beta\alpha} = \underline{r}_\alpha \cdot \underline{r}_\beta$. Dann gilt $\hat{a} = \det \hat{a}_{ij} = a\vartheta^2 > 0$, wobei $a = \det a_{\alpha\beta}$ die Diskriminante der metrischen Fundamentalform der Fläche S ist, während $\vartheta = (1-\kappa_1 x^3)(1-\kappa_2 x^3)$ ist. Mit κ_1 und κ_2 bezeichnen wir die beiden Hauptkrümmungen der Fläche S, während $K = \kappa_1\kappa_2$ ihre Gauß'sche Krümmung und $H = \frac{1}{2}(\kappa_1+\kappa_2)$ ihre mittlere Krümmung bezeichnen.

2. Die Gleichung für das statische Gleichgewicht des Kontinuums lautet:

$$\frac{1}{\sqrt{\hat{a}}} \partial_i (\sqrt{\hat{a}}\, \underline{\hat{P}}^i) + \underline{\hat{\Phi}} = 0 \qquad \underline{\hat{P}}^i = \hat{P}^{ij}\, \underline{r}_j , \tag{2}$$

wobei $\hat{P}^{ij} = \hat{P}^{ji}$ die kontravarianten Komponenten des Spannungstensors sind, während $\underline{\hat{\Phi}}$ die Volumenkraft (vgl. [3]) ist.
Wenn wir mit $\hat{e}^{ij}$ bzw. $\hat{a}^{ij}$ die kontravarianten Komponenten des Deformationstensors bzw. des metrischen Raumtensors bezeichnen, dann kann man mit Hilfe des HOOKEschen Gesetzes: $\hat{P}^{ij} = \lambda\hat{e}^k_k\hat{a}^{ij} + 2\mu\hat{e}^{ij}$, wobei λ und μ die Laméschen Konstanten sind, sehr leicht die Formeln

$$\hat{P}^{ij} = \hat{T}^{ij} + \hat{Q}^{ij} , \tag{3}$$

$$\hat{T}^{\alpha\beta} = \hat{T}^{\beta\alpha} = E(1-\sigma^2)^{-1}\left[\sigma\hat{e}^{\gamma}_{\gamma}a^{\alpha\beta} + (1-\sigma)\hat{e}^{\alpha\beta}\right] , \quad \hat{T}^{3i} = \hat{T}^{i3} = 0 , \tag{4}$$

$$\hat{Q}^{\alpha\beta} = \hat{Q}^{\beta\alpha} = \sigma(1-\sigma)^{-1}\,\hat{P}^{33}\,\hat{a}^{\alpha\beta} , \quad \hat{Q}^{3i} = \hat{Q}^{i3} = \hat{P}^{3i} \tag{5}$$

herleiten, wobei E der YOUNGsche Modul und σ der POISSONsche Koeffizient ist, $0 < \sigma < 0{,}5$. Wenn man die Bezeichnung: $\underline{T}^{\alpha} = \vartheta\hat{T}^{\alpha\beta}\underline{r}_{\beta}$, $\underline{Q}^{i} = \vartheta\hat{Q}^{ij}\underline{r}_{j}$ benutzt, dann nimmt die Gleichung (2) die Form

$$\frac{1}{\sqrt{a}}\,\partial_{\alpha}(\sqrt{a}\,\underline{T}^{\alpha}) + \frac{1}{\sqrt{a}}\,\partial_{i}(\sqrt{a}\,\underline{Q}^{i}) + \underline{\Phi} = 0 , \quad \underline{\Phi} = \vartheta\underline{\hat{\Phi}} \tag{6}$$

an. Weil $\underline{n}\,\underline{T}^{\alpha} = 0$ ist, deshalb bezeichnet man den Spannungszustand, der durch den Tensor $\hat{T}^{\alpha\beta}$ definiert wird, als Tangential-Spannungsfeld (bezüglich S) oder kurz als Feld $\underline{T}$. In den Beziehungen (4) hängt dieses von den Tangentialkomponenten $\hat{e}_{\alpha\beta}$ des Deformationstensors ab. Aus diesen Beziehungen erhalten wir die äquivalenten Gleichungen

$$\hat{e}_{\alpha\beta} = E^{-1}\left[-\sigma\,\hat{T}^{\gamma}_{\gamma}\,\hat{a}_{\alpha\beta} + (1+\sigma)\hat{T}_{\alpha\beta}\right] . \tag{7}$$

Das durch den Tensor Q^{ij} definierte Spannungsfeld (kurz: das Feld $\underline{Q}$) hängt nur vom Spannungsvektor $\underline{\hat{P}}^{3}$ ab, der auf der Fläche mit der Normalen $\underline{n}$ angreift. Wir bezeichnen dieses als bedingtes transversales Spannungsfeld, wenn es auch für $P^{33} \neq 0$ eine von null verschiedene Tangentialkomponente besitzt.

3. Die Darstellung des allgemeinen Spannungsfeldes in der Gestalt einer Summe der Felder $\underline{T}$ und $\underline{Q}$, und auch die Schreibweise der Gleichgewichtsbedingung in der Form (6), gestatten eine umfangreiche Klasse von Problemen aus der Theorie der elastischen Schalen zu reduzieren auf ein System von partiellen Differentialgleichungen erster Ordnung für zwei gesuchte Funktionen, die von den Gauß'schen Parame-

tern x^1, x^2 der Fläche S abhängen.
Zu der Zahl derartiger Probleme gehören zum Beispiel auch die Randwertprobleme für Zapfenverbindungen (vergleiche [2], Kap. 5, § 8). Diese Verbindungen sind verwirklicht, wenn die Schale mit ihren Randflächen auf absolut feste glatte Wände stößt; zum Beispiel in Löchern und Spalten der Schale seien absolut harte glatte Zapfen (Pfropfen) oder Stempel fest eingesetzt. Dann sind die Tangentialspannungen und die Normalverschiebungen auf den Randflächen der Schale, deren Punktmenge wir mit $\partial\Omega$ bezeichnen, gleich null. Wenn man mit $\underline{1}$ und $\underline{\bar{\bar{S}}}$ die Einheitsvektoren der Tangentennormale bzw. der Tagente des Randes der Fläche x^3=const. bezeichnet, dann erhalten die Randbedingungen der Zapfverbindung die Gestalt

$$\hat{P}^{\alpha\beta} 1_\alpha \bar{\bar{S}}_\beta = 0 \quad , \quad \hat{P}^{3\alpha} 1_\alpha = 0 \ , \ \hat{U}^\alpha 1_\alpha = 0 \ (\text{auf } \partial\Omega) \ , \tag{8}$$

wobei $\hat{U}^\alpha$ die kontravarianten Komponenten des Verschiebungsvektors $\underline{U}$, während 1_α bzw. $\bar{\bar{S}}_\alpha$ die kovarianten Komponenten der Einheitsvektoren $\underline{1}$ bzw. $\underline{\bar{\bar{S}}}$ sind. Leicht ersichtlich ist, daß die physikalischen Randbedingungen $(8_{1,2})$ die Gestalt

$$T_{(1\bar{\bar{S}})} \equiv T^{\alpha\beta} 1_\alpha \bar{\bar{S}}_\beta \equiv 0 \ , \ Q_{(13)} \equiv Q^{3\alpha} 1_\alpha = 0 \ (\text{auf } \partial\Omega) \tag{9}$$

besitzen. Daher haben wir für das Tangential- und für das Transversalspannungsfeld voneinander unabhängige Randbedingungen (9_1) bzw. (9_2). In einer Reihe von Fällen gelingt es deshalb das Problem: "den Spannungszustand der Schale zu bestimmen" in zwei Teilprobleme zu entkoppeln; nämlich einmal sucht man das Feld $\underline{T}$ und zum anderen muß das Feld $\underline{Q}$ bestimmt werden. Eine solche Zerlegung des Problems ist gewöhnlich durch die Anwendung des halbinversen Prinzips von St. Venant verwirklicht, das auf gewissen physikalischen und geometrischen Annahmen beruht.

4. Als Beispiel betrachten wir das Problem des statischen Gleichgewichtes einer dünnen Schale der konstanten Dicke 2h, wenn an ihren Außenflächen, $x^3 = h$ und $x^3 = -h$, die vorgegebenen Spannungen $\overset{(+)}{\underline{P}}$ bzw. $\overset{(-)}{\underline{P}}$ angreifen. Dann kann man im Innern der Schale die Spannung $\underline{\hat{P}}^3$ angenähert darstellen in der Form

$$\underline{\hat{P}}^3 = \frac{1}{2}(\overset{(+)}{\underline{P}} + \overset{(-)}{\underline{P}}) + \frac{x^3}{2h}(\overset{(+)}{\underline{P}} - \overset{(-)}{\underline{P}}) \ . \tag{10}$$

Wenn außerdem die Randbedingungen: $\overset{(+)}{P}{}^{\alpha} l_{\alpha} = \overset{(-)}{P}{}^{\alpha} l_{\alpha} = 0$ (auf $\partial\Omega$) erfüllt sind, dann wird das entsprechende Feld $\underline{Q}$ die Bedingung (9_2) erfüllen. Diese Bedingungen sind offensichtlich erfüllt, wenn an den Außenflächen nur Normalkräfte (zum Beispiel: Staudruck) angreifen. Dann ist $\overset{(+)}{P}{}^{\alpha} = \overset{(-)}{P}{}^{\alpha} = 0$.

Im allgemeinen, wenn wir annehmen, daß das Feld $\underline{Q}$ gegeben ist und die Bedingungen auf den Außenflächen und die Randbedingung (9_2) erfüllt sind, erhalten wir zur Bestimmung des Feldes $\underline{T}$ die Gleichung

$$\frac{1}{\sqrt{a}}\, \partial_{\alpha}(\sqrt{a}\, \underline{T}^{\alpha}) + \underline{X} = 0 \;, \quad \underline{X} = \frac{1}{\sqrt{a}}\, \partial_{i}(\sqrt{a}\, \underline{Q}^{i}) + \underline{\Phi} \;. \tag{11}$$

Wenn wir nun zu dieser Gleichung noch die Bedingung (9_1) hinzufügen, dann erhalten wir ein Randwertproblem, das wir kurz als Problem T benennen werden. Angenommen, dieses Problem sei gelöst (auf die Bedingungen seiner Lösbarkeit kommen wir weiter unten noch zurück), dann kann das Verschiebungsfeld mittels des Gleichungssystems (7) bestimmt werden, nachdem noch die Randbedingung (8_3) hinzugefügt wurde. Dieses Randwertproblem werden wir kurz als Problem T' bezeichnen.
Der Vektor $\underline{X}$ heiße äußere Last oder rechte Seite des Problems T . Wenn also das Feld $\underline{Q}$ gegeben und die Bedingung (9_2) erfüllt ist, dann ist das Problem, das Tangentialspannungsfeld $\underline{T}$ und den Verschiebungsvektor $\underline{U}$ zu bestimmen, die die Randbedingungen (9_1) bzw. (8_3) erfüllen, äquivalent zur Vereinigung der Probleme T und T' . Die beiden Probleme in ihrer Gesamtheit löst man wie folgt:

Zuerst ermitteln wir die Lösungen des Problems T, falls sie existieren, und dann werden wir die Lösungen des Problems T' finden, dessen rechte Seite eine lineare Funktion der Lösungen des Problems T ist. Im weiteren werden wir dieses allgemeine Problem als Problem (T, T') bezeichnen, während seine Lösung durch (T, U) bezeichnet werde.

Die skalare Koordinate x^3 ist in den Problemen T und T' als Parameter enthalten; von ihr hängen der Vektor $\underline{X}$ und folglich auch die gesuchten Felder $\underline{T}$ und $\underline{U}$ ab. Aber man kann auch eine beliebige Fläche $x^3 = \text{const.}$ als Basis für die Parametrisierung des Bereiches Ω nehmen. Deshalb kann man die allgemeine Untersuchung der Probleme T und T' stets zurückführen auf den Fall $x^3 = 0$. In diesem Fall - wenn wir die Darstellungen: $\underline{T}^{\alpha} = T^{\alpha\beta}\underline{r}_{\beta}$, $\underline{X} = X^{\alpha}\underline{r}_{\alpha} + X\,\underline{n}$, benutzen und außerdem noch annehmen, daß der Deformationstensor $e_{\alpha\beta}$ linear vom Verschiebungsvektor $\underline{U}$ abhängt - führen wir die Gleichungen (11) und (7) über in das System

$$\nabla_\alpha T^{\alpha\beta} + X^\beta = 0 \ , \ b_{\alpha\beta} T^{\alpha\beta} + X = 0 \ , \tag{12}$$

$$\frac{1}{2}(\nabla_\alpha U_\beta + \nabla_\beta U_\alpha) - b_{\alpha\beta} U_3 = E^{-1}(-\sigma T^\gamma_\gamma a_{\alpha\beta} + (1-\sigma) T_{\alpha\beta}) , \tag{13}$$

wobei ∇_α das Symbol für die kovariante Ableitung auf S und $b_{\alpha\beta} = -\underline{r}_\alpha \underline{n}_\beta$ ist.

Wir wenden nun die Formel

$$\iint_S \frac{1}{\sqrt{a}} \partial_\alpha (\sqrt{a}\, \underline{T}^\alpha) \underline{U} dS + \iint_S \underline{T}^\alpha \partial_\alpha \underline{U} dS = \int_{\partial S} \underline{U}\, \underline{T}^\alpha 1_\alpha d\bar{\bar{S}} \tag{14}$$

an, die für beliebige Tensoren $T^{\alpha\beta}$ und für beliebige Vektoren $\underline{U}$ gilt, die auf $S + \partial S$ stetig und in S stetig differenzierbar sind. Damit erhalten wir die folgenden notwendigen Bedingungen für die Lösbarkeit der Probleme T bzw. T':

$$\iint_S \underline{\tilde{U}}\, \underline{X}\, dS = 0 \ , \ \iint_S (\sigma \tilde{T}^\alpha_\alpha T^\beta_\beta - (1+\sigma) \tilde{T}^\alpha_\beta T^\beta_\alpha) dS = 0 \ , \tag{15}$$

wobei $\underline{\tilde{U}}$ und $\tilde{T}^{\alpha\beta}$ beliebige Lösungen des homogenen Problems T'_0 bzw. T_0 sind. Speziell für konvexe Schalen wurde auch die Hinlänglichkeit dieser Bedingungen (vergleiche [2] , Kap. 6, § 5) bewiesen. Folglich sind die Probleme T und T' für konvexe Schalen zueinander konjugierte Randwertprobleme. Im folgenden werden wir gerade diesen Fall betrachten.

Wir bemerken nun, daß die äußere Last $\underline{X} = \underline{Y} + \underline{Z}$ ist, wobei gesetzt wurde

$$\underline{Y} = \frac{1}{\sqrt{a}} \partial_\alpha (\sqrt{a}\, Q^{\alpha\beta} \underline{r}_\beta) \ , \ \underline{Z} = \partial_3 (\vartheta \hat{\underline{P}}^3) + \frac{1}{\sqrt{a}} \partial_\alpha (\sqrt{a}\, \hat{P}^{\alpha 3} \underline{n}) + \underline{\Phi} \ . \tag{15 a}$$

Man überzeugt sich jetzt leicht davon, daß stets die Bedingung

$$\iint_S \underline{Y}\, \underline{\tilde{U}}\, dS = 0 \ , \ \underline{Y} = \frac{1}{\sqrt{a}} \partial_\alpha (\sqrt{a}\, \underline{\tilde{Q}}^\alpha) \ , \ \underline{\tilde{Q}}^\alpha = Q^{\alpha\beta} \underline{r}_\beta \tag{15 b}$$

gilt. Das folgt aus der Formel (14), wenn man in ihr $\underline{T}^\alpha = \underline{\tilde{Q}}^\alpha$, $\underline{U} = \underline{\tilde{U}}$ substituiert und wenn man außerdem noch die Gleichungen

$$\underline{\tilde{U}}\,\underline{\tilde{Q}}^{\alpha}\,1_{\alpha} = \sigma(1-\sigma)^{-1}\,\hat{P}^{33}\,\tilde{U}_{(1)} = 0 \qquad \text{(auf } \partial S)$$

$$\underline{\tilde{Q}}^{\alpha}\partial_{\alpha}\,\underline{\tilde{U}} = \sigma(1-\sigma)^{-1}\,\hat{P}^{33}\,\underline{r}^{\alpha}\partial_{\alpha}\underline{\tilde{U}} = 0 \qquad \text{(in } S)$$

berücksichtigt. Die letzte Gleichung folgt aus der Bedingung: $d\,\underline{\tilde{U}}\,d\,\underline{r} = 0$ und diese ist die Differentialgleichung der unendlich kleinen Verbiegung der Fläche S in Vektorform (siehe [2], Kap. 5, §1). Aufgrund der Gleichung (15 b) nimmt die Bedingung (15_1) nun die Gestalt

$$\iint_S \underline{Z}_0 \underline{\tilde{U}} dS = 0 \;, \quad \underline{Z}_0 = (\underline{Z})_{x^3=0} \qquad (15\text{ c})$$

an. Die äußere Last $\underline{Y}$, die durch die erste von den beiden Formeln (15 a) definiert ist, nennen wir Gleichgewichtslast. Das Problem T für eine konvexe Schale ist offenbar lösbar, wenn seine rechte Seite eine Gleichgewichtslast, d. h. $\underline{Z} = 0$ ist. Wenn $\overset{(+)}{P}{}^{\alpha} = \overset{(-)}{P}{}^{\alpha} = 0$ gilt, dann haben wir für die Spannung $\underline{\hat{P}}^3$ gemäß Formel (10) die Gleichung:

$$\underline{Z} = \frac{1}{2h}(1-2hH)\overset{(+)}{\underline{P}} + \frac{1}{2h}(1+2hH)\overset{(-)}{\underline{P}} + \underline{\Phi}\;. \qquad (15\text{ d})$$

Folglich kann man annehmen, daß bei dünnen Schalen Gleichgewicht herrscht, falls die Bedingungen:

$$\frac{1}{2h}(1-2hH)\overset{(+)}{\underline{P}} + \frac{1}{2h}(1+2hH)\overset{(-)}{\underline{P}} + \underline{\Phi} = 0 \;, \quad \overset{(+)}{P}{}^{3\alpha} = \overset{(-)}{P}{}^{3\alpha} = 0 \qquad (15\text{ e})$$

erfüllt sind.

5. Falls $K > 0$ auf S ist, kann man als Gauß'sche Parameter die konjugierten isometrischen Koordinaten x und y nehmen, die die Bedingungen: $b_{11} = b_{22} = \sqrt{a\,K} > 0$, $b_{12} = b_{21} = 0$ erfüllen (siehe [2], Kap. 2, § 6). Dann nimmt die Gleichung (12_2) die Gestalt $\sqrt{a\,K}\,(T^{11} + T^{22}) + X = 0$ an. In unsere Betrachtung bringen wir nun die komplexen Funktionen für die Spannung und für die Verschiebung:

$$w = \frac{1}{2} K^{-\frac{1}{4}} a(T^{11} - T^{22} - 2iT^{12}) \equiv 2K^{-\frac{3}{4}}\,\underline{T}(\underline{n}_z \otimes \underline{n}_z) \qquad (16)$$

$$w' = \frac{U_1 + iU_2}{\sqrt{a\sqrt{K}}} \equiv \frac{2}{\sqrt{a\sqrt{K}}}\,\underline{\tilde{U}}\,\underline{r}_{\bar{z}} \quad \left(()_{\bar{z}} \equiv \partial_{\bar{z}} = \frac{1}{2}(\partial_x + i\partial_y)\right) \qquad (17)$$

und zeigen, daß die Systeme (12) und (13) äquivalent sind zu den entsprechenden Gleichungen

$$\partial_{\bar z} w - \bar B \bar w = F \quad \text{und} \quad \partial_{\bar z} w' + B \bar w' = F' , \tag{18}$$

$$B = \frac{2i}{K\sqrt{a}} \underline{n}\, \underline{n}_{\bar z}\, \underline{n}_{\bar z \bar z} , \quad F = K^{-\frac{1}{4}} \sqrt{a}\left[\underline{X}\, \underline{n}_z + \frac{1}{2} K \partial_z (K^{-1} X)\right],$$

$$F' = \frac{E^{-1}}{\sqrt{a\sqrt{K}}}\left[-\sigma T^{\gamma}_{\gamma}(a_{11} - a_{22} + 2ia_{12}) + (1-\sigma)(T_{11} - T_{22} + 2iT_{12})\right]. \tag{19}$$

Aus den Formeln (16) und (17) erhalten wir die Gleichungen

$$T_{(\ell\ell)} \equiv T^{\alpha\beta} l_\alpha l_\beta = -K^{-\frac{1}{4}} \operatorname{Re}\left[w\left(\frac{dz}{ds}\right)^2\right] - \frac{\bar{\bar K}_s}{2K} X ,$$

$$T_{(\ell s)} \equiv T^{\alpha\beta} l_\alpha s_\beta = K^{-\frac{1}{4}} \operatorname{Re}\left[w \frac{dz}{ds} \frac{dz}{dl}\right] - \frac{\tau_s}{2K} X , \tag{20}$$

$$U_{(1)} \equiv U^{\alpha} l_\beta = \sqrt{a\sqrt{K}} \operatorname{Re}\left(w' \frac{dz}{dl}\right) ,$$

$$U_{(s)} \equiv U^{\alpha} \bar{\bar S}_\beta = \sqrt{a\sqrt{K}} \operatorname{Re}\left(w' \frac{dz}{ds}\right) , \tag{21}$$

wobei $\bar{\bar K}_s$ und τ_s die Normalkrümmung bzw. die geodätische Windung der Fläche S in der Richtung des Bogens $d\bar{\bar S}$ sind. Diese Gleichungen drücken die Normalkomponenten (zur Fläche Σ_1 mit der Normalen 1) $T_{(11)}$ und $U_{(1)}$ und die Tangentialkomponenten (zur Fläche Σ_1 in der Richtung $\underline{\bar{\bar S}}$) $T_{(1s)}$ und $U_{(s)}$ des Tangentialspannungsfeldes $\underline{T}$ und des Verschiebungsvektors $\underline{U}$ durch die Lösungen w und w' der Gleichungen (18_1) und (18_2) aus; $\underline{1}$ und $\underline{\bar{\bar S}}$ sind die Einheitsvektoren der Tangentennormale bzw. der Tangente an eine Kurve der Fläche S : $\underline{1} \times \underline{\bar{\bar S}} = \underline{n}$.

Die Gleichungen (18) sind zueinander adjungierte verallgemeinerte inhomogene Cauchy-Riemannsche Differentialgleichungen. Insbesondere, wenn S eine algebraische Fläche zweiter Ordnung mit positiver Krümmung (Ellipsoid, elliptisches Paraboloid, zweischaliges Hyperboloid) ist, dann ist $B = 0$ und wir haben die inhomogenen Cauchy-Riemannschen Diferentialgleichungen: $\partial_{\bar z} w = F$ und $\partial_{\bar z} w' = F'$ (siehe [2], Kap. 5, § 4). Nun sei ein Gebiet $\mathcal{D}$ bzw. sein Rand Γ topologisches Bild der Fläche S bzw. seines Randes $L = \partial S$ in der komplexen Zahlenebene $\mathbb{C}_z$, $z = x + iy$. Aufgrund der Formeln (20_2) und (20_1) nehmen dann die Randbedingungen (9_1) und (8_3) die Form

$$\operatorname{Re}\left[w \frac{dz}{ds} \frac{dz}{dl}\right] = \frac{1}{2} \tau_s K^{-\frac{3}{4}} X , \quad \operatorname{Re}\left(w' \frac{dz}{dl}\right) = 0 \tag{22}$$

an. Auf diese Weise erkennt man die Probleme T und T' als Spezialfälle des verallgemeinerten Riemann-Hilbert'schen Randwertproblems [2] (Kap. 4). Nach der Anwendung der bekannten allgemeinen Sätze untersuchen wir jetzt die Lösbarkeit der Probleme T und T'.

6. Die Schale habe $m + 1$ Löcher, deren Ränder einfache Ljapunow-Kurven seien. Dann ist der Index des Problems T bzw. T' gleich $n = 2(m-1)$ bzw. $n' = (1-m)$ (siehe [2], Kap. 6, § 5). Jetzt betrachten wir vier Fälle.

I. Für die Schale mit einem Loch, d. h. $m = 0$ und $n = -2$, $n' = 1$ erhalten wir: Das homogene Problem T_o besitzt dann keine (nichttrivialen) Lösungen. Aber das homogene Problem T_o' besitzt drei linear unabhängige Lösungen $\overset{(1)}{\underline{U}}$, $\overset{(2)}{\underline{U}}$, $\overset{(3)}{\underline{U}}$. Das inhomogene Problem T' ist immer lösbar und seine Lösung kann man in der Gestalt

$$\underline{U} = \overset{(0)}{\underline{U}} + c_1\overset{(1)}{\underline{U}} + c_2\overset{(2)}{\underline{U}} + c_3\overset{(3)}{\underline{U}} \tag{23}$$

darstellen, wobei c_1, c_2, c_3 beliebige reelle Konstanten sind, während $\overset{(0)}{\underline{U}}$ eine partikuläre Lösung des Problems T' ist. Das inhomogene Problem T besitzt eine Lösung und überdies ist sie die einzige, wenn die drei notwendigen und hinreichenden Bedingungen:

$$\iint_S \underline{Z}_o \overset{(i)}{\underline{U}} \, dS = 0 \ , \quad i = 1, 2, 3 \tag{24}$$

erfüllt sind. Die in der Formel (23) vorkommenden Konstanten c_i sind eindeutig bestimmt, wenn in drei beliebigen, aber festen Punkten auf dem Rande des Schalenloches die Werte der Tangentialkomponenten $U_{(s)}$ des Verschiebungsvektors $\underline{U}$ (siehe [2], Kap. 4, § 6) gegeben sind. Solche Verbindungen lassen sich mechanisch zuverlässig realisieren mittels sehr dünnen, ideal harten Dornen auf den Lochwandungen der Schale, die stramm eingepaßt werden in dazu entsprechende kleine Nuten im Zapfen (oder umgekehrt) längs dreier fester Normalen zu $\partial\overline{\overline{S}}$. Diese Vorsprünge stellen Hindernisse dar, die tangentielle Verschiebungen senkrecht zu den Dornen unmöglich machen, aber dabei die normalen (senkrecht zur tangentiellen Richtung) Verschiebungen (parallel der Dornen)nicht behindern. Auf diese Weise kann man im Falle einer konvexen Schale mit einer Öffnung, deren Rand glatt ist, die kinematische Verbindung, die durch die Bedingung $U_{(1)} = 0$ zum Ausdruck kommt, verstärken, indem man die Fixierung (zum Beispiel durch die Forderung des Null-Werdens) auch der Tangentialkomponente $U_{(s)}$ des Verschiebungsvektors $\underline{U}$ längs dreier fester Normalen zum Rand der Schale gewährleistet. In diesem Fall ist das Problem (T, T') lösbar und besitzt eine einzige Lösung, falls das Problem T lösbar ist. Insbesondere ist das letztere Problem immer lösbar, falls entweder seine rechte Seite eine äußere Gleichgewichtslast (d. h. $\underline{Z}_o = 0$) darstellt, oder wenn die Verschiebungsfelder $\overset{(i)}{\underline{U}}$ die Bewegung eines ideal starren Körpers beschreiben. Dann stimmen die Gleichungen (24), die die notwendigen und hinreichenden Bedingungen für die Lösbarkeit des Problems T ausdrücken, mit drei von den sechs statischen Gleichgewichtsbedingungen einer starren Fläche S überein. Wir zeigen, daß dies gilt, wenn die Fläche S längs des Randes der Schalenöffnung von einem Kegel berührt wird. In der Tat, wenn man die Spitze des Kegels in den Ursprung des kartesischen Koordinatensystems legt, dann erfüllt der Vektor $\underline{U} = \underline{C} \times \underline{r}$, wobei $\underline{C}$ ein beliebiges konstantes Vektorfeld $\underline{C} \neq 0$ ist, die Randbedingung: $U_{(1)} = 0$ und die Gleichung (24) beschreibt dann drei der sechs statischen Gleichgewichtsbedingungen des starren Körpers. Das ist klar, weil $\underline{1}$ der Einheitsvektor der entsprechenden Erzeugenden des Kegels ist.

II. Für die Schale mit zwei Löchern, d. h. $m = 1$ und $n = n' = 0$ können wir zwei Fälle unterscheiden:
(II a) die homogenen Probleme T_o und T'_o besitzen keine Lösungen oder (II b) sie besitzen eine normierte Lösung $\overset{(1)}{T}{}^{\alpha\beta}$ bzw. $\overset{(1)}{\underline{U}}$. Als Normierungsbedingung für $\overset{(1)}{T}{}^{\alpha\beta}$ kann man

$$\iint_S (\sigma \overset{(1)}{T}{}^{\alpha}_{\alpha} \overset{(1)}{T}{}^{\beta}_{\beta} - (1+\sigma) \overset{(1)}{T}{}^{\alpha}_{\beta} T^{\beta}_{\alpha}) dS = 1 \tag{25}$$

nehmen. Im Falle (II a) sind die Probleme T und T' stets lösbar und jedes der Probleme besitzt eine einzige Lösung. Im Falle (II b) gestatten die Probleme T und T' Lösungen , nur wenn die Bedingungen

$$\iint_S \underline{Z}_o \overset{(1)}{\underline{U}} \, dS = 0 \quad \text{bzw.} \quad \iint_S (\sigma T^{\alpha}_{\alpha} \overset{(1)}{T}{}^{\beta}_{\beta} - (1+\sigma) T^{\alpha}_{\beta} \overset{(1)}{T}{}^{\beta}_{\alpha}) dS = 0 \tag{26}$$

erfüllt sind. Dann lassen sich die gesuchten Lösungen mit Hilfe der Formeln

$$T^{\alpha\beta} = c \overset{(1)}{T}{}^{\alpha\beta} + \overset{(0)}{T}{}^{\alpha\beta} , \quad \underline{U} = c' \overset{(1)}{\underline{U}} + \overset{(0)}{\underline{U}} \tag{27}$$

ausdrücken, wobei c und c' beliebige Konstanten sind, während

$\overset{(0)}{T}{}^{\alpha\beta}$ und $\overset{(0)}{\underline{U}}$ partikuläre Lösungen des Problems T bzw. des Problems T' sind. Setzt man nun den Ausdruck (27_1) auf der linken Seite in die Formel (26_2) ein, dann erhalten wir vermöge der Bedingung (25) sofort

$$c = \iint_S \left(\sigma \overset{(0)}{T}{}^{\alpha}_{\alpha} \overset{(1)}{T}{}^{\beta}_{\beta} - (1+\sigma) \overset{(0)}{T}{}^{\alpha}_{\beta} \overset{(1)}{T}{}^{\beta}_{\alpha}\right) dS. \tag{28}$$

Somit ist im Falle (II b) das Feld $\underline{T}$, falls es existiert, eindeutig definiert, das Problem T' ist immer lösbar und das zugehörige Verschiebungsfeld $\underline{U}$ erhält man aus der Formel (27_2) bis auf einen additiven Summanden $c'\overset{(1)}{\underline{U}}$. Die in diese Formel eingehende Konstante c' ist eindeutig bestimmt, falls in einem festen Punkt des Schalenrandes der Wert der Tangentialkomponente $U_{(s)}$ des Vektors $\underline{U}$ (siehe [2], Kap. 4, § 6) vorgegeben ist. Das kann man verwirklichen mittels eines dünnen harten querstehenden Dorns auf einer der zwei Lochwandungen der Schale, den man stramm in eine kleine Nute einpaßt, der sich auf dem entsprechenden Zapfen (oder umgekehrt) befindet. In diesem Fall ist das Problem (T, T') eindeutig auflösbar, wenn man das Problem T gelöst hat. Das letztere Problem ist immer dann lösbar, wenn $\underline{Z}_o = 0$ ist, oder $\overset{(1)}{\underline{U}}$ die Bewegung eines starren Körpers beschreibt. Dann stellt die Gleichung (26_1) eine von den sechs statischen Gleichgewichtsbedingungen des Körpers dar. Man kann zeigen, daß dieser Fall realisiert ist, wenn die Fläche S längs der Ränder der Löcher, die sich in der Schale befinden, von Kegeln berührt wird. In der Tat, wenn der Ursprung des Koordinantensystems in die Spitze von einem der Kegel gelegt wird und das konstante Feld $\underline{C} \neq 0$ parallel zur Geraden ist, die durch die Spitzen der Kegel geht, dann erfüllt der Vektor $\underline{C} \times \underline{r}$ die Randbedingungen des homogenen Problems T'_o.

III. Für Schalen mit drei oder mehr Löchern ist $m > 1$ und $n > m-1$, $n' < 0$. Dann besitzt das homogene Problem T'_o keine Lösung, während das homogene Problem T_o $\kappa = 3m-3$ linear unabhängige Lösungen $\overset{(1)}{T}{}^{\alpha\beta}, \ldots, \overset{(\kappa)}{T}{}^{\alpha\beta}$ besitzt, denen man die folgenden Normierungsbedingungen

$$\iint_S \left(\sigma \overset{(p)}{T}{}^{\alpha}_{\alpha} \overset{(q)}{T}{}^{\beta}_{\beta} - (1+\sigma) \overset{(p)}{T}{}^{\alpha}_{\beta} \overset{(q)}{T}{}^{\beta}_{\alpha}\right) dS = \delta_{pq} \tag{29}$$

auferlegen kann. Wir zeigen jetzt, daß in diesem Fall beide Probleme T und T' lösbar sind und jedes von ihnen eindeutig. In der Tat, die allgemeine Lösung des Problems T kann man in der Form

$$T^{\alpha\beta} = \sum_{j=1}^{\kappa} c_j \overset{(j)}{T}{}^{\alpha\beta} + \overset{(0)}{T}{}^{\alpha\beta} \tag{30}$$

darstellen, wobei c_j beliebige Konstante und $\overset{(0)}{T}{}^{\alpha\beta}$ eine partikuläre Lösung des Problems T sind, die immer existiert. Setzt man nun den Ausdruck (30) in die rechte Seite der Gleichung (13) ein, dann haben wir

$$\frac{1}{2}(\nabla_\alpha U_\beta + \nabla_\beta U_\alpha) - b_{\alpha\beta} U_3 = \sum_{j=1}^{\kappa} c_j \overset{(j)}{e}_{\alpha\beta} + \overset{(0)}{e}_{\alpha\beta} ; \tag{31}$$

$$\overset{(j)}{e}_{\alpha\beta} = E^{-1}(-\sigma \overset{(j)}{T}{}^{\gamma}_{\gamma} a_{\alpha\beta} + (1+\sigma)\overset{(j)}{T}_{\alpha\beta}), \; j = 0, 1, \ldots, \kappa. \tag{32}$$

Wenn wir jetzt die notwendigen und hinreichenden Lösbarkeitsbedingungen für das Problem T' :

$$\sum_{j=1}^{\kappa} c_j \iint_S \overset{(i)}{T}{}^{\alpha\beta} \overset{(j)}{e}_{\alpha\beta} dS + \iint_S \overset{(i)}{T}{}^{\alpha\beta} \overset{(0)}{e}_{\alpha\beta} dS = 0, \; i=1,\ldots,\kappa \tag{33}$$

hinschreiben, dann erhalten wir aufgrund der Normierungsbedingungen (29) die Gleichungen

$$c_j = E \iint_S \overset{(j)}{T}{}^{\alpha\beta} \overset{(0)}{e}_{\alpha\beta} dS, \; j = 1, \ldots, \kappa , \tag{34}$$

durch die die in der Formel (30) vorkommenden Konstanten $c_1, \ldots, c_\kappa$; $\kappa = 3m - 3$; festgelegt werden.

Somit stellen wir fest, daß für $m > 1$ das Problem (T, T') stets lösbar ist und eine eindeutig bestimmte Lösung besitzt, wobei sie stetig von der äußeren Last $\underline{X}$ abhängt.

Weil die Indizes der Probleme T und T', die gleich $n = 2(m-1)$ bzw. gleich $n' = 1-m$ sind, nicht von der Koordinate x^3 abhängen, gelten die oben für das Problem (T, T') formulierten Ergebnisse auch für beliebige Koordinatenflächen $\hat{S} : x^3 = c = \text{const}, -h \leq c \leq h$.

7. Zum Abschluß betrachten wir den Fall einer geschlossenen konvexen Schale. Dann kann man das Problem (T, T') überführen in das Aufsuchen der Lösung der Gleichungssysteme (12) und (13), die die Felder

$\underline{T}$ und $\underline{U}$ definieren, die stetig auf der geschlossenen Schale sind. In diesem Fall kann man die geschlossene Fläche S topologisch abbilden auf die erweiterte komplexe Zahlenebene $(z = x + iy)$, wobei in der Nähe des unendlichfernen Punktes die Bedingungen (siehe [2], Kap. 2, § 6; Kap. 5, § 4; Kap. 6, § 5) erfüllt sind:

$$B = O(|z|^{-2}) , \quad w = O(|z|^{-4}), \quad w' = O(|z|^{+2}) . \tag{35}$$

Nach dem verallgemeinerten Satz von LIOUVILLE besitzt die homogene Gleichung $\partial_{\bar{z}} w - \bar{B}\,\bar{w} = 0$ beim Erfülltsein der Bedingungen $(35_{1,2})$ keine nichttriviale reguläre Lösung in der z-Ebene. Jedoch besitzt die homogene Gleichung $\partial_{\bar{z}} w' + B\,\bar{w}' = 0$ beim Erfülltsein der Bedingung $(35_{1,3})$ sechs linear unabhängige Lösungen, die das Bewegungsfeld eines absolut starren Körpers beschreiben: $\underline{U} = \underline{C}_o + \underline{C} \times \underline{r}$, wobei $\underline{C}_o$ und $\underline{C}$ beliebige konstante Vektorfelder sind.
Für die Auflösbarkeit des Problems T ist notwendig und hinreichend das Erfülltsein der sechs Gleichungen

$$\iint_S \underline{Z}_o \overset{(i)}{\underline{U}} \, dS = 0 , \quad i = 1, \ldots, 6 , \tag{36}$$

wobei $\overset{(i)}{\underline{U}}$ die Bewegungsfelder sind. Diese Bedingungen sind die sechs Bedingungen für das statische Gleichgewicht einer absolut starren Fläche S .
Daher ist das Problem (T, T') für die geschlossene konvexe Schale stets lösbar.

(Anm.: Die Ergebnisse der vorliegenden Arbeit wurden vom Autor auch auf der All-Unions-Konferenz über Partielle Differentialgleichungen anläßlich des 75. Geburtstages (Januar 1976) des Akademie-Mitglieds I. G. Petrowsky in Moskau vorgetragen und sollen in russisch in den zugehörigen Proceedings erscheinen).

Zitierte Literatur

[1] I. N. Vekua: Grundlagen der Tensoranalysis, Verlag der Tbilisser Universität, Tbilissi 1967 (russ.).

[2] I. N. Vekua: Verallgemeinerte analytische Funktionen, Akademie-Verlag, Berlin 1963.

[3] I. N. Vekua: Theorie der dünnen, gewölbten Schalen mit veränderlicher Dicke, Trudy Tbilisskogo matematičeskogo instituta, im. A. M. Rasmadse, Tom 30, 1965.

[4] E. Kreyszig: Differentialgeometrie, Akadem. Verlagsgemeinschaft, Leipzig, 1957.

ZUR DARSTELLUNG PSEUDOANALYTISCHER FUNKTIONEN

Karl Wilhelm Bauer

1. Im folgenden wird die Differentialgleichung

(1) $$w_{\bar{z}} = c\bar{w} \quad \text{mit} \quad c = \frac{\gamma_{\bar{z}}}{\gamma}$$

behandelt, wobei $\gamma(z,\bar{z})$ eine im betrachteten einfach zusammenhängenden Gebiet G nicht verschwindende, reellwertige, zweimal stetig differenzierbare Funktion bezeichnet.
Die Lösungen der Differentialgleichung (1) sind in verschiedener Hinsicht ausgezeichnet. Während für pseudoanalytische Funktionen allgemein nur ein "unscharfes" Maximumprinzip gilt (vgl. [6]), konnte in [10] für die Lösungen von (1) ein "scharfes" Maximumprinzip bewiesen werden, falls γ^{-2} eine in G subharmonische Funktion bezeichnet. Zum anderen gilt, daß die Lösungen von (1) auf Grund der Differenzierbarkeitsvoraussetzung für den Koeffizienten c Funktionen darstellen, die auch Lösungen der elliptischen Differentialgleichung

$$w_{z\bar{z}} - \frac{c_z}{c} w_{\bar{z}} - c\bar{c}w = 0$$

sind (vgl. [18], S.140). Im Fall $\gamma = (\alpha+\bar{\alpha})^m$, $m \in \mathbb{N}$ [1], war es möglich, die Lösungen von (1) mit Hilfe von Differentialoperatoren darzustellen ([5]). [2] Schließlich erhält man mit $\gamma = (z+\bar{z})^{1/2}$ eine Differentialgleichung, deren Lösungen bei geeigneter Transformation Lösungen der Ernst-Gleichung liefern, die in der allgemeinen Relativitätstheorie auftritt (vgl. [7, 8]).

2. Leitet man die Differentialgleichung (1) nach z ab, und verwendet man $w = u+iv$, u und v reellw., so erhält man für den Realteil u und den Imaginärteil v von w die elliptischen Differentialgleichungen

(2) $$u_{z\bar{z}} - \frac{\gamma_{z\bar{z}}}{\gamma} u = 0,$$

---

1) Mit $\mathbb{N}$, $\mathbb{Z}$, $\mathbb{R}$ bzw. $\mathbb{C}$ wird die Menge der natürlichen, ganzen, reellen bzw. komplexen Zahlen bezeichnet. $\mathbb{N}_0 = \mathbb{N} \cup \{0\}$.

2) Die Darstellung der Lösungen von (1) im Fall $c = n(z+\bar{z})^{-1}+i\,\psi(z-\bar{z})$, $n \in \mathbb{N}$, ψ reellw. und reellanalytisch, wurde in [9] behandelt.

(3) $$v_{z\bar z} + \left[\frac{\gamma_{z\bar z}}{\gamma} - \frac{2\gamma_z\gamma_{\bar z}}{\gamma^2}\right] v = 0.$$

Dabei gilt, wie man leicht verifiziert, daß die Differentialgleichung (2) durch die linearen Bäcklund-Transformationen [3)]

(4a,b) $$(u-v)_z = \frac{\gamma_z}{\gamma}(u+v), \qquad (u+v)_{\bar z} = \frac{\gamma_{\bar z}}{\gamma}(u-v)$$

in die Differentialgleichung (3) transformiert wird. Das System (4) kann man jedoch sofort integrieren. Geht man von (4a) aus und setzt das Ergebnis in (4b) ein, so erhält man

(5) $$v = -u + \frac{1}{\gamma}\Phi, \qquad \Phi_z = 2\gamma u_z, \qquad \Phi_{\bar z} = 2\gamma_{\bar z} u.$$

Geht man von (4b) aus, so folgt

(6) $$v = u - \frac{1}{\gamma}\Psi, \qquad \Psi_z = 2\gamma_z u, \qquad \Psi_{\bar z} = 2\gamma u_{\bar z}.$$

Kennt man also eine partikuläre Lösung oder ein allgemeines Integral u von (2), so erhält man durch (5) oder (6) eine partikuläre Lösung bzw. ein allgemeines Integral der Differentialgleichung (3).

3. In diesem Abschnitt wird angenommen, daß der Realteil u der betrachteten pseudoanalytischen Funktion w der Differentialgleichung

(7) $$\eta^2 RSu - n(n+1)u = 0, \qquad n \in \mathbb{N},$$

genügt. Dabei gilt $R = \frac{1}{\alpha'}\frac{\partial}{\partial z}$, $S = \frac{1}{\overline{\alpha'}}\frac{\partial}{\partial \bar z}$ und $\eta = \alpha + \bar\alpha$, wobei $\alpha(z)$ eine in G holomorphe Funktion mit $(\alpha+\bar\alpha)\alpha' \neq 0$ in G bezeichnet. Die Differentialgleichung (7) stellt den Spezialfall einer in [3] behandelten Differentialgleichung dar, für die wir im nachstehenden Satz einige Resultate zusammenstellen, die im folgenden benötigt werden.

Satz 1

a) Zu jeder in G definierten Lösung u der Differentialgleichung (7) gibt es in G holomorphe Funktionen g(z) und h(z), so daß

---

3) Bezüglich der Anwendung von Bäcklund-Transformationen in der Theorie ultra-kurzer optischer Impulse und in der Theorie des Josephson-Effektes wird auf [1,2,11,12,13] bzw. [17] verwiesen. Eine Anwendung bei hyperbolischen Differentialgleichungen im Zusammenhang mit der infinitesimalen Deformation von Flächen findet sich in [4].

(8) $$u = Hg + \overline{Hh}$$

mit

(9) $$H = \sum_{k=0}^{n} \frac{A_k^n}{\eta^{n-k}} R^k , \qquad A_k^n = \frac{(-1)^{n-k}(2n-k)!}{k!(n-k)!} .$$

b) Umgekehrt stellt (8) für jedes Paar von in G holomorphen Funktionen $g(z)$ und $h(z)$ eine Lösung von (7) in G dar.

c) Bei vorgegebener Lösung u sind die Funktionen $R^{2n+1}g$ und $S^{2n+1}\bar{h}$ eindeutig gemäß ($P = \eta^2 R$, $Q = \eta^2 S$)

$$R^{2n+1}g = \frac{P^{n+1}u}{\eta^{2n+2}} , \qquad S^{2n+1}\bar{h} = \frac{Q^{n+1}u}{\eta^{2n+2}}$$

bestimmt. Die Funktionen $g(z)$ und $h(z)$ sind bei Vorgabe von u nicht eindeutig festgelegt. Man erhält die allgemeinsten Funktionen $\tilde{g}(z)$ und $\tilde{h}(z)$ dieser Art durch

$$\tilde{g}(z) = g(z) + \sum_{\mu=0}^{2n} a_\mu \alpha^\mu , \qquad \tilde{h}(z) = h(z) - \sum_{\mu=0}^{2n} (-1)^\mu \overline{a_\mu}\, \alpha^\mu, \qquad a_\mu \in \mathbb{C}.$$

d) Jede reellwertige Lösung u von (7) läßt sich in der Form $w=Hg+\overline{Hg}$ mit einer geeigneten in G holomorphen Funktion $g(z)$ darstellen.

Gleichung (2) zeigt, daß γ eine beliebige reellwertige Lösung der Differentialgleichung (7) darstellen kann, während der Imaginärteil v in diesem Fall der Differentialgleichung

(10) $$\eta^2 RSv + \left[n(n+1) - \frac{2\eta^2 R\gamma S\gamma}{\gamma^2} \right] v = 0$$

genügen muß. Verwendet man hier einmal $\alpha(z) = z$, so erhält man

(11) $$(z+\bar{z})^2 v_{z\bar{z}} + \left[n(n+1) - \frac{2(z+\bar{z})^2 \gamma_z \gamma_{\bar{z}}}{\gamma^2} \right] v = 0 .$$

Mit (11) liegt aber eine Differentialgleichung vor, die - bedingt durch das Auftreten des ersten bzw. zweiten Beltrami-Operators bei γ bzw. v - durch gewisse Invarianzeigenschaften bei Automorphismen der rechten Halbebene ausgezeichnet ist. Damit sind die komplexwertigen Lösungen in der Theorie der automorphen Funktionen von Interesse. Mit Rücksicht auf diese Zusammenhänge wird im folgenden zunächst angenommen, daß γ eine beliebige komplexwertige Lösung der Differentialgleichung (7) bezeichnet. Es gelte $\gamma = H\varphi + \overline{H\psi}$, $\varphi(z)$, $\psi(z)$ hol. in G . Geht man nun von den Relationen (6) aus und verwendet $u = Hg$, $g(z)$ hol. in G , so erhält man mit $\Psi_{\bar{z}} = 2\gamma u_{\bar{z}}$ zunächst

$$(12) \qquad \Psi = \sum_{k=0}^{n-1} \left[p_k(\varphi) + \overline{q_k(\psi)} \right] \frac{A_k^n R^k g}{\eta^{n-k}} + r(z)$$

mit

$$p_k(\varphi) = \sum_{s=0}^{n} \frac{2(n-k)A_s^n}{2n-k-s} \frac{R^s\varphi}{\eta^{n-s}} , \qquad q_k(\psi) = \sum_{s=0}^{n-1} B_s^k \frac{R^s\psi}{\eta^{n-s}} ,$$

$$B_s^k = 2(k-n) \sum_{\mu=0}^{n-s-1} \frac{(2n-s-k-1)!}{(n+\mu-k)!} A_{n-\mu}^n ,$$

während $r(z)$ eine vorerst beliebige in G holomorphe Funktion bezeichnet. Zur Bestimmung von $r(z)$ setzt man (12) in $\Psi_z = 2u\gamma_z$ ein und erhält damit $v = Hg - \frac{1}{\gamma} K_{\varphi\psi} g$, $g(z)$ hol. in G, mit

$$K_{\varphi\psi}\, g = \sum_{k=0}^{n-1} \left[p_k(\varphi) + \overline{q_k(\psi)} \right] \frac{A_k^n R^k g}{\eta^{n-k}} + 2 \int \alpha' R^{n+1}\varphi \; R^n g \, dz .$$

Da sich jede Lösung der Form $v = u - \frac{1}{\gamma}\Psi$ wegen $v = -u + \frac{1}{\gamma}(2\gamma u - \Psi)$ auch in der Form $v = -u + \frac{1}{\gamma}\Phi$ darstellen läßt, ist es vorteilhaft, zur Bestimmung des zweiten Lösungsanteils von den Relationen (5) auszugehen. Hier folgt unter Verwendung von $u = -\overline{Hh}$ sofort $v = \overline{Hh} - \frac{1}{\gamma}\overline{K_{\psi\varphi} h}$. Berücksichtigt man noch die Aussagen in Satz 1c,d, so erhält man den folgenden

Satz 2

G sei ein einfach zusammenhängendes Gebiet der komplexen Zahlenebene. $\alpha(z)$ sei holomorph in G, und es gelte $\eta = \alpha + \bar{\alpha}$, $\eta\, \alpha' \neq 0$ in G. γ sei eine in G definierte und dort nicht verschwindende Lösung der Differentialgleichung (7) mit der Darstellung $\gamma = H\varphi + \overline{H\psi}$. Dann gelten die folgenden Aussagen:

a) Zu jeder in G definierten Lösung v der Differentialgleichung (10) gibt es in G holomorphe Funktionen $g(z)$ und $h(z)$, so daß

$$(13) \qquad v = Hg + \overline{Hh} - \frac{1}{\gamma}[K_{\varphi\psi}g + \overline{K_{\psi\varphi}h}] .$$

b) Umgekehrt stellt (13) eine Lösung von (10) dar, falls $g(z)$ und $h(z)$ zwei in G holomorphe Funktionen bezeichnen.

c) Ist eine Lösung v von (10) vorgegeben, so sind für eine Darstellung der Form (13) die Größen $R[\gamma^{-1}(Hg-\overline{Hh})]$ und $S[\gamma^{-1}(Hg-\overline{Hh})]$ eindeutig gemäß

$$(14) \qquad R[\gamma^{-1}(Hg-\overline{Hh})] = \gamma^{-2}R(\gamma v), \qquad S[\gamma^{-1}(Hg-\overline{Hh})] = -\gamma^{-2}S(\gamma v)$$

bestimmt. Die Funktionen g(z) und h(z) sind bei Vorgabe von v nicht eindeutig festgelegt. Man erhält das allgemeinste Funktionenpaar $\tilde{g}(z)$ und $\tilde{h}(z)$ durch $\tilde{g} = g+a\varphi+g_o$, $\tilde{h} = h-\bar{a}\psi+h_o$ mit

$$g_o = \sum_{\mu=0}^{2n} a_\mu \alpha^\mu, \quad h_o = \sum_{\mu=0}^{2n} (-1)^\mu \overline{a_\mu} \alpha^\mu, \quad a, a_\mu \in \mathbb{C} \quad , \quad 2\gamma H g_o = K_{\varphi\psi} g_o + \overline{K_{\psi\varphi} h_o} \; .$$

d) Bezeichnet γ eine reellwertige Lösung von (7) mit $\gamma = H\varphi+\overline{H\varphi}$, so erhält man die in $\mathfrak{G}$ definierten reellwertigen Lösungen von (10) durch $v = Hf+\overline{Hf}-\frac{1}{\gamma}[K_{\varphi\varphi}f+\overline{K_{\varphi\varphi}f}]$, f(z) hol. in $\mathfrak{G}$.

4. Die in Satz 2 genannten Darstellungen für die Lösungen der Differentialgleichung (10) vereinfachen sich erheblich, wenn sich die Funktionen φ und ψ in der Darstellung von γ auf Polynome in α vom Grad 2n reduzieren. In diesem Fall kann man γ unter Berücksichtigung von Satz 1c auch in der Form $\gamma = H\varphi = \overline{H\psi}$ darstellen, wodurch man erreichen kann, daß die Terme $q_k(\psi)$ im Operator $K_{\varphi\psi}$ verschwinden. Man erhält damit

Satz 3

γ sei eine in $\mathfrak{G}$ definierte und dort nicht verschwindende Lösung von (7) mit der Darstellung

$$(15) \qquad \gamma = H\varphi = \overline{H\psi}, \quad \varphi = \sum_{\mu=0}^{2n} b_\mu \alpha^\mu, \quad \psi = \sum_{\mu=0}^{2n} \overline{b_\mu}(-\alpha)^\mu, \quad b_\mu \in \mathbb{C}.$$

a) Zu jeder in $\mathfrak{G}$ definierten Lösung der Differentialgleichung (10) mit γ gemäß (15) gibt es eine Konstante $C \in \mathbb{C}$ und zwei in $\mathfrak{G}$ holomorphe Funktionen g(z) und h(z), so daß

$$(16) \qquad v = Hg+\overline{Hh} - \frac{1}{\gamma}[K_\varphi g+\overline{K_\psi h}+C]$$

mit

$$K_\varphi = \sum_{k=0}^{n-1} T_k^n(\varphi)R^k, \quad T_k^n(\varphi) = \sum_{s=0}^{n} \frac{2(n-k)A_s^n A_k^n R^s \varphi}{(2n-k-s)_\eta{}^{2n-k-s}} + 2(-1)^{n-1-k} R^{2n-k}\varphi \quad .$$

b) Umgekehrt stellt (16) für jede Konstante $C \in \mathbb{C}$ und jedes Paar von in $\mathfrak{G}$ holomorphen Funktionen g(z) und h(z) eine Lösung von (10) in $\mathfrak{G}$ dar.

Aus der Menge der in Satz 3 behandelten Funktionen γ seien zwei Fälle wegen ihres Zusammenhanges mit den Legendre'schen Polynomen besonders hervorgehoben. Es gilt

$$\gamma = H\left[\frac{(-i\alpha)^n}{n!}\right] = i^n P_n(\sigma), \qquad \sigma = \frac{\alpha-\bar{\alpha}}{\alpha+\bar{\alpha}},$$

$$\gamma = H\left[\frac{(\alpha^2-1)^n}{2^n n!}\right] = P_n(\omega), \qquad \omega = \frac{1+\alpha\bar{\alpha}}{\alpha+\bar{\alpha}},$$

wobei $P_n(\tau)$ das Legendre'sche Polynom n-ten Grades in τ bezeichnet. Darüber hinaus ergeben sich auch für andere Funktionen γ, die nicht mehr durch Polynome in α vom Grad 2n erzeugt werden können, vereinfachte Darstellungen. So gelingt es z.B. im Fall $\gamma = D_1\eta^{n+1} + D_2\eta^{-n}$, $D_1, D_2 \in \mathbb{C}$, die Lösungen der Differentialgleichung (10) wieder mit Hilfe von Operatoren des Typs H_n [4] darzustellen. Man erhält nach geeigneter Umformung und Zusammenfassung

$$(17) \qquad v = \frac{1}{\gamma}\left[D_1\eta^{n+1}[H_{n+1}g+\overline{H_{n+1}h}]+D_2\eta^{-n}[H_{n-1}(R^2g)+\overline{H_{n-1}(R^2h)}]\right],$$

wobei g(z) und h(z) wieder beliebige in $\mathfrak{G}$ holomorphe Funktionen bezeichnen.

5. Mit Hilfe der in Satz 1 und Satz 2 zusammengefaßten Resultate lassen sich nun Darstellungen für die pseudoanalytischen Funktionen ermitteln, die als Lösungen der entsprechenden Differentialgleichung (1) auftreten.
γ sei in diesem Abschnitt eine reellwertige, in $\mathfrak{G}$ nicht verschwindende Lösung der Differentialgleichung (7) und habe die Darstellung $\gamma = H\varphi + \overline{H\varphi}$, $\varphi(z)$ hol. in $\mathfrak{G}$. Dann haben die in $\mathfrak{G}$ definierten Lösungen der Differentialgleichung (1) notwendig die Form

$$(18) \qquad w = u+iv = Hg+\overline{Hg}+i\left[Hf+\overline{Hf}-\frac{K_{\varphi\varphi}f+\overline{K_{\varphi\varphi}f}}{\gamma}\right].$$

Setzt man (18) in (1) ein, so folgt mit (14) zunächst $H(g-if)+\overline{H(g-if)} = C_1\gamma$, $C_1 \in \mathbb{R}$, und damit

$$(19) \qquad w = C_1\gamma + 2iHf - \frac{i}{\gamma}(K_{\varphi\varphi}f+\overline{K_{\varphi\varphi}f}).$$

Da man die partikuläre Lösung $w_1 = C_1\gamma$, $C_1 \in \mathbb{R}$, auch durch $2iHf - \frac{i}{\gamma}(K_{\varphi\varphi}f+\overline{K_{\varphi\varphi}f})$ mit $f = -iC_1$ erzeugen kann, darf die Konstante C_1 in (19) ohne Beschränkung der Allgemeinheit zu Null normiert werden.

[4] Hier wird zur Unterscheidung verschiedener Operatoren dieser Art der in Definition (9) auftretende Parameter n als Index angegeben.

Für $w = 2iHf - \frac{i}{\gamma}(K_{\varphi\varphi}f+\overline{K_{\varphi\varphi}f})$ folgt sodann mit (14) $S(\gamma w)=2iS\gamma(Hf-\overline{Hf})$, womit sich die Null-Lösungen bestimmen lassen. Schließlich erhält man mit $f \equiv 0$ die partikuläre Lösung $w_2 = iC_2\gamma^{-1}$, $C_2 \in \mathbb{R}$. Die Lösungen w_1 und w_2 sind die einzigen reellen bzw. imaginären Lösungen der Differentialgleichung (1), was unmittelbar aus der Darstellung der pseudoanalytischen Funktionen in der Nähe von Nullstellen folgt (vgl.[6], S.22-23). Setzt man $C_1 = C_2 = 1$, so erhält man mit $F = \gamma$, $G = i\gamma^{-1}$ partikuläre Lösungen, die den Bedingungen $FG \neq 0$ und $\operatorname{Im}\frac{G}{F} > 0$ in G genügen. Diese Funktionen stellen also im Sinne von L. Bers ein Erzeugendenpaar für die Lösungen von (1) in G dar. Zusammenfassend erhält man damit den folgenden

Satz 4

G sei ein einfach zusammenhängendes Gebiet der komplexen Zahlenebene. $\alpha(z)$ sei holomorph in G, und es gelte $(\alpha+\bar{\alpha})\alpha' \neq 0$ in G. γ sei eine reellwertige, in G nicht verschwindende Lösung der Differentialgleichung (7) und habe die Darstellung $\gamma = H\varphi + \overline{H\varphi}$, $\varphi(z)$ hol. in G.
Dann gelten die folgenden Aussagen:
a) Zu jeder in G definierten Lösung der Differentialgleichung

$$w_{\bar{z}} = \frac{\gamma_{\bar{z}}}{\gamma}\,\bar{w} \tag{20}$$

gibt es eine in G holomorphe Funktion f(z), so daß

$$w = 2iHf - \frac{i}{\gamma}(K_{\varphi\varphi}f+\overline{K_{\varphi\varphi}f}) \,. \tag{21}$$

b) Umgekehrt stellt (21) für jede in G holomorphe Funktion f(z) eine Lösung von (20) in G dar.
c) Ist eine Lösung w von (20) vorgegeben, so ist für eine Darstellung der Form (21) die Größe $Hf-\overline{Hf}$ eindeutig gemäß $2i(Hf-\overline{Hf})S\gamma=S(\gamma w)$ bestimmt. Die Funktion f ist bei Vorgabe von w nicht eindeutig festgelegt. Man erhält die allgemeinste Funktion $\tilde{f}(z)$ durch $\tilde{f} = f+f_0$ mit

$$f_0 = \sum_{\mu=0}^{2n} a_\mu \alpha^\mu, \quad a_\mu \in \mathbb{C}\,, \quad a_\mu-(-1)^\mu\overline{a_\mu} = 0, \quad K_{\varphi\varphi}f_0+\overline{K_{\varphi\varphi}f_0} = 2\gamma Hf_0 \,.$$

d) Die partikulären Lösungen $w_1 = C_1\gamma$, $w_2 = iC_2\gamma^{-1}$, $C_1, C_2 \in \mathbb{R}$, stellen die einzigen reellen bzw. imaginären Lösungen von (20) in G dar und liefern mit $C_1 = C_2 = 1$ ein Erzeugendenpaar für die Lösungen der Differentialgleichung (20) in G.
Um den Zusammenhang mit einem früher gewonnenen Resultat herzustellen, sei noch der Spezialfall $\gamma = D_1\eta^{n+1}+D_2\eta^{-n}$, $D_1, D_2 \in \mathbb{R}$, herausgegriffen. Unter Verwendung der Darstellung (17) erhält man

$$w = H_n g + \overline{H_n g} + \frac{i}{\gamma}\left[D_1\eta^{n+1}[H_{n+1}h + \overline{H_{n+1}h}] + D_2\eta^{-n}[H_{n-1}(R^2h) + \overline{H_{n-1}(R^2h)}]\right].$$

Setzt man in die Differentialgleichung (1) ein, so folgt $g = R(ih)$. Unter Verwendung von

$$Q_n f = \sum_{k=0}^{n} \frac{b_k^n}{\eta^{n-k}}[nR^k f - (n-k)\overline{R^k f}], \qquad b_k^n = \frac{(-1)^{n-k}(2n-1-k)!}{k!(n-k)!}$$

erhält man sodann mit $f = 2ih$ nach geeigneter Umformung und Zusammenfassung $w = \frac{1}{\gamma}[D_1\eta^{n+1}Q_{n+1}f - iD_2\eta^{-n}Q_n(iRf)]$. Mit Hilfe von $S(\gamma w) = [H_n(Rf) + \overline{H_n(Rf)}]S\gamma$ lassen sich sodann unter Verwendung von Satz 1c die Null-Lösungen bestimmen, und man erhält

Satz 5

a) Zu jeder in G definierten Lösung der Differentialgleichung

$$(22) \qquad w_{\bar z} = \frac{\gamma_{\bar z}}{\gamma}\bar w, \qquad \gamma = D_1\eta^{n+1} + D_2\eta^{-n}, \qquad D_1, D_2 \in \mathbb{R},$$

$$\eta = \alpha + \bar\alpha, \qquad \gamma\eta\alpha' \neq 0 \text{ in } G,$$

gibt es eine in G holomorphe Funktion $f(z)$, so daß

$$(23) \qquad w = \frac{1}{\gamma}[D_1\eta^{n+1}Q_{n+1}f - iD_2\eta^{-n}Q_n(iRf)].$$

b) Umgekehrt stellt (23) für jede in G holomorphe Funktion $f(z)$ eine Lösung von (22) in G dar.

c) Bei Vorgabe einer Lösung w von (22) ist für eine Darstellung der Form (23) die Größe $H_n(Rf) + \overline{H_n(Rf)}$ eindeutig gemäß $[H_n(Rf) + \overline{H_n(Rf)}]S\gamma = S(\gamma w)$ bestimmt. Die Funktion $f(z)$ ist bei Vorgabe von w nicht eindeutig bestimmt; man erhält die allgemeinste Funktion $\tilde f(z)$ durch

$$\tilde f(z) = f(z) + \sum_{\mu=0}^{2n+1} a_\mu \alpha^\mu, \qquad a_\mu \in \mathbb{C},$$

mit $a_\mu - (-1)^\mu \overline{a_\mu} = 0$, $\mu = 1, 2, \ldots, 2n+1$, $\binom{2n}{n}D_1(a_0 - \overline{a_0}) + (-1)^{n+1}D_2 a_{2n+1} = 0$.
Verwendet man in (22) $D_2 = 0$ und ersetzt man $n+1$ durch m, so erhält man die in [5] behandelte Differentialgleichung, und (23) geht in die in [5], Satz 4, genannte Lösungsdarstellung über.

6. Wir nehmen nun an, daß die Funktion γ in (1) die Form $\gamma = (\alpha+\bar\alpha)^\lambda$, $\lambda \in \mathbb{R}$, $\lambda \notin \mathbb{Z}$, hat, und erhalten damit die Differentialgleichung

$$(24) \qquad w_{\bar z} = \frac{\lambda\overline{\alpha'}}{\alpha+\bar\alpha}\bar w.$$

Unter Verwendung von (2) bzw. (3) folgen sodann als Differentialgleichungen für den Real- bzw. Imaginärteil von w

$$\eta^2 u_{z\bar{z}} - \lambda(\lambda-1)\alpha'\overline{\alpha'}u = 0, \tag{25}$$

$$\eta^2 v_{z\bar{z}} - \lambda(\lambda+1)\alpha'\overline{\alpha'}v = 0, \tag{26}$$

wenn wir wieder $\eta = \alpha+\bar{\alpha}$ setzen. Allgemein erhalten wir also Differentialgleichungen der Form

$$\eta^2 u_{z\bar{z}} + C\alpha'\overline{\alpha'}u = 0 \quad \text{mit} \quad C \neq -n(n+1), \quad n \in \mathbb{N}_0, \tag{27}$$

für deren Lösungen kein allgemeiner Darstellungssatz mit Differentialoperatoren bekannt ist.[5] Eine partikuläre Lösung von (25) erhält man mit $u = \eta^\lambda$. Ersetzt man in (25) λ durch $1-\lambda$, so geht die Differentialgleichung in sich über. Mit $u = \eta^{1-\lambda}$, $\lambda \neq \frac{1}{2}$, liegt also eine weitere Lösung vor, die wir zusammen mit $\gamma = \eta^\lambda$ in den Bäcklund-Transformationen (4) verwenden. Mit Hilfe von (6) erhalten wir sodann die Lösung $v = \eta^{-\lambda}(\alpha-\bar{\alpha})$ von (26). Ersetzt man in (26) λ durch $-\lambda$, so erhält man die Differentialgleichung (25). Mit $u = \eta^\lambda(\alpha-\bar{\alpha})$ liegt also eine dritte Lösung dieser Differentialgleichung vor.
Verwendet man nun $u = \eta^{1-\lambda}(\alpha-\bar{\alpha})$ zusammen mit $\gamma = \eta^\lambda$ in (6), so folgt wie vorher $u = \eta^\lambda[\alpha^2 - 2\lambda(1+\lambda)^{-1}\alpha\bar{\alpha} + \bar{\alpha}^2]$ als Lösung von (25). In dieser Weise fortfahrend erhält man sukzessiv die Lösungen

$$u = \alpha^m \eta^\lambda F(\lambda, -m, 1-\lambda-m; \frac{-\bar{\alpha}}{\alpha}), \quad m \in \mathbb{N}_0, \tag{28}$$

der Differentialgleichung (25), wobei $F(a,b,c;x)$ die hypergeometrische Funktion bezeichnet, und auch $\lambda = \frac{1}{2}$ zulässig ist. Ersetzt man in (28) wiederum λ durch $-\lambda$, so liegen mit

$$v = \alpha^m \eta^{-\lambda} F(-\lambda, -m, 1+\lambda-m; \frac{-\bar{\alpha}}{\alpha}), \quad m \in \mathbb{N}_0, \tag{29}$$

Lösungen von (26) vor. Damit besteht die Möglichkeit, Lösungen der Differentialgleichung (24) zu ermitteln. Berücksichtigt man, daß die Differentialgleichung (26) in sich übergeht, wenn man λ durch $-\lambda-1$ ersetzt, und daß die Koeffizienten in den auftretenden hypergeometrischen Funktionen der Relation

---

[5] Setzt man in (27) $\alpha = iz$, so erhält man mit $(z-\bar{z})^2 u_{z\bar{z}} - Cu = 0$ eine Differentialgleichung, die von verschiedenen Autoren behandelt worden ist (vgl. z.B. [14,15,16]).

$$\frac{(\lambda)_s(-m)_s}{s!(1-\lambda-m)_s} = \frac{(\lambda)_{m-s}(-m)_{m-s}}{(m-s)!(1-\lambda-m)_{m-s}}$$

genügen, so erhält man mit $u = (i\alpha)^m \eta^\lambda F(\lambda,-m,1-\lambda-m;\ \frac{-\bar{\alpha}}{\alpha})$, $m \in \mathbb{N}_0$, und $v = (i\alpha)^n \eta^{\lambda+1} F(\lambda+1,-n,-\lambda-n;\ \frac{-\bar{\alpha}}{\alpha})$, $n \in \mathbb{N}_0$, reellwertige Lösungen der Differentialgleichung (25) bzw. (26). Zur Bestimmung von Lösungen der Differentialgleichung (24) verwenden wir mit Rücksicht auf die auftretenden Potenzen von α und $\bar{\alpha}$ den Ansatz

$$w = (i\alpha)^m \eta^\lambda F(\lambda,-m,1-\lambda-m;\frac{-\bar{\alpha}}{\alpha}) + iC(i\alpha)^{m-1}\eta^{\lambda+1}F(\lambda+1,1-m,1-\lambda-m;\frac{-\bar{\alpha}}{\alpha}),\ m \in \mathbb{N},$$

wobei $C \in \mathbb{R}$ eine vorerst beliebige Konstante bezeichnet. Setzt man in (24) ein, so erhält man mit $C = \frac{m}{2\lambda+m}$ eine Lösung, die sich nach geeigneter Zusammenfassung und Normierung in der Form $w_1 = (i\alpha)^m \eta^\lambda F(\lambda,-m,-\lambda-m;\frac{-\bar{\alpha}}{\alpha})$ schreiben läßt, wobei auch $m = 0$ zulässig ist. Ersetzt man in w_1 den Parameter λ durch $-\lambda$, so liegt eine Lösung der Differentialgleichung $w^*_{\bar{z}} = \frac{(-\lambda)\overline{\alpha'}}{\eta}\overline{w^*}$ vor, die durch die Transformation $w = iw^*$ wieder in (24) übergeht. Man erhält also mit $w_2 = i(i\alpha)^n \eta^{-\lambda} F(-\lambda,-n,\lambda-n;\frac{-\bar{\alpha}}{\alpha})$, $n \in \mathbb{N}_0$, eine weitere Lösung der Differentialgleichung (24). Verwendet man bei w_1, w_2 $m=0$ bzw. $n=0$, so liegt mit $F = \eta^\lambda$, $G = i\eta^{-\lambda}$ ein Erzeugendenpaar im Sinne von L. Bers vor.

<u>Satz 6</u>

$\mathfrak{G}$ sei ein einfach zusammenhängendes Gebiet. $\alpha(z)$ sei holomorph in $\mathfrak{G}$, und es gelte $\eta = \alpha+\bar{\alpha} \neq 0$ in $\mathfrak{G}$.

a) Dann stellt

$$w = C_1(i\alpha)^m \eta^\lambda F(\lambda,-m,-\lambda-m;\frac{-\bar{\alpha}}{\alpha}) + iC_2(i\alpha)^n \eta^{-\lambda} F(-\lambda,-n,\lambda-n;\frac{-\bar{\alpha}}{\alpha})$$

mit $C_1, C_2, \lambda \in \mathbb{R}$ und $m,n \in \mathbb{N}_0$ eine Lösung der Differentialgleichung (24) in $\mathfrak{G}$ dar.

b) Die Funktionen $F = \eta^\lambda$, $G = i\eta^{-\lambda}$ bilden ein Erzeugendenpaar und sind bis auf reelle Faktoren die einzigen reellen bzw. rein imaginären Lösungen der Differentialgleichung (24) in $\mathfrak{G}$.

Literatur

[1] Ames, W. F.: Non Linear Partial Differential Equations in Engineering. Vol. II. London-New York, Academic Press, 1972.

[2] Barnard, T. W.: 2Np Ultrashort Light Pulses. Phys. Rev., A 7, 1, 373-376 (1973).

[3] Bauer, K. W. und G. Jank: Differentialoperatoren bei einer inhomogenen elliptischen Differentialgleichung. Rend. Ist. Mat. Univ. Trieste, Heft II, 140-169 (1971).

[4] --- und C. Rogers: Zur infinitesimalen Deformation von Flächen. Math.-stat. Sektion, Forsch.-Z. Graz, Ber. Nr. 31, 1-16 (1975).

[5] --- und St. Ruscheweyh: Ein Darstellungssatz für eine Klasse pseudoanalytischer Funktionen. Ber. d. Ges. f. Math. u. Datenv., Bonn, Nr. 75, 3-15 (1973).

[6] Bers, L.: Theory of Pseudo-Analytic Functions. New York University, 1953.

[7] Bitsadze, A. V. und V. I. Paškovskiĭ: On the Theory of the Maxwell-Einstein Equations. Dokl. Akad. Nauk SSSR, Tom 216, 762-764 (1964).

[8] Ernst, F. J.: New Formulation of the Axially Symmetric Gravitational Field Problem. Phys. Rev., 167, 1175-1178 (1968).

[9] Jank, G. und St. Ruscheweyh: Eine Bemerkung zur Darstellung gewisser pseudoanalytischer Funktionen. Ber. d. Ges. f. Math. u. Datenv., Bonn, Nr. 75, 17-19 (1973).

[10] --- und K.-J. Wirths: Über eine Abschätzungsmethode bei gewissen Klassen pseudoanalytischer Funktionen. Erscheint in Kürze.

[11] Lamb, G. L. Jr.: π Pulse Propagation in a Lossless Amplifier. Phys. Letters, 29 A, 507-508 (1969).

[12] --- Higher Conservation Laws in Ultrashort Optical Pulse Propagation. Phys. Letters, 32 A, 251-252 (1970).

[13] --- Analytic Descriptions of Ultrashort Optical Pulse Propagation in a Resonant Medium. Rev. Mod. Phys., 43, 99-124 (1971).

[14] Maaß, H.: Über eine neue Art von nichtanalytischen automorphen Funktionen und die Bestimmung Dirichlet'scher Reihen durch Funktionalgleichungen. Math. Ann., 121, 141-183 (1949).

[15] Roelcke, W.: Über die Wellengleichung bei Grenzkreisgruppen erster Art. Sitz.-Ber. Heidelberger Akad. Wiss., Math.-natw. Kl., Heidelberg, 1956.

[16] Ruscheweyh, St.: Hardy Spaces of λ-harmonic Functions. Erscheint in Kürze.

[17] Scott, A. C.: Propagation of Magnetic Flux in a Long Josephson Junction. Nuovo Cimento, 69 B, 241-261 (1970).

[18] Vekua, I. N.: Verallgemeinerte analytische Funktionen. Berlin, Akademie-Verlag, 1963.

Über das Randwert-Normproblem für ein nichtlineares elliptisches System *)

von

Heinrich Begehr

I. Math. Institut
Freie Universität Berlin

Robert P. Gilbert +)

Department of Math.
University of Delaware

An die Ergebnisse von Bers [3] und Vekua [16] über lineare elliptische Differentialgleichungssysteme erster Ordnung, in Hilbertscher Normalform gegeben durch

$$u_x - v_y = au + bv + c$$

$$u_y + v_x = \alpha u + \beta v + \gamma$$

bzw. in komplexer Schreibweise

$$w_{\bar{z}} = Aw + B\bar{w} + C,$$

haben sich viele Untersuchungen und Verallgemeinerungen (vgl. [4], [5] [6], [10], [13]) angeschlossen. Die funktionentheoretischen Eigenschaften der Lösungen solcher Systeme sind bis hin zur Nevanlinnaschen Wertverteilungstheorie [2], [3], [11], [12], [16] entwickelt und entspringen dem auf Bers und Vekua zurückgehenden Ähnlichkeitsprinzip.

---

*) Herrn Professor Rolf Nevanlinna zum 80. Geburtstag gewidmet.

+) Diese Arbeit entstand, während sich der zweitgenannte Verfasser durch die Alexander von Humboldt-Stiftung mit dem "Senior U.S. Scientist Award" ausgezeichnet im Sommersemester 1975 an der Freien Universität Berlin aufhielt.

Randwertprobleme für obige Systeme werden ausführlich in [10], [16] und in [9], [17], [18] behandelt. Hier sollen wie in [10] und [17] die Greenschen Funktionen erster und zweiter (Neumannsche Funktion) Art benutzt werden, um das Randwert-Normproblem für eine nichtlineare Gleichung der Form

$$(1) \qquad w_{\bar{z}} = f(z,w)$$

zu lösen. Existenz und Eindeutigkeit der Lösung dieses Problems wird mit Hilfe einer allgemeineren Bedingung gesichert, als es die Lipschitzbedingung ist. Mit anderen Methoden (Einbettungsmethode) und andersartigen Voraussetzungen (zweimalige stetige Differenzierbarkeit von f nach w und $\bar{w}$) ist das Problem in [9] behandelt worden. Neben den verallgemeinerten analytischen Funktionen sind die approximativ analytischen Funktionen (vgl. [3], [14], [1]) Lösungen von Differentialgleichungen des Typ (1).

<u>1. Vorbereitende Betrachtungen.</u> Ist ϕ eine konforme Abbildung des einfach zusammenhängenden Gebietes D der komplexen Ebene $\mathbb{C}$ mit mehr als einem Randpunkt auf den Einheitskreis, so sind

$$G^I(\zeta,z) := -\frac{1}{2\pi} \log \left| \frac{\phi(\zeta)-\phi(z)}{1-\overline{\phi(\zeta)}\phi(z)} \right| \qquad (\zeta,z\in D)$$

$$G^{II}(\zeta,z) := -\frac{1}{2\pi} \log \left| (\phi(\zeta)-\phi(z))\,(1-\overline{\phi(\zeta)}\phi(z)) \right| \qquad (\zeta,z\in D)$$

die Greenschen Funktionen erster und zweiter Art für D. Hat D einen glatten Rand ∂D, so existiert eine Konstante c, die durch

$$c = 4 \sup_{\zeta,z\in D} \left| \frac{\phi'(z)\;(\zeta-z)}{\phi(\zeta)-\phi(z)} \right| \geq 4$$

festgelegt werden kann, so daß

$$(2) \quad |G^k_\zeta(\zeta,z)| \leq \frac{1}{2\pi} \left| \frac{\phi'(\zeta)}{\phi(\zeta)-\phi(z)} \right| \leq \frac{c}{4|\zeta - z|} \qquad (\zeta,z\in D; k = I,II).$$

Charakteristische Eigenschaften auf ∂D sind

$$G^{I}(\zeta,z) = 0,\ d_n G^{II}(\zeta,z) = -\frac{1}{2\pi}|d\phi(\zeta)| \quad (\zeta\in\partial D, z\in D),$$

$$\int_{\partial D} G^{II}(\zeta,z)\ |d\phi(\zeta)| = 0 \ (z\in D).$$

Mit Hilfe einer auf (glattem Rand) ∂D gegebenen reellen stetigen bzw. Hölder-stetigen Funktion φ wird durch

$$\tilde{\varphi}(z) := -\int_{\partial D} \varphi(\zeta)[d_n G^{I}(\zeta,z) - i\, dG^{II}(\zeta,z)] \quad (z\in D),$$

wo

$$d := \frac{\partial}{\partial\zeta}d\zeta + \frac{\partial}{\partial\bar{\zeta}}d\bar{\zeta},\ d_n := -i\left[\frac{\partial}{\partial\zeta}d\zeta - \frac{\partial}{\partial\bar{\zeta}}d\bar{\zeta}\right],$$

in D eine holomorphe Funktion $\tilde{\varphi}$ definiert, die den Randbedingungen

$$\mathrm{Re}\tilde{\varphi}|_{\partial D} = \varphi,\ \int_{\partial D} \mathrm{Im}\ \tilde{\varphi}(\zeta)|d\phi(\zeta)| = 0$$

genügt, und unter Hinzunahme ihrer Randwerte in $\hat{D} = D\cup\partial D$ stetig bzw. Hölder-stetig ist (vgl. [10], 9.4 oder [17]). Für in D stetige Funktionen w mit verallgemeinerten ersten Ableitungen in $L_p(\hat{D})$ $(2<p)$ gilt in D folgende Integraldarstellung ([10], 10.4):

$$(3)\quad w(z) = -\Theta(z) + i\int_D \{w_{\bar{\zeta}}(\zeta)[G^{I}_{\zeta}(\zeta,z)+G^{II}_{\zeta}(\zeta,z)] + \overline{w_{\bar{\zeta}}(\zeta)}[G^{I}_{\bar{\zeta}}(\zeta,z)-G^{II}_{\bar{\zeta}}(\zeta,z)]\}d\zeta d\bar{\zeta},$$

$$(4)\qquad \Theta(z) := \int_{\partial D} \{\mathrm{Re}\ w(\zeta)[d_n G^{I}(\zeta,z) - i\,dG^{II}(\zeta,z)] + i\ \mathrm{Im}\ w(\zeta) d_n G^{II}(\zeta,z)\}.$$

Setzt man anstelle von (4)

$$(5)\qquad \Theta(z) = \int_{\partial D} \mathrm{Re}\ w(\zeta)[d_n G^{I}(\zeta,z) - i\,dG^{II}(\zeta,z)] - iC$$

mit einer willkürlichen reellen Konstanten C, so ist die "Randnorm" von Im w,

$$\frac{1}{2\pi}\int_{\partial D} \mathrm{Im}\ w(\zeta)\ |d\phi(\zeta)|$$

gleich dieser Konstanten (vgl. [10], 4.8). Man erkennt dies durch Integration von (3) in Verbindung mit (5).

<u>Hilfssatz:</u> D sei ein beschränktes Gebiet von $\mathbb{C}$ und $g(z,x)$ eine in $\hat{D} \times [0, +\infty)$ nichtnegative, in x stetige Funktion mit den Eigenschaften

i. $g(z,o) = 0, \quad g(z,x) \leq g(z,y) \quad (z \in D,\ 0 \leq x \leq y)$,

ii. $g(z,x(z)) \in L_p(\hat{D})\,(2<p)$ für jede in $\hat{D}$ stetige, nichtnegative Funktion x,

iii. Es existiert ein K>0, so daß

$$\int_D g(\zeta,K) \frac{d\xi\, d\eta}{|\zeta - z|} \leq K, \tag{6}$$

iv. Für den auf der Menge der auf $\hat{D}$ stetigen, nichtnegativen, durch K nach oben beschränkten Funktionen betrachteten Integraloperator $\underset{\sim}{I}$,

$$(\underset{\sim}{I}x)\ (z) := \int_D g(\zeta,x(\zeta)) \frac{d\xi\, d\eta}{|\zeta - z|} \quad (z \in D,\ x \in C(\hat{D}),\ \zeta = \xi + i\eta),$$

ist 1 nicht Eigenwert.

Dann hat die Integralungleichung

$$\delta(z) \leq (\underset{\sim}{I}\delta)(z) \quad (z \in D) \tag{7}$$

in der Menge der auf $\hat{D}$ stetigen Funktionen mit Wertevorrat in [0,K] nur die Nullfunktion als Lösung.

<u>Beweis:</u> Es sei Δ_o eine stetige Lösung von (7) mit $0 \leq \Delta_o(z) \leq K$. Die durch

$$\Delta_n := \underset{\sim}{I}\ \Delta_{n-1} \quad (n \in \mathbb{N})$$

gegebene Funktionenfolge von stetigen, nichtnegativen, durch K nach oben beschränkten Funktionen ist monoton wachsend und beschränkt, so daß

$$\Delta = \lim_{n \to \infty} \Delta_n$$

existiert. Wegen der monotonen Konvergenz und der Stetigkeit von g in der zweiten Veränderlichen gilt

$$\Delta = \underset{\sim}{I}\Delta .$$

Damit ist Δ eine stetige Funktion aus der Definitionsmenge von $\underset{\sim}{I}$ und damit die Nullfunktion.

Wegen $0 \le \Delta_0 \le \Delta$ gilt also $\Delta_0(z) \equiv 0$ in $\hat{D}$.

Ein Beispiel für eine Funktion g mit den Eigenschaften dieses Hilfssatzes ist

$$(8) \qquad g(z,x) = g(z)x$$

mit

$$0 \le g(z) \quad (z \in \hat{D}), \quad g \in L_p(\hat{D}) \quad (2<p), \quad \int_D g(\zeta) \frac{d\xi\, d\eta}{|\zeta - z|} < 1 \quad (z \in \hat{D}).$$

Hier braucht Bedingung iv. nicht gefordert zu werden. Vielmehr ergibt sie sich ebenso wie die Behauptung des Hilfssatzes in diesem Fall sogleich mit einem indirekten Beweisschluß aus der Ungleichungskette

$$\delta(z) \le \int_D g(\zeta)\delta(\zeta)\frac{d\xi\, d\eta}{|\zeta - z|} \le \operatorname*{Max}_{z \in \hat{D}} \delta(z) \int_D g(\zeta) \frac{d\xi\, d\eta}{|\zeta - z|} < \operatorname*{Max}_{z \in \hat{D}} \delta(z).$$

<u>Lemma:</u> Ist D ein beschränktes Gebiet von $\mathbb{C}$ mit Durchmesser d(D) und $w \in L_p(\hat{D})$ $(2<p)$, so gilt

$$\int_D |f(\zeta)| \frac{d\xi\, d\eta}{|\zeta - z|} \le M\, L_p(f,\hat{D}) \quad (z \in \mathbb{C})$$

mit

$$L_p(f,\hat{D}) := \left(\int_D |f(\zeta)|^p\, d\xi\, d\eta\right)^{\frac{1}{p}}$$

und

$$M = M(p,D) := \left(\frac{2\pi}{\alpha q}\right)^{\frac{1}{q}} d^{\alpha}(D) \quad \left(\alpha = \frac{p-2}{p}, \frac{1}{p} + \frac{1}{q} = 1\right).$$

Der Beweis ergibt sich durch Anwendung der Hölderschen Ungleichung.

<u>2. Die erste Randwertaufgabe.</u> Es sei D ein einfach zusammenhängendes, beschränktes Gebiet von $\mathbb{C}$ mit stetig gekrümmtem Rand ∂D, φ eine auf ∂D stetige, reelle Funktion und f eine in $\hat{D} \times \mathbb{C}$ gegebene komplexe

Funktion mit folgenden Eigenschaften:

v. $f(z,w(z)) \in L_p(\hat{D})$ $(2<p)$ für alle in $\hat{D}$ stetigen, komplexen Funktionen w.

vi. Es existiert eine in $\hat{D} \times [0, +\infty)$ definierte, in der zweiten Variablen stetige und in x = o gleichgradig bezüglich $z\in\hat{D}$ stetige Funktion g(z,x) mit den Eigenschaften i. - iv. des Hilfssatzes, so daß mit c aus (2)

$$|f(z,w) - f(z,\omega)| \leq \frac{1}{c}\, g(z,|w-\omega|) \quad (z\in\hat{D}; w,\omega\in\mathbb{C}).$$

Unter diesen Voraussetzungen wird das Randwertproblem

$$w_{\bar{z}} = f(z,w), \quad \mathrm{Re}\, w|_{\partial D} = \varphi \tag{9}$$

untersucht. Es wird sich zeigen, daß dieses Problem eindeutig lösbar ist, wenn die Randnorm von Im w

$$\frac{1}{2\pi}\int_{\partial D} \mathrm{Im}\, w(\zeta)\,|d\phi(\zeta)| = C \tag{10}$$

vorgegeben wird und die Konstante K aus (6) genügend groß ist. Um dies zu sehen, ist zu beachten, daß sich eine Lösung von (9), (10) nach (3), (4) durch

$$w(z) = -\,\Theta(z) + (\underset{\sim}{P}w)(z) \quad (z\in D) \tag{11}$$

mit

$$\Theta(z) := \int_{\partial D} \varphi(\zeta)[d_n G^{I}(\zeta,z) - i\,dG^{II}(\zeta,z)] - iC \quad (z\in D)$$

und für $w\in C(\hat{D})$ und $z\in D$

$$(\underset{\sim}{P}w)(z) := 2\int_D \{f(\zeta,w(\zeta))[G^{I}_{\zeta}(\zeta,z)+G^{II}_{\zeta}(\zeta,z)] + \overline{f(\zeta,w(\zeta))}[G^{I}_{\bar{\zeta}}(\zeta,z)-G^{II}_{\bar{\zeta}}(\zeta,z)]\}\,d\xi\, d\eta$$

darstellen läßt. Der Integraloperator $\underset{\sim}{P}$ ist wegen

$$|(\underset{\sim}{P}w - \underset{\sim}{P}\omega)(z)| \leq \int_D g(\zeta,|w(\zeta) - \omega(\zeta)|)\frac{d\xi\, d\eta}{|\zeta-z|}$$

mit Rücksicht auf die gleichgradige Stetigkeit von g(z,x) in x=o auf dem Banachraum

$$B := \{w\colon w\in C(\hat{D}),\ ||w|| := \max_{z\in\hat{D}} |w(z)|\}$$

der in $\hat{D}$ stetigen Funktionen mit Maximumnorm stetig. $\underset{\sim}{P}w$ ist in D beschränkt und Hölder-stetig mit dem von w unabhängigen Hölder-Exponenten $\alpha = \frac{p-2}{p}$ (vgl. [16] (I,§6.1)),da $f(z,w(z)) \in L_p(\hat{D})$ $(2<P)$.

Satz 1: Erfüllt die Konstante K die ((6) einschließende) Bedingung

$$(12) \qquad ||\Theta|| + M[L_p(f_o,\hat{D}) + L_p(g_{2K},\hat{D})] \leq K$$

mit

$$f_o(z): = cf(z,o), g_{2k}(z): = g(z,2K)\ (z\in\hat{D}),$$

so exisitert in

$$A: = \{w: w\in C(\hat{D}), ||w|| \leq K,\ \mathrm{Re}w|_{\partial D}=\varphi,\ \frac{1}{2\pi}\int_{\partial D} \mathrm{Im}\ w(\zeta)|d\phi(\zeta)| = C\} \subset B$$

nur eine Lösung w der Integralgleichung (11); für sie gelten

$$\mathrm{Re}w|_{\partial D} = \varphi,\ \frac{1}{2\pi}\int_{\partial D} \mathrm{Im}\ w(\zeta)|d\phi(\zeta)| = C.$$

Beweis: Da $-\Theta$ zu A gehört, ist A nicht leer. Wegen (12) bildet $-\Theta + \underset{\sim}{P}$ die kompakte, konvexe Menge A von B in sich ab, so daß nach dem Schauderschen Fixpunktsatz in A eine Lösung der nichtlinearen Integralgleichung (11) existiert. Diese Lösung ist sogar Hölder-stetig, wenn die Randvorgaben φ Hölder-stetig sind. Daß die Randbedingungen erfüllt sind, läßt sich wie in [10] zeigen.

Um die Eindeutigkeit der Lösung in A nachzuweisen, nehme man die Existenz zweier Lösungen w und ω an. Dann gilt für $z\in D$

$$|w(z)-\omega(z)| = |\underset{\sim}{P}w-\underset{\sim}{P}\omega)\ (z)| \leq \int_D g(\zeta,|w(\zeta)-\omega(\zeta)|)\ \frac{d\xi\ d\eta}{|\zeta-z|} \leq ML_p(g_{2K},\hat{D}) \leq K.$$

Da demnach $|w(z)-\omega(z)|$ der Integralungleichung (7) genügt und die Funktion $g(z,x)$ die Voraussetzungen des Hilfssatzes erfüllt, gilt $w = \omega$.

Da $\underset{\sim}{P}w$ unter der Voraussetzung $f(z,w(z)) \in L_p(\hat{D})$ verallgemeinerte Ableitungen erster Ordnung hat (siehe etwa [16]), und die verallgemeinerte Ableitung nach $\bar{z}$ durch

$$\frac{\partial}{\partial \bar{z}}\ (\underset{\sim}{P}w)(z) = f(z,w(z))\ \ (z\in D)$$

gegeben wird, stimmt die Lösung von (11) mit der verallgemeinerten

Lösung des Randwert-Normproblems (9), (10) überein. Ist $f(z,w(z))$ für die Lösung von (11) beschränkt und Hölder-stetig, so existiert die Ableitung von w im klassischen Sinn und man erhält eine Lösung im klassischen Sinn. Zum Beweis der Existenz einer Lösung wird für $g(z,x)$ nur die bezüglich $z \in D$ gleichgradige Stetigkeit in $x = o$ benötigt. Die Voraussetzungen für $g(z,x)$ aus dem Hilfssatz sind nur für den Eindeutigkeitsbeweis benötigt worden.

Satz 2: Das Randwert-Normproblem (9), (10) hat eine in der Menge A eindeutig bestimmte verallgemeinerte Lösung, wenn f den Bedingungen v. und vi. genügt. Ist φ Hölder-stetig auf ∂D und $f(z,w(z))$ beschränkt und Hölder-stetig für Hölder-stetige Funktionen w, so ist die Lösung im klassischen Sinn zu verstehen.

3. Anmerkungen und Folgerungen. Es genügt, an Stelle von (12) nur

$$|\Theta(z)| + \int_D (|f_o(\zeta)| + g(\zeta,K)) \frac{d\xi\, d\eta}{|\zeta - z|} \leq K \quad (z \in \hat{D}) \tag{13}$$

und

$$\int_D g(\zeta,2K) \frac{d\xi\, d\eta}{|\zeta - z|} \leq K \quad (z \in \hat{D}) \tag{14}$$

zu verlangen.

Bedingung (12) ist für den bereits erwähnten Spezialfall der Lipschitz-Bedingung (8) erfüllbar. Man hat nur K gemäß

$$2||\Theta|| \leq K, \quad 2ML_p(f_o,\hat{D}) \leq K(1-4ML_p(g,\hat{D})) \tag{15}$$

unter der Voraussetzung

$$4M\, L_p(g,\hat{D}) < 1,$$

die für hinreichend kleine Gebiete D erfüllbar ist, zu wählen. Da K hier beliebig groß festgelegt werden kann, ist die Lösung des Randwert-Normproblems (9), (10) unter (8) und (15) eindeutig in B bestimmt.

Satz 3: Das Randwert-Normproblem (9), (10) hat im Raum $C(\hat{D})$ eine eindeutig bestimmte verallgemeinerte Lösung, wenn

$$f(z,w(z)) \in L_p(\hat{D})\,(2<p,\ w \in C(\hat{D}),\ z \in \hat{D}),$$

$$|f(z,w) - f(z,\omega)| \le \frac{1}{c} g(z)|w - \omega|\,(z \in \hat{D}; w,\omega \in \mathbb{C}),$$

$$g \in L_p(\hat{D}),\ 2\int_D g(\zeta)\,\frac{d\xi\, d\eta}{|\zeta - z|} \le 1.$$

Die Bedingungen (6), (12), (13), (14) sind für hinreichend kleines Gebiet D für vorgegebenes K(gegebenenfalls $K > 2||\Theta||$) erfüllbar.

Satz 4: Das Randwert-Normproblem

$$w_{\bar{z}} = f(z,w),\ \operatorname{Re} w|_{\partial D} = \varphi,\ \frac{1}{2\pi}\int_{\partial D} \operatorname{Im} w(\zeta)|d\phi(\zeta)| = C$$

für ein einfach zusammenhängendes beschränktes Gebiet D mit stetig gekrümmtem Rand ∂D, auf ∂D stetiger reeller Funktion φ und reeller Konstanten C ist in $C(\hat{D})$ eindeutig lösbar im verallgemeinerten Sinn, wenn f in $\hat{D} \times \mathbb{C}$ folgenden Bedingungen genügt:

1. $f(z,w(z))$ gehört für alle in $\hat{D}$ stetigen Funktionen zu $L_p(\hat{D})$ $(2<p)$,
2. $|f(z,w) - f(z,\omega)| \le \frac{1}{c}\, g(z,|w-\omega|)\ (z \in \hat{D};\ w,\omega \in \mathbb{C})$

mit einer in $\hat{D} \times [0,+\infty)$ nichtnegativen, in x monoton wachsenden und stetigen, in x = o gleichgradig bezüglich $z \in \hat{D}$ stetigen und dort für x = o identisch in z verschwindenden Funktion g(z,x), für die für jede in $\hat{D}$ stetige, nichtnegative Funktion x(z)

1. $g(z,x(z)) \in L_p(\hat{D})$ $(2<p)$,
2. $x(z) \neq \int_D g(\zeta,x(\zeta))\,\frac{d\xi\, d\eta}{|\zeta - z|}$ wenigstens für ein $z \in \hat{D}$, falls $x(z) \not\equiv 0$,

erfüllt sind.

Zu den betrachtenden Funktionen gehören unter den zusätzlichen Voraussetzungen

$$f(z,o) \equiv 0,\ g(z,x) = g(z)x\ (g \in L_p(\hat{D}))$$

die approximativ analytischen Funktionen.

Literatur

[1] Begehr, H. Zur Wertverteilung approximativ analytischer Funktionen. Arch. Math. (Basel) 23(1972), 41 - 49.

[2] Begehr, H. Die logarithmische Methode in der Wertverteilungstheorie pseudoanalytischer Funktionen. Ann.Acad.Sci.Fenn. AI 549 (1973).

[3] Bers, L. Theory of pseudo-analytic functions. Vorlesungsausarb. New York University 1953.

[4] Bojarski, B.B. Die Theorie des verallgemeinerten analytischen Vektors. Ann.Pol.Math. 17(1966), 281-320 (Russisch).

[5] Gilbert, R.P. Constructive methods for elliptic partial differential equations. Lecture Notes in Mathematics. 365, Springer, Berlin-Heidelberg-New York, 1974. 397 S.

[6] Gilbert, R.P. - Hile, G. Generalized hyperanalytic function theory. Trans. Amer.Math.Soc. 195(1974), 1-29.

[7] Gilbert, R.P. - Hsiao, G.C. On Dirichlet's problem for quasilinear elliptic equations. In: Constructive and computational methods for differential and integral equations.
Herausgegeben von D.L. Colton und R.P. Gilbert. Lecture Notes in Mathematics 430, Springer, Berlin-Heidelberg-New York, 1974, 184-236.

[8] Gilbert, R.P.-Weinacht, R.J. Interative schemes for elliptic systems. Ibid. 253-260.

[9] Wendland, W.L. An integral equation method for generalized analytic functions. Ibid. 414-542.

[10] Haack, W. - Wendland, W. Vorlesungen über partielle und Pfaffsche Differentialgleichungen. Birkhäuser, Basel-Stuttgart, 1969, 555 S.

[11] Habetha, K. Über die Werteverteilung pseudoanalytischer Funktionen. Ann.Acad.Sci.Fenn.AI 406 (1967).

[12] Habetha, K. On zeros of elliptic systems of first order in the plane. Erscheint demnächst.

[13] Tutschke, W. Konstruktion von globalen Lösungen mit vorgeschriebenen Singularitäten bei partiellen komplexen Differentialgleichungssystemen. S. - ber. Sächsische Akad. Wiss. Leipzig, Math. Nat. Kl. 109, 7.

[14] Tutschke, W. Ein Differenzierbarkeitssatz für approximativ analytische Funktionen. Math. Z. 142 (1975), 27-31.

[15] Tutschke, W. Theorie und Anwendungen morpher Funktionen. Beiträg zur Analysis 4(1972), 167-175.

[16] Vekua, I.N. Verallgemeinerte analytische Funktionen. Akademie Verlag, Berlin, 1963, 538 S.

[17] Wendland, W. Über Ähnlichkeitsprinzip und Randwertaufgaben für verallgemeinerte analytische Funktionen. Appl. Anal. 2 (1972), 101-110.

[18] Wendland, W. On boundary value problems of generalized analytic functions. Conf. theory ordinary partial diff. equ., Dundee/ Scottland 1972, Lecture Notes in Mathematics. Springer, Berlin-Heidelberg-New York 280 (1972), 190-201.

STABILITY OF MINIMAL SURFACES

Reinhold Böhme

The purpose of the present paper is to explain the following result: Let us consider the set M of all 2-dimensional minimal surfaces in $\mathbb{R}^3$ of topological type of the 2-disc and of the class C^∞. We define E as the set of all C^∞-embeddings of S^1 into $\mathbb{R}^3$. If $g \in E$, we denote by M(g) the set of all $x \in M$ such that the surface x is bounded by g.
If $x \in M(g)$, what happens if one varies g slightly? If h is near g in E, will M(h) contain a surface y, which is near x in M? We will show:
There is a C^∞-open and C^∞-dense subset N in M, such that all $x \in N$ are stable in the above sense. We give a complete exposition, for details cf. [1].
Apart from this stability result we will give here a formal description of the space of all minimal surfaces (without branch points on the boundary) as a kind of algebraic variety in a space of infinite dimensions, where the variety has an infinite dimension and an infinite codimension (§2).
This result may be of interest in itself.
There exists a preprint of A.J.Tromba [8] with interesting related results. He is however dealing mainly with minimal surfaces without branch points at all.

§1. Sobolev spaces of holomorphic functions

Definition 1.1: Let B denote the open unit disc of $\mathbb{R}^2$. For nonnegative integers m we define $A^m(B) = A^m(B,\mathbb{C})$ as the space of all holomorphic functions f defined on B, such that the derivatives of f up to the order m are square integrable on B.

It is easy to see that $A^m(B)$ is a Hilbert space with the norm $\|f\|^2_{A^m} := \sum_{0\le\mu\le m} \|(\frac{d}{dz})^\mu f\|^2_{L_2(B)}$.

We define the Frechet space $A^\infty(B) := \varprojlim A^m(B,\mathbb{C})$ of the functions holomorphic on B and C^∞ on $\overline{B}$ having the topology which is induced by the family $\{\|.\|_{A^m}\}_m$ of norms on this space.

If $z_0 \in B$, $f \in A^0(B,\mathbb{C})$ and $k \geq 0$, then we denote the k-jet of f at z_0 by $j_{k,z_0}(f)$.Its a k-order Taylor polynomial of f.

Lemma 1.2: For all $m \geq 0$, $k \geq 0$ and all $z_0 \in B$ the mapping $j_{k,z_0} : A^m(B) \to \mathbb{C}^{k+1}$ is continuous. Its kernel is closed and therefore a Hilbert space.

Now the Sobolev embedding theorem implies: If $m \geq 2$, then the pointwise multiplication of functions defines a continuous product on $A^m(B,\mathbb{C})$, and $A^m(B,\mathbb{C})$ is an algebra. Less obvious is the following continuity result about pointwise division of functions:

Proposition 1.3: Assume $m \geq 2$, let $h \in A^m(B)$ denote a function which is nowhere zero on ∂B. Let $W_h{}^m$ denote the subspace of all functions in A^m that vanish in all points

of B, where h vanishes, with at least the same multiplicity. Then W_h^m is a Hilbert space, and the mapping
$\hat{H} : W_h^m \to A^m(B)$,
defined by $\hat{H}(f)(z) := \frac{f(z)}{h(z)}$ for all $z \in B$,
is a continuous isomorphism.

<u>Proof:</u> We can identify the space $A^m(B)$ with a certain Sobolev space $\overline{H}^m(\partial B)$ of complex valued functions on ∂B using the Poisson formula. Since $h|_{\partial B}$ is nonzero on ∂B, we see that $h|_{\partial B}$ and $h^{-1}|_{\partial B}$ are continuous multiplicators on $\overline{H}^m$. Then we observe, that the singularities of H(f) are removable and 1.3 follows.

§2. The Weierstrass representation for minimal surfaces

The purpose of this section is a study of the properties of a representation formula for minimal surfaces which was published by K. Weierstrass in 1866 [9]:

<u>Definition 2.1:</u> Assume that (f,g) is a pair of complex valued functions defined on B. The pair (f,g) will be called compatible iff f is holomorphic on B, g is mermorphic on B and fg^2 is holomorphic on B too.

<u>Definition 2.2:</u> A C^2-mapping $x : B \to \mathbb{R}^3$ is called a (parametrized 2-dimensional) minimal surface (of the topological type of the disc) iff x is harmonic and the complex gradient $F := x_u - ix_v$ satisfies the relation
$F.F := (F_1)^2 + (F_2)^2 + (F_3)^2 \equiv 0$ on B.

<u>Theorem 2.3:</u> Assume that x is minimal surface as above and define $f := \frac{1}{2}(F_1 - iF_2)$ and $g := \frac{F_3}{F_1 - iF_2}$, then the pair (f,g) is compatible.

Conversely, if (f,g) is a compatible pair, define

$F_1 := f - fg^2$, $F_2 := if + ifg^2$ and $F_3 := 2fg$, choose an arbitrary vector $q \in \mathbb{R}^3$ and define

$x_j := \mathrm{Rea} \int_0^z F_j(t)dt + q_j$ for $j = 1,2,3$, then

$x : B \to \mathbb{R}^3$ is a minimal surface with $x(0) = q$.

For a proof see [4], §8. We call (f,g) the Weierstrass representation of x.

The definition of a minimal surface is invariant under the orthogonal group of $\mathbb{R}^3$. We may use this invariance to simplify the representation.

<u>Definition 2.4:</u> Assume $x \in C^2(\bar{B}, \mathbb{R}^3)$ is a minimal surface. A point $z_0 \in \bar{B}$ is a singular point of x iff the rank of the differential $Dx(z_0)$ is less than two. Such a point is called a branch point of order s iff there exist $a_0 \in \mathbb{C}^3$ and a representation near z_0 of the following form:

$\frac{d}{dz}x(z) = a_0(z-z_0)^s + o(|z-z_0|^s)$.

It is easy to see, that z_0 is a singular point of x iff f and fg^2 vanish both at z_0.

<u>Lemma 2.5:</u> Assume that $x \in H^4(B, \mathbb{R}^3)$ is a minimal surface without a singular point on ∂B. Then one can choose a coordinate system in $\mathbb{R}^3$ such that the function $f = \frac{1}{2}(F_1 - iF_2)$ is nowhere zero on ∂B.

Such a coordinate system will be called appropriate for x.

Definition 2.6: Modifying 1.1 we define for all $m \geq 2$

$A_*^m(B,\mathbb{C}) := \{f \in A^m(B,\mathbb{C}) \mid f(z) \neq 0 \text{ for all } z \in \partial B\}$.

We define

$\Pi := \{P \in \mathbb{C}[z] \mid P(z) = \prod_{\nu=0}^{n} (z-a_\nu),\ a_\nu \in B \text{ for all } \nu\}$ and

$\Pi^* := \{P \in \Pi \mid \text{all zeros of } P \text{ are simple}\}$,

$\Pi_n := \{P \in \Pi \mid \text{degree of } P \text{ is } n\}$ and

$\Pi_n^* := \Pi_n \cap \Pi^*$.

We think of Π_n and of Π_n^* as finite dimensional complex manifolds.

Theorem 2.7: Assume $m \geq 5$ and $x \in H^m(B, \mathbb{R}^3)$ is a minimal surface without singular points on ∂B and given in an appropriate coordinate system. We define $F := \frac{d}{dz} x$.

Then there exists a H^m-neighbourhood U of x such that all minimal surfaces $\tilde{x} \in U$ have representations as follows:

$\tilde{F} := \frac{d}{dz} \tilde{x}$ and

$\tilde{F} =: (\tilde{Q}^2\tilde{R}\tilde{f} - \tilde{R}\tilde{f}\tilde{g}^2,\ i\tilde{Q}^2\tilde{R}\tilde{f} + i\tilde{R}\tilde{f}\tilde{g}^2,\ 2\tilde{Q}\tilde{R}\tilde{f}\tilde{g})$.

Here are $\tilde{Q} \in \Pi$, $\tilde{R} \in \Pi_*$, $\tilde{f} \in A_*^{m-1}$ and $\tilde{g} \in A^{m-1}$.

In fact $\tilde{Q}$, $\tilde{R}$, $\tilde{f}$ and $\tilde{g}$ are uniquely determined.

The proof is a straightforward computation, essentially using 1.3 and 2.3. The polynomials $\tilde{Q}$ and $\tilde{R}$ are introduced only in order to avoid a discussion of meromorphic functions which enter the classical representation formula.

Corollary 2.8: Assume that $F = \frac{d}{dz} x$ has a representation $F = (Q^2Rf - Rfg^2,\ iQ^2Rf + iRfg^2,\ 2QRfg)$. Suppose that the functions Q, R, g have no common zero in B (f has none). Then you can choose the neighbourhood U of x so small that for all $\tilde{x} \in U$, represented as above, the following holds:

$\tilde{Q}$, $\tilde{R}$, $\tilde{g}$ have no common zero in B, and deg $\tilde{Q}$ = deg Q = : q_o, deg $\tilde{R}$ = deg R = : r_o.

It is apparent that 2.8 applies for all minimal surfaces which have only simple branch points and no branch points on the boundary ∂B. Then we have:

Theorem 2.9: Assume $x \in H^m(B, \mathbb{R}^3)$ is a minimal surface with only simple branch points in B. Then there exists an open neighbourhood V^m of x in $H^m(B, \mathbb{R}^3)$ such that the set $W^m \subset V^m$ of all minimal surfaces in V^m is a Hilbert manifold and a Hilbert submanifold of $H^m(B, \mathbb{R}^3)$ with the model space $\mathbb{R}^3 \times \mathbb{C}^{q_o} \times \times \mathbb{C}^{r_o} \times A^{m-1}(B, \mathbb{C}^2)$.

Proof: Represent x as above. It is not difficult to show that the mapping
$\Psi : \tilde{x} \mapsto (\tilde{Q}, \tilde{R}, \tilde{f}, \tilde{g})$ defining the representation 2.7 in a neighbourhood of x, is continuous and injective as a mapping
$\Psi : W^m \to \Pi_{q_o} \times \Pi_{r_o} \times A^{m-1}(B, \mathbb{C}^2)$.
Its inverse Ψ^{inv} exists on the open set $\Psi(W^m)$ and is a differentiable embedding.

Obviously there are complications when considering a surface x, that gives common zeros of Q, R, g, i.e. if x has multiple branch points. What really happens can be described easily using the concept of stratification of a topological space.

Definition 2.10 (J. Cerf [3]): Assume E is a topological space. A sequence $E^o, E^1, \dots E^i, \dots$ of subsets of E is called a stratification of E iff the sets E^i are pairwise disjoint, their union is E and the set $E^o \cup E^1 \cup \dots \cup E^i$ is open in E for all $i \geq 0$.

Very well known is the stratification of an algebraic set (H. Whitney). In fact, in some sense the theorem 2.12 below shows that the set of all minimal surfaces of class H^m and without branch points on the boundary can be understood as an algebraic subset of $H^m(B, \mathbb{R}^3)$ with infinite dimension and infinite codimension. We will denote the set of these surfaces by M^m.

<u>Definition 2.11:</u> Given $m \geq 5$, $p \geq 0$ and $q \geq 0$, then for all weakly monotonic sequences $\underline{l} = (l_1, \ldots, l_p)$ of odd natural numbers and for all weakly monotonic sequences $\underline{m} = (m_1, \ldots, m_q)$ of even natural numbers we define

$$N^{q,\underline{m}}_{p,\underline{l}} := N^{q;(m_1, \ldots, m_q)}_{p;(l_1, \ldots, l_p)} := \{x \in H^m(B, \mathbb{R}^3) \mid x \text{ minimal}$$

surface with exactly p branch points in B with odd multiplicities $(l_1, \ldots, l_p)$ and exactly q branch points in B with even orders $(m_1, \ldots, m_q)$ in B and no others$\}$.

We denote by $\mathcal{N}$ the set of all these $N^{q,\underline{m}}_{p,\underline{l}}$. Then we have:

<u>Theorem 2.12:</u>

(i) The sets $N^{q,\underline{m}}_{p,\underline{l}}$ define a partition of M^m.

(ii) Every $N \in \mathcal{N}$ is a real analytic submanifold of $H^m(B,\mathbb{R}^3)$ without boundary.

(iii) For all $N \in \mathcal{N}$ the relative boundary of N in M^m is a countable union of elements $N' \in \mathcal{N}$.

(iv) $\tilde{N}$ is a boundary component of N iff there is a reordering of the indices such that in the new order $\underline{\tilde{m}}-\underline{m}$ and $\underline{\tilde{l}}-\underline{l}$ are nonnegative and have only even components.

(v) M^m consists of exactly two connected components. These are the surfaces of the odd and of the even total numbers of branch points.

(vi) The total number of branch points is constant for all surfaces in a fixed set $N \in \mathcal{N}$. This number may be called the order of N.

(vii) The union of all $N \in \mathcal{N}$ of order less or equal to k, $k \geq 0$, is an open subset of M^m. The partition $\mathcal{N}$ defines a stratification of M^m.

(viii) The union of all $N \in \mathcal{N}$ with arbitrary p, with $\underline{1} = (1,1, \ldots, 1)$ and $q = 0$ is H^m-dense in M^m. These are the strata of surfaces with exactly p simple branch points in the interior, referred to in 2.7.

The proof of 2.12 is an immediate interpretation of Weierstrass' formula using the results of §1.

Corollary 2.13: Let $M^m{}_i$ denote the manifold of all minimal surfaces in M^m with exactly i simple branch points. Define $M^\infty := \bigcap_m M^m$ and $M^\infty{}_i := \bigcap_m M^m{}_i$. Then $M^\infty{}_i$ is a Frechet manifold with model spaces $\mathbb{C}^n \times \mathbb{A}^\infty(B, \mathbb{C}^2)$. The union $\bigcup_i M^\infty{}_i$ is H^5-open and C^∞-dense in M^∞.

Remark: It can be important to consider minimal surfaces with branch points on the boundary too, but there a more refined argument is needed.

§3. Spaces of embeddings of S^1 into $\mathbb{R}^3$.

We denote the unit circle of $\mathbb{R}^2$ by S^1 or by ∂B. If $g : \partial B \to \mathbb{R}^3$ is a continuous mapping, then there exists a unique harmonic mapping $\overline{g} : \overline{B} \to \mathbb{R}^3$ with boundary values g. It is very convenient to introduce norms for g by defining $\|g\|_{\overline{H}^m} := \|\overline{g}\|_{H^m(B,\mathbb{R}^3)}$.

<u>Definition 3.1:</u> For $m \geq 4$ we will denote by G^m the set of all C^1-embeddings $g : \partial B \to \mathbb{R}^3$ with the property that the norm $\|g\|_{\overline{H}^m}$ is finite. G^m is open in the space $\overline{H}^m(\partial B, \mathbb{R}^3)$. We define $G^\infty := \bigcap_m G^m$ with the induced Frechet topology.

<u>Lemma 3.2:</u> Given g_1 and g_2 in G^m, suppose there exists a homeomorphism ω of ∂B such that $g_1 = g_2 \circ \omega$. Then ω is a C^1-diffeomorphism of ∂B and is of class $\overline{H}^m(\partial B, \mathbb{R}^2)$.

These diffeomorphisms together constitute a topological group, which will be denoted by $\mathrm{Diff}^m(\partial B)$. Therefore we can define equivalence classes in the space G^m with respect to right action of $\mathrm{Diff}^m(\partial B)$ and equivalence classes in G^∞ with respect to $\mathrm{Diff}^\infty(\partial B)$. The spaces of equivalence classes will be denoted by E^m resp. E^∞.

Now assume $g \in G^\infty$. Then there exists a normal bundle N_g of g in $\mathbb{R}^3$ and we can define an $\overline{H}^m$-topology on the space Γ^m of $\overline{H}^m$-sections in this bundle.

We can identify a $\overline{H}^m$-neighbourhood V_m of the 0-section in Γ^m with a subset of some $\overline{H}^m$-neighbourhood U_m of g in G^m.

<u>Lemma 3.3:</u> If U_m and V_m are appropriately chosen, then there exists a continuous projection $\rho_m : U_m \to V_m$, idempotent and

surjective, such that the fibres of the projection are equivalence classes with respect to $\mathrm{Diff}^m(\partial B)$. V_m has a structure of a Hilbert manifold with model space $\overline{H}^m(\partial B,\mathbb{R}^2)$.

This lemma can be generalized in a natural manner in order to obtain the following theorem.

Theorem 3.4: The space E^∞ has a structure as a Frechet manifold with model space $C^\infty(\partial B, \mathbb{R}^2)$. There exists a continuous projection ρ of G^∞ onto E^∞, whose fibers are the equivalence classes with respect to the group $\mathrm{Diff}^\infty(\partial B)$. The mapping ρ extends to a continuous projection ρ_m of G^m onto E^m. The mapping $\rho : G^\infty \to E^\infty$ is weakly differentiable up to all orders, but it can be extended to give a 2-times continuously differentiable mapping between Hilbert spaces only if considered as a mapping $\rho_{m+2,m} : G^{m+2} \longrightarrow E^m$.

Remark 3.5: The mapping $\rho_{m+2,m}$ is apparently not surjective, nor does it define a local surmersion. The loss of 2 derivatives here is very natural and cannot be avoided. A very clear idea about what happens here may be obtained from the paper [6] of Sergeraert, who introduces the concepts of $\mathcal{L}$-space and $\mathcal{L}$-mappings. We will actually apply his implicit function theorem in the sequel.

§4. Jacobi fields of minimal surfaces

Assume that $g : \partial B \to \mathbb{R}^3$ is an embedding of class C^∞. Let $F^m = F^m(g)$ denote the $\overline{H}^m$-equivalence class of g in the space G^m. If $f \in F^m(g)$, we can identify f with its harmonic extension to $\overline{B}$.

<u>Lemma 4.1:</u> F^m is a differentiable Hilbert manifold with the model space $\overline{H}^m(\partial B, \mathbb{R})$. If we identify F^m with a space of harmonic functions, then the Dirichlet integral $\mathbb{D}$ defines a differentiable mapping $\mathbb{D} : F^m \to \mathbb{R}$.

The gradient of $\mathbb{D}$ vanishes in $x \in F^m$ iff x is a minimal surface in F^m.

If $x \in F^m$ is a minimal surface of F^m, then the Hessian $D^2\mathbb{D}(x)$ is defined as an equivalence class of symmetric bilinear forms on the model space. There exists a well defined extension of $D^2\,\mathbb{D}(x)$ inducing a projective equivalence class of quadratic forms on $\overline{H}^1(\partial B, \mathbb{R})$. This class can be represented as a compact perturbation of the standard metric form of this space.

The dimension of the space, on which $D^2\,\mathbb{D}(x)$ is negative definite, is finite and defines the Morse index of x.

There exists a finite dimensional subspace $J(x)$ of $\overline{H}^1$ on which the form $D^2\,\mathbb{D}(x)$ is identically zero.

The space $J(x)$ shall be called the space of Jacobi fields of x, in a natural analogy to the classical concept.

For a proof see [2].

<u>Proposition 4.2:</u> Assume that x is a minimal surface such that x restricted to ∂B is an embedding of ∂B into $\mathbb{R}^3$ of class $\overline{H}^m$. (This is equivalent to saying: $x \in F^m$). Then the space $J(x)$ has a natural splitting

$J(x) = J_0(x) \oplus J_1(x) \oplus J_2(x)$

with the following properties:

(i) $\dim J_0(x) = 3$. J_0 is exactly the space of degeneracy of Dirichlet's integral coming from its invariance under the group of conformal automorphisms of B.

(ii) dim $J_1(x)$ = 2 times the order of x. This space is a space of variations of x inducing infinitesimal movements of the branch points of x but inducing no first order change in the boundary curve.

(iii) dim $J_2(x) \geq 0$. The vectors of this space only should be considered for 'true' Jacobi fields. If the order of x is zero, they can be identified with the classical Jacobi fields for minimal immersions. See [7].

The proof of 4.2 is found in [2].

<u>Theorem 4.3:</u> There exists an H^m-open and H^m-dense subset N^m in the space of all minimal surfaces of class H^m such that for all $x \in N^m$ the following holds:

(i) x restricted to ∂B is an embedding of class $\bar{H}^m$.

(ii) x has only simple branch points, and an H^m-neighbourhood V^m of x in the space of all minimal surfaces is a submanifold of $H^m(B, \mathbb{R}^3)$.

(iii) The space $J_2(x)$ has dimension 0; i.e. there exist no true Jacobi fields for x.

(iv) The space $K(x) := T_x V^m \cap J(x) = J_0(x)$: the intersection of the tangent space of V^m at x with the space of Jacobi fields of x is exactly the space $J_0(x)$.

(v) There is a natural projection r_m of V^m into a neighbourhood X^m of $x|_{\partial B}$ in the space of boundary curves E^m. r_m is nothing else than the restriction mapping to boundary followed by the mapping ρ_m of §3.

(vi) If x is of class C^∞ and contained in N^m then the differential of the mapping r_m evaluated at x is a

Fredholmoperator of index +3. Since its kernel is exactly the space $J_0(x)$, which is a consequence of (iv), and since the dimension of $J_0(x)$ is 3, therefore this differential is surjective.

Proof: The proof is rather complicated and can be found in §5 of [1]. The essential point seems to be the following: The minimal surface equation with an additional boundary condition, given by an element of E, can be seen as a non-linear elliptic system. This system has a linearization at x as above, which is actually a linear elliptic system of index zero. This system defines in a precise sense a 'formal' tangent space for the manifold V^m of all vectors satisfying the equations which define V^m up to the first order. This formal tangent space is useless for us, and we have explicitely calculated the difference between the formal and the true tangent space. It turns out, that in a generic case as above, this complement is exactly $J_1(x)$. Now one can also calculate the fredholm index of $Dr_m(x)$ defined on T_xV^m.

§5. Stability of minimal surfaces against perturbation of the boundary curve

Assume that $g : S^1 \to \mathbb{R}^3$ isa C^∞-embedding, i.e. $g \in G^\infty$.

Definition 5.1: If $g \in G^\infty$, then we define
$\mathcal{M}^\infty(g) := \{x \in C^\infty(\bar{B}, \mathbb{R}^3) \mid x$ is a minimal surface and there exists a homeomorphism $u : S^1 \to S^1$, such that $x|_{\partial B} = g \circ u\}$.
A surface $x \in C^\infty(\bar{B}, \mathbb{R}^3)$ is called a solution of the Plateau problem for $g \in G^\infty$, iff $x \in \mathcal{M}^\infty(g)$.

Remark 5.2: There are classical theorems which ensure that $\mathcal{M}^{\infty}(g)$ is not void. It is clear that $\mathcal{M}^{\infty}(g)$ depends only on the equivalence class [g] of g in E^{∞}

We may define $\mathcal{M}^{\infty} := \bigcup_{g \in G^{\infty}} \mathcal{M}^{\infty}(g)$. $\mathcal{M}^{\infty}$ is a subset of $H^{\infty}(B,\mathbb{R}^3)$ and will have its Frechet topology. There is a natural mapping r of $\mathcal{M}^{\infty}$ into E^{∞} such that $r(x) = [g]$ if $x \in \mathcal{M}^{\infty}(g)$. Assume $x \in \mathcal{M}^{\infty}$ has no branch points on the boundary, i.e. x resticted to the boundary is itself an embedding. Now theorem 3.4 implies: There exists an open neighbourhood V^{∞} of x in $\mathcal{M}^{\infty}$, such that the mapping $r : V^{\infty} \to E^{\infty}$ is continuous. More precisely, there exist continuous extensions $r_m : V^m \to E^m$ of r for all m. See §3.

Definition 5.3: Assume $x \in \mathcal{M}^{\infty}$ is a minimal surface without branch points on the boundary. Then x is stable against perturbation of the boundary iff the mapping $r : V^{\infty} \to E^{\infty}$ is open at x. This definition is natural. Assume $x \in \mathcal{M}^{\infty}(g)$ and h near g in E^{∞}, then x is stable as above iff there exists $y \in \mathcal{M}^{\infty}(h)$ such that y is near x in $\mathcal{M}^{\infty}$.

Theorem 5.3: There exists a C^{∞}-open and C^{∞}-dense subset N^{∞} in the set of all minimal surfaces of class C^{∞} such that all minimal surfaces in N^{∞} are stable against perturbation of the boundary curve.

Proof: We define $N^{\infty} := \bigcap_m N^m$ exactly as in 4.3. N^{∞} is open and dense in M^{∞}. The mapping r has a differential $Dr(x)$ at x which extends by continuity to a linear isomorphism $Dr(x):T_xN^m \longrightarrow E^m$; but the mapping r is not continuously differentiable as a mapping $r_m : N^m \longrightarrow E^m$.

There is a 'loss of derivatives' as discussed in 3.5.

This loss makes it necessary to apply a hard version of the implicit function theorem. We may refer to [5] or [6]. But such a theorem implies that r is open at x if its differential is open and some estimates hold. See [1].

References:

[1] Böhme, R.: Über Isoliertheit und Stabilität der Lösungen des klassischen Plateauproblems. 1976. Submitted to Math.Z.

[2] Böhme, R.: Die Jacobifelder zu Minimalflächen im $\mathbb{R}^3$. Manuscr.math. 16, 51-73, 1975.

[3] Cerf, J.: La stratification naturelle des espaces de fonctions différentiables réelles et le théorème de la pseudoisotopie. Publ.Math. IHES, 41, 5-173, 1970.

[4] Osserman, R.: A survey of minimal surfaces. New York, van Nostrand, 1969.

[5] Schwartz, J.T.: Nonlinear functional analysis. New York London, Paris. Gordon and Breach, 1969.

[6] Sergeraert, F.: Un théorème des fonctions implicités dans certains espaces de Frechet et quelques applications. Ann.Sci.Ec.Norm.Sup. (3), 5, 559-660, 1972.

[7] Simons, J.: Minimal varieties in Riemannian manifolds. Ann.Math. 88, 52-105, 1968.

[8] Tromba, A.J.: On the number of simply connected minimal surfaces spanning a curve in $\mathbb{R}^3$. Subm. to Trans.AMS.1975.

[9] Weierstrass, K.: Untersuchungen über die Flächen, deren mittlere Krümmung überall Null ist. Werke Bd.3, S.39 ff, 1894.

NON-(k)-MONOGENIC POINTS OF FUNCTIONS OF A QUATERNION VARIABLE

F.F. Brackx

Seminarie voor Wiskundige Analyse
Rijksuniversiteit Gent, J. Plateaustraat 22,
B-9000 GENT (Belgium)

Abstract

The concept of (k)-monogeneity for functions with values in the algebra $\mathcal{H}$ of real quaternions has been introduced in [1]. Its definition and some results concerning Cauchy's Formula, the Taylor expansion and Weierstrass's convergence theorem are recalled in a first paragraph; for the proofs we refer the reader to [1] and [2].

We then continue the study of such functions by examining their behaviour in the neighbourhood of singular points. This gives rise to Laurent's expansion (§3), the notions of (k)-pole, essential-non-(k)-monogenic point and removable singularity (§§4 and 6), Mittag-Leffler's theorem on (k)-meromorphic functions (§5) and a finite residue theory (§7).

§1. INTRODUCTION

In the sequel Ω will always denote an open, non-void subset of $\mathbb{R}^4$, and $(\mathcal{H},+,\times)$ the algebra of real quaternions, with basis $\{e_0,e_1,e_2,e_3\}$. The functions considered here have their argument in $\mathbb{R}^4$ and take values in $\mathcal{H}$; they are of the form :

$$f = \sum_{\alpha=0}^{3} e_\alpha f_\alpha : x = (x_0,x_1,x_2,x_3) \mapsto f(x) = \sum_{\alpha=0}^{3} e_\alpha f_\alpha(x) ,$$

where the f_α ($\alpha=0,1,2,3$) are real valued. The notion of (k)-monogeneity ($k\in\mathbb{N}$) is now defined by means of the differential operator $D = \sum_{\alpha=0}^{3} e_\beta \frac{\partial}{\partial x_\beta}$, which can act from the left or the right upon a function f, namely $Df = \sum_{\alpha,\beta=0}^{3} e_\beta e_\alpha \frac{\partial f_\alpha}{\partial x_\beta}$ and $fD = \sum_{\alpha,\beta=0}^{3} e_\alpha e_\beta \frac{\partial f_\alpha}{\partial x_\beta}$.

<u>Definition 1.1</u>. A function $f : \mathbb{R}^4 \to \mathcal{H}$ is called left-(k)-monogenic in Ω iff (i) $f \in C_k(\Omega)$, this means $f_\alpha \in C_k(\Omega)$ for all $\alpha=0,1,2,3$;
(ii) $D^k f = D(D^{k-1}f) = 0$ in Ω.

<u>Theorem 1.1</u>. If $f \in C_{2k}(\Omega)$ is left-(k)-monogenic in Ω, then f is polyharmonic in Ω (this means $\Delta^k f = 0$ in Ω with $\Delta = \sum_{\alpha=0}^{3} e_\alpha \frac{\partial^2}{\partial x_\alpha^2}$) and belongs automatically to $C_\infty(\Omega)$.

<u>Theorem 1.2</u>. (Stokes) Let M be a four dimensional, differentiable, manifold contained in Ω, C a four chain on M, and $f,g \in C_1(\Omega)$. Then

$$\int_{\partial C} f d\sigma g = \int_C (fD.g + f.Dg)dx^4$$

where $d\sigma = \sum_{\alpha=0}^{3} e_\alpha (-1)^\alpha d\hat{x}_\alpha$, $d\hat{x}_\alpha = dx_0 \wedge \ldots \wedge dx_{\alpha-1} \wedge dx_{\alpha+1} \wedge \ldots \wedge dx_3$

and $dx^4 = dx_0 \wedge dx_1 \wedge dx_2 \wedge dx_3$.

Theorem 1.3. The function g_k, defined by $g_k(x) = \frac{1}{\omega_4} \cdot \frac{\overline{x}}{\rho_x^4} \cdot \frac{x_0^{k-1}}{(k-1)!}$ for $x \neq 0$ (where $x = \sum_{\alpha=0}^{3} e_\alpha x_\alpha$, $\overline{x} = \sum_{\alpha=0}^{3} \varepsilon_\alpha e_\alpha x_\alpha$, $\varepsilon_0 = +1$, $\varepsilon_i = -1$ for $i = 1,2,3$, $\rho_x^2 = \sum_{\alpha=0}^{3} x_\alpha^2$ and $\omega_4 = 2\pi^2$ denotes the area of the four dimensional unit sphere), is a two-sided fundamental solution of the operator D^k, i.e. :

(i) g_k is analytic in co{0} and $D^k g_k = g_k D^k = 0$ in co{0} (see also def. 1.2.) ;

(ii) $g_k \in L_1^{loc}(\mathbb{R}^4, \mathcal{H})$;

(iii) $D^k g_k = g_k D^k = \delta$ (δ : Diracmeasure) where the derivatives are taken in the sense of distributions.

Theorem 1.4. (Cauchy) Let f be left-(k)-monogenic in Ω and $S \subset \Omega$ a four dimensional, compact, differentiable, oriented manifold-with-boundary. Then for any $x \in \mathring{S}$,

$$f(x) = \frac{1}{\omega_4} \int_{\partial S} \sum_{j=0}^{k-1} (-1)^j \frac{\overline{u-x}}{\rho^4} \cdot \frac{(u_0-x_0)^j}{j!} . d\sigma_u . D^j f(u)$$

where ρ is the euclidean distance between u and x.

Notation : By $p^{(a)}_{\ell_1 \ldots \ell_n}$ we denote the two-sided-(1)-monogenic (in the sequel also called regular) polynomials of degree n, already introduced by R. Fueter in [3], and given by

$$p^{(a)}_{\ell_1 \ldots \ell_n}(x) = \frac{1}{n!} \sum_{\pi(\ell_1, \ldots, \ell_n)} (z_{\ell_1} - a'_{\ell_1}) \ldots (z_{\ell_n} - a'_{\ell_n})$$

where the sum runs over all possible permutations with repetition of the sequence $(\ell_1, \ldots, \ell_n) \in \{1,2,3\}^n$, and the $z_\ell - a'_\ell$ are hypercomplex variables defined by $z_\ell - a'_\ell = (x_\ell - a_\ell)e_0 - (x_0 - a_0)e_\ell$, $\ell = 1,2,3$.

Definition 1.2. A function $f : \mathbb{R}^4 \to \mathcal{H}$ is called analytic in Ω iff the components f_α, $\alpha = 0,1,2,3$ are real-analytic in Ω (see e.g. [4], p.473).

Theorem 1.5. (1st Taylor expansion for an analytic function)
If f is analytic in a certain region Λ containing the origin, there exists an open neighbourhood Ω of the origin in which

$$f(x) = \sum_{n=0}^{\infty} \frac{1}{n!} \sum_{\ell_1=0}^{3} \cdots \sum_{\ell_n=0}^{3} x_{\ell_1} \cdots x_{\ell_n} \left. \frac{\partial^n f}{\partial x_{\ell_1} \cdots \partial x_{\ell_n}} \right|_{x=0}$$

where the series, considered as a multiple power series, converges absolutely and uniformly on each compact subset of Ω.

Theorem 1.6. (2nd Taylor expansion for a monogenic function)
If f is left-(k)-monogenic in Ω, then

(i) f is analytic in Ω ;

(ii) for each $a \in \Omega$ there exists an open neighbourhood Ω_a of a, in which

$$f(x) = \sum_{n=0}^{\infty} \sum_{m=0}^{k-1} \sum_{(\ell_1,\ldots,\ell_{n-m})} \frac{(x_0-a_0)^m}{m!} \cdot p^{(a)}_{\ell_1\cdots\ell_{n-m}}(x) \cdot \left. \frac{\partial^{n-m} D^m f}{\partial x_{\ell_1} \cdots \partial x_{\ell_{n-m}}} \right|_{x=a}$$

where the third sum runs over all possible combinations with repetition of the elements 1,2,3 in groups of (n-m) ;

(iii) this expansion is unique.

Theorem 1.7. (Weierstrass) If the sequence $\{f_n\}_{n=1}^{\infty}$ of left-(k)-monogenic functions in Ω, converges uniformly on the compact subsets of Ω to a function f, then f is left-(k)-monogenic in Ω.

§2. THE CONTINUATION OF FUNCTIONS TO (k)-MONOGENIC FUNCTIONS

Theorem 2.1. If $g : \mathbb{R}^4 \to \mathcal{H}$ is (k-1)-times continuously differentiable on the boundary ∂B of the sphere B(0,R), and f is defined in $\overset{\circ}{B}$ by

$$f(x) = \frac{1}{\omega_4} \int_{\partial B(0,R)} \sum_{j=0}^{k-1} (-1)^j \frac{\overline{u}-\overline{x}}{\rho^4} \cdot \frac{(u_0-x_0)^j}{j!} \cdot d\sigma_u \cdot D^j g(u) \qquad (2.1)$$

then, (i) f is left-(k)-monogenic in $\overset{\circ}{B}$;

(ii) $\overset{\circ}{B}{}^*_1 = \overset{\circ}{B}(0,(\sqrt{2}-1)R)$ is an open neighbourhood of 0 wherein f can be expanded into its 2nd Taylor series, the coefficients of which are given by

$$\left. \frac{\partial^{n-m} D^m f}{\partial x_{\ell_1} \cdots \partial x_{\ell_{n-m}}} \right|_{x=0} = \frac{1}{\omega_4} \int_{\partial B(0,R)} \left. \frac{\partial^{n-m}}{\partial x_{\ell_1} \cdots \partial x_{\ell_{n-m}}} \right|_{x=0} \left(\frac{\overline{u}-\overline{x}}{\rho^4}\right) \cdot \sum_{j=m}^{k-1} \frac{(-u_0)^{j-m}}{(j-m)!} \cdot d\sigma_u \cdot D^j g(u) .$$

Proof. (i) Differentiation under the integral sign in (2.1) being allowed, we have : $f \in C_\infty(\overset{\circ}{B})$ and

$$D^k f(x) = \int_{\partial B} \sum_{j=1}^{k-1} (-1)^j D^k_x[g_{j+1}(u-x)] \cdot d\sigma_u \cdot D^j g(u) = 0 \text{ in } \overset{\circ}{B}, \text{ because of}$$

$D^k_x[g_{j+1}(u-x)] = 0$ as long as $x \neq u$ and $j+1 \leq k$.

(ii) It is known that $\frac{1}{\rho} = \sum_{n=0}^{\infty} \frac{1}{n!}[\sum_{\alpha=0}^{3} x_\alpha \frac{\partial}{\partial x_\alpha}]^n (\frac{1}{\rho})$ for all $u \in \partial B$ and $x \in \mathring{B}_1^*$ (see e.g. [5], p.136), and so

$$\frac{\overline{u}-\overline{x}}{\rho^4} = \sum_{n=0}^{\infty} \frac{1}{n!} \sum_{\ell_1,\ldots,\ell_n=0}^{3} x_{\ell_1}\ldots x_{\ell_n} \frac{\partial^n}{\partial x_{\ell_1}\ldots\partial x_{\ell_n}}\Big|_{x=0} (\frac{\overline{u}-\overline{x}}{\rho^4}),\ u \in \partial B,\ x \in \mathring{B}_1^*.$$

As $\frac{\overline{u}-\overline{x}}{\rho^4}$ is regular for $u \neq x$, we obtain in view of Theorem 1.6.

$$\frac{\overline{u}-\overline{x}}{\rho^4} = \sum_{n=0}^{\infty} \sum_{(\ell_1,\ldots,\ell_n)} p^{(0)}_{\ell_1\ldots\ell_n}(x).U^{(0)}_{\ell_1\ldots\ell_n}(u) \quad , \quad x \in \mathring{B}_1^* \text{ and } u \in \partial B \tag{2.2}$$

where the functions $U^{(0)}_{\ell_1\ldots\ell_n}(u) = \frac{\partial^n}{\partial x_{\ell_1}\ldots\partial x_{\ell_n}}\Big|_{x=0} (\frac{\overline{u}-\overline{x}}{\rho^4})$ are two-sided-regular in co{0}.

Substitution of the uniformly convergent power series (2.2) into (2.1) yields

$$f(x) = \sum_{n=0}^{\infty} \sum_{m=0}^{k-1} \sum_{(\ell_1,\ldots,\ell_{n-m})} \frac{x_0^m}{m!}.p^{(0)}_{\ell_1\ldots\ell_{n-m}}(x).a^{(m)}_{\ell_1\ldots\ell_{n-m}} \quad , \quad x \in \mathring{B}_1^*$$

where, in view of the uniquess of the 2nd Taylor expansion,

$$a^{(m)}_{\ell_1\ldots\ell_{n-m}} = \frac{1}{\omega_4} \int_{\partial B} U^{(0)}_{\ell_1\ldots\ell_{n-m}}(u). \sum_{j=m}^{k-1} \frac{(-u_0)^{j-m}}{(j-m)!} d\sigma_u D^j(u) = \frac{\partial^{n-m} D^m f}{\partial x_{\ell_1}\ldots\partial x_{\ell_{n-m}}}\Big|_{x=0}$$

Q.E.D.

In an analogous way one can prove

<u>Theorem 2.2.</u> If g is (k-1)-times continuously differentiable on the boundary ∂B of the sphere $B(0,R)$, and f' is defined in $co\overline{B}$ by

$$f'(x) = \frac{1}{\omega_4} \int_{\partial B} \sum_{j=0}^{k-1} (-1)^j \frac{\overline{u}-\overline{x}}{\rho^4}.d\sigma_u.\frac{(u_0-x_0)^j}{j!}.D^j g(u) \quad ,$$

then

(i) f' is left-(k)-monogenic in $co\overline{B}$;

(ii) in $co\overline{B}_2^* = co\overline{B}(0,(\sqrt{2}+1)R)$ f' can be expanded into

$$f'(x) = \sum_{n=0}^{\infty} \sum_{m=0}^{k-1} \sum_{(\ell_1,\ldots,\ell_{n-m})} \frac{x_0^m}{m!}.X^{(0)}_{\ell_1\ldots\ell_{n-m}}(x).b^{(m)}_{\ell_1\ldots\ell_{n-m}}$$

where the $X^{(0)}_{\ell_1\ldots\ell_{n-m}}(x) = \frac{\partial^{n-m}}{\partial u_{\ell_1}\ldots\partial u_{\ell_{n-m}}}\Big|_{u=0} (\frac{\overline{u}-\overline{x}}{\rho^4})$ are two-sided-regular in co{0}, and

$$b^{(m)}_{\ell_1\ldots\ell_{n-m}} = \frac{1}{\omega_4} \int_{\partial B} p^{(0)}_{\ell_1\ldots\ell_{n-m}}(u). \sum_{j=m}^{k-1} \frac{(-u_0)^{j-m}}{(j-m)!}.d\sigma_u.D^j g(u).$$

§3. LAURENT'S EXPANSION

Consider the open set $G = \mathring{B}(0,R)\setminus\overline{B}(0,r)$ $(r<R)$ which satisfies the following conditions (S) : there exists an open subset $G_1 = \mathring{B}(0,R_1)\setminus\overline{B}(0,r_1)$, with $0 < r < r_1 < R_1 < R$, so that $(\sqrt{2}+1)r_1 = r_1' < R_1' = (\sqrt{2}-1)R_1$. If we put $B_1 = B(0,R_1)$, $B_2 = B(0,r_1)$, $B_1' = B(0,R_1')$, $B_2' = B(0,r_1')$, then we call $G^* = \mathring{B}_1'\setminus\overline{B}_2'$ a Laurent domain associated with G.

<u>Theorem 3.1</u>. If f is left-(k)-monogenic in an open annulus G which satisfies the conditions (S), then in an associated Laurent domain G^*,

$$f(x) = \sum_{n=0}^{\infty}\sum_{m=0}^{k-1}\sum_{(\ell_1,\ldots,\ell_{n-m})}\frac{x_0^m}{m!}.P^{(0)}_{\ell_1\ldots\ell_{n-m}}(x).a^{(m)}_{\ell_1\ldots\ell_{n-m}}$$

$$-\sum_{n=0}^{\infty}\sum_{m=0}^{k-1}\sum_{(\ell_1,\ldots,\ell_{n-m})}\frac{x_0^m}{m!}.X^{(0)}_{\ell_1\ldots\ell_{n-m}}(x).b^{(m)}_{\ell_1\ldots\ell_{n-m}}$$

where $$\left\{{a \atop b}\right\}^{(m)}_{\ell_1\ldots\ell_{n-m}} = \frac{1}{\omega_4}\int_{\partial B(0,R^*)}\left\{{U \atop P}\right\}^{(0)}_{\ell_1\ldots\ell_{n-m}}(u).d\sigma_u.\sum_{j=m}^{k-1}\frac{(-u_0)^{j-m}}{(j-m)!}D^j f(u)$$

and R^* is choosen so that $(\sqrt{2}+1)r_1 < R^* < (\sqrt{2}-1)R_1$.

<u>Proof</u>. In view of Cauchy's formula (Theorem 1.3) we have for all $x \in G = \mathring{B}_1\setminus\overline{B}_2$, $f(x) = f_1(x)+f_2(x)$, with

$$f_i(x) = (-1)^{i-1}\int_{\partial B_i}\sum_{j=0}^{k-1}(-1)^j g_{j+1}(u-x).d\sigma_u.D^j f(u) \qquad (i=1,2).$$

Now Theorems 2.1 and 2.2 enable us to develop f_1 and f_2, both left-(k)-monogenic respectively in $\mathring{B}_1$ and $co\overline{B}_2$, in the following way :

$$\left\{{f_1(x) \atop f_2(x)}\right\} = \left\{{1 \atop -1}\right\}.\sum_{n=0}^{\infty}\sum_{m=0}^{k-1}\sum_{(\ell_1,\ldots,\ell_{n-m})}\frac{x_0^m}{m!}\left\{{P \atop X}\right\}^{(0)}_{\ell_1\ldots\ell_{n-m}}(x).\left\{{a \atop b}\right\}^{(m)}_{\ell_1\ldots\ell_{n-m}}$$

If we choose R^* so that $r_1' < R^* < R_1'$, and notice that $U^{(0)}_{\ell_1\ldots\ell_n}(u)$ is right-regular in $co\{0\}$, $P^{(0)}_{\ell_1\ldots\ell_n}(u)$ is right-regular in $\mathbb{R}^4$ and

$g(u) = \sum_{j=m}^{k-1}\frac{(-u_0)^{j-m}}{(j-m)!}.D^j f(u)$, then the constants $a^{(m)}_{\ell_1\ldots\ell_{n-m}}$ and

$b^{(m)}_{\ell_1\ldots\ell_{n-m}}$ can be expressed by $\left\{{a \atop b}\right\}^{(m)}_{\ell_1\ldots\ell_n} = \frac{1}{\omega_4}\int_{\partial B(0,R^*)}\left\{{U \atop P}\right\}^{(0)}_{\ell_1\ldots\ell_n} d\sigma_u g(u)$

in virtue of Stokes' formula (see Theorem 1.2). Q.E.D.

Theorem 3.2. If f is left-(k)-monogenic in an open annulus which satisfies the conditions (S), then the Laurent expansion of f in an associated Laurent domain may be "D-derived" term by term.

Outline of the proof : It is shown that the formal term by term D-derivative of the Laurent expansion, coincides with the Laurent series of Df, which is left-(k-1)-monogenic in the same annulus, and thus converges itself uniformly to Df in the associated Laurent domain. Q.E.D.

Theorem 3.3. The Laurent expansion of a left-(k)-monogenic function in an associated Laurent domain is unique.

Outline of the proof : Given two series which converge uniformly, together with the first (k-1) D-derived series, in $\mathring{B}(0,R)$, respectively $co\overline{B}(0,r)$:

$$\left\{ \begin{matrix} f_1'(x) \\ -f_2'(x) \end{matrix} \right\} = \sum_{n=0}^{\infty} \sum_{(\ell_1,\dots,\ell_n)} \left\{ \begin{matrix} P \\ X \end{matrix} \right\}^{(0)}_{\ell_1\dots\ell_n}(x) \cdot \sum_{n=0}^{k-1} \frac{x_0^m}{m!} \left\{ \begin{matrix} a' \\ b' \end{matrix} \right\}^{(m)}_{\ell_1\dots\ell_n} ,$$

we put $f = f_1'+f_2'$, so obtaining a left-(k)-monogenic function in $G = \mathring{B}(0,R)\setminus\overline{B}(0,r)$. From Theorem 3.1 it follows that f can be written as $f = f_1+f_2$, f_1 and f_2 being series of the same form as f_1' and f_2'. Straightforward computations show that the occuring coefficients in the corresponding series are equal to each other, so $f_1 = f_1'$ and $f_2 = f_2'$ in G. Q.E.D.

§4. DEFINITIONS OF NON-(k)-MONOGENIC POINTS

Definition 4.1. A point $a \in \mathbb{R}^4$ is called a non-left-(k)-monogenic point of f iff there is no open neighbourhood of a in which f is left-(k)-monogenic.

Definition 4.2. The non-left-(k)-monogenic point a of f is called isolated iff there exist r and R $(0<r<R<+\infty)$ so that f is left-(k)-monogenic in $\mathring{B}(a,R)\setminus\overline{B}(a,r)$, how small r may be.

Suppose now that a is an isolated non-left-(k)-monogenic point of f. Then there exist r and R so that f is left-(k)-monogenic in $G = \mathring{B}(a,R)\setminus\overline{B}(a,r)$, and in an associated Laurent domain is $f(x) = f_1(x)+f_2(x)$ (see Theorem 3.1). We call f_1 and f_2 the first, respectively the second series of f.

Definition 4.3. A point a is a left-(k)-pole of order m of f iff

(i) a is an isolated non-left-(k)-monogenic point of f ;

(ii) the second series of f breaks off from $n \geq m$; this finite second series is called the principal part of a and denoted by $G(\frac{1}{x-a})$.

Definition 4.4. An isolated non-left-(k)-monogenic point of f is called essential iff the second series of f does not break off.

§5. LEFT-(k)-MEROMORPHIC FUNCTIONS AND MITTAG-LEFFLER'S THEOREM

Definition 5.1. The function f is called (k)-meromorphic in Ω iff there exists a subset $S \subset \Omega$ so that

(i) S does not contain an accumulation point in Ω ;
(ii) f is (k)-monogenic in $\Omega \setminus S$;
(iii) f has a (k)-pole in each point of S.

Note that (i) implies that S is at most countable.

Now consider a left-(k)-meromorphic function f in $\mathbb{R}^4$.

CASE A : f has no left-(k)-poles.
Then f is left-(k)-monogenic in the whole $\mathbb{R}^4$; we call it a left-(k)-entire function.

Theorem 5.1. If f is left-(k)-entire, then in $\mathbb{R}^4$,

$$f(x) = \sum_{n=0}^{\infty} \sum_{(\ell_1,\dots,\ell_n)} p^{(0)}_{\ell_1\dots\ell_n}(x) \cdot \sum_{s=0}^{k-1} \frac{x_0^s}{s!} a^{(s)}_{\ell_1\dots\ell_n} .$$

Proof. Take $R > 0$; f is left-(k)-monogenic in $\mathring{B}(0,R)$, and in $\mathring{B}(0,(\sqrt{2}-1)R)$ is

$$f(x) = \sum_{n=0}^{\infty} \sum_{(\ell_1,\dots,\ell_n)} p^{(0)}_{\ell_1\dots\ell_n}(x) \cdot \sum_{s=0}^{k-1} \frac{x_0^s}{s!} a^{(s)}_{\ell_1\dots\ell_n} .$$

As the coefficients $a^{(s)}_{\ell_1\dots\ell_n}$ are independent of R, and R is arbitrary, this expansion holds in $\mathbb{R}^4$. Q.E.D.

CASE B : f has a finite number of left-(k)-poles $a_1,\dots,a_i$.
If $G_j(\frac{1}{x-a_j})$ $(j=1,\dots,i)$ is the principal part of a_j, then $f - \sum_{j=1}^{i} G_j$ is a left-(k)-entire function. So in $\mathbb{R}^4$,

$$f(x) = \sum_{n=0}^{\infty} \sum_{(\ell_1,\dots,\ell_n)} p^{(0)}_{\ell_1\dots\ell_n}(x) \sum_{s=0}^{k-1} \frac{x_0^s}{s!} a^{(s)}_{\ell_1\dots\ell_n}$$

$$- \sum_{j=1}^{i} \sum_{n=0}^{m_j-1} \sum_{(\ell_1,\dots,\ell_n)} X^{(aj)}_{\ell_1\dots\ell_n}(x) \cdot \sum_{s=0}^{k-1} \frac{(x_0-a_{j,0})^s}{s!} b^{(s)(j)}_{\ell_1\dots\ell_n} .$$

CASE C : f has a countable set of left-(k)-poles.

Theorem 5.2. (Mittag-Leffler) Let $S = \{a_i \in \mathbb{R}^4 : i \in \mathbb{N}\}$ be given so that S has no accumulation point in $\mathbb{R}^4$ and so that with each $a_i \in S$ there corresponds a $m(a_i) \in \mathbb{N}$ and a function

$$G_i\left(\frac{1}{x-a_i}\right) = -\sum_{n=0}^{m(a_i)-1} \sum_{(\ell_1,\dots,\ell_n)} X^{(a_i)}_{\ell_1\dots\ell_n}(x) \cdot \sum_{s=0}^{k-1} \frac{(x_0-a_{i,0})^s}{s!} \cdot b^{(s)(i)}_{\ell_1\dots\ell_n}.$$

Then there exists at least one function which is left-(k)-meromorphic in $\mathbb{R}^4$ and has a_i as a left-(k)-pole of order $m(a_i)$ with principal part $G_i\left(\frac{1}{x-a_i}\right)$, for all $i \in \mathbb{N}$.

Proof. We order the elements a_i of S so that their euclidean norm form a non-decreasing sequence of positive numbers : $0 \leq |a_1| \leq \dots \leq |a_n| \leq \dots$ Choose fixed α and α_1 so that $0 < \alpha, \alpha_1 < 1$. The principal part $G_i\left(\frac{1}{x-a_i}\right)$ is certainly left-(k)-monogenic in $\mathring{B}_i(0,\alpha|a_i|)$, so in $\mathring{B}^*_i = \mathring{B}_i(0,\alpha_1\alpha(\sqrt{2}-1)a_i)$ holds $G_i\left(\frac{1}{x-a_i}\right) = \sum_{n=0}^{\infty} g_n^{(i)}$ with

$$g_n^{(i)} = \sum_{(\ell_1,\dots,\ell_n)} P^{(0)}_{\ell_1\dots\ell_n}(x) \cdot \sum_{s=0}^{k-1} \frac{x_0^s}{s!} a^{(s)(i)}_{\ell_1\dots\ell_n}.$$ So for each $\varepsilon > 0$ there

exists a $N_i(\varepsilon) \in \mathbb{N}$ so that $|G_i - V_{i,N_i}| < \frac{\varepsilon}{2^i}$ for all $x \in \mathring{B}^*_i$, if we put

$$V_{i,N_i} = \sum_{n=0}^{N_i} g_n^{(i)} \quad \text{(a polynomial of degree } N_i+k-1).$$

Take now $R > 0$ arbitrarily and associate with $\overline{B}(0,R)$ the sphere $\overline{B}(0,R')$, where $R' = \alpha_1^{-1}\alpha^{-1}(\sqrt{2}+1)R$. In the compact set $\overline{B}(0,R')$ at most a finite number of a_i occur, say $a_1,a_2,\dots,a_t$. For each $a_m \in co\overline{B}(0,R')$ is $\alpha_1\alpha(\sqrt{2}-1)|a_m| > R$ and so $\mathring{B}^*_m \supset \mathring{B}(0,R)$ for $m > t$. Now define two functions:

1°) $f_1(x) = \sum_{i=1}^{t} [V_{i,N_i} - G_i]$. This function f_1 is left-(k)-monogenic in $\mathring{B}(0,R)$ except in $\{a_i : 0 \leq i \leq t\} \subset \mathring{B}(0,R)$, which consists of left-(k)-poles of f_1 with respective principal parts $G_i\left(\frac{1}{x-a_i}\right)$.

2°) $f_2(x) = \sum_{m=t+1}^{\infty} [V_{m,N_m} - G_m]$. As $\mathring{B}^*_m \supset \mathring{B}(0,R)$ for $m > t$, we have $|V_{m,N_m} - G_m| < \frac{\varepsilon}{2^m}$ for all $x \in \mathring{B}(0,R)$, $m = t, t+1, \dots$

The series $\sum_{m=t+1}^{\infty} \frac{\varepsilon}{2^m}$ converges, hence $\sum_{m=t+1}^{\infty} (V_{m,N_m} - G_m)$ will converge uniformly in $\mathring{B}(0,R)$. Moreover, each term $V_{m,N_m} - G_m$ being left-(k)-monogenic in $\mathring{B}(0,R)$, we conclude with Weierstrass (Theorem 1.7) that f_2 is also left-(k)-monogenic in $\mathring{B}(0,R)$. Consequently, the function

$$f_1+f_2 = \sum_{m=1}^{\infty} [V_{m,N_m} - G_m]$$

is left-(k)-monogenic in $\mathring{B}(0,R)$ except in $\{a_i : 0 \leq i \leq t\} \subset \mathring{B}(0,R)$ which consists of left-(k)-poles with respective principal parts $G_i(\frac{1}{x-a_i})$. Let $R \to +\infty$, then we obtain a function $f = f_1+f_2$, which is left-(k)-meromorphic in $\mathbb{R}^4$ and has the prescribed points a_i and the prescribed functions G_i, as (k)-poles and principal parts respectively.

Q.E.D.

Remark. If to the constructed function f_1+f_2 an arbitrary left-(k)-entire function is added, the new function remains satisfying the prescribed conditions. The structure of the most general left-(k)-meromorphic function in $\mathbb{R}^4$ with prescribed (k)-poles and principal parts is thus

$$f = f_1 + f_2 + \sum_{n=0}^{\infty} \sum_{(\ell_1,\ldots,\ell_n)} p^{(0)}_{\ell_1\ldots\ell_n}(x) . \sum_{s=0}^{k-1} \frac{x_0^s}{s!} a^{(s)}_{\ell_1\ldots\ell_n} .$$

§6. REMOVABLE SINGULARITIES

Definition 6.1. Let f be left-(k)-monogenic in $\Omega\setminus\{a\}$. The point a is called a removable non-left-(k)-monogenic point of f if there exists an extension h of f so that h is left-(k)-monogenic in Ω.

Theorem 6.1. Let f be (k)-monogenic in $\Omega\setminus\{0\}$. If $|D^j f(x)| < M_j$, $j = 0,1,\ldots,k-1$ (all $M_j > 0$) for all $x \in B(0,R)\setminus\{0\} \subset \Omega$, then 0 is a removable non left-(k)-monogenic point.

Proof. Choose $0 < r < R_1 < R$ so that with $G = B(0,R_1)\setminus\overline{B}(0,r)$ a Laurent domain G^* can be associated ; then in G^*

$$f = \sum_{n=0}^{\infty} \sum_{(\ell_1,\ldots,\ell_n)} p^{(0)}_{\ell_1\ldots\ell_n}(x) \sum_{m=0}^{k-1} \frac{x_0^m}{m!} a^{(m)}_{\ell_1\ldots\ell_n}$$

$$- \sum_{n=0} \sum_{(\ell_1,\ldots,\ell_n)} X^{(0)}_{\ell_1\ldots\ell_n}(x) \sum_{m=0}^{k-1} \frac{x_0^m}{m!} b^{(m)}_{\ell_1\ldots\ell_n}$$

where a.o. $b^{(m)}_{\ell_1\ldots\ell_n} = \frac{1}{\omega_4} \int_{\partial B(0,r)} p^{(0)}_{\ell_1\ldots\ell_n}(u) . d\sigma_u . \sum_{j=m}^{k-1} \frac{(-u_0)^{j-m}}{(j-m)!} . D^j f(u)$.

As the $p^{(0)}_{\ell_1\ldots\ell_n}$ are homogeneous polynomials of degree n, they can be written as $p^{(0)}_{\ell_1\ldots\ell_n}(u) = \sum_{\alpha=0}^{3} P^{(\alpha)}_{\ell_1\ldots\ell_n}(u) e_\alpha$, the $P^{(\alpha)}_{\ell_1\ldots\ell_n}$ being real valued homogeneous polynomials of degree n. So, taking into account that $d\hat{u}_\alpha = (-1)^\alpha n_\alpha dS$ (n_α : the α-th component of the external surface normal ; dS : the elementary surface area) and $n_\alpha = u_\alpha/r$ for the sphere B(0,r), we get :

$$|b^{(m)}_{\ell_1\ldots\ell_n}| \leq \frac{1}{\omega_4} \sum_{\alpha,\beta,\gamma=0}^{3} \int_{\partial B(0,r)} |P^{(\alpha)}_{\ell_1\ldots\ell_n}(u)| . \sum_{j=m}^{k-1} \frac{u_0^{j-m}}{(j-m)!} |(D^j f)_\beta(u)| \frac{|u_\gamma|}{r} dS$$

The functions $P^{(\alpha)}_{\ell_1\ldots\ell_n}$ are continuous in $\mathbb{R}^4$; hence there exist constants $C^{(\alpha)}_{\ell_1\ldots\ell_n}$ so that $|P^{(\alpha)}_{\ell_1\ldots\ell_n}(u)| \leq C^{(\alpha)}_{\ell_1\ldots\ell_n}$, for all $u \in \overline{B}(0,R_1)$.

Put $C_{\ell_1\ldots\ell_n} = \max\limits_{\alpha} C^{(\alpha)}_{\ell_1\ldots\ell_n}$. Further more it follows from the assumptions made that $|(D^j f)_\beta(u)| < M_j$ for all $u \in \partial B(0,r)$ and all j and β. Hence,

$$|b^{(m)}_{\ell_1\ldots\ell_n}| \leq 4^3 . C_{\ell_1\ldots\ell_n} . r^3 . \sum_{j=m}^{k-1} \frac{r^{j-m}}{(j-m)!} . M_j .$$

As r can be choosen arbitrarily small, it follows that $|b^{(m)}_{\ell_1\ldots\ell_n}| = 0$ for all $(\ell_1,\ldots,\ell_n) \in \{1,2,3\}^n$ and all m = 0,1,...,k-1. Consequently in G* :

$$f(x) = \sum_{n=0}^{\infty} \sum_{(\ell_1,\ldots,\ell_n)} P^{(0)}_{\ell_1\ldots\ell_n}(x) . \sum_{m=0}^{k-1} \frac{x_0^m}{m!} a^{(m)}_{\ell_1\ldots\ell_n} .$$

Now define the function h as follows : $h(0) = a_0^{(0)}$ and h(x) = f(x) for $x \in \Omega\setminus\{0\}$. It is clear that h is an extension of f, which is moreover left-(k)-monogenic in Ω.

Q.E.D.

§7. RESIDUE THEORY

Suppose that O is a left-(k)-pole of order m (m>0) or an isolated essential-non-left-(k)-monogenic point for the function f. Then the first term in the second series of Laurent's expansion round the origin is $-X_0^{(0)}(x).b_0^{(0)}$ or $\frac{\overline{x}}{\rho^4}.b_0^{(0)}$.

<u>Definition 7.1</u>. $b_0^{(0)}$ is called the residue of f at the non-left-(k)-monogenic point O. (Notation : $Res(O) = b_0^{(0)}$).

<u>Theorem 7.1</u>. If O is an isolated non-left-(k)-monogenic point of f, then

$$Res(O) = \frac{1}{\omega_4} \int_{\partial B} d\sigma_u . \sum_{j=0}^{k-1} \frac{(-u_0)^j}{j!} D^j f(u) .$$

Proof. This formula follows readily from the general expression of the coefficients $b^{(m)}_{\ell_1\ldots\ell_n}$ in the Laurent series of f.

Q.E.D.

Theorem 7.2. If 0 is a left-(k)-pole of the first order of f, then

$$\mathrm{Res}(0) = \lim_{x\to 0} (\rho_x^2 . x . f(x)) .$$

Proof. In the associated Laurent domain we have

$$f(x) = \sum_{n=0}^{\infty} \sum_{(\ell_1,\ldots,\ell_n)} P^{(0)}_{\ell_1\ldots\ell_n}(x) \sum_{m=0}^{k-1} \frac{x_0^m}{m!} a^{(m)}_{\ell_1\ldots\ell_n} + \frac{\bar{x}}{\rho^4} . \sum_{m=0}^{k-1} \frac{x_0^m}{m!} b_0^{(m)} .$$

Hence, $\rho_x^2 . x . f(x) = \rho_x^2 . x . f_1(x) + b_0^{(0)} + \sum_{m=1}^{k-1} \frac{x_0^m}{m!} b_0^{(m)}$, from which it follows that $\lim_{x\to 0} \rho_x^2 . x . f(x) = b_0^{(0)} = \mathrm{Res}(0)$.

Q.E.D.

Theorem 7.3. Let S be a four dimensional, compact, differentiable, oriented manifold-with-boundary, contained in Ω. Assume f to be left-(k)-monogenic in $\Omega \setminus \{a_1,\ldots,a_t\}$, with $\{a_1,\ldots,a_t\} \subset \mathring{S}$. Then

$$\int_{\partial S} d\sigma_x \sum_{j=0}^{k-1} \frac{(-x_0)^j}{j!} D^j f(x) = \omega_4 . \sum_{s=1}^{t} \mathrm{Res}(a_s) .$$

Proof. Since all a_s $(s = 1,\ldots,t)$ belong to $\mathring{S}$, there exist spheres $B(a_s,r_s)$ so that $\bar{B}(a_s,r_s) \subset \mathring{S}$ for all $s = 1,\ldots,t$ and $\bar{B}(a_s,r_s) \cap \bar{B}(a_{s'},r_{s'})$ $= \phi$ for $s \neq s'$. The function f is left-(k)-monogenic in $\mathring{S} \setminus [\bigcup_{s=1}^{t} \bar{B}(a_s,r_s)]$, hence

$$g = \sum_{j=0}^{k-1} \frac{(-x_0)^j}{j!} D^j f$$

is left-(k)-monogenic in the same region.

So $\int_{\partial S} d\sigma g = \sum_{s=1}^{t} \int_{\partial B_s} d\sigma g$ (Theorem 1.2). But by definition is

$$\int_{\partial B_s} d\sigma \sum_{j=0}^{k-1} \frac{(-x_0)^j}{j!} D^j f(x) = \omega_4 . \mathrm{Res}(a_s) \qquad (s = 1,\ldots,t) .$$

Consequently

$$\int_{\partial S} d\sigma \sum_{j=0}^{k-1} \frac{(-x_0)^j}{j!} D^j f(x) = \omega_4 . \sum_{s=1}^{t} \mathrm{Res}(a_s) .$$

Q.E.D.

REFERENCES

[1] F. BRACKX, On (k)-monogenic functions of a quaternion variable (to appear)

[2] F. BRACKX, On the space of left-(k)-monogenic functions of a quaternion variable and an associated quaternion Hilbert space with reproducing kernel (to appear)

[3] R. FUETER, Über die analytische Darstellung der regulären Funktionen einer Quaternionenvariablen, Comm. Math. Helv., 8 (1935), 371-378

[4] H.G. GARNIR, Fonctions de variables réelles, II, Gauthier-Villars, Paris, 1965

[5] O. KELLOG, Foundations of potential theory, Springer, Berlin, 1929

ACKNOWLEDGEMENT

I wish to thank Prof. Dr. R. Delanghe for his aid and guidance in connection with this work.

ON THE THEORY OF LINIAR EQUATIONS WITH SPATIAL DERIVATIVES

Mariana Coroi-Nedelcu
Polytechnical Institut
Bucharest

Starting from the notion of areolar derivative given by Dimitrie Pompeiu [22] in 1912, Gr.C.Moisil and N.Teodorescu [4],[5],[25] - [30] have generalized it to R^m.

Thus, in the matrix writing, we say that a matrix field $\varphi = \{\varphi^i_j\} \in C^0(\Delta \subset R^m)$ is of class $\Lambda^1_\gamma(\Delta)$ it there exists a matrix of summable functions Φ on Δ, so that

$$\int_\sigma (\vec{n}_M \cdot \gamma)\varphi(M)d\sigma_M + \int_\omega \Phi(Q)d\omega_Q = 0$$

for any regular pair $(\sigma,\omega) \subset \Delta \subset R^m$, where $\vec{n}_M$ is the scalar product between the interior normal to the hypersurface σ limiting the volume ω and γ the vector of constant square matrices with 2^s lines and columns, $\gamma^1,\ldots,\gamma^m$ which have the property that

$$\gamma^i\bar{\gamma}^j + \gamma^j\bar{\gamma}^i = 2\delta^{ij}e, \qquad i,j = 1,\ldots,m$$

e is a unit matrix and $\bar{\gamma}^i$ the transposed matrix of γ^i.

Matrix Φ was called spatial derivative (in a wide sense) of φ and was denoted $\Phi = D\varphi$. In this casè is given the representation formula Pompeiu-Teodorescu-Moisil:

$$\frac{1}{s_m}\int_\Sigma \frac{(\vec{MP}\cdot\bar{\gamma})}{MP^m}(\vec{n}_M\cdot\gamma)\varphi(M)d\sigma_M + \frac{1}{s_m}\int_\Omega \frac{(\vec{QP}\cdot\bar{\gamma})}{QP^m}D\varphi(Q)d\omega_Q = \begin{cases} \varphi(P), & P\in\Omega\subset\Delta \\ 0, & P\notin\Omega\cup\Delta \end{cases}$$

where s_m is the area of the unite hypersphere in R^m, Σ is the hypersurface with a continuous normal in each point, which bounds the volume $\Omega\subset\Delta$, $\vec{n}_M$ is the interior normal to Σ in M, and $\bar{\gamma}$ the vector of components the matrix $\bar{\gamma}^1,\bar{\gamma}^2,\ldots,\bar{\gamma}^m$. If $\varphi\in C^1(\Delta)$ then

$$D\varphi = \sum_{i=1}^m \gamma^i\frac{\partial\varphi}{\partial x^i}, \quad \bar{D}\varphi = \sum_{i=1}^m \bar{\gamma}^i\frac{\partial\varphi}{\partial x^i}, \quad \text{and } D\bar{D} = \bar{D}D = e\Delta$$

and D is called the spatial derivative in a restricted sense. The matrix of functions $\varphi \in C^1(\Delta)$, which satisfy the partial differential equations system $D\varphi = 0$ in Δ, shall admit a representation by an surface integral in Δ and they define the holomorphic functions in space.

The solutions of the partial differential equations'system $D^n\psi(P) = \varphi(P)$ with φ holomorphic in space, define the spatial polynomial of degree n. The representation of such a polynomial and its extension to series has been given in [7] - [12], [15] - [17]. These have put into evidence the existence of some matrices $\xi^k(M,P)$ given in [6], [9], [10] which have the property $D^{k+1}\frac{(\overrightarrow{MP}\cdot\bar{\gamma})}{MP^m}\xi^k(M,P) = D^k\frac{(\overrightarrow{MP}\cdot\bar{\gamma})}{MP^m}\xi^{k-1}(M,P) = D\frac{(\overrightarrow{MP}\cdot\bar{\gamma})}{MP^m} = 0$ for $P \neq M$. If m is uneven, ξ^k presents algebraic singularity, for m even presents algebraic - logarithmic singularity and $\frac{1}{s_m}\frac{(\overrightarrow{MP}\cdot\bar{\gamma})}{MP^m}\xi^k(M,P)$ represents an elementary solution for the operator D^{k+1} and in particular for polyharmonic equations'systems. The determination of operators $\xi^k(M,P)$ was made starting from the elemetary polyharmonic functions supposing that

$$\bar{\gamma}^i\bar{\gamma}^j + \bar{\gamma}^j\bar{\gamma}^i = \gamma^i\gamma^j + \gamma^j\gamma^i, \qquad i,j = 1,2,\dots,m.$$

They have permitted the extension to the space of the elementary functions (exponentials, trigometric functions).

Written problems with areolar or spatial derivative have been studied also by I.N.Vecua [31], A.V.Bitadze [1], B.V.Bojarski [2], V.Iftimie [3], J.Ridder [23], Marcel Roşculeţ [24].

In [5], Muravèv has used an operator alike the spatial derivative, defined on commutative algebra and has constructed a theory of linear equations defined by this operator, by analogy with that of differential equations.

Let in $\Delta \subset R^m$ be the system of partial differential equations, defined by partial differential equation of order n

$$L\varphi = D^n\varphi(P) + \alpha_1(P)D^{n-1}\varphi(P) + \dots + \alpha_{n-1}(P)D\varphi(P) + \alpha_n(P)\varphi(P) = f(P) \qquad (1)$$

with $\alpha_i \in C^0(\Delta)$, i=1,...,n, matrices with 2s lines and 2s columns, with the non identic determinant null in Δ, and the matrix $f \in C^0(\Delta)$.

Let in Δ be the matrices of functions $\varphi_i(P)$, i=1,...,p of the form

$$\varphi_i(P) = k_i(P)c_i(P) \tag{2}$$

where the matrices k_i admit spatial derivatives until the order n in and the matrices of arbitrary functions $c_1,\dots,c_p \in C^1(\Delta)$ are supposed to satisfy the following system of equations with partial derivatives:

$$\sum_{i=1}^{p}\sum_{j=1}^{m} \gamma^i k_i(P)\frac{\partial c_i(P)}{\partial x^j}=0,\ \sum_{i=1}^{p}\sum_{j=1}^{m} \gamma^j D^{\ell} k_i(P)\frac{\partial c_i(P)}{\partial x^j}=0,\ \ell=1,2,\dots,n-1 \tag{3}$$

in any point $P\in\Delta$. The class of matrices $c_1,\dots,c_p$ shall be denoted with $(A^n_{k_1,\dots,k_p})$. A particular case is that of the constant matrices. We suppose that matrices φ_i, k_i, c_i have 2s lines and 2s columns, even if they are completed by zero.

We observe that <u>L is a linear operator with respect to matrices of the class</u> $(A^n_{k_1,\dots,k_p})$, that is:

$$L\Big(\sum_{i=1}^{p} k_i(P)c_i(P)\Big) = \sum_{i=1}^{p} L(k_i(P))\cdot c_i(P) \tag{4}$$

for any $c_1,\dots,c_p \in (A^n_{k_1,\dots,k_p})$.

Results immediatly the following:

Theorem 1. *If the nonhomogeneous equation (1) admits a solution* $\varphi_0(P)$, *for* $P\in\Delta$, *then the general solution of equation (1) is written as follows*

$$\varphi(P) = \Phi(P) + \varphi_0(P) \tag{5}$$

for any $P\in\Delta$ *with* $\Phi(P)$ *the general solution of the homogeneous equation (1).*

Theorem 2. *Let* $k_1(P),\dots,k_p(P)$ *be a system of solutions of the homogeneous equation (1). Then* $\sum_{i=1}^{p}\varphi_i(P)$ *with* $\varphi_i(P)$ *given by (2), is a solution of the homogeneous equation (1) depending of p matrices of functions of the class* $(A^n_{k_1,\dots,k_p})$ *in* Δ.

We shall call *the matrices of functions* $k_i(P)$, $i=1,\dots,p$ *linearly independent in* Δ, *if there exists no combination of the form*

$$\sum_{i=1}^{p} k_i(P)c_i(P) = 0$$

with $c_1,\ldots,c_p$ matrices of class $(A^n_{k_1,\ldots,k_p})$ only if all matrices c_i are null for any $P\in\Delta$ and are strictly linearly independent in Δ, if also

$$W(k_1,\ldots,k_p) = \begin{vmatrix} k_1 & k_2 & \ldots k_p \\ Dk_1 & Dk_2 & \ldots Dk_p \\ \vdots & \vdots & \vdots \\ D^{p-1}_{k_1} & D^{p-1}_{k_2} & \ldots D^{p-1}_{k_p} \end{vmatrix} \neq 0$$

in any point $P\in\Delta$.

The matrices $k_1(P),\ldots,k_p(P)$ form a fundamental system of solutions of homogeneous equation (1), if they are solution of the equations $L\varphi = 0$ in Δ and are strictly liniarly independent for any $P\in\Delta$.

In [18] is given that:

Theorem 3. The general solution of the homogeneous equation (1) $L\varphi = 0$, in Δ, can be written in the form

$$\Phi(P) = \sum_{i=1}^{n} k_i(P)c_i(P) \tag{6}$$

where $k_i(P)$, $i=1,\ldots,n$, form a fundamental system of solutions of the homogeneous equations (1), and $c_1,\ldots,c_n \in (A^n_{k_1,\ldots,k_n})$ in Δ.

A particular solution of the nonhomogeneous equation (1), $L\varphi = f$, is given by

$$\varphi_0 = \sum_{k=1}^{n} k_i(P)C_i(P) \tag{7}$$

where $C_1,\ldots,C_n \in (A^{n-1}_{k_1,\ldots,k_n})$ and are a particular solution of the equation

$$\sum_{j=1}^{m}\sum_{i=1}^{m} \gamma^j D^{n-1}k_i(P)\,\frac{\partial C_i(P)}{\partial x^j} = f(P) \tag{8}$$

Theorem 4. The integration of equation (1) is equivalent with the solution of the Fredholm integral equation

$$D^n\varphi(P) + T_{n_P} D^n\varphi = f(P) - \sum_{i=1}^{n} T_{i-1_P} h_i \tag{9}$$

where was denoted with

$$\begin{aligned}
T_{oP} &= \alpha_n(P)\\
T_{i-1P} &= \sum_{k=0}^{i-2} \alpha_{n-k}(P)\Pi_P^{i-1-k} + \alpha_{n-i+1}(P), \quad i=1,2,\dots,n-1\\
T_{nP} &= T_{n-1P}\Pi_P\\
\Pi_P\chi &= \frac{1}{s_m}\int_{\Omega} \frac{(\overrightarrow{QP}\cdot\overline{\gamma})}{QP^m}\chi(Q)d\omega_Q
\end{aligned} \tag{10}$$

and the volume $\Omega \subset \Delta$, $P \in \Omega$, $\chi \in C^o(\Delta)$, and the matrices of arbitrary functions $h_i(P)$ are holomorphic in the space thus satisfy $D\,h_i(P) = 0$, $i=1,\dots,n$.

Demonstration. Taking into account the notation (10), for $P \in \Omega \subset \Delta$, introducing the arbitrary holomorphic matrices $h_1(P),\dots,h_n(P)$, we obtain:

$$\begin{aligned}
D^n\varphi(P) &= D^n\varphi(P)\\
D^{n-1}\varphi(P) &= \Pi_P D^n\varphi + h_n(P)\\
D^{n-2}\varphi(P) &= \Pi_P D^{n-1}\varphi + h_{n-1}(P) = \Pi_P^2 D^n\varphi + \Pi_P h_n + h_{n-1}(P)\\
&\dots\dots\dots\dots\dots\dots\dots\dots\\
D^2\varphi(P) &= \Pi_P^{n-1} D^n\varphi + \Pi_P^{n-2} h_n + \dots + \Pi_P h_3 + h_2(P)\\
\varphi(P) &= \Pi_P^n D^n\varphi + \Pi_P^{n-1} h_n + \dots + \Pi_P h_2 + h_1(P).
\end{aligned} \tag{11}$$

Introducing in (1) is obtained (9).

Theorem 5. The solution of the liniar equation with spatial derivatives (1) with α_i, $f \in C^o(\Delta)$, $\bar{\Omega} \subset \Delta$, $i=1,2,\dots,n$, is given by the development in series absolutely and uniformly convergent in $\Omega_o \subset \Omega$

$$\begin{aligned}
\varphi(P) = &\left\{ e - \Pi_P^n\Big[e + \sum_{k=1}^{\infty} (-1)^k T_n^k\Big] T_o\right\} h_1 +\\
&+ \sum_{\ell=1}^{n-1}\Pi_P^{\ell}\left\{ e - \Pi^{n-\ell}\Big[e + \sum_{k=1}^{\infty} (-1)^k T_n^k\Big] T_{\ell}\right\} h_n +\\
&+ \Pi_P^n\left\{ e + \sum_{k=1}^{\infty} (-1)^k T_n^k\right\} f .
\end{aligned} \tag{12}$$

with $h_1(P),\dots,h_n(P)$ arbitrary holomorphic functions in Ω.

The solution is valid in $\mathcal{R}_0 \subset \mathcal{R}$ defined by condition $M_1 e^{MR} < 1$ with $M_1 = \max\{\|\alpha_1(P)\|, \ldots, \|\alpha_n(P)\|\}$, $R = \max_{Q,P \in \mathcal{R}_0} QP$ and M constant, the matrix's norm being considered the maximum of modules of its components in $\mathcal{R}_0$.

Demonstration. From theorem 4 results that the solutions of equation (1) are given by the solutions of equation (9). The unicity of the solution is obtained for $\|T_n\| < 1$, condition which defines the domain $\mathcal{R}_0$.

From [16] are obtained the increases

$$\|\Pi_P^n(e)\| \leq \frac{M^n R^n}{n!} \tag{13}$$

with M constant, $R = \max_{Q,P \in \mathcal{R}_0} QP$, $\mathcal{R}_0 \subset \mathcal{R}$. Let M given in the ennunciation of the theorem. From the expression of T_n from (10), results

$$\|T_n\| \leq M_1 \left\{ \frac{M^n R^n}{n!} + \ldots + \frac{M^2 R^2}{2!} + \frac{MR}{1} \right\} \leq M_1 e^{MR} < 1 \tag{14}$$

Supposing that α_i, i=1,...,n, are such as there exists a domain $\mathcal{R}_0$ defined by (14). Then, in $\mathcal{R}_0$ can be determined the solution of equation (9), because

$$D^n \varphi(P) = \left[e + \sum_{k=1}^{\infty} (-1)^k T_{n_P}^k \right] \left\{ f - \left[\sum_{i=1}^{n} T_{i-1} h_i \right] \right\}$$

and from the last relation (11), results (12).

The series (12) is absolutely and uniformly convergent because from (13), denoting $M_2 = \max_{P \in \mathcal{R}_0} |f_{ij}(P)|$ because $f = \{f_{ij}\} \in C^0(\bar{\mathcal{R}})$, we obtain

$$\|\Pi^n T_n^k f\| \leq M_1^R M_2 \|\Pi^n e\| \, \|(\Pi(e) + \Pi^2(e) + \ldots + \Pi^n(e))^k\| \leq$$

$$\leq M_1^R M_2 \frac{M^n R^n}{n!} \frac{n^k R^k M^{nk}}{k!}$$

and similar increases for the other terms of the series (12).

Because α_i, $f \in C^0(\Delta)$, results that the integrals $\Pi_P \chi$ shall have for χ combinations of coefficients and f or combinations of coefficients and h_i, thus $\chi \in C^0(\Delta)$ and because $\Pi_P \chi$ represents the gradient of a potential of continuous density volume, it results that $\Pi_P \chi$ admits only spatial derivatives in the sense of the large, definition given at the beginning of the paper, from which $D\Pi_P\chi = \chi(P)$ and thus, in the sense of this definition (12) admits succesive spatial derivatives defined by absolute and uniform convergent series in Ω_0. Thus (12) satisfies equation (1) in the sense of the large definition of the spatial derivative.

If α_i, $f \in C^1(\Delta)$, $i=1,2,...,n$, then $\Pi_P\chi$ admits spatial derivatives in restricted sense in Ω_0 and the solution (12) becomes common solution of equation (1).

It is observed that the solution depends of n holomorphic matrices in the space, arbitrary.

The particular case, when the coefficients of equation (1) are constant matrices has been approached in [18] and [19].

BIBLIOGRAPHY

1. A.V.Bitzadze, Prostransvenîi analog integrala tipa Cauchy i nekotorîe evo primeneniia. Izvestiia A.N.SSSR, Seriia Matematiceskaia, 1953, 17, 6.
2. B.V.Bojarski, Ob odnoi kraievîi zadace dlia sistemi uravnenii v ciastnîh proizvodnîh pervovo poriadka ellipticeskovo tipa, Dok.SSSR, 1955, 102, 2, 201-204.
3. Eftimie Viorel, Functions Hypercomplexes. Bul. Math. Soc., Sci.Math. RSR, 9, 279-332 (1965).
4. Gr.C.Moisil, Sur une classe de systemes d'équations aux dérivées partielles de la Physique Mathématique, Bucarest, I Göbl et fils, 1931.
5. Gr.C.Moisil, N.Teodorescu, Fonctions holomorphes dans éspace, Bul. Soc. de Şt. Cluj, 1931, 6, 177-194.
6. A.P.Muraviev, Analog formule Taylor dlia tonnotztnîh ellipticeskih funktzii, An.St. ale Univ.Iaşi, t.9, 1963, 357-368.
7. Mariana Nedelcu, Polinomul areolar de ordinul n în spaţiul cu trei dimensiuni, Studii şi cercetări matematice, Acad.R.P.R., 1956,7,1-3.
8. Mariana Nedelcu, Formule de mediaţie pentru polinomul areolar de ordinul n în spaţiul cu 3 dimensiuni, Bul.Ştiinţ. Acad. R.P.R., Seria Şt. mat. şi fizice, 1956, 8, 1.
9. Mariana Nedelcu, Le théorème de Morera pour le polynôme aréolaire d'ordre n dans l'espace à 3 dimensions. Bull.Math. de la Sc. Mat.de la R.P.R., 1957, 1,
10. Mariana Nedelcu, La théorie des polynômes aréolaire dans l'espace à n dimensions I. Le polynôme aréolaire d'ordre I. Rev.de Math. t. 4, 1959, nr.4, 693-723.
11. Mariana Nedelcu, The theory of the areolar polynomials in the m-dimensional space II. The n-order areolar polynomial. Rev. de Math. t. 5, 1960, nr.1, 121-168.
12. Mariana Nedelcu, La théorie des polynmes aréolaires dans l'espace à n dimensions III. Théorème de Morera et applications du polynômes aux équations aux dérivées aréolaires et aux problèmes aus limites. Bull. Math. t.3,(51) 1959, nr.2, 163-207.
13. Mariana Nedelcu, Une application de la derivée spatial aux équations de l'élasticité dans l'espace. Rev. de Math. pures et appl., t. 5, 1960, nr.3-4, 549-572.

14. Mariana Coroi-Nedelcu, Asupra soluţiei unui sistem de ecuaţii cu derivate parţiale de tip eliptic în E în analog ecuaţiei lui Beltrami în complex. St. şi Cerc. nr.7, 1965, 679-706.
15. Mariana Coroi-Nedelcu, Unele funcţii elementare în clasa funcţiilor olomorfe (α) în spaţiu. St. şi Cercet., nr.10, 1965, 1541-1567.
16. Mariana Coroi-Nedelcu, Representations of spatial polinoms and their properties. Bull. Math., t.11, 1967, nr.1, 63-85.
17. Mariana Coroi-Nedelcu, Development in series for a holomorphic functions in the space. Rev. de Math., t.12, 1967, nr.10, 1429-1451.
18. Mariana Coroi-Nedelcu, Asupra teoriei ecuaţiilor liniare cu derivate parţiale. Anale Univ. Bucureşti, 1968, 1, 15-37.
19. Mariana Coroi-Nedelcu, Equations aux derivées spatiales. Rev. Roum. de Math. Pures et Appl., t.13, 9, 1968, 1285-1291.
20. Mariana Coroi-Nedelcu, O problemă Dirichlet pentru un sistem de ecuaţii cu derivate parţiale de tip eliptic. Studii şi Cercet. nr.7, t.22, 1970, 997-1008.
21. Miron Nicolescu, Recherches sur les fonctions polyhormoniques. Annales scientifiques de l'Ecole Normale Supérieure, 52.
22. Dimitrie Pompeiu, Sur une classe de fonctions d'une variable complexe. Rend. Circolo Mat. Palermo, 1912, 33.
23. J.Ridder, Über areolar harmonische Funktionen , Acta matematica, 1946, 76.
24. Marcel Roşculeţ, Funcţii monogene pe algebre comutative. Edit. Academiei R.S.Romania, Bucureşti, 1975.
25. N.Teodorescu, Sur l'emploi de conditions globales dans quelques problèmes physiques, Annali di Matematica, 1933, 11, p.325-326.
26. N.Teodorescu, La dérivée aréolaire, Bucarest, 1936.
27. N.Teodorescu, La dérivée aréolaire et ses applications à la Physique Mathématique, Thèse, Paris, 1931.
28. N.Teodorescu, Dérivée aréolaire globale et dérivée généralisée.Bull. Math. de la Soc. de Sc. Math. de la R.P.R., t.6, nr.3-4, 1962.
29. N.Teodorescu, Théorie des primitives aréolaires bornées, Mem. Acad. Royale de Belgique, t. XXXII, 1962.
30. N.Teodorescu, Dérivées spatiales et opérateurs différentiels généralisés. Proc.International Congr. of Mathematiciens in Edinburg, 1959, Cambridge University, Press, 1960.
31. I.N.Vekua, Obobscenie analiticeskie funktzii, Moscova, 1959.

ON HILBERT MODULES WITH REPRODUCING KERNEL

R. Delanghe
Seminar of Higher Analysis, State University of Ghent,
Krijgslaan 271, B-9000 Gent, Belgium.

Introduction

This lecture is a report on part of the research which has been carried out by the author in his study on topological modules over certain Banach algebras. In this study, special interest has been paid to the theory of Hilbert modules over finite dimensional H^*-algebras, a theory which has essentially grown out of considerations in hypercomplex function theory. In collaboration with Dr. F. Brackx, a unified treatment of this subject, including new spaces of functions associated with systems of linear partial differential equations, is in preparation.
This paper will be divided into two parts :

I. Hilbert modules over finite dimensional H^*-algebras

II. Hilbert modules with reproducing kernel.

The first part is intended to give the necessary background material for Part II. The essential result concerns a characterization of the dual of a Hilbert module which furnishes a generalization of the classical Riesz representation theorem for Hilbert spaces.
In the second part we deal with Hilbert modules with reproducing kernel and prove a generalization of the Aronszajn-Bergman theorem. We also work out an example of such a Hilbert module arising from hypercomplex function theory.

Part I. HILBERT MODULES OVER FINITE DIMENSIONAL H^*-ALGEBRAS

I.1. Preliminary results

In [1], W. Ambrose has given the following definition of an H^*-algebra A :

A is a complex algebra (not necessarily commutative) provided with an inner product (,) and an involution $x \to \bar{x}$ such that

(i) $(xy,z) = (y,\bar{x}z)$

(ii) $(yx,z) = (y,z\bar{x})$

for all $x,y,z \in A$ and

(iii) A is a Banach algebra for the norm $|.|_0$ induced by the

inner product.
The same author has proved that, if A is a proper H^*-algebra (i.e. $xA = \{0\}$ implies that $x=0$), then it becomes an involutive Banach algebra.
Throughout this paper we assume that A is a proper finite dimensional H^*-algebra from which it follows that A contains an identity element, denoted by e with $|e|_0 \geq 1$.
In [4], F.F. Bonsall and A.W. Goldie have shown that a proper H^*-algebra furnishes an example of a trace algebra with non singular trace τ, defined by

$$\tau(a,b) = \mathrm{Re}(b,\bar{a}) \quad , \quad a,b \in A .$$

The non singularity of τ means that if for $a \in A$ fixed, $\tau_a(b) = \tau(ab) = 0$ for all $b \in A$, then $a = 0$.
Following [4], τ so represents the real linear functionals on A ; i.e. if T is a real linear functional on A, then there exists a unique element $a \in A$ such that

$$T(x) = \tau_a(x) = \tau(ax) \quad , \quad x \in A .$$

In what follows, an important role will be played by the real linear functional $\tau_e(\ .\ )$ on A. Moreover, when we consider right A-modules X, we always assume that $xe = x$, $\forall x \in X$, so that X becomes a real or complex vector space.

Theorem 1.1. If X is a right A-module and T a right A-linear functional on X, then $\tau_e T$ is a real linear functional on X. Moreover $T = 0$ if and only if $\tau_e T = 0$.
Proof. It is clear that $\tau_e T$ is a real linear functional on X and that $T = 0$ implies $\tau_e T = 0$.
Conversely, assume that T is a right A-linear functional on X such that $\tau_e T = 0$ and choose $x_0 \in X$. Then for any $\lambda \in A$,

$$\tau_{T(x_0)}(\lambda) = \tau(T(x_0)\lambda) = \tau_e(T(x_0)\lambda) = \tau_e T(x_0\lambda) = 0 .$$

Since τ is a non singular trace on A, $T(x_0) = 0$. Q.E.D.

If X is a right A-module, then we call a function $\|\cdot\| : X \to \mathbb{R}$ a proper norm on X iff for all $x,y \in X$ and $\lambda \in A$,

(i) $\|x\| \geq 0$ and $\|x\| = 0 \Leftrightarrow x = 0$

(ii) (a) $\|x\lambda\| \leq \|x\| \, |\lambda|_0$

(b) $\|x\lambda\| = \|x\| . |\lambda|$ if $\lambda \in \mathbb{C}$

(iii) $\|x+y\| \leq \|x\| + \|y\|$.

As usual, a right A-linear functional T on X, provided with a proper norm, is said to be bounded iff there exists a constant $C > 0$ such that

$$|T(x)|_0 \leq C\|x\| \quad , \quad \forall x \in X .$$

Theorem 1.2. Let X be a right A-module provided with a proper norm. Then a right A-linear functional T on X is bounded iff $\tau_e T$ is bounded on X.

Proof. If T is bounded, then from

$$|\tau_e T(x)| = |\mathrm{Re}(T(x),e)| \leq |(Tx,e)| \leq |e|_0 |T(x)|_0$$

it follows that $\tau_e T$ is bounded on X.

Let now T be a right A-linear functional on X such that $\tau_e T$ is bounded on X. Then there exists a constant $C > 0$ such that

$$|\tau_e T(y)| \leq C\|y\| \quad , \quad \forall y \in X .$$

Since

$$\tau_e(T(x).\overline{T(x)}) = |T(x)|_0^2$$

we find, on the one side, putting $y = x\overline{T(x)}$, that

$$|\tau_e T(x\overline{T(x)})| = |\tau_e(T(x).\overline{T(x)})| = |T(x)|_0^2$$

while on the other hand

$$\begin{aligned} |\tau_e T(x\overline{T(x)})| &\leq C\|x\overline{T(x)}\| \\ &\leq C|T(x)|_0\|x\| . \end{aligned}$$

Hence, for all $x \in X$,

$$|T(x)|_0^2 \leq C|T(x)|_0\|x\|$$

or

$$|T(x)|_0 \leq C\,\|x\| . \qquad \text{Q.E.D.}$$

I.2. Hilbert modules

Let A be a proper finite dimensional (not necessarily commutative) H^*-algebra with identity e. Then we call a right A-module $\mathcal{H}$ a right Hilbert A-module iff

(1) A function $(\ .\ ) : \mathcal{H}\times\mathcal{H} \to A$, called innerproduct on $\mathcal{H}$, is defined such that for all $x,y,z \in \mathcal{H}$ and $\lambda \in A$,

(i) $(x,y+z) = (x,y) + (x,z)$

(ii) $(x,y\lambda) = (x,y)\lambda$

(iii) $(x,y) = \overline{(y,x)}$

(iv) $\tau_e(x,x) \geq 0$ and $\tau_e(x,x) = 0 \Leftrightarrow x = 0$.

If moreover we put $\|x\|^2 = \tau_e(x,x)$, then

(v) $\|x\lambda\|^2 \leq |\lambda|_0^2 \|x\|^2$

(2) $\mathcal{H}$ is complete for the norm topology induced by the inner product on $\mathcal{H}$.

Remark that $\|\cdot\|$ is a proper norm on $\mathcal{H}$ and that $\mathcal{H}$, considered as a real vector space, is a real Hilbert space with inner product $\tau_e(\ ,\ )$.

Just as in the classical case, a family $(x_i)_{i\in I}$ of elements in $\mathcal{H}$ is said to be orthonormal iff $(x_i,x_j) = 0$, $(i \neq j)$ and $\|x_i\|^2 = 1$, $\forall i \in I$.

Theorem 1.3. (Bessel's inequality) If $(x_i)_{i=1}^m$ is a finite orthonormal family in $\mathcal{H}$, then for all $x \in \mathcal{H}$,

$$\sum_{i=1}^m |(x_i,x)|_0^2 \leq \|x\|^2 .$$

Proof. Working out $\|x - \sum_{i=1}^m x_i(x_i,x)\|^2$, we find :

$$0 \leq \|x - \sum_{i=1}^m x_i(x_i,x)\|^2$$

$$= \|x\|^2 - \sum_{i=1}^m |(x_i,x)|_0^2 - \sum_{i=1}^m |(x_i,x)|_0^2 + \sum_{i=1}^m \|x_i(x_i,x)\|^2 .$$

But as for all i, $\|x_i(x_i,x)\|^2 \leq |(x_i,x)|_0^2 \|x_i\|^2 = |(x_i,x)|_0^2$, the result follows immediately. Q.E.D.

Corollary (Inequality of Cauchy-Schwarz) For all $x,y \in \mathcal{H}$,

$$|(x,y)|_0 \leq \|x\| \|y\| .$$

We now give a characterization of the dual of a right Hilbert A-module.

Theorem 1.4. (The Riesz representation theorem) T is a bounded right A-linear functional on the right Hilbert A-module $\mathcal{H}$ if and only if there exists a unique element $y \in \mathcal{H}$ such that $T(x) = (y,x)$ for all $x \in \mathcal{H}$.

Proof. It is clear that if $y \in \mathcal{H}$ is fixed, then the A-functional $T_y(x) = (y,x)$, $x \in \mathcal{H}$, is bounded and right A-linear on $\mathcal{H}$. Let now T be a bounded right A-linear functional on $\mathcal{H}$. Then $\tau_e T$ is a bounded real linear functional on $\mathcal{H}$, considered as a real Hilbert space with inner product $\tau_e(\ ,\ )$.

In virtue of the classical Riesz representation theorem, there exists a unique element $y \in \mathcal{H}$ such that $\tau_e T(x) = \tau_e(x,y)$ for all $x \in \mathcal{H}$.

As $\tau_e(a) = \tau_e(\overline{a})$ for any $a \in A$, $\tau_e T(x) = \tau_e(y,x)$, $x \in \mathcal{H}$. Consider the function $T_y : \mathcal{H} \to A$ with $T_y(x) = (y,x)$, $x \in \mathcal{H}$. Then T_y is a bounded right A-linear functional on $\mathcal{H}$ and $\tau_e T_y(x) = \tau_e T(x)$ for all $x \in \mathcal{H}$, so that, by means of Theorem 1.1., $T_y = T$ or $T(x) = (y,x)$, $\forall x \in \mathcal{H}$. Q.E.D.

Part II. HILBERT MODULES WITH REPRODUCING KERNEL

II.1. Preliminary results

Let A be a finite dimensional (real or complex) H^*-algebra and $\mathcal{H}$ be a right Hilbert A-module consisting of functions which are defined on a certain set E and take on values in A.

Definition 2.1. A function $K : E\times E \to A$ is said to be a reproducing kernel for $\mathcal{H}$ iff for every $t \in E$ fixed

(i) $K(x,t)$, as a function of x, belongs to $\mathcal{H}$

(ii) $f(t) = (K(x,t),f(x))_x$ for all $f \in \mathcal{H}$.

Of course, any Hilbert space with reproducing kernel is a Hilbert module with reproducing kernel by taking $A = \mathbb{C}$.

Theorem 2.1. (Aronszajn-Bergman) Let $\mathcal{H}$ be a right Hilbert A-module consisting of functions $f : E \to A$. Then $\mathcal{H}$ possesses a reproducing kernel K iff for any $t \in E$, there exists a constant $C(t) > 0$ such that for all $f \in \mathcal{H}$,

$$|f(t)|_0 \leq C(t)\|f\| \ .$$

Proof. If K exists, then for any $t \in E$,

$$f(t) = (K(x,t),f(x))_x$$

so that

$$|f(t)|_0 \leq \|K(x,t)\| \ \|f\| \ .$$

Conversely, assume that a constant $C(t) > 0$ may be found such that $|f(t)|_0 \leq C(t)\|f\|$ for all $f \in \mathcal{H}$.
Then the right A-linear functional $T_t : \mathcal{H} \to A$ defined as

$$T_t(f) = f(t) \quad , \qquad f \in \mathcal{H} \ ,$$

is bounded on $\mathcal{H}$ which implies, by means of Theorem 1.4., that there exists a unique element $h_t \in \mathcal{H}$ for which

$$f(t) = T_t(f) = (h_t(x),f(x))_x \quad , \quad \forall f \in \mathcal{H}.$$

Put $K(x,t) = h_t(x)$. Then K is a reproducing kernel for $\mathcal{H}$.

Q.E.D.

Theorem 2.2. If the right Hilbert A-module $\mathcal{H}$ possesses a reproducing kernel, then it is unique.

Proof. Suppose that K and K' are two different reproducing kernels of $\mathcal{H}$. Then for at least one $y \in E$, $K(x,y) \neq K'(x,y)$ or

$$0 < \| K(x,y)-K'(x,y)\|^2 = \tau_e(K-K',K-K')_x$$
$$= \tau_e(K,K-K')_x - \tau_e(K',K-K')_x$$
$$= 0$$

which yields a contradiction. Q.E.D.

II.2. The right Hilbert A-module $Lr_2(\Omega;A)$

Let A be the Clifford algebra constructed over a quadratic n-dimensional real or complex vector space V with orthogonal basis $e=(e_1,\ldots,e_n)$. Then it is wellknown that A admits as basis e_0 ; $e_1,\ldots,e_n$; $e_1e_2,\ldots, e_{n-1}e_n$; ... ; $e_1e_2\ldots e_n$, so that an arbitrary basic element may be written as $e_A = e_{\alpha_1}e_{\alpha_2}e_{\alpha_h}$ with $A = \{\alpha_1,\ldots,\alpha_h\} \subset N = \{1,\ldots,n\}$ and $1 \leq \alpha_1 < \alpha_2 < \ldots < \alpha_h \leq n$ $(e_\phi = e_0)$.
It then follows that in A, $e_ie_j + e_je_i = 0$ $(i\neq j$; $i,j = 1,\ldots,n)$, $e_0e_i = e_ie_0 = e_i$ $(i=1,\ldots,n)$ and we require that $e_i^2 = e_0$, $i=1,\ldots,n$.
Hence, an arbitrary element $a \in A$ may be put into the form $a = \sum_A a_Ae_A$, $a_A \in \mathbb{R}$ or $\mathbb{C}$.
Call $[\cdot]_0 : A \to \mathbb{R}$ or $\mathbb{C}$ the function which assigns to any element $a \in A$, its coefficient a_0 of e_0 and define for any basic element $e_A=e_{\alpha_1}e_{\alpha_2}\ldots e_{\alpha_h}$ of A, $\overline{e}_A = e_{\alpha_h}e_{\alpha_{h-1}}\ldots e_{\alpha_1}$. Then it is clear that if for

$a = \sum_A a_Ae_A \in A$, $\overline{a} = \sum_A \overline{a}_A\overline{e}_A$, the function $a \to \overline{a}$ is an involution on A.
Define for any $a,b \in A$,

$$(a,b) = 2^n [a\overline{b}]_0 .$$

Then one may check that (,) is an inner product on A, turning A into a proper, non commutative and finite dimensional H^*-algebra with

$$|a|_0 = 2^{n/2}\sqrt{\sum_A |a_A|^2} , \quad a \in A .$$

Let now A be the <u>real</u> Clifford algebra of dimension 2^n and $\Omega \subset \mathbb{R}^n$ be open. Then in [6] we have studied the so-called left regular functions with values in A : $f \in C_1(\Omega;A)$ is said to be left regular in Ω iff $M(f) = 0$ in Ω where

$M = \sum_{i=1}^n e_i \frac{\partial}{\partial x_i}$. Note that $M^2 = \Delta = \sum_{i=1}^n \frac{\partial^2}{\partial x_i^2}$.

As examples we mention :

(1) Let $n = 3$, $\vec{f} = (f_1, f_2, f_3) \in C_1(\Omega; R^3)$ and put $f = \sum_{i=1}^{3} f_i e_i$; then $M(f) = 0$ in Ω is equivalent to $\operatorname{div} \vec{f} = 0$ and $\operatorname{rot} \vec{f} = 0$ in Ω.

(2) Let again $n = 3$ and consider the system of equations occurring in the theory of elasticity and defined by

$$(\Delta + \beta \operatorname{grad} \operatorname{div})\vec{f} = 0, \ \vec{f} = (f_1, f_2, f_3) \in C_2(\Omega, R^3), \ \beta = \frac{a^2 - b^2}{b^2} \in R .$$

Then, if we put $f = \sum_{i=1}^{3} f_i e_i$ and call $\psi = (\operatorname{div} \vec{f})e_0$, $\phi = M(f) + \psi$ satisfies $M\phi = 0$ in Ω.

Let $Lr(\Omega;A)$ be the right A-module of left regular functions in Ω and put for any $K \subset \Omega$ compact and $f \in Lr(\Omega;A)$,

$$p_K(f) = \sup_{x \in K} |f(x)|_0 \quad ;$$

then we obtain that $P = \{p_K : K \subset \Omega, K \text{ compact}\}$ is a proper system of seminorms on $Lr(\Omega;A)$ which is equivalent to a countable proper system of seminorms (see also [7]).

Theorem 2.3. $(Lr(\Omega;A), P)$ is a right Fréchet A-module.

Proof. We only need to prove that if $(f_m)_{m \in N}$ is a Cauchy sequence in $Lr(\Omega;A)$, then it converges to some element of $Lr(\Omega;A)$. But this immediately follows from the fact that for each $m \in N$ and $A \subset \{1,\ldots,n\}$, $\Delta f_{m,A} = 0$ in Ω and that $Lr(\Omega;A)$ is provided with the proper system of seminorms P. Q.E.D.

Definition 1.2. We call $Lr_2(\Omega;A)$ the set of functions $f \in Lr(\Omega;A)$ such that $[2^n \int_\Omega \bar{f} f dx^N]_0 < +\infty$ and we define for any $f, g \in Lr_2(\Omega;A)$,

$$(f,g) = \int_\Omega \bar{f} g dx^N ,$$

where $dx^N = dx_1 \ dx_2 \ \ldots \ dx_n$.

It is clear that $Lr_2(\Omega;A)$ is a submodule of $Lr(\Omega;A)$ and that for any $f \in Lr_2(\Omega;A)$,

$$\|f\|^2 = \tau_{e_0}(f,f) = \int_\Omega \tau_{e_0}(\bar{f}f) dx^N$$
$$= \int_\Omega |f|_0^2 dx^N = [2^n \int_\Omega \bar{f} f dx^N]_0$$

We of course wish to show that $Lr_2(\Omega;A)$ is a right Hilbert A-module with reproducing kernel.

We therefore first prove :

Theorem 2.4. Let $t \in \Omega$ be fixed. Then there exists a constant $C(t) > 0$

such that for all $f \in Lr_2(\Omega;A)$,

$$|f(t)|_0 \leq C(t) \; \|f\| \; .$$

Proof. Call $R(t) = d(t, co\Omega)$ and consider a ball $B(t,r)$ where $0 < r = R(t)-\varepsilon$.

Then in view of Cauchy's Formula (see [6]),

$$f(t) = \frac{1}{A_{n-1}} \int_{\partial B} \frac{\overline{x}-\overline{t}}{\rho^n} \, d\sigma_x \; f(x)$$
$$= \frac{1}{A_{n-1}\cdot r^n} \int_{\partial B} (\overline{x}-\overline{t}) \, d\sigma_x \; f(x)$$

But, since $\int_{\partial B} (\overline{x}-\overline{t}) d\sigma_x f(x) = n \int_B f(x) dx^N$,

$$f(t) = \frac{n}{r^n A_{n-1}} \sum_A e_A \int_{B(t,r)} f_A(x) dx^N$$

so that

$$|f(t)|_0^2 = \frac{2^n n^2}{r^{2n} A_{n-1}^2} \sum_A \left(\int_B f_A(x) dx^N\right)^2 .$$

In virtue of the inequality of Cauchy-Schwarz for integrals,

$$|f(t)|_0^2 \leq \frac{n}{r^n A_{n-1}} \int_B |f|_0^2 dx^N$$

$$\leq \frac{n}{(R(t)-\varepsilon)^n A_{n-1}} \|f\|^2$$

If $\varepsilon \to 0+$, then

$$|f(t)|_0^2 \leq \frac{n}{R^n(t) A_{n-1}} \|f\|$$

or

$$|f(t)|_0 \leq C(t) \|f\| \text{ with } C(t) = n^{1/2}(R^n(t).A_{n-1})^{-1/2}$$

Q.E.D.

Corollary. For any compact subset $K \subset \Omega$, there exists $C(K) > 0$ such that $p_K(f) = \sup_{x \in K} |f(x)|_0 \leq C(K).\|f\|$, $f \in Lr_2(\Omega;A)$.

Theorem 2.5. $Lr_2(\Omega;A)$ is a right Hilbert A-module.

Proof. Let $(f_i)_{i=1}^{\infty}$ be a Cauchy sequence in $Lr_2(\Omega;A)$. Then for any $\varepsilon>0$, there exists an $N(\varepsilon)$ such that $\|f_r - f_s\| \leq \varepsilon$ whenever $r,s \geq N(\varepsilon)$.

Let $(K_m)_{m\in \mathbb{N}}$ be a compact exhaustion of Ω. Then in view of the Corollary to Theorem 2.4., to each $m \in \mathbb{N}$ there corresponds a constant $C_m > 0$ such that for $r,s \geq N(\varepsilon)$,

$$p_{K_m}(f_r - f_s) = \sup_{x \in K_m} |f_r(x) - f_s(x)|_0 \leq C_m \|f_r - f_s\| \leq C_m \varepsilon$$

which of course means that $(f_i)_{i=1}^{\infty}$ is a Cauchy sequence in $Lr(\Omega;A)$. Hence, by means of Theorem 2.3., there exists an $f \in Lr(\Omega;A)$ such that $f_m \to f$ in $Lr(\Omega;A)$.

We now show that $f \in Lr_2(\Omega;A)$ and that $f_m \to f$ in $Lr_2(\Omega;A)$. Indeed, as $(\|f_i\|)_{i=1}^{\infty}$ is bounded, there exists a constant $L > 0$ such that $\|f_i\| \leq L$ for all $i \in N$.

Consequently, for any m,

$$\int_{K_m} |f_i|_0^2 dx^N \leq \int_{\Omega} |f_i|_0^2 dx^N \leq L^2$$

and as the sequence $(f_i)_{i=1}^{\infty}$ converges uniformly to f on each compact subset of Ω,

$$\int_{K_m} |f|_0^2 dx^N \leq L^2 \ .$$

so that, as $K_m \uparrow \Omega$,

$$\int_{\Omega} |f|_0^2 dx^N = \lim_{m\to\infty} \int_{K_m} |f|_0^2 dx^N \leq L^2$$

or $\qquad \|f\|^2 \leq L^2$ which implies that $f \in Lr_2(\Omega;A)$.

Let now $s_0 \geq N(\varepsilon)$ be fixed. Then in view of Fatou's Lemma, for all $m \in N$,

$$\begin{aligned}\int_{K_m} |f-f_{s_0}|_0^2 dx^N &= \int_{K_m} \liminf |f_r - f_{s_0}|_0^2 dx^N \\ &\leq \liminf \int_{K_m} |f_r - f_{s_0}|_0^2 dx^N \\ &\leq \liminf \int_{\Omega} |f_r - f_{s_0}|_0^2 dx^N \\ &\leq \varepsilon^2\end{aligned}$$

from which it follows that

$$\lim_{m\to\infty} \int_{K_m} |f-f_{s_0}|_0^2 dx^N = \int_{\Omega} |f-f_{s_0}|_0^2 dx^N \leq \varepsilon^2$$

or $\qquad \|f-f_{s_0}\| \leq \varepsilon \quad$ whenever $\quad s_0 \geq N(\varepsilon)$. Q.E.D.

In view of the Theorems 2.1, 2.4. and 2.5. we may conclude with

<u>Theorem 2.6.</u> $Lr_2(\Omega;A)$ is a right Hilbert A-module with reproducing kernel.

If we call $K(x,y)$ the reproducing kernel of $Lr_2(\Omega;A)$, we so have that for any $f \in Lr_2(\Omega;A)$ and $t \in \Omega$,

$$f(t) = \int_\Omega \overline{K(x,t)}\, f(x)\,dx^N .$$

In some way, $Lr_2(\Omega;A)$ and $K(x,y)$ may thus be regarded as generalizations respectively of the classical HL^2-space and S. Bergman's kernel function. As a further result concerning the space $Lr_2(\Omega;A)$, we have obtained that, if Ω is connected and bounded, it contains an orthonormal family of elements $(\psi_k)_{k\in N}$ such that $Lr_2(\Omega;A) = c(sp_A\{\psi_k : k\in N\})$, where c denotes the closure operator in $Lr_2(\Omega;A)$ and $sp_A\{\psi_k : k\in N\}$ stands for the right A-span of the set $\{\psi_k : k\in N\}$.
To that end, we proceed as follows.

Let $k \to \alpha_k = (\alpha_k^{(2)},\ldots,\alpha_k^{(n)})$ be a bijection of N onto N^{n-1} and consider the differential operator D^{α_k} on $Lr_2(\Omega;A)$ with

$$D^{\alpha_k}f = \frac{\partial^{|\alpha_k|}}{\partial x_2^{\alpha_k^{(2)}}\cdots\partial x_n^{\alpha_k^{(n)}}}\, f .$$

Fix $t \in \Omega$ and call for each $m = 0,1,2,\ldots,$

$$E_m(\Omega) = \{f \in Lr_2(\Omega;A) : D^{\alpha_k}f(t) = \delta_{k,m}e_0,\ k \leq m\}.$$

Then it is clear that $0 \notin E_m(\Omega)$. Moreover, $E_m(\Omega) \neq \phi$ since, if $\alpha_m = (s_2,\ldots,s_n)$ and $p = \sum_{i=2}^{n} s_i$, the homogeneous polynomial $V^{(t)}_{k_1\ldots k_p} \in E_m(\Omega)$ where $(k_1,\ldots,k_p) \in \{2,\ldots,n\}^p$ is such that the elements $2,3,\ldots,n$ occur respectively $s_2,\ldots,s_n$ times.

Recall that (see [6]), $V^{(t)}_{k_1\ldots k_p} = \frac{1}{p!} \sum_{(k_1,\ldots,k_p)} z^*_{k_1}\cdots z^*_{k_p}$

where $\qquad z^*_k = (x_k-t_k)e_0 - (x_1-t_1)e_k \quad , \quad k = 2,\ldots,n .$

As $E_m(\Omega)$ is a closed convex subset of $Lr_2(\Omega;A)$, we obtain that there exists a unique element $g_m \in E_m(\Omega)$ for which $\|g_m\| = \inf_{f\in E_m(\Omega)} \|f\|$.

As to the sequence $(g_m)_{m\in N}$, we have obtained the following results.

Lemma 2.1. If for $f \in Lr_2(\Omega;A)$, $D^{\alpha_k}f(t) = 0$, $0 \leq k \leq m$, then $(g_m,f) = 0$.

Proof. For each $\lambda \in R$ and $A \in PN$, the elements $g_m+\lambda f\overline{e_A}$ belong to $E_m(\Omega)$. Hence,

$$\begin{aligned}\|g_m\|^2 &\leq \|g_m+f\lambda\overline{e_A}\|^2\\ &= \tau_{e_0}(g_m+f\lambda\overline{e_A},\ g_m+f\lambda\overline{e_A})\\ &= \|g_m\|^2 + 2\lambda\tau_{e_0}(g_m,f\overline{e_A}) + \lambda^2\|f\overline{e_A}\|^2 .\end{aligned}$$

Put $$\lambda = - \frac{\tau_{e_0}(g_m, f\overline{e_A})}{\| f\overline{e_A}\|^2} .$$

Then
$$0 \leqslant -(\tau_{e_0}(g_m, f\overline{e_A}))^2 . \| f\overline{e_A}\|^2$$

from which it follows that $\tau_{e_0}(g_m, f\overline{e_A}) = 0$.
As for any $a \in A$, $\tau_{e_0}(a) = \tau(e_0 a) = 2^n [a]_0$, $[(g_m, f\overline{e_A})]_0 = 0$ for all $A \in PN$ so that, since $(g_m, f) = \sum_A e_A [(g_m, f\overline{e_A})]_0$, $(g_m, f) = 0$.

Q.E.D.

Corollary. If $i \neq j$, then $(g_i, g_j) = 0$. If furthermore $\psi_i = g_i \| g_i \|^{-1}$, $i \in N$, then $(\psi_i)_{i=0}^{\infty}$ is an orthonormal family in $Lr_2(\Omega;A)$.

Lemma 2.2. For any $k \in N$, (ψ_k, ψ_k) is invertible in A.

Proof. Suppose that for some k, (ψ_k, ψ_k) is not invertible. As $(\psi_k, \psi_k) \neq 0$ and an element $a \in A$ is either invertible or a zero divisor, there exists $b \in A$ with $b \neq 0$ such that $(\psi_k, \psi_k).b = 0$ and hence $\overline{b}(\psi_k, \psi_k)b = 0$ or $(\psi_k . b, \psi_k . b) = 0$.
Consequently,
$$\tau_{e_0}(\psi_k b, \psi_k b) = \| \psi_k . b \|^2 = 0$$
or $\psi_k . b \equiv 0$ in Ω and so $D^{\alpha_k}\psi_k(x).b \equiv 0$ in Ω.
But for $x = t$, $D^{\alpha_k}\psi_k(t) = e_0 \| g_k \|^{-1}$, which implies that $b = 0$.

Q.E.D.

Lemma 2.3. If $f \in Lr_2(\Omega;A)$ and $(\psi_k, f) = 0$ for all $k = 0,1,\ldots,m$, then $D^{\alpha_k}f(t) = 0$, $0 \leqslant k \leqslant m$.

Proof. We first show that elements $a_k^{(m)} \in A$ may be found such that for the function $f_m = \sum_{k=0}^{m} \psi_k . a_k^{(m)}$, $D^{\alpha_r}f_m(t) = D^{\alpha_r}f(t)$, $0 \leqslant r \leqslant m$.

Take $r = 0$. Then the condition becomes :
$$D^{\alpha_0}f_m(t) = D^{\alpha_0}f(t) = a_0^{(m)} \| g_0 \|^{-1} .$$
If $r = 1$, we obtain
$$D^{\alpha_1}f_m(t) = D^{\alpha_1}f(t) = D^{\alpha_1}\psi_0(t).a_0^{(m)} + a_1^{(m)} \| g_1 \|^{-1}$$
or
$$a_1^{(m)} = \| g_1 \| (D^{\alpha_1}f(t) - \| g_0 \| D^{\alpha_1}\psi_0(t).D^{\alpha_0}f(t)) .$$

Proceeding in this way, we indeed find elements $a_k^{(m)} \in A$ with the desired property.

Consequently, $D^{\alpha_k}(f-f_m)(t) = 0$, $0 \leq k \leq m$, so that, in view of Lemma 2.1., $(\psi_k, f-f_m) = 0$ and hence $(\psi_k, f) = (\psi_k, f_m) = 0$.
But

$$(\psi_k, f_m) = (\psi_k, \psi_k) a_k^{(m)} ,$$

so that, by means of Lemma 2.2., $a_k^{(m)} = 0$, $0 \leq k \leq m$.
Hence $f_m \equiv 0$ in Ω and so $D^{\alpha_k} f_m(t) = D^{\alpha_k} f(t) = 0$, $0 \leq k \leq m$.

Q.E.D.

Lemma 2.4. If $f \in Lr_2(\Omega;A)$ and $(\psi_k, f) = 0$ for all $k \in N$, then $f = 0$.

Proof. In virtue of [6], Th.13, there exists an $r > 0$ such that in $\mathring{B}(t,r) \subset \overline{B}(t,r) \subset \Omega$,

$$f(x) = \sum_{p=0}^{\infty} \sum_{(k_1,\ldots,k_p)} V^{(t)}_{k_1\ldots k_p}(x) . D^p_{x_{k_1}\ldots x_{k_p}} f(t) .$$

As $(\psi_k, f) = 0$ for all $k \in N$, $D^{\alpha_k} f(t) = 0$ and so $f \equiv 0$ in $\mathring{B}(t,r)$.
Since Ω is connected, $f \equiv 0$ in Ω.

Q.E.D.

If we put for any subset $W \subset Lr_2(\Omega;A)$,

$$sp_A W = \{ \sum_{(i)} w_i \lambda_i : w_i \in W, \lambda_i \in A \} ,$$

then we obtain

Theorem 2.8. $Lr_2(\Omega;A) = c(sp_A\{\psi_k : k \in N\})$.

Proof. As $M = c(sp_A\{\psi_k : k \in N\})$ is a closed submodule of $Lr_2(\Omega;A)$, we easily find that

$$Lr_2(\psi;A) = M \oplus M^{\perp} .$$

Take $g \in M^{\perp}$; then $g \perp \psi_k$, $k \in N$, so that in view of Lemma 2.4., $g = 0$ or $M^{\perp} = \{0\}$.

Q.E.D.

Corollary. $Lr_2(\Omega;A)$ is separable.

II.3. Further examples

In [5], Dr. F. Brackx has studied the theory of (k)-monogenic functions of a quaternionvariable, i.e. the theory of functions $f \in C_k(\Omega;H)$ ($k \in N$ fixed) satisfying $D^k f = 0$ in Ω, where

(i) Ω is an open non empty subset of R^4

(ii) H is the algebra of real quaternions with basis $\{e_0, e_1, e_2, e_3\}$

$$\text{(iii)} \quad D = \sum_{i=0}^{3} e_i \frac{\partial}{\partial x_i} .$$

If $F_k(\Omega;H)$ is the right vector space of (k)-monogenic functions of a quaternionvariable, call $A_k(\Omega;H)$ the subspace of $F_k(\Omega;H)$ consisting of those elements $f \in F_k(\Omega;H)$ for which

$$\int_\Omega |D^j f|^2 dx^4 < +\infty \quad , \quad j = 0,1,\ldots,k-1,$$

where $\quad dx^4 = dx_0 \quad dx_1 \quad dx_2 \quad dx_3$.

Then, if for $f,g \in A_k(\Omega;H)$,

$$(f,g) = \int_\Omega \sum_{j=0}^{k-1} \overline{D^j f} \, D^j g \, dx^4 ,$$

$A_k(\Omega;H)$ becomes a right Hilbert H-module with reproducing kernel $K_k(x,t)$, $x,t \in \Omega$.

Moreover, in the case where Ω is bounded and connected, an orthonormal family $(\psi_i^{(j)})$, $j = 1,2,\ldots,k$; $i = 0,1,2,\ldots$, may be found such that

$$K_k(x,t) = \sum_{j=1}^{k} \sum_{i=0}^{\infty} \psi_i^{(j)}(x) \, \overline{\psi_i^{(j)}(t)} .$$

For more details, we refer the reader to [5].

Bibliography

[1] W. Ambrose, Structure theorems for a special class of Banach algebras, Trans.Amer.Math.Soc. 57 (1945) 364-386.

[2] N. Aronszajn, Theory of reproducing kernels, Trans.Amer.Math.Soc. 68 (1950) 337-404.

[3] S. Bergman, Sur les fonctions orthogonales de plusieurs variables complexes avec les applications à la théorie des fonctions analytiques (Mémorial des Sciences Mathématiques, Fasc. CVI, Gauthier-Villars, Paris, 1947).

[4] F.F. Bonsall and A.W. Goldie, Algebras which represent their linear functionals, Proc.Cambridge Phil.Soc. 49 (1953) 1-14.

[5] F. Brackx, Theory of (k)-monogenic functions of a quaternion-variable, Ph.D. thesis, Ghent(1973) (in Dutch).

[6] R. Delanghe, On regular-analytic functions with values in a Clifford algebra, Math.Ann. 185 (1970) 91-111.

[7] H.G. Garnir, M. De Wilde, J. Schmets, Analyse fontionnelle, Tome I (Birkhäuser Verlag 1968).

[8] H. Meschkowski, Hilbertsche Räume mit Kernfunktion (Springer Verlag, Berlin Göttingen Heidelberg, 1962).

A priori Abschätzungen für eine Klasse elliptischer Pseudo-Differentialoperatoren im Raum $L^p(R^n)$

J. Donig

(Darmstadt)

Einleitung

In der vorliegenden Arbeit betrachten wir Pseudo-Differentialoperatoren positiv ganzzahliger Ordnung m, welche erzeugt werden von polabhängigen singulären Michlin-Integraloperatoren S_α und gewissen verallgemeinerten Faltungsoperatoren K_α gemäß

$$A := \sum_{|\alpha|\leq m} (S_\alpha + K_\alpha)\tilde{D}^\alpha .$$

Hierbei wird im Gegensatz zu zahlreichen anderen Arbeiten über allgemeinere Klassen von Pseudo-Differentialoperatoren (vgl. J. J. Kohn - L. Nirenberg [13], R. T. Seeley [18], L. Hörmander [11], V. V. Gruzin [10] u. a.) weitgehend auf Glattheitsvoraussetzungen über die Daten von S_α und K_α bzw. das Symbol von A verzichtet. Einen Operator A der obigen Gestalt fassen wir auf als unbeschränkte Abbildung im Lebesgueschen Raum $L^p(R^n)$, wobei für p alle reellen Zahlen mit $1 < p < \infty$ zugelassen sind, für die die Operatoren S_α existieren. Es wird ferner generell $n \geq 2$ vorausgesetzt; Operatoren vom Typ A im eindimensionalen Fall wurden vom Verfasser in [8] zuerst gesondert betrachtet.

Ziel der Arbeit ist die Angabe notwendiger und hinreichender Bedingungen für die Gültigkeit geeigneter a priori Abschätzungen, welche mit der Fredholmeigenschaft von A äquivalent sind. Diese Abschätzungen müssen sich notwendig von den Abschätzungen unterscheiden, die in Funktionenräumen über beschränkten, offenen Teilmengen des R^n oder kompakten C^∞-Mannigfaltigkeiten ohne Rand bekannt sind (vgl. M. S. Agranowitsch [2], S. 76, Theorem 12.1 für speziellere Operatoren $A_0 := \sum_{|\alpha|\leq m} S_\alpha D^\alpha$). Die Ursache dafür liegt darin, daß der für offene, beschränkte $\Omega\subset R^n$ gültige Rellichsche Satz über die Kompaktheit der Einbettung $H_0^{m,p}(\Omega) \to L^p(\Omega)$ für $\Omega := R^n$ falsch wird. Wir werden zeigen, daß der zitierte Satz von Agranowitsch so modifiziert werden kann, daß auch im hier vorliegenden Falle allgemeinerer Operatoren und der nichtkompakten Mannigfaltigkeit R^n eine entsprechende Aussage gilt. Da die verwendeten Beweismethoden global sind, gelingt es insbesondere, für Fredholmoperatoren A die Regularisatoren explizit anzugeben.

Einen lokalen Zugang zur Charakterisierung der Fredholmeigenschaft der Operatoren A im Hilbertraum $L^2(R^n)$, $n \geq 2$, basierend auf den Arbeiten von I. B. Simonenko [19], [20], wählte F.-O. Speck [21]. A priori Abschätzungen und globale Regularisatoren werden in dieser Arbeit nicht angegeben.

1. Bezeichnungen und Definitionen

Es sei $n \in N$, $n \geq 2$. Wir bezeichnen wie üblich mit $S(R^n)$ den Schwartzschen Raum der C^∞-Funktionen φ auf R^n, für die

$$\left(\frac{\partial}{\partial x}\right)^\alpha \varphi(x) = o(|x|^{-t}) \quad \text{für} \quad |x| \to \infty \quad \text{und}$$
$$\text{für alle} \quad \alpha \in N_o^n, \; t \in N_o$$

($N_o := N \cup \{0\}$). $S(R^n)$ wird durch Auszeichnung eines abzählbaren Halbnormensystems lokalkonvex. Der topologische Dualraum von $S(R^n)$ wird mit $S'(R^n)$ bezeichnet. Mit F bezeichnen wir die Fouriertransformation in $S(R^n)$:

$$F\varphi := \hat{\varphi} := \int_{R^n} e^{-2\pi i(\cdot,y)}\varphi(y)dy, \quad \varphi \in S(R^n),$$

die in bekannter Weise nach $S'(R^n)$ fortgesetzt werden kann. Für diese Fortsetzung von F verwenden wir ebenfalls das Symbol F.

Für $\ell \in N$ und reelle r mit $1 < r < \infty$ bezeichne $W^{\ell,r}(R^n)$ den Sobolewraum der Funktionen $u \in L^r(R^n)$, deren schwache Ableitungen $D^\alpha u$ für $0 < |\alpha| := \alpha_1 + \ldots + \alpha_n \leq \ell$ existieren und in $L^r(R^n)$ liegen. $W^{\ell,r}(R^n)$ wird durch die Norm

$$||u||_{\ell,r} := \left(\sum_{|\alpha| \leq \ell} \int_{R^n} |D^\alpha u|^r dx\right)^{\frac{1}{r}}, \quad u \in W^{\ell,r}(R^n),$$

eine Banachraumstruktur aufgeprägt. Mit Hilfe des Besselpotentialoperators $(I+\tilde{\Delta})^{-\ell/2}$ der Ordnung ℓ, definiert durch

$$(I+\tilde{\Delta})^{-\ell/2} v := [(1+|.|^2)^{-\frac{\ell}{2}}]^{\wedge} * v$$

$$= F^{-1}(1+|\cdot|^2)^{-\frac{\ell}{2}} F v, \quad v \in S'(R^n)$$

(I : Identität, $\tilde{\Delta} := -\frac{1}{4\pi^2}\Delta$: normierter Laplaceoperator), kann man bekanntlich auf $W^{\ell,r}(R^n)$ eine zu $\|\cdot\|_{\ell,r}$ äquivalente Norm einführen gemäß

$$|||u|||_{\ell,r} := \|(I+\tilde{\Delta})^{\frac{\ell}{2}} u\|_r, \quad u \in W^{\ell,r}(R^n).$$

$W^{\ell,r}(R^n)$ mit dieser Norm und $L^r(R^n)$ sind isometrisch isomorph. Für beliebige reelle ℓ definiert man allgemein:

$$W^{\ell,r}(R^n) := \{u \in S'(R^n) : |||u|||_{\ell,r} < \infty\}.$$

Es sei im folgenden M eine kompakte C^∞-Mannigfaltigkeit ohne Rand und $\mathcal{A} := \{(U_i, \varphi_i)\}_{i\in I}$ ein Atlas auf M. Aufgrund der Kompaktheit von M können wir eine endliche Teilüberdeckung $\{U_{i_1},\dots,U_{i_k}\} \subset \{U_i\}_{i\in I}$ auswählen. Bezüglich dieser Teilüberdeckung existiert eine Zerlegung der Einheit gemäß $1 = \gamma_1+\dots+\gamma_k$ auf M mit $\gamma_j \in C_o^\infty(U_{i_j})$, $j = 1,\dots,k$. Für reelle Zahlen ℓ, r mit $\ell \geq 0$ und $1 < r < \infty$ definiert man dann $W^{\ell,r}(M) := \overline{C^\infty(M)}^{\|\ \|^M_{\ell,r}}$ mit

$$\|v\|^M_{\ell,r} := \Sigma_{j=1}^k |||(\gamma_j v)\circ \varphi_{i_j}^{-1}|||^{R^n}_{\ell,r}, \quad v \in C^\infty(M).$$

Die durch diese Norm in $W^{\ell,r}(M)$ definierte Topologie ist bekanntlich unabhängig sowohl von der speziellen Wahl der endlichen Teilüberdeckung von $\{U_i\}_{i\in I}$, als auch von der Art der Zerlegung der Einheit. In der vorliegenden Arbeit ist stets $M := S^{n-1}$ die Einheitskugelfläche im R^n.

Es sei Ω eine Teilmenge des R^n und f sei eine beliebige meßbare, komplexwertige Funktion auf $R^n \times \Omega$. Dann bezeichnen wir mit f' bzw. f'' die vektorwertige Funktion $x \mapsto f'(x,\cdot) := f(x,\cdot)$ auf R^n

bzw. $y \mapsto f''(\cdot,y) := f(\cdot,\cdot-y)$ auf Ω.

Es sei schließlich A ein beliebiger dicht definierter, unbeschränkter Operator in $L^r(R^n)$. Wir nennen wie üblich A Semi-Fredholmoperator, wenn er abgeschlossen ist und ferner sein Nullraum $N(A)$ endlichdimensional und sein Bildraum $R(A)$ abgeschlossen ist. A heiße Fredholmoperator, wenn er ein Semi-Fredholmoperator ist und wenn $\dim R(A)^{\perp} < \infty$. Die Zahl $\kappa(A) := \dim N(A) - \dim R(A)^{\perp}$ heiße Index von A.

Nach diesen einführenden Erklärungen kommen wir zur Definition der zu untersuchenden Pseudo-Differentialoperatoren. Es seien $n, m \in N$, $n \geq 2$, und $q, p \in R$ mit $1 < q, p < \infty$ und $p \geq \frac{q}{q-1}$. Für $\alpha \in N_o^n$, $|\alpha| \leq m$, wählen wir meßbare, komplexwertige Funktionen a_α, s_α und k_α auf R^n, $R^n \times S^{n-1}$ und $R^n \times R^n$ resp., welche folgende Eigenschaften haben:

i) $a_\alpha \in C(\overline{R^n})$ $\quad (\overline{R^n} := R^n \cup \{\infty\})$.

ii) $s'_\alpha \in C(\overline{R^n}, L^q(S^{n-1}))$ und

$$\int_{S^{n-1}} s'_\alpha(x,\cdot)do = o \qquad \text{für alle } x \in R^n$$

(d. h. s' hat den Mittelwert Null über S^{n-1}).

iii) $k'_\alpha \in C(\overline{R^n}, L^1(R^n))$, $k''_\alpha \in L^\infty(R^n, L^1(R^n))$.

Mit Hilfe dieser Funktionen definieren wir im gesamten Raum $L^p(R^n)$ Integraloperatoren S_α bzw. K_α durch

$$S_\alpha v := a_\alpha(\cdot)v(\cdot) + \lim_{\varepsilon \to 0} \int_{|y| \geq \varepsilon} \frac{s_\alpha(\cdot, \frac{y}{|y|})}{|y|^n} v(\cdot - y)dy$$

bzw.

$$K_\alpha v := \int_{R^n} k_\alpha(\cdot,y)v(\cdot-y)dy \ , \quad v \in L^p(R^n).$$

Wir nennen S_α singulären Michlin-Integraloperator mit der Charakteristik s_α und K_α (regulären) verallgemeinerten Faltungsoperator mit dem Kern k_α. Es ist bekannt, daß unter den obigen Voraussetzungen über die Funktionen a_α, s_α und k_α die Operatoren S_α und K_α Endomorphismen sind (bzgl. S_α s. A. P. Calderon und A. Zygmund [4], S. 290, Theorem 2).
Wir setzen noch:

$$\tilde{D}^\alpha := (2\pi i)^{-|\alpha|} D^\alpha \; ; \quad D^\alpha := D_1^{\alpha_1} \dots D_n^{\alpha_n} .$$

Definition 1.1. Es sei $m \in N$. Für $|\alpha| \le m$ seien S_α und K_α die unter den obigen Voraussetzungen i) - iii) definierten Integraloperatoren, wobei zusätzlich $S_\alpha \neq 0$ sei für mindestens ein $|\alpha| = m$.

Dann definieren wir den Pseudo-Differentialoperator A der Ordnung m als singulären Integro-Differentialoperator in $L^p(R^n)$ mit dem Definitionsbereich $D(A) := W^{m,p}(R^n)$:

$$Au := \sum_{|\alpha| \le m} (S_\alpha + K_\alpha) \tilde{D}^\alpha u \, , \quad u \in D(A) .$$

Wir nennen $A_1 := \sum_{|\alpha| = m} S_\alpha \tilde{D}^\alpha$: Hauptteil und $A_2 := A - A_1$ Nebenteil von A.
Die Funktionen a_α, s_α und k_α, welche S_α und K_α festlegen, nennen wir Daten von A.

Es sei bemerkt, daß A ein dicht definierter, abgeschlossener und unbeschränkter Operator in $L^p(R^n)$ ist.

Definition 1.2. Für den Pseudo-Differentialoperator A definieren wir das Symbol σ_A bzw. das Hauptsymbol σ_{A_1} durch

$$\sigma_A(x,\xi) := \sum_{|\alpha| \le m} \{ a_\alpha(x) + \left[pv \frac{s_\alpha(x, \frac{\cdot}{|\cdot|})}{|\cdot|^n} \right]^\wedge (\xi) + [k_\alpha(x,\cdot)]^\wedge(\xi) \} \xi^\alpha$$

bzw.

$$\sigma_{A_1}(x,\xi) := \sum_{|\alpha|=m} \{a_\alpha(x) + \left[pv\, \frac{s_\alpha\left(x, \frac{\cdot}{|\cdot|}\right)}{|\cdot|^n}\right]^\wedge(\xi)\}\xi^\alpha ;$$

$$x \in R^n, \quad \xi \in R^n \setminus \{0\}.$$

Hierbei ist für festes $x \in R^n$:

$pv\, \frac{s_\alpha\left(x, \frac{\cdot}{|\cdot|}\right)}{|\cdot|^n} \in S'(R^n)$ definiert durch den Cauchyschen Hauptwert:

$$\langle pv\, \frac{s_\alpha\left(x, \frac{\cdot}{|\cdot|}\right)}{|\cdot|^n}, \varphi\rangle :=$$

$$= \lim_{\varepsilon\to 0} \int_{|y|\geq 0} \frac{s_\alpha\left(x, \frac{y}{|y|}\right)}{|y|^n} \varphi(y)dy, \quad \varphi \in S(R^n).$$

Nach A. P. Calderon und A. Zygmund (vgl. [5], S. 312, Theorem 2) gilt für alle $x \in R^n$: $\sigma_{A_1}(x,\cdot) \in C(R^n \setminus \{0\})$.

Definition 1.3. Ein Pseudo-Differentialoperator A mit dem Hauptsymbol σ_{A_1} heiße elliptisch in $\overline{R^n}$, wenn gilt:

i) $\sigma_{A_1}(\cdot,\xi) \in C(\overline{R^n})$ für alle $\xi \in R^n \setminus \{0\}$.

ii) $\sigma_{A_1}(x,\xi) \neq 0$ für alle $x \in \overline{R^n}$ und alle $\xi \in R^n \setminus \{0\}$.

2. Formulierung des Hauptsatzes

Aufgrund der Voraussetzungen i) - iii), Seite 174, über die Daten a_α, s_α und k_α von A gelten die eindeutigen Zerlegungen:

(1) $a_\alpha = a_\alpha^\infty + a_\alpha^o, \quad s_\alpha = s_\alpha^\infty + s_\alpha^o, \quad k_\alpha = k_\alpha^\infty + k_\alpha^o$

mit

$$a_\alpha^\infty \in \mathbb{C},\ s_\alpha^\infty \in L^q(S^{n-1}),\ k_\alpha^\infty \in L^1(R^n);$$

$$a_\alpha^o \in C_o(R^n),\ (s_\alpha^o)' \in C_o(R^n, L^q(S^{n-1})),\ (k_\alpha^o)' \in C_o(R^n, L^1(R^n))^{1)},$$

1) Für einen Banachraum $(Y, \|\cdot\|)$ sei $C_o(R^n,Y) := \{f \in C(R^n,Y) : \|f(x,\cdot)\| = o(1)$ für $|x| \to \infty$

welche eine entsprechende eindeutige Zerlegung des Symbols σ_A nach sich ziehen:

$$\sigma_A = \sigma_A^\infty + \sigma_A^0 .$$

Satz 2.1 (Hauptsatz) Es sei $n \geq 2$. Seien q, p reelle Zahlen mit $1 < q < \infty$, $p \geq \frac{q}{q-1}$, sei

$$\ell_0(q) := \begin{cases} \frac{n-1}{q}, & 1 < q \leq 2 \\ \frac{n-1}{2}, & 2 < q < \infty \end{cases},$$

und sei ℓ ebenfalls reell mit $\ell > \ell_0(q)$. Seien a_α, s_α und k_α ($|\alpha| \leq m$) Funktionen mit den Eigenschaften i) - iii), Seite 174; zusätzlich sei

ii') $s_\alpha{}' \in C(\overline{R^n}, W^{\ell,q}(S^{n-1}))$,

und es sei $(\sigma_A^\infty)^{-1} . (1+|\cdot|)^2)^{\frac{m}{2}}$ ein Fourier-Multiplikator für $L^p(R^n)$.

Dann sind die folgenden Aussagen äquivalent:

j) A ist elliptisch in $\overline{R^n}$, und $\inf_{\xi \in R^n} |\sigma_A^\infty(\xi)| > 0$.

jj) A ist koerzitiv über $W^{m,p}(R^n)$; d. h. es existieren eine Konstante $\gamma > 0$ und eine kompakte Halbnorm ϕ auf $W^{m,p}(R^n)$, so daß folgende a priori Abschätzung gilt:

(2) $\gamma \| u \|_{m,p} \leq \| Au \|_p + \langle \phi, u \rangle$, $u \in W^{m,p}(R^n)$.

jjj) A ist ein Fredhomoperator vom Index $\kappa(A) = \kappa(\Sigma_{|\alpha|=m} S_\alpha R^\alpha)$ (R^α : Riesz-Transformation, vgl. (8), (9) unten).

Korollar. Der Rechts- und Linksregularisator B von A hat die Gestalt:

$$B := B^\infty \cdot S^\infty \cdot S_1 ,$$

wobei S_1, S^∞ und B^∞ gemäß (14) bzw. (15) bzw. (16) unten definierte Pseudo-Differentialoperatoren sind.

3. Beweis des Hauptsatzes

Wir wenden uns zunächst der Behauptung jj) <=> jjj) zu und beweisen für einen beliebigen unbeschränkten, abgeschlossenen linearen Operator

A in $L^p(R^n)$ mit dem Definitionsbereich $D(A) := W^{m,p}(R^n)$:

Satz 3.1 Die folgenden Aussagen sind äquivalent:

i) A ist koerzitiv über $W^{m,p}(R^n)$, und es gilt $\beta(A) := \dim R(A)^{\perp} < \infty$.

ii) Es existieren endlichdimendionale Projektoren P und Q in $W^{m,p}(R^n)$ bzw. $L^p(R^n)$, sowie ein beschränkter, linearer Operator B von $L^p(R^n)$ in $W^{m,p}(R^n)$, so daß

$$BA = I-P, \; AB = I-Q.$$

iii) A ist ein Fredholmoperator.

Beweis i) $\Rightarrow$ ii). Aus (2) folgt zunächst mit Hilfe eines Kompaktheitsschlusses, daß $\alpha(A) := \dim N(A) < \infty$, und daraus schließt man auf die Existenz eines beschränkten Projektors P in $W^{m,p}(R^n)$ mit $R(P)=N(A)$ und $W^{m,p}(R^n) = N(P) \oplus N(A)$. Wir definieren $A_o := A|_{N(P)}$, so daß $A = A_o(I-P)$ und $R(A) = R(A_o)$. A_o ist aufgrund der Abgeschlossenheit von $N(P)$ nach dem Satz vom beschränkten Inversen invertierbar, falls nur $R(A_o)$ abgeschlossen ist. $R(A_o)$ ist aber abgeschlossen. Zum Beweis ist es hinreichend, daß für $\phi \neq 0$ in (2) ein $\gamma' > 0$ existiert, so daß

$$(3) \qquad \gamma' < \phi, u > \leq \| A_o u \|_p , \quad u \in N(P).$$

Wir schließen indirekt. Wäre (3) falsch, so gäbe es eine Folge $\{u_r\} \subset N(P)$ mit $< \phi, u_r > = 1$ für alle $r \in N$ und $\lim_{r\to\infty} \| A_o u_r \|_p = 0$. Daraus und aus (2) folgt, daß für $\eta > 0$ und hinreichend großes r gilt: $\gamma \| u_r \|_{m,p} \leq \eta + 1$. Als beschränkte Folge enthält $\{u_r\}$ eine **schwach konvergente** Teilfolge $\{u_{r_i}\}$ mit einem Grenzwert $u_o \in N(P)$. Da andererseits $A_o u_{r_i} \to A_o u_o = o$ schwach konvergiert, hat man $u_o \in N(A_o) \cap N(P)$; d. h. $u_o = 0$ im Widerspruch zu $< \phi, u_{r_i} > = 1$ für alle $i \in N$. Also ist $R(A_o)$ abgeschlossen.

Aus $\beta(A) < \infty$ schließlich folgt die Existenz eines beschränkten, endlichdimensionalen Projektors Q in $L^p(R^n)$ mit $N(Q) = R(A_o)$ und $L^p(R^n) = R(A_o) \oplus N(I-Q)$. Mit $B := A_o^{-1}(I-Q)$ verifiziert man dann die Behauptung ii).

Die Umkehrung ii) $\Rightarrow$ i) ist offensichtlich, ii $\Leftrightarrow$ iii) bekannt (vlg. z.B. K. Jörgens |12|, Seite 60, Satz 5.5).

Satz 3.2 Es sei A koerzitiv über $W^{m,p}(R^n)$.
Dann existiert ein $\gamma_1 > 0$, so daß

$$\gamma_1 \| u \|_{m,p} \leq \| Au \|_p + \| u \|_p, \quad u \in W^{m,p}(R^n).$$

Beweis. Es sei $u \in W^{m,p}(R^n)$. Wir zerlegen u in die direkte Summe $u = u_1 + u_2$ mit $u_1 \in N(P)$, $u_2 \in N(A)$ (vgl. Satz 3.1). Dann folgt aus (2) mit (3) und wegen der Äquivalenz der Normen $\| \cdot \|_p$ und $\| \cdot \|_{m,p}$ auf dem endlichdimensionalen Raum $N(A)$:

$$(4) \qquad \gamma \| u \|_{m,p} \leq \| Au \|_p + \langle \phi, u_1 \rangle + \langle \phi, u_2 \rangle =$$

$$\leq (1 + \frac{1}{\gamma'}) \| Au \|_p + \gamma'' \| \phi \| \, \| u_2 \|_p, \quad \gamma'' > 0 .$$

Da $N(A)$ auch in $L^p(R^n)$ einen Komplementärraum $N(\tilde{P})$ mit einem in $L^p(R^n)$ beschränkten Projektor $\tilde{P}$ der Eigenschaft $R(\tilde{P}) = N(A)$ besitzt, hat man insbesondere

$$(5) \qquad \| u_2 \|_p \leq \| \tilde{P} \| \cdot \| u \|_p, \quad u \in W^{m,p}(R^n) .$$

Mit (5) folgt aus (4) die Behauptung.

Die Umkehrung dieses Satzes gilt i. a. nicht, was man am Gegenbeispiel $A := \tilde{D}^2$ in $L^2(R^n)$ leicht demonstrieren kann. Jedoch ist die Umkehrung z. B. für abgeschlossene, lineare Operatoren A in $L^p(\Omega)$ mit $D(A) := H_o^{m,p}(\Omega) := \overline{C_o^\infty(\Omega)}^{\| \, \|_{m,p}}$, $\Omega \subset R^n$, offen und beschränkt, richtig aufgrund des Rellichschen Einbettungssatzes.
Zum vollständigen Beweis von Satz 2.1 haben wir noch zu zeigen:
j) $\Leftrightarrow$ jj) und j) $\Rightarrow$ $\beta(A) < \infty$. Dazu benötigen wir folgende Hilfssätze:

Satz 3.3 Es sei $n \geq 2$. Seien q und ℓ reelle Zahlen mit $\ell \geq \frac{n-1}{2}$, $1 < q \leq 2$.
Dann gilt:

$$W^{\ell,q}(S^{n-1}) \rightarrow W^{\ell - \frac{n-1}{q} + \frac{n-1}{2}, 2}(S^{n-1});$$

d. h. der links vom Pfeil stehende Raum ist ein linearer Teilraum des rechts stehenden, und der Einbettungsoperator ist stetig).
Vgl. R. A. Adams [1], S. 218, Theorem 7.58.

Satz 3.4 Es sei p reell mit $1 < p < \infty$. Sei k eine meßbare, komplexwertige Funktion auf $R^n \times R^n$, so daß $k' \in C_o(R^n, L^1(R^n))$,

$k'' \in L^\infty(R^n, L^1(R^n))$. K sei ein verallgemeinerter Faltungsoperator in $L^p(R^n)$ mit den Kern k:

(6) $$Kv := \int_{R^n} k(\cdot,y)v(\cdot-y)dy \ , \quad v \in L^p(R^n).$$

Dann ist A kompakt.

Man beweist die Behauptung des Satzes leicht mit Hilfe des Kriteriums von A. Weil (vgl. R. E. Edwards [9], S 269, Theorem 4.20.1) durch Verifikation folgender Abschätzungen:

i) $$\sup_{\|v\|_p=1} \|Kv\|_p < \infty$$

ii) Für alle $\varepsilon > 0$ existiert ein $r_0 \in N$, so daß für alle $r \geqq r_0$ gilt:

$$\sup_{\|v\|_p=1} \int_{|y|>r} |Kv(y)|^p \, dy \leqq \varepsilon.$$

iii) Für alle $\varepsilon > 0$ existiert ein $\delta > 0$, so daß für alle $h \in R^n$, $|h| < \delta$, gilt[1]:

$$\sup_{\|v\|_p=1} \|\tau_{-h} Kv - Kv\|_p \leqq \varepsilon .$$

Im folgenden sei $\{y_{r,t}\}_{r\in N_0,\ t=1,2,\ldots d_r}$

mit

$$d_r := \binom{r+n-1}{n-1} - \binom{r-2+n-1}{n-1}$$

das maximale, orthonormierte System der Kugelflächenfunktionen in $L^2(S^{n-1})$ bzgl. des Skalarproduktes

$$(u,v)_2^{S^{n-1}} := \int_{S^{n-1}} u\cdot\bar{v} \, do \ ; \ u,v \in L^2(S^{n-1}).$$

---

[1] Hier und im folgenden bez. wir für $v \in L^p(R^n)$ und $h \in R^n$ mit τ_h die Translation: $\tau_h v(x) = v(x-h)$, $x \in R^n$.

Satz 3.5 Es sei $n \geq 2$, seien q, p, $\ell_0(q)$ wie in Satz 2.1 gewählt, und sei $\ell > \ell_0(q)$. Sei s eine meßbare, komplexwertige Funktion auf $R^n \times R^n$ mit $s' \in C_0(R^n, W^{\ell,p}(S^{n-1}))$, und s' habe den Mittelwert Null über S^{n-1}. Ferner sei $k \in L^1(R^n)$. S und K seien Operatoren in $L^p(R^n)$: S sei definiert als singulärer Michlin-Operator der Gestalt

$$Sv := \lim_{\varepsilon \to 0} \int_{|y| \geq \varepsilon} \frac{s(\cdot, \frac{y}{|y|})}{|y|^n} v(\cdot - v)dy \ , \ v \in L^p(R^n) \ ,$$

und K sei ein verallgemeinerter Faltungsoperator mit dem Kern k (vgl. (6)).
Dann ist SK kompakt.

Beweis. Wir betrachten zunächst den Fall $1 < q \leq 2$. Nach dem Einbettungssatz 3.3 gilt wegen $\ell > \frac{n-1}{q} \geq \frac{n-1}{2}$ für $g := \ell - \frac{n-1}{q} + \frac{n-1}{2}$: $s' \in C_0(R^n, W^{g,2}(S^{n-1}))$.
Sei $x \in R^n$. Dann ist $s'(x,\cdot) \in L^2(S^{n-1})$, und $s'(x,\cdot)$ hat den Mittelwert Null über S^{n-1}. Folglich besitzt $s(x,\cdot)$ die $L^2(S^{n-1})$-konvergente Fourierreihenentwicklung:

$$s(x,.) = \sum_{r=1}^{\infty} \sum_{t=1}^{d_r} a_{rt}(x)\, y_{rt},$$

wobei

$$a_{rt}(x) := (s'(x,\cdot), y_{rt})_2^{S^{n-1}} \ .$$

Durch die Vorschrift $x \mapsto a_{rt}(x)$ werden für die obig. r und t komplexwertige Funktionen a_{rt} auf R^n definiert. Im folgenden seien stets r, t beliebig mit $r \in N$, $t \in \{1,\dots,d_r\}$. Dann gilt:

i) $a_{rt} \in C_0(R^n)$

ii) Es existiert ein $C_1 > 0$, unabhängig von $\{a_{rt}\}_{r \in N, t \in \{1,\dots,d_r\}}$ und s, so daß

$$\sup_{x \in R^n} \sum_{r=1}^{\infty} \sum_{t=1}^{d_r} r^{2g} |a_{rt}(x)|^2 \leq C_1 \max_{x \in R^n} \| s'(x,\cdot) \|_{g,2}^{S^{n-1}} \ .$$

iii) Es existiert ein $C_1' > 0$, unabhängig von a_{rt}, so daß

$$\max_{x \in R^n} |a_{rt}(x)| \leq C_1' \, r^{-2g} \ .$$

Beweis i), ii). Der Beweis von i) folgt sofort aus der Stetigkeitseigenschaft von s'; ii) folgt aus einem Satz von M. S. Agranowitsch (s. [2], S. 32, Proposition 6.2).

Beweis iii). Es sei Δ_1 der Laplace-Beltramioperator in $L^2(S^{n-1})$ mit dem Definitionsbereich $D(\Delta_1) := W^{2,2}(S^{n-1})$. Wählt man in R^n ein Kugelkoordinatensystem $\{0; f_o, f_1, \dots, f_{n-1}\}$, so daß jeder Punkt von S^{n-1} festgelegt ist durch die Kugelkoordinaten $(\theta_1, \dots, \theta_{n-1})$ $(-\pi \leq \theta_1 \leq \pi,\ o \leq \theta_2; \dots, \theta_{n-1} \leq \pi)$, so hat man für Δ_1 bekanntlich die Darstellung:

$$\Delta_1 := \Sigma_{j=1}^{n-1} \frac{1}{q_j \sin^{n-j-1} \theta_j} \frac{\partial}{\partial \theta_j} (\sin^{n-j-1} \theta_j \frac{\partial}{\partial \theta_j})$$

mit

$$q_j := \begin{cases} 1 & , \quad j = 1 \\ (\sin \theta_1 \dots \sin \theta_{j-1})^2 & , \quad j > 1 \end{cases}$$

(vgl. Triebel [24], S. 420, (31.34)).

Es ist bekannt, daß $\Delta_1^{g/2}$ eine selbstadjungierte Friedrichsche Fortsetzung in $L^2(S^{n-1})$ mit dem Definitionsbereich $W^{g,2}(S^{n-1})$ besitzt. Bezeichnet man diese Fortsetzung ebenfalls mit dem Symbol $\Delta_1^{g/2}$, so gilt also:

$$(\Delta_1^{g/2} u, v)_2^{S^{n-1}} = (u, \Delta_1^{g/2} v)_2^{S^{n-1}} \quad ; \ u, v \in L^2(S^{n-1}).$$

Nach Agranowitsch (vgl. [2], S. 32 ob.) besitzt $\Delta_1^{g/2}$ die Eigenwerte

$$\gamma_r := (-r)^g (r+n-2)^g \ ;$$

die zugehörigen Eigenfunktionen sind y_{rt}.

Aufgrund der vorstehenden Eigenschaften von $\Delta_1^{g/2}$ können wir jetzt a_{rt} abschätzen. Es gilt für alle $x \in R^n$:

$$a_{rt}(x) = \lambda_r (s'(x,\cdot), \Delta_1^{g/2} y_{rt})_2^{S^{n-1}} = \lambda_r (\Delta_1^{g/2} s'(x,\cdot), y_{rt})_2^{S^{n-1}} ,$$

und folglich

$$|a_{rt}(x)| \leq r^{-2g} \| \Delta_1^{g/2} s'(x,\cdot) \|_2^{S^{n-1}} ,$$

wobei die Cauchy-Schwarzsche Ungleichung verwendet und die Orthonormalität des Systems $\{y_{rt}\}_{r \in N, t \in \{1,\dots,d_r\}}$ beachtet wurde. Da $\Delta_1^{g/2}$

beschränkt ist, hat man weiter

$$|a_{rt}(x)| \leq C\, r^{-2g} \| s'(x,\cdot)\|_{g,2}^{S^{n-1}} \quad , \quad C > 0 \ ,$$

und nach dem Einbettungssatz 3.3:

$$|a_{rt}(x)| \leq C \cdot r^{-2g} \| s'(x,\cdot)\|_{\ell,q}^{S^{n-1}} \ .$$

Daraus folgt wegen $s' \in C_0(R^n, W^{\ell,q}(S^{n-1}))$ die Behauptung ii) mit $C_1' := C \cdot \max_{x\in R^n} \| s'(x,\cdot)\|_{\ell,q}^{S^{n-1}}$.

Es gilt ferner:

iv) Die Reihe $\Sigma_{r=1}^{\infty} \Sigma_{t=1}^{d_r} a_{rt}(x)\, y_{rt}(\xi)$ konvergiert absolut und gleichmäßig bzgl. $x \in R^n$ und $\xi \in S^{n-1}$.

Beweis iv). Es ist zu zeigen, daß für $r_2 > r_1 \geq 0$

$$\sum_{r=r_1+1}^{r_2} \sum_{t=1}^{d_r} |a_{rt}(x)|\ |y_{rt}(\xi)| \to 0$$

für $r_1, r_2 \to \infty$ gleichmäßig in x und ξ. Nach der Cauchy-Schwarzschen Ungleichung hat man

$$\sum_{r=r_1+1}^{r_2} \sum_{t=1}^{d_r} |a_{rt}(x)|\,|y_{rt}(\xi)| \leq$$

$$\leq \left[\sum_{r=1}^{\infty} \sum_{t=1}^{d_r} r^{2g} |a_{rt}(x)|^2\right]^{\frac{1}{2}} \cdot \left[\sum_{r=r_1+1}^{r_2} r^{-2g} \sum_{t=1}^{d_r} |y_{rt}(\xi)|^2\right]^{\frac{1}{2}} \leq$$

$$\leq \left[\max_{x\in R^n} \| s'(x,\cdot)\|_{g,2}^{S^{n-1}}\right]^{\frac{1}{2}} \cdot \left[\sum_{r=r_1+1}^{r_2} r^{-2g} \frac{d_r}{\omega_{n-1}}\right]^{\frac{1}{2}} ,$$

wobei in der zweiten Abschätzung die Aussage ii) und die Identität

$$\sum_{t=1}^{d_r} |y_{rt}(\xi)|^2 = \frac{d_r}{\omega_{n-1}}$$

berücksichtigt wurden (ω_{n-1} : Flächeninhalt von S^{n-1}). Beachtet man noch, daß

(7) $$d_r \leq C_3 r^{n-1} \quad , \quad C_3 > 0 \ ,$$

so resultiert:

$$\sum_{r=r_1+1}^{r_2} \sum_{t=1}^{d_r} |a_{rt}(x)||y_{rt}(\xi)| \leq$$

$$\leq \left[\max_{x\in R^n} \| s'(x,\cdot)\|_{g,2}^{S^{n-1}}\right]^{\frac{1}{2}} \cdot \frac{C_3^{1/2}}{\omega_{n-1}^{1/2}} \cdot \left[\sum_{r=r_1+1}^{r_2} r^{-2(g+1)+n}\right]^{\frac{1}{2}}.$$

Hier strebt wegen

$$g = \ell - \frac{n-1}{q} + \frac{n-1}{2} > \frac{n-1}{2}$$

wie gewünscht die rechte Seite gegen Null für $r_2, r_1 \to \infty$, so daß die Behauptung iv) bewiesen ist.

Es seien Y_{rt} die Giraudoperatoren in $L^p(R^n)$; diese sind definiert durch

$$Y_{rt}v := \lim_{\varepsilon\to 0} \int_{|y|\geq\varepsilon} \frac{Y_{rt}(\frac{y}{|y|})}{|y|^n} v(\cdot - y)dy, \quad v\in L^p(R^n).$$

Dann gilt:

v) Es existiert ein $C_4 > 0$, unabhängig von r und t, so daß

$$\sup_{\|v\|_p\leq 1} \|Y_{rt}v\|_p \leq C_4.$$

Beweis v). Nach einem Satz von A. P. Calderon und A. Zygmund (vgl. [4], S. 290, Theorem 2) hat man für alle $v\in L^p(R^n)$:

$$\|Y_{rt}v\|_p \leq C\cdot \|y_{rt}\|_q^{S^{n-1}} \|v\|_p$$

mit $C > 0$ unabhängig von y_{rt}.
Daraus folgt zusammen mit der Abschätzung

$$\|y_{rt}\|_q^{S^{n-1}} \leq C'\|y_{rt}\|_2^{S^{n-1}} \leq C', \quad C' > 0,$$

die Behauptung v) mit $C_4 := C\,C'$.

Wir definieren:

$$S' := \Sigma_{r=1}^{\infty} \Sigma_{t=1}^{d_r} a_{rt} Y_{rt} .$$

Dann gilt:

vi) Es ist $S = S'$ in $L^p(R^n)$, und S' konvergiert gleichmäßig im Raum der stetigen Endomorphismen von $L^p(R^n)$.

Beweis vi). $S = S'$ folgt sofort aus iv). Zum Beweis der Konvergenzaussage bzgl. S' ist es aufgrund der Vollständigkeit von $L^p(R^n)$ hinreichend, zu zeigen, daß die Folge $\{S_j'\}$ der Partialsummen von S' fundamental ist. Seien $r_2 > r_1 \geq 0$. Mit Hilfe der Minkowskischen Integralungleichung erhalten wir zunächst für alle $v \in L^p(R^n)$:

$$\| (S'_{r_2} - S'_{r_1})v \|_p \leq \Sigma_{r=r_1+1}^{r_2} \Sigma_{t=1}^{d_r} \| a_{rt} Y_{rt} v \|_p ,$$

und daraus mit iii), v) und (7):

$$\| S'_{r_2} - S'_{r_1} \|_{L^p(R^n) \to L^p(R^n)} \leq C_1' C_4 \cdot \Sigma_{r=r_1+1}^{r_2} r^{-2(g+1)+n} .$$

Wie im Beweis von iv) strebt hier die rechte Seite gegen Null für $r_1, r_2 \to \infty$, w. z. b. war.
Es gilt:

vii) Die Operatoren $a_{rt} Y_{rt} K$ sind kompakt in $L^p(R^n)$.

Beweis vii). Man hat wegen $Y_{rt} : F^{-1} \gamma_r y_{rt} F$ und $K = F^{-1} \hat{k} F$,

wobei

$$\gamma_r := i^r \pi^{\frac{n}{2}} \frac{\Gamma(\frac{r}{2})}{\Gamma(\frac{r+n}{2})} ,$$

die Identität:

$$a_{rt} Y_{rt} K = a_{rt} F^{-1} \gamma_r y_{rt} \hat{k} F = (a_{rt} K) \cdot Y_{rt} .$$

Hier ist nach i) und Satz 3.4 $a_{rt} K$ kompakt, und da Y_{rt} beschränkt ist, ist die Behauptung vii) bewiesen.

Die Behauptung des Satzes im betrachteten Fall $1 < q \leq 2$ folgt jetzt aus vi) und vii). Im Fall $2 < q < \infty$ hat man $g = \ell$ zu setzen, und der Beweis verläuft analog zum ersten Fall.

Beweis von j) ⇒ jj), Satz 2.1

Der Beweis besteht wieder aus mehreren Teilen. Man hat zunächst, da der Besselpotentialoperator $(I+\tilde{\Delta})^{-\frac{m}{2}}$ den Raum $L^p(R^n)$ isometrisch isomorph auf $W^{m,p}(R^n)$ abbildet:

i) Die Operatoren $\tilde{A} := A(I+\tilde{\Delta})^{-\frac{m}{2}}$ (mit $D(\tilde{A}) := L^p(R^n)$) und A (als Abbildung von $W^{m,p}(R^n)$ in $L^p(R^n)$) haben dieselben algebraischen und topologischen Eigenschaften; $\tilde{A}$ ist ein linearer, beschränkter Operator in $L^p(R^n)$.

Ferner gilt:

ii) $\tilde{A}$ ist, modulo kompakter Endomorphismen V von $L^p(R^n)$, ein Integraloperator der Gestalt (11) unten.

Beweis ii). Wir definieren auf R^n eine komplexwertige Funktion j_m durch Angabe ihres Fourier-Bildes gemäß

$$\hat{j}_m(y) := |y|^m (1+|y|^2)^{-\frac{m}{2}} - 1 , \qquad y \in R^n ;$$

für $|\alpha| \leq m-1$ definieren wir weitere komplexwertige Funktionen auf R^n durch

$$j_\alpha(x) := \tilde{D}^\alpha \left[(1+|\cdot|^2)^{-\frac{m}{2}} \right]^\wedge (x), \quad x \in R^n .$$

Man sieht leicht, daß j_m, $j_\alpha \in L^1(R^n)$. Wir bezeichnen mit J_m bzw. J_α die Faltungsoperatoren mit den Kernen j_m bzw. j_α. Nach einem Satz von Young sind J_m und $J_{\bar{\alpha}}$ in $L^p(R^n)$ stetig. Es sei ferner für $k = 1,\dots,n$, R_k die Riesz-Transformation in $L^p(R^n)$, definiert durch

$$R_k\, v := \lim_{\varepsilon\to 0} \frac{\Gamma(\frac{n+1}{2})}{\pi^{\frac{n+1}{2}}} \cdot i \int_{|y|\geq\varepsilon} \frac{y_k}{|y|^{n+1}} v(\cdot - y)dy , \quad v \in L^p(R^n).$$

Nach Cordes (vgl. [6], S. 1010, Lemma 5) hat man die Fourierdarstellung

(8) $$R_k = F^{-1} \frac{x_k}{|x|} F , \; k = 1,\dots,n .$$

Wir definieren für beliebiges $\alpha = (\alpha_1,\dots,\alpha_n) \in N_o^n$:

(9) $$R^\alpha := R_1^{\alpha_1} \dots R_n^{\alpha_n} .$$

Wegen (8) gilt:

$$R^\alpha = F^{-1} \frac{x^\alpha}{|x|^{|\alpha|}} F ,$$

und daher für $|\alpha| = m$:

(10) $$\tilde{D}^\alpha \left[(1+|\cdot|^2)^{-\frac{m}{2}} \right]^\wedge = R^\alpha(\delta + j_m)$$

(δ : Deltadistribution). Beachtet man die Definition von J_m und J_α, sowie (10), so erhält man:

(11) $$\tilde{A} = \Sigma_{|\alpha|=m} (S_\alpha + K_\alpha) R^\alpha (I + J_m) + \Sigma_{|\alpha| \leq m-1} (S_\alpha + K_\alpha) J_\alpha .$$

Wir definieren für $|\alpha| \leq m$ Operatoren S_α^∞ und K_α^∞ in $L^p(R^n)$ als singulären Michlin-Operator mit den Daten a_α^∞ und s_α^∞ bzw. als Faltungsoperator mit dem Kern k_α^∞ (vgl. die Zerlegung (1) der Daten a_α, s_α und k_α). Nach Satz 3.5 und 3.4 sind dann $S_\alpha - S_\alpha^\infty$ und $K_\alpha - K_\alpha^\infty$ kompakt. Berücksichtigen wir das in (11), so resultiert die Darstellung:

(12) $$\tilde{A} = \Sigma_{|\alpha|=m} S_\alpha R^\alpha + \Sigma_{|\alpha|=m} \{ K_\alpha^\infty R^\alpha + (S_\alpha^\infty + K_\alpha^\infty) R^\alpha J_m \} + \Sigma_{|\alpha| \leq m-1} (S_\alpha^\infty + K_\alpha^\infty) J_\alpha \qquad (\text{mod } V).$$

Es sei bemerkt, daß in (12) lediglich singuläre Michlin-Operatoren und Faltungsoperatoren mit integrierbaren Kernen auftreten. Es gilt:

iii) Es existiert ein beschränkter Regularisator S_1 von $S := \Sigma_{|\alpha|=m} S_\alpha R^\alpha$; d. h.

$$S_1 S = S S_1 = I \quad (\text{mod } V).$$

Beweis iii). Nach Konstruktion ist S ein singulärer Michlin-Operator, dem wir das Symbol $(x,\xi) \to \sigma_S(x,\xi) := \sigma_{A_1}(x,\xi)|\xi|^{-m}$ zuordnen.

Dieses Symbol hat folgende Eigenschaften:

a) Für alle $x \in \overline{R^n}$ ist $\sigma_S(x,\cdot)$ homogen vom Grade Null, so daß $\sigma_S(x,\cdot)$ identifiziert werden kann mit der Einschränkung $\sigma_{A_1}(x,\cdot)$ auf S^{n-1}.

b) $\sigma_S' \in C(\overline{R^n}, C(S^{n-1}) \cap W^{g+\frac{n}{2},2}(S^{n-1}))$

mit

$$g := \begin{cases} \ell - \frac{n-1}{q} + \frac{n-1}{2}, & 1 < q \leq 2 \\ \ell, & 2 < q < \infty \end{cases}$$

c) $\min_{x \in \overline{R^n}} \min_{\xi \in S^{n-1}} |\sigma_S(x,\xi)| > 0$.

Beweis a) Fundamentaleigenschaft des Symbols .

Beweis b) Die Stetigkeit von $\sigma_S'(x,\cdot)$ für alle $x \in \overline{R^n}$ wurde bereits früher bemerkt; $\sigma_S' \in C(\overline{R^n}, W^{g+\frac{n}{2},2}(S^{n-1}))$ folgt nach Agranowitsch (vgl. [2], S. 43, Theorem 7.12).

Beweis c) Folgerung aus der Elliptizität von A in $\overline{R^n}$. Die Elliptizitätsvoraussetzung i), Def. 1.3 , über A , aus der $\sigma_S(\cdot,\xi) \in C(\overline{R^n})$ für $\xi \in S^{n-1}$ folgt, bedeutet keine zusätzliche Glattheitsvoraussetzung über die Daten von A , da wegen b) $\sigma_S(\cdot,\xi)$ in eine absolut und gleichmäßig konvergente Reihe von $C(\overline{R^n})$-Funktionen entwickelt werden kann, was man analog zum Beweisschritt iv) von Satz 3.5 verifiziert.
Mit a), b), c) folgt für das zu σ_S inverse Symbol σ_S^{-1} : $(\sigma_S^{-1})' \in C(\overline{R^n}, C(S^{n-1}) \cap W^{g+\frac{n}{2},2}(S^{n-1}))$. Wir entwickeln σ^{-1} in eine Fourierreihe

$$\sigma^{-1}(x,\xi) = \Sigma_{r=1}^{\infty} \Sigma_{t=1}^{d_r} b_{rt}(x)\, y_{rt}(\xi) \;;\; x \in \overline{R^n}, \quad \xi \in S^{n-1} ,$$

welche sowohl für jedes x in $L^2(S^{n-1})$ konvergiert, als auch bzgl. x und ξ absolut und gleichmäßig konvergent ist. Dann hat die Reihe

$$s_1(x,\xi) := \Sigma_{r=1}^{\infty} \Sigma_{t=1}^{d_r} \frac{b_{rt}(x)}{\gamma_r}\, y_{rt}(\xi) \;;\; x \in \overline{R^n}, \quad \xi \in S^{n-1}$$

die gleichen Konvergenzeigenschaften wie die σ^{-1}-Reihe, es ist

(13) $s_1' \in C(\overline{R^n}, L^q(S^{n-1}) \cap W^{g,2}(S^{n-1}))$,

und s_1' hat den Mittelwert Null über S^{n-1}. Wir definieren einen singulären Michlin-Operator S_1 durch

(14) $$S_1 v := \lim_{\varepsilon \to 0} \int_{|y| \geq \varepsilon} \frac{s_1(\cdot, |\frac{y}{y}|)}{|y|^n} v(\cdot - y)dy, \quad v \in L^p(R^n).$$

S_1 ist ein beschränkter Operator in $L^p(R^n)$, da die S_1-Reihe gleichmäßig im Raum der stetigen Endomorphismen von $L^p(R^n)$ konvergiert, und man hat $\sigma_{S_1} = \sigma_S^{-1}$. Analog zum Beweis eines Satzes von S. G. Michlin (vgl. [14], S. 151, Theorem 1.33) kann man zeigen, daß gilt:

$$S_1 S = S S_1 = I \pmod V,$$

w.z.b. war.

Definiert man jetzt

(15) $$S^\infty := \Sigma_{|\alpha|=m} S_\alpha^\infty R^\alpha, \quad A^\infty := \Sigma_{|\alpha| \leq m} (S_\alpha^\infty + K_\alpha^\infty)\tilde{D}^\alpha,$$

so folgt aus (12) wegen (13), (15) und Satz 3.5:

$$S^\infty S_1 \tilde{A} = A^\infty (I + \tilde{\Delta})^{-\frac{m}{2}}$$

(16) $$= \bar{F}^1 \sigma_A^\infty (1 + |\cdot|^2)^{-\frac{m}{2}} F \pmod V$$

$$=: (B^\infty)^{-1} .$$

Aufgrund der Voraussetzungen über σ_A^∞ erhält man daraus die gewünschte a priori Abschätzung.

Beweis von jj) => j)

Die Elliptizität von A in $\overline{R^n}$ folgt mit Hilfe der Abschätzung von Satz 3.2 analog zum Beweis der Elliptizität aus der Gårdingschen Ungleichung für Differentialoperatoren.

Zum Beweis von $\inf_{\xi \in R^n} |\sigma_A^\infty(\xi)| > 0$ benötigt man folgende Hilfsaussagen auf deren Beweis wir hier aus Platzgründen verzichten müssen:

i) $\tilde{A}^\infty := A^\infty (I + \tilde{\Delta})^{-\frac{m}{2}}$ ist koerzitiv über $W^{m,p}(R^n)$.

ii) $\inf_{\xi \in R^n} |\sigma_{\tilde{A}^\infty}(\xi)| > 0.$

Beweis von j) ⇒ $\beta(A) < \infty$

Die Behauptung folgt sofort aus der Existenz eines Rechtsregularisators von A ; dieser ist identisch mit dem Linksregularisator von A.

Die Indexaussage in jjj) folgt aus (16).

Literatur

[1] Adams, R. A.:
Sobolev spaces, New York: Academic Press 1975.

[2] Agranovich, M. S.:
Elliptic singular integro-differential operators.
RMS 20, 1 - 121 (1969).

[3] Atkinson, F. V.:
The normal solubility of linear equations in normed spaces.
Mat. Sbornik, N. S. 28, 3 - 14 (1951).

[4] Calderon, A. P. u. Zygmund, A.:
On singular integrals.
Amer. J. of Math. 78, 289 - 309 (1956).

[5] ---:
Algebras of certain singular operators.
Amer. J. of Math. 78, 310 - 320 (1956).

[6] Cordes, H. O.:
The algebra of singular integral operators in R^n.
J. of Math. Mech. 14, No. 6, 1007 - 1032 (1965).

[7] Cordes, H. O., Herman, E. A.:
Gel'fand theory of pseudodifferential operators.
Amer. J. of Math. 90, 681 - 717 (1968).

[8] Donig, J.:
Zur Theorie einer Klasse elliptischer singulärer Integro-Differentialoperatoren in Grund- und Distributionenräumen.
Dissertation, Tübingen 1973.

[9] Edwards, R. E.:
Functional analysis.
Chicago: Holt, Rinehart and Winston 1965.

[10] Gruzin, V. V.:
Pseudodifferential operators on R^n with bounded symbols.
Funktsional'nyi Analiz i ego Prilozheniya 4, 37 - 50 (1970).

[11] Hörmander, L.:
Pseudodifferential operators.
Moscow: Mir 1967.

[12] Jörgens, K.:
Lineare Integraloperatoren.
Stuttgart: Teubner-Verlag 1970.

[13] Kohn, J. J. und Nirenberg, L.:
On the algebra of pseudodifferential operators.
Comm. Pure and Appl. Math. 18, 269 - 305.

[14] Mikhlin, S. G.:
Multidimensional singular integrals and integral equations.
Oxford: Pergamon Press 1965.

[15] Neri, U.:
Singular integrals.
Berlin: Springer-Verlag 1971.

[16] Palais, R. S.:
Seminar on the Atiyah-Singer index theorem.
Princeton: Princeton University Press 1965.

[17] Prößdorf, S.:
Einige Klassen singulärer Gleichungen.
Basel: Birkhäuser Verlag 1974.

[18] Seeley, R. T.:
Integro-differential operators on vector bundles.
I. Trans. Amer. Math. Soc. 117, 167 - 204 (1965).

[19] Simonenko, I. B.:
A new general method of investigating linear operator equations of the type of singular integral equations.
Soviet Math. Dokl. 5, 1323 - 1326 (1964).

[20] ---:
Singular integral equations with a continuous or piecewise continuous symbol.
SMD 8, 1320 - 1323 (1967).

[21] Speck, F.-O.:
Über verallgemeinerte Faltungsoperatoren und eine Klasse von Integrodifferentialgleichungen.
Dissertation, Darmstadt 1974.

[22] Stein, E.:
Singular integrals and differentiability properties of functions.
Princeton: Princeton University Press 1970.

[23] Stummel, F.:
Rand- und Eigenwertaufgaben in Sobolewschen Räumen.
Berlin: Springer-Verlag 1969.

[24] Triebel, H.:
Höhere Analysis.
Berlin: VEB Deutscher Verlag d. Wissenschaften 1972.

A Solution of the Biharmonic Dirichlet Problem by means of Hypercomplex Analytic Functions

J. Edenhofer

Technical University of Munich

1. Introduction

This paper deals with an extension of the theory of a special class of hypercomplex analytic functions, playing a similar part for the solution of the biharmonic Dirichlet problem as classical function theory for the solution of Laplace equation. Among others we give a generalization of the Cauchy integral formula for hypercomplex analytic functions and derive a Riemann mapping theorem. This finally leads to a solution of the biharmonic Dirichlet problem similar to that of Laplace equation by classical function theory.

I am very much obliged to my teacher E. Lammel, supporting me by word and deed to write this paper.

2. An Algebra of Hypercomplex Numbers

The algebra used in this paper is the same as in [7]. We therefore confine ourselves to a short summary of definitions and results.

Let A be the commutative and associative algebra of dimension 4 over the field $\mathbb{R}$ of real numbers with unity e_o and basis $e_o, e_1, e_2 = e_1^2$, $e_3 = e_1^3$, satisfying the relation

$$e_o + 2e_1^2 + e_1^4 = 0 . \tag{1}$$

Besides $a = \sum_{o}^{3} \alpha_i e_i$, $\alpha_i \in \mathbb{R}$, we write $a = (\alpha_o, \alpha_1, \alpha_2, \alpha_3)^t = (\alpha_i)^t$ (t= = transposed) for a hypercomplex number $a \in A$, yielding a geometric interpretation of hypercomplex numbers as points of $\mathbb{R}^4$.

The product (γ_i) of two hypercomplex numbers (α_i) and (β_i) is given by

$$(\gamma_i) = \Gamma(a) \cdot (\beta_i), \tag{2}$$

where

$$\Gamma(a) = \begin{pmatrix} \alpha_0 & -\alpha_3 & -\alpha_2 & 2\alpha_3-\alpha_1 \\ \alpha_1 & \alpha_0 & -\alpha_3 & -\alpha_2 \\ \alpha_2 & \alpha_1-2\alpha_3 & \alpha_0-2\alpha_2 & 3\alpha_3-2\alpha_1 \\ \alpha_3 & \alpha_2 & \alpha_1-2\alpha_3 & \alpha_0-2\alpha_2 \end{pmatrix} \qquad (3)$$

is the matrix of a.

$a \in A$ is a zero-divisor if and only if

$$\det \Gamma(a) = ((\alpha_0 - \alpha_2)^2 + (\alpha_1 - \alpha_3)^2)^2 = 0. \qquad (4)$$

The set of zero-divisors is a 2-dimensional linear subspace of A, which we denote by N.

A hypercomplex function is a mapping

$$f: D \to A \; ; \quad D \subset A.$$

Let A be normed by the Euclidean vector norm and f defined in a neighbourhood U(z) of $z \in D$. f is called hypercomplex differentiable in z, if there is a hypercomplex number f'(z), independent of h, such that

$$f(z+h) - f(z) = h \cdot f'(z) + \omega_f(z,h); \; z+h \in U(z); \qquad (5)$$

and

$$\frac{\|\omega_f(z,h)\|}{\|h\|} \to 0 \qquad \text{if} \quad h \to 0.$$

f is analytic in a domain $G \subset D$, if f is hypercomplex differentiable for all $z \in G$.

f is hypercomplex differentiable in $z = (\xi_0, \xi_1, \xi_2, \xi_3) \in D$ if and only if the components of f are totally differentiable in z and satisfy the system of generalized Cauchy - Riemann differential equations

$$\frac{\partial f(z)}{\partial \xi_j} = e_j \frac{\partial f(z)}{\partial \xi_0} \; ; \qquad j = 1,2,3 \; . \qquad (6)$$

Let $G \subset D$ be a simply-connected (any simple closed Jordan curve in G may be continuously contracted in G to a point of G) domain of $\mathbb{R}^4$ and $C \subset G$ a simple closed rectifiable Jordan curve. If f is analytic in G, we have the Cauchy integral theorem

$$\int_C f(z)\, dz = 0 \ . \tag{7}$$

Moreover every component f_i of f satisfies the biharmonic equation

$$\left(\frac{\partial^4}{\partial \xi_0^4} + 2\,\frac{\partial^4}{\partial \xi_0^2 \partial \xi_1^2} + \frac{\partial^4}{\partial \xi_1^4}\right) f_i(\xi_0,\xi_1,\xi_2,\xi_3) = 0, \tag{8}$$

where ξ_2, ξ_3 play the part of real parameters.

The behaviour of convergence of a hypercomplex power series

$$\sum_{0}^{\infty} a_n z^n \ , \quad a_n = (\alpha_i^{(n)}) \ , \quad z = (\xi_i) \ , \tag{9}$$

was solved by E. Lammel [6].

Accordingly, (9) is convergent for

$$(\xi_0 - \xi_2)^2 + (\xi_1 - \xi_3)^2 < \rho^2 \ , \tag{10}$$

where ρ is the smaller of the numbers $\left(\overline{\lim_{n\to\infty}} \sqrt[n]{(\alpha_0^{(n)} - \alpha_2^{(n)})^2 + (\alpha_1^{(n)} - \alpha_3^{(n)})^2}\right)^{-1}$ and $\left(\overline{\lim_{n\to\infty}} \sqrt[n]{(\alpha_1^{(n)} - 3\alpha_3^{(n)})^2 + 4(\alpha_2^{(n)})^2}\right)^{-1}$.

Finally we join [3], calling a hypercomplex function of the variable $(\xi_0,\xi_1,0,0)$ reduced. All given definitions and results hold analogously for reduced hypercomplex functions.

3. The Cauchy Integral Formula

In [2], by a detailed study of the set of zero divisors, we derived generalized Cauchy integral formulas for a wide class of algebras, special cases of them were given among others by L. Sobrero [7] and W. Eichhorn [3] in the case of reduced analytic functions and by R. Fueter [4] for the quaternions. The integral formula used here is the same as in [7], but we drop its restriction to reduced hypercomplex functions thus gaining a deeper insight in the structure of hypercomplex analytic functions defined in 2.

Let G be a simply-connected domain of $\mathbb{R}^4$, f a hypercomplex function analytic in G and $C \subset G$ a simple closed rectifiable Jordan curve, whose projection parallel the set of zero divisors onto $\xi_0\xi_1$-subspace of A, is assumed to be again a simple closed Jordan curve $\overline{C}$. We denote the interior of $\overline{C}$ by $G_{\overline{C}}$. Then the integral

$$\oint_C \frac{f(\xi)}{\xi - z}\, d\xi$$

is defined for all z of the cylinder

$$Z_{\overline{C}} = \{z = \overline{z} + h \mid \overline{z} \in G_{\overline{C}}\ ;\ h \in N\},$$

since $\xi - z$ is no zero divisor for $\xi \in C$, $z \in Z_{\overline{C}}$ and one can establish the Cauchy integral formula

$$f(z) = -\frac{1}{4\pi}\begin{pmatrix}0\\3\\0\\1\end{pmatrix}\oint_C \frac{f(\xi)}{\xi - z}\, d\xi, \quad z \in G \cap Z_{\overline{C}}\ . \tag{11}$$

Admitting $z \in Z_{\overline{C}} \setminus G$, f can be continued analytically on the whole of $Z_{\overline{C}}$ by virtue of (11), which will be tacitly assumed in the following. From (11) we conclude that f can be differentiated arbitrarily often and that

$$f^{(n)}(z) = -\frac{n!}{4\pi}\begin{pmatrix}0\\3\\0\\1\end{pmatrix}\oint_C \frac{f(\xi)}{(\xi - z)^{n+1}}\, d\xi\ ;\ n \in \mathbb{N}\,. \tag{12}$$

We now choose especially $C = \partial K_r(z_0)$ where

$$K_r(z_0) = \{z = z_0 + \rho(\cos\varphi\, e_0 + \sin\varphi\, e_1) \mid 0 \le \rho < r;\ 0 \le \varphi < 2\pi\},$$

$z_0 = (\xi_i^{(o)}) \in G$ and $r > 0$ such that the closure of $K_r(z_0)$ is contained in G. Then

$$f(z) = \sum_0^\infty \frac{f^{(n)}(z_0)}{n!}(z - z_0)^n \tag{13}$$

for all z of the cylinder $Z_{K_r(z_0)} := \{z = \overline{z} + h \mid \overline{z} \in K_r(z_0);\ h \in N\}$.

For $(z - z_0) = h \in N$ the series (13) breaks off and we get the fundamental relation

$$f(z_0 + h) = f(z_0) + h \cdot f'(z_0). \tag{14}$$

4. Definition of Hypercomplex Analytic Functions by means of Holomorphic Functions

In this section we describe a procedure to attach to any complex-valued function, holomorphic (in the classical sense) on a domain G of the complex plane a hypercomplex function, analytic on the cylinder

$$Z_G = \{z = \bar{z} + h | \bar{z} \in G \ ; \quad h \in N\}$$

of $\mathbb{R}^4$. We apply this to obtain a generalization of the Riemann mapping theorem for hypercomplex functions.

Let f be holomorphic on a domain G of the complex plane and $z_o = \xi_o^{(o)} + i\xi_1^{(o)} \in G$. f may be expanded in a power series

$$\sum_0^\infty a_k (z - z_0)^k ; \tag{15}$$

$$a_k = \alpha_0^{(k)} + i\alpha_1^{(k)} \in \mathbb{C} \ ; \quad z = \xi_0 + i\xi_1 \ ;$$

at least convergent in the largest circle $K_r(z_0)$ with center z_0 and radius r contained in G. Starting from (15) we define a hypercomplex analytic function by

$$F(z) = \sum_0^\infty a_k (z - z_0)^k \ ; \tag{16}$$

$$a_k = (\alpha_0^{(k)}, \alpha_1^{(k)}, 0, 0); \ z_0 = (\xi_0^{(o)}, \xi_1^{(o)}, 0, 0); \ z = (\xi_i);$$

for all z of the cylinder $Z_{K_r(z_0)} = \{z = \bar{z} + h | \bar{z} \in K_r(z_0); \ h \in N\}$ and call F a "hypercomplex analytic extension" of f (For simplicity we denote complex and hypercomplex variables with the same letter if no confusion is to be feared).

The convergence of the series (16) for $z \in Z_{K_r(z_0)}$ is a consequence of (10).

We should notice, that the hypercomplex analytic function F, defined by (16) is not uniquely determined by the holomorphic function f, but depends on the manner, f is written down.

Writing for example $f(z) = i^3 z = -iz$, we may define $F(z) = e_1^3 z$ or $F(z) = -e_1 z$. But one can prove, that the difference of two such functions is always an analytic function, whose range consists of nothing but zero divisors.

For let

$$F_o(z) := \sum_{o}^{\infty} b_k(z - z_o)^k \qquad (17)$$

be another hypercomplex function, defined like F starting from f. Since the residue class algebra A/N is isomorphic to $\mathbb{C}$, the "residue class power series" corresponding to (16) and (17) are identical and equal to (15). Thus the coefficients of (16) and (17) only differ by zero divisors. Admitting for z only the reduced variable $\overline{z} = (\xi_o,\xi_1,0,0)$ and applying the first of the Cauchy - Riemann equations (6) to $F(\overline{z}) - F_o(\overline{z})$, we conclude, that

$$F(\overline{z}) - F_o(\overline{z}) = (u(\overline{z}),\ v(\overline{z}),\ u(\overline{z}),\ v(\overline{z}))^t, \qquad (18)$$

where $u(\overline{z}) = u(\xi_o,\xi_1)$, $v(\overline{z}) = v(\xi_o,\xi_1)$ are conjugate harmonic functions. Thus any hypercomplex analytic extension F of a holomorphic function f ba (14) may we written

$$F(z) = F_o(z) + (u(\overline{z}),v(\overline{z}),u(\overline{z}),v(\overline{z}))^t;\ z = \overline{z} + h;\ h \in N; \qquad (19)$$
$$\overline{z} = (\xi_o,\xi_1,0,0);$$

where F_o is a special hypercomplex analytic extension of f and u, v are suitable conjugate harmonic functions.

6. Riemann Mapping Theorem for Hypercomplex Functions

In this section we generalize the Riemann mapping theorem for hypercomplex functions, which will be applied in the sequel to solve the biharmonic Dirichlet boundary value problem.

Let G be a simply-connected domain of the complex plane with more than one boundary point and w = f(z) a holomorphic function, mapping G conformally onto the interior of the unit circle $K_1(0)$.

The inverse function $z = f^{-1}(w)$ can be expanded in a power series, convergent for $|w| < 1$. Its hypercomplex analytic extension

$$z = \overset{-1}{F}(w) = \sum_{0}^{\infty} a_k w^k; \; w = (\omega_i); \tag{20}$$

converges in the cylinder

$$Z_{K_1(0)} = \{w = \overline{w} + k \mid \overline{w} \in K_1(0); \; k \in N\}.$$

By $\overset{-1}{F}$ $K_1(0)$ (= unit circle of $\omega_0\omega_1$ - subspace of A) will be mapped one-to-one onto a 2-parametric manifold M. We even show, that $\overset{-1}{F}$ is a bijective mapping of $Z_{K_1(0)}$ onto the cylinder $Z_M = \{z = z_M + h \mid z_M \in M; \; h \in N\}$ Every $z \in Z_M$ respectively $w \in Z_{K_1(0)}$ may be decomposed uniquely

$$z = z_M + h; \; z_M \in M; \; h \in N \tag{21}$$

respectively

$$w = \overline{w} + k; \; \overline{w} \in K_1(0); \; k \in N. \tag{22}$$

For an arbitrary $z = z_M + h \in Z_M$ there exists exactly one $\overline{w} \in K_1(0)$ with $z_M = \overset{-1}{F}(\overline{w})$, thus $z = \overset{-1}{F}(\overline{w}) + h$. If there is a $k \in N$ such that $h = \overset{-1}{F}{}'(\overline{w})\cdot k$, we conclude by (14) $z = \overset{-1}{F}(\overline{w}) + \overset{-1}{F}{}'(\overline{w})\cdot k = \overset{-1}{F}(\overline{w} + k)$ and have shown the existence of an inverse image $w = \overline{w} + k$ of z and thus proved the surjectivity of $\overset{-1}{F}$. This number k exists, since $\overset{-1}{F}{}'(\overline{w})$ has an inverse for every $\overline{w} \in K_1(0)$. Otherwise by differentiation of (20) and change to residue class algebra A/N we would obtain

$$0 = \overset{-1}{F}{}'(\overline{w}) = \sum_{1}^{\infty} k \, \tilde{a}_k \, \tilde{\overline{w}}^k = \overset{-1}{f}{}'(w)$$

contradicting $\overset{-1}{f}{}' \neq 0$ in $K_1(0)$. ($\tilde{a}$ residue class of $a \in A$).
To prove the injectivity of $\overset{-1}{F}$, we suppose, that $\overset{-1}{F}(w_1) = \overset{-1}{F}(w_2)$ for $w_1 \neq w_2$. By (14) and (22) we conclude

$$\overset{-1}{F}(\overline{w}_1) + k_1\overset{-1}{F}{}'(\overline{w}_1) = \overset{-1}{F}(\overline{w}_2) + k_2\overset{-1}{F}{}'(\overline{w}_2).$$

In case $\overline{w}_1 = \overline{w}_2$ we get $k_1 = k_2$ contradicting $w_1 \neq w_2$.
In case $\overline{w}_1 \neq \overline{w}_2$, the right-hand side of

$$\overset{-1}{F}(\overline{w}_1) - \overset{-1}{F}(\overline{w}_2) = k_2\overset{-1}{F}{}'(\overline{w}_2) - k_2\overset{-1}{F}{}'(\overline{w}_1)$$

is a zero divisor contrary to the left.

Thus F^{-1} is a bijective mapping of $Z_{K_1(0)}$ onto Z_M.

The projection of M parallel N onto $\xi_0\xi_1$-subspace of A is exactly G. We conclude this from (20) by transition to residue class algebra A/N. The inverse mapping F of F^{-1} maps G onto a 2-parametric manifold $\hat{G}$, whose projection parallel N onto $\omega_0\omega_1$-subspace of A is exactly $K_1(0)$. Furthermore $(F'(z_G))^{-1}$ exists for $z_G \in G$, which can be proved the same as above the existence of $(F^{-1'}(\overline{w}))^{-1}$ for $\overline{w} \in K_1(0)$.

Collecting results, we get the

Riemann mapping theorem for hypercomplex functions:

Let G be a simply-connected domain of $\xi_0\xi_1$-subspace of A with more than one boundary point. Then there is a hypercomplex function F, analytic on the cylinder $Z_G = \{z = \overline{z} + h | \overline{z} \in G;\ h \in N\}$ of $\mathbb{R}^4$ and $(F'(\overline{z}))^{-1}$ existing for $\overline{z} \in G$, mapping G one-to-one onto a 2-parametric manifold $\hat{G}$ of $w = (\omega_i)$-space, whose projection parallel N onto $\omega_0\omega_1$-subspace is the interior of the unit circle $K_1(0)$. Furthermore F is a bijective mapping of Z_G onto the circular cylinder $Z_{K_1(0)} = \{w = \overline{w} + h | \overline{w} \in K_1(0);\ h \in N\}$.

7. Solution of the Biharmonic Dirichlet Problem

Let G be a simply-connected bounded domain of $\xi_0\xi_1$-subspace of A and its boundary of class $C^{1+\alpha}$; $0 < \alpha < 1$. We try to solve the biharmonic Dirichlet problem for G, i.e. we try to determine a function $u_0(\xi_0, \xi_1)$, biharmonic in G and of class $C^1(G \cup \partial G)$ with

$$u_0(P) = g(P);\quad \frac{\partial u_0(P)}{\partial n} = h(P);\quad P \in \partial G; \tag{23}$$

where $g \in C^1(\partial G)$, $h \in C(\partial G)$ are prescribed real-valued functions.

By the hypercomplex Riemann mapping function F of G, we map G onto a 2-parametric manifold $\hat{G}$ with parametric equation

$$w(r,\alpha) = r(\cos\alpha, \sin\alpha, 0, 0)^t + (h_0(r,\alpha), h_1(r,\alpha) h_0(r,\alpha), h_1(r,\alpha))^t; \tag{24}$$
$$0 \le \alpha < 2\pi;\quad 0 \le r \le 1;$$

contained in the closure of the cylinder $Z_{K_1(0)} = \{w = \overline{w} + h | \overline{w} \in K_1(0);\ h \in N\}$

and consider the "transformed" Dirichlet problem: Find a hypercomplex function U, analytic on $Z_{K_1(0)}$, whose first component U_o, satisfies

$$U_o(\hat{P}) = g(P) \;; \qquad P \in \partial G \;;$$
$$\frac{\partial U_o(\hat{P})}{\partial r} = \frac{1}{|f'(P)|} h(P) \;;\; \hat{P} = F(P); \tag{25}$$

where f is the classical Riemann mapping function. By Kellog [5] $f'(P)$ is a continuous nonvanishing function of $P \in \partial G$.

The existence of U can be shown in the following way: Starting from u_o, by a result of Eichhorn [3], one can show the existence of "conjugate" functions u_1, u_2, u_3, such that $u = (u_i)$ is an analytic reduced hypercomplex function in G. Applying Cauchy's integral formula (11) with $C \subset G$ u can be continued analytically on the whole of the cylinder Z_G. Then the compositum $u \circ F$ can evidently be taken for U.

In order to obtain explicit formulas for U_o, we start from the most general hypercomplex function analytic on $Z_{K_1(0)}$, which can be written

$$U(w) = \sum_o^\infty a_k w^k . \tag{26}$$

Writing $a_k = (\alpha_i^{(k)})$, $\overline{w} = r(\cos\alpha, \sin\alpha, 0, 0)^t$, $h = (h_o, h_1, h_o, h_1)^t$ and using (14) and a formula of Sobrero [7] for $\overline{w}^k$, we obtain

$$U(w) = \sum_o^\infty a_k w^k = \sum_o^\infty a_k \overline{w}^k + h \sum_1^\infty k a_k \overline{w}^{k-1} = \tag{27}$$

$$= \sum_o^\infty r^k (\alpha_i^{(k)}) \begin{pmatrix} \cos k\alpha + \frac{k}{2}\sin\alpha\sin(k-1)\alpha \\ \frac{3}{2}\sin k\alpha - \frac{k}{2}\sin\alpha\cos(k-1)\alpha \\ \frac{k}{2}\sin\alpha\sin(k-1)\alpha \\ \frac{1}{2}\sin k\alpha - \frac{k}{2}\sin\alpha\cos(k-1)\alpha \end{pmatrix} +$$

$$+ \begin{pmatrix} h_o \\ h_1 \\ h_o \\ h_1 \end{pmatrix} \cdot \sum_1^\infty k\, r^{k-1} (\alpha_i^{(k)}) \begin{pmatrix} \cos(k-1)\alpha + \frac{k-1}{2}\sin\alpha\sin(k-2)\alpha \\ \frac{3}{2}\sin(k-1)\alpha - \frac{k-1}{2}\sin\alpha\cos(k-2)\alpha \\ \frac{k-1}{2}\sin\alpha\sin(k-2)\alpha \\ \frac{1}{2}\sin k-1)\alpha - \frac{k-1}{2}\sin\alpha\cos(k-2)\alpha \end{pmatrix} .$$

The first component of this equation is

$$U_0(w) = \sum_{0}^{\infty} r^k[(\alpha_o^{(k)} - \tfrac{k}{4}\alpha_k)\cos k\alpha + ((\tfrac{1}{2}+\tfrac{k}{4})\beta_k - \alpha_1^{(k)})\sin k\alpha + \tag{28}$$

$$+ \tfrac{k}{4}\,\alpha_k \cos(k-2)\alpha - \tfrac{k}{4}\,\beta_k \sin(k-2)\alpha] +$$

$$+ h_0 \cdot \sum_{1}^{\infty} kr^{k-1}(\alpha_k \cos(k-1)\alpha - \beta_k \sin(k-1)\alpha) +$$

$$+ h_1 \cdot \sum_{1}^{\infty} kr^{k-1}(\alpha_k \sin(k-1)\alpha - \beta_k \cos(k-1)\alpha) ,$$

where $\alpha_k = \alpha_o^{(k)} - \alpha_2^{(k)}$, $\beta_k = \alpha_1^{(k)} - \alpha_3^{(k)}$.

Substituting $k\alpha_k = c_{k-1}$, $-k\beta_k = d_{k-1}$ we finally get

$$U_0(w) = \omega(r,\alpha) + \frac{r^2-1}{4}\sum_{0}^{\infty} r^k(c_{k+1}\cos k\alpha + d_{k+1}\sin k\alpha) + \tag{29}$$

$$+ h_0 \cdot \sum_{0}^{\infty} r^k\,(c_k \cos k\alpha + d_k \sin k\alpha)$$

$$+ h_1 \sum_{0}^{\infty} r^k\,(c_k \sin k\alpha + d_k \cos k\alpha),$$

where ω is a harmonic function.

Inversely,it is not difficult proving that any function (29) with harmonic ω can be taken as the first component of a hypercomplex function U, analytic on $Z_{K_1(0)}$. Some further easy calculation and the substitution $\varphi(r,\alpha) = \sum_{0}^{\infty} r^k(c_{k+1}\cos k\alpha + d_{k+1}\sin k\alpha)$ yield

$$U_0(w) = \omega(r,\alpha) + \frac{r^2-1}{4}\,\varphi(r,\alpha) + \tag{30}$$

$$+ h_0(r\cos\alpha\,\varphi(r,\alpha) - r\sin\alpha\,\psi(r,\alpha)) +$$

$$+ h_1(r\sin\alpha\,\varphi(r,-\alpha) - r\cos\alpha\,\psi(r,-\alpha)) + c_0 h_0 + d_0 h_1,$$

where ψ is the conjugate harmonic of φ.

Substituting $w = w(r,\alpha)$ (= parametric equation (24) of $\hat{G}$) in (29) and (30), or what is the same, putting $h_0 = h_0(r,\alpha)$, $h_1 = h_1(r,\alpha)$ in (29) and (30), we get representation formulas for the first component U_0 on $\hat{G}$ of the most general hypercomplex function analytic on $Z_{K_1(0)}$, where ω and φ are arbitrary harmonic functions on $K_1(0)$. We now suppose to have $\omega, \varphi, c_0, d_0$ such that U_0 satisfies (25). The compositum $U \circ F$ is an analytic function on Z_G. Every of its components is a biharmonic function for $z \in G$. Moreover its first component satisfies boundary conditions (23) and therefore is a solution of the biharmonic Dirichlet problem for G.

References

[1] J. Horvath, A Generalization of the Cauchy - Riemann Equations, Contributions to Differential Equations, Volume I, Interscience Publishers, 1963, 39-58.

[2] J. Edenhofer, Analytische Funktionen auf Algebren, TU München, 1973, Dissertation.

[3] W. Eichhorn, Funktionentheorie in Algebren über dem reellen Zahlenkörper und ihre Anwendung auf partielle Differentialgleichungen, Würzburg, 1961, Dissertation.

[4] R. Fueter, Über die analytische Darstellung der regulären Funktionen einer Quaternionenvariablen, Comment. Math. Helvet., 8,1935 ,371.

[5] O.D. Kellogg, Harmonic functions and Greens integral, Transactions of the Americ. Math. Society 13, 1912, 109-132.

[6] E. Lammel, Über eine zur Differentialgleichung $(a_0 \frac{\partial^n}{\partial x^n} + a_1 \frac{\partial^n}{\partial x^{n-1}\partial y} + \ldots + \frac{\partial^n}{\partial x^n}) U(x,y) = 0$ gehörige Funktionentheorie I, Math. Ann. 122, 1950/51, 109-126.

[7] L. Sobrero, Theorie der ebenen Elastizität, 1934, B.G. Teubner.

Existenz- und Eindeutigkeitsproblem bei der Abstrahlung ebener Wellen aus einem angeströmten Ringkanal

Norbert FRIEDRICH

Mathematisches Institut der Universität
Saarbrücken

Einleitung

Behandelt wird ein idealisiertes Modell für den Einlauf eines Strahltriebwerkes. Der Triebwerksmantel wird dabei repräsentiert durch einen halbunendlich langen Zylinder vom Radius a. Die Triebwerksnabe wird dargestellt durch einen koaxial angeordneten beidseitig ins Unendliche ausgedehnten Zylinder vom Radius b, $0<b<a$. Diese Anordnung der beiden Zylinder wird in axialer Richtung angeströmt und durchströmt durch eine homogene Unterschallgrundströmung eines reibungsfreien kompressiblen Gases. Repräsentativ für den vom Triebwerk erzeugten Lärm werde ganz hinten im Ringkanal eine ebene Welle angeregt. Der Fall der Anregung durch umlaufende Wellen ist in [2] behandelt. Gesucht ist das durch diese Anregung im ganzen Raum induzierte Geschwindigkeitsfeld in der akustischen Näherung. Die Lösung erfolgt durch Anwendung der Fouriertransformation in axialer Richtung und mit Hilfe der Wiener-Hopf-Methode. Durch Anwendung der Fouriertransformation für temperierte Distributionen gelingt die Diskussion der Eindeutigkeitsfrage in einem Arbeitsgang. Solange die sogenannten cut-off Frequenzen gemieden werden, erhält man Eindeutigkeit, andernfalls treten neue Lösungsanteile mit Resonanzcharakter auf. Die physikalische Bedeutung dieser Lösungsanteile ist noch zu untersuchen.

Der mathematischen Formulierung des Problems werden noch einige Vorbemerkungen und Bezeichnungen vorangestellt:

a_0 Schallgeschwindigkeit bei ungestörter Grundströmung der Geschwindigkeit - U,

p_0, d_0 Druck und Dichte bei ungestörter Grundströmung,

$M := \frac{U}{a_0}$ Machzahl bei ungestörter Grundströmung, $0<M<1$.

(r,ϑ,z) bezeichnet das auf den mit der Geschwindigkeit $-U$ mitgeführten Beobachter bezogene Zylinderkoordinatensystem. Gleichzeitig ist dabei der Radius des äußeren Zylinders auf 1 normiert, und der Radius des inneren Zylinders ergibt sich als $\frac{b}{a}$.

Da nur ebene Wellen betrachtet werden, genügt eine 2-dimensionale Formulierung des Problems:

$E := \{(r,z) \mid r \geq \frac{b}{a},\ z \in \mathbb{R}\} \subseteq \mathbb{R}^2$, so daß $E \times [o,2\pi]$ den 3-dimensionalen Raum ohne den inneren Zylinder darstellt.

$Z_1 := \{\frac{b}{a}\} \times \mathbb{R}$ repräsentativ für die Wand des inneren Zylinders,

$Z_l := \{1\} \times]-\infty,0[$ repräsentativ für die Wand des äußeren Zylinders,

$Z_r := \{1\} \times]0,\infty[$ gedachte Verlängerung von Z_l nach rechts,

$K := \{1\} \times \{0\}$ Vorderkante von Z_l.

Zur mathematischen Formulierung des Problems sei noch vorausgeschickt, daß nur solche Geschwindigkeitsfelder untersucht werden, die sich aus einem Geschwindigkeitspotential gewinnen lassen. Außerdem werden nur eingeschwungene Zustände betrachtet. In der akustischen Näherung bedeutet dies: $(v_r,v_z) = \operatorname{grad} \varphi\, e^{-i\omega t} - (0,U)$, wobei grad φ als Störung des Ruhezustandes stets hinreichend klein zu sein hat. Die Störung des Druckes aus dem Ruhezustand ergibt sich dabei als

$p(r,z) = \frac{d_o U}{a\sqrt{1-M^2}} \left(i\frac{k}{M} + \frac{\partial}{\partial z}\right) \varphi(r,z)$. φ genügt der Schwingungsgleichung $\left(\frac{\partial^2}{\partial z^2} + \frac{\partial^2}{\partial r^2} + \frac{1}{r}\frac{\partial}{\partial r} + k^2\right)\phi = o$ mit $k = \frac{\omega a}{a_o\sqrt{1-M^2}}$

<u>Mathematische Formulierung</u>: Betrachtet wird folgende Klasse von Geschwindigkeitspotentialen $\varphi : E \setminus (Z_l \cup K) \to \mathbb{C}$ mit den Eigenschaften:

1. φ ist stetig auf $E \setminus (Z_l \cup K)$. Außerdem existieren für alle $z<0$:

   $\lim\limits_{r\to 1+} \varphi(r,z) =: \varphi(1+,z)$ und $\lim\limits_{r\to 1-} \varphi(r,z) =: \varphi(1-,z)$.

2. Auf $E \setminus (Z_1 \cup Z_l \cup K)$ ist φ zweimal partiell stetig differenzierbar und erfüllt die Schwingungsgleichung $\left(\frac{\partial^2}{\partial z^2} + \frac{\partial^2}{\partial r^2} + \frac{1}{r}\frac{\partial}{\partial r} + k^2\right)\phi = o$

3. Auf Z_1 und Z_l verschwindet die Normalgeschwindigkeit, also

   $\lim\limits_{r\to \frac{b}{a}+} \frac{\partial}{\partial r}\varphi(r,z) = 0$ für alle $z \in \mathbb{R}$ und $\lim\limits_{r\to 1\pm} \frac{\partial}{\partial r}\varphi(r,z) = 0$

für alle $z \in \,]-\infty,0[$.

Physikalisch bedeutet das, daß die Rohrwände schallhart sind.

4. Der Druck $p(r,z) = \frac{d_0 U}{a\sqrt{1-M^2}} (i\frac{k}{M} + \frac{\partial}{\partial z})\varphi(r,z)$ ist stetig in $E \setminus (Z_l \cup K)$ und von beiden Seiten auf Z_l fortsetzbar in dem Sinne: $\lim_{r\to 1+} p(r,z)$ und $\lim_{r\to 1-} p(r,z)$ existieren für alle $z<0$.

5. An der Kante K gilt folgende Kantenbedingung für den Druck:
$p(r,z) = O(|(r-1,z)|^{-\frac{1}{2}})$ für $(r,z)\to(1,0)$ mit der Verschärfung
$\varphi(r,z) = O(|(r-1,z)|^{-\frac{1}{2}})$ und $\frac{\partial}{\partial z}\varphi(r,z) = O(|(r-1,z)|^{-\frac{1}{2}})$.
Die Kantenbedingung trägt wesentlich zur Sicherung der Eindeutigkeit der Lösung bei. Eine Konsequenz der Kantenbedingung ist die Beschränktheit der Druckkraft an der Vorderkante. Die physikalische Beobachtung aber fordert noch mehr, das Beschränktbleiben des Geschwindigkeitsfeldes im ganzen Raum. Im Rahmen der hier verwendeten reibungsfreien Theorie ist diese Forderung aber nicht erfüllbar.

6. Im Ringkanal, also für $(r,z) \in [\frac{b}{a},1]\times]-\infty,0[$, hat φ folgende Zerlegung mit $A_0 \in \mathbb{C}$:
$$\varphi(r,z) = A_0 u_\nu(r) e^{i\varkappa_\nu z} + \varphi_{ind}(r,z).$$
$u_\nu(r)e^{i\varkappa_\nu z}$ ist dabei eine Hauptschwingung im Ringkanal, so daß die Energietransportrichtung von $u_\nu(r)e^{i\varkappa_\nu z}\, e^{-i\omega t}$ von $-\infty$ in Richtung der Mündung des Ringkanals verläuft. Die u_ν sind durch die Forderung $u_\nu(1)=1$ normiert. Explizit ergibt sich
$$u_0(r)\equiv 1,\quad u_\mu(r) = \frac{J_1(\lambda_\mu\frac{b}{a})Y_0(\lambda_\mu r) - J_0(\lambda_\mu r)Y_1(\lambda_\mu\frac{b}{a})}{J_1(\lambda_\mu\frac{b}{a})Y_0(\lambda_\mu) - J_0(\lambda_\mu)Y_1(\lambda_\mu\frac{b}{a})} \quad \mu\in\mathbb{N}$$
λ_μ ist dabei die μ-te positive Lösung von
$$J_1(\lambda\tfrac{b}{a})Y_1(\lambda) - J_1(\lambda)Y_1(\lambda\tfrac{b}{a}) = 0;\quad \varkappa_0 := k,\quad \varkappa_\mu := \sqrt{|k^2-\lambda_\mu^2|}\ .$$
Mit J bzw. Y werden die entsprechenden Bessel- bzw. Neumannfunktionen bezeichnet wie bei [1o](Watson).
Sei $\nu_0 - 1 \in \mathbb{N}$ der Index mit $\lambda_\mu \le k$ für alle $\mu \le \nu_0 - 1$, so ergibt sich, daß von den aus den Hauptschwingungen abgeleiteten Wellen genau

die Wellen $u_\mu(r)e^{i\varkappa_\mu z}e^{-i\omega t}$ $0\le\mu\le\nu_0-1$ Energie von $-\infty$ in Richtung der Rohrmündung transportieren. Im cut-off Fall $\lambda_{\nu_0-1}=k$ ist der Index $\mu=\nu_0-1$ von dieser Aussage auszuschließen.

φ_{ind} ist als der von $A_0u_\nu(r)e^{i\varkappa_\nu z}$ induzierte Anteil des Geschwindigkeitspotentials aufzufassen. φ_{ind} hat die Sommerfeldsche Ausstrahlungsbedingung zu erfüllen in dem Sinn, daß φ_{ind} keine Energie aus dem Unendlichen ins Endliche transportiert, siehe [8],[2]. Offen ist zunächst noch, ob eine Anregung durch ebene Wellen nicht auch noch umlaufende Wellen induziert. Das führt auf das Eindeutigkeitsproblem bei Anregung durch umlaufende Wellen, siehe [2]. Dabei ergibt sich, daß bis auf die noch offenen Fälle bei den cut-off Frequenzen durch ebene Wellen wieder ebene Wellen angeregt werden.

7. Außerhalb des Ringkanals erfüllt φ die Sommerfeldsche Ausstrahlungsbedingung. Alle vorkommenden Wellenkomponenten entsprechen daher auslaufenden Wellen. Dabei wird gefordert:

$\lim\limits_{r\to\infty}\varphi(r,z)=0$ für jedes $z\in\mathbb{R}$, $\lim\limits_{z\to\infty}\varphi(r,z)=0$ für jedes $r\ge\frac{b}{a}$ und

$\lim\limits_{z\to-\infty}\varphi(r,z)=0$ für jedes $r\ge1$.

Speziell für z=0 wird gefordert

$$\varphi(r,0) = A_1\frac{e^{ikr}}{r} + B_1\frac{e^{ikr}}{r^{3/2}} + O(\frac{1}{r^2}) \quad \text{und}$$

$$\frac{\partial}{\partial z}\varphi(r,0) = A_2\frac{e^{ikr}}{r} + B_2\frac{e^{ikr}}{r^{3/2}} + O(\frac{1}{r^2}) \quad \text{mit } A_1,B_1,A_2,B_2 \in \mathbb{C} .$$

Diese Forderungen lassen sich plausibel machen mit Hilfe des Huygensschen Prinzips, wonach von jedem Kantenelement der Kante K des Ringkanals auslaufende Kugelwellen in den Außenraum induziert werden. Für z=0 wird eine stärkere Forderung benötigt, die durch die angewandte Beweisführung bedingt ist.

Nach diesen an der Physik orientierten Bedingungen folgen nun drei mathematisch bedingte Forderungen, die das Problem der Behandlung durch die Fouriertransformation und Wiener-Hopf-Methode zugänglich machen.

8. Es gibt lokal integrierbare, im Unendlichen polynombeschränkte Funktionen g_1 und g_2 mit den Eigenschaften:

$|\varphi(r,z)| \leq g_1(z)$

$|\frac{\partial}{\partial z} \varphi(r,z)| \leq g_2(z)$ auf dem ganzen Definitionsbereich.

9. Es gibt eine lokal integrierbare, im Unendlichen polynombeschränkte Funktion h_1 mit der Eigenschaft:

$|\frac{\partial}{\partial r} \varphi(r,z)| \leq h_1(z)$ auf dem ganzen Definitionsbereich.

10. Zu jedem $r>\frac{b}{a}$, $r\neq 1$, gibt es eine Umgebung $U(r)$ und eine lokal integrierbare Funktion $h_U : \mathbb{R} \to \mathbb{R}$, welche für große $|z|$ polynombeschränkt ist, so daß gilt:

$|\frac{\partial^2}{\partial r^2} \varphi(r',z)| \leq h_U(z)$ für alle $(r',z) \in U(r) \times \mathbb{R}$.

Die letzten drei Bedingungen sind sehr allgemein gehalten und bieten einen wesentlichen Fortschritt gegenüber Formulierungsversuchen im Rahmen der klassischen Fouriertransformation.

Lösungsgang: Zunächst wird die Existenz einer Lösung entsprechend den Bedingungen 1. bis 1o. angenommen. Davon ausgehend wird eine Lösungsformel gewonnen, deren Lösungseigenschaften anschließend zu verifizieren sind. Offenbar läßt sich $\varphi(r,z)$ als vom Parameter r abhängige Schar temperierter Distributionen auffassen. Für $r\neq 1$ sind $\varphi(r,\cdot), \frac{\partial}{\partial z}\varphi(r,\cdot), \frac{\partial^2}{\partial z^2}\varphi(r,\cdot), \frac{\partial}{\partial r}\varphi(r,\cdot), \frac{\partial^2}{\partial r^2}\varphi(r,\cdot)$ stetige, für $|z|\to\infty$ polynombeschränkte Funktionen. Also existiert die distributionelle Fouriertransformation entlang der z-Achse für $r>\frac{b}{a}$, $r\neq 1$.
Für $f\in\mathfrak{S}$ (siehe [7]) sei $\mathfrak{F}f(\alpha) := \int_{\mathbb{R}} f(x)e^{i\alpha x}dx$ für $\alpha\in\mathbb{R}$. $\mathfrak{F}$ sei wie üblich fortgesetzt auf $\mathfrak{S}'$. Da kaum Verwechslungen zu befürchten sind, setze man $\varphi(r,\alpha) := (\mathfrak{F}\varphi(r,\cdot))(\alpha)$.
Bei dieser Schreibweise ist vorweggenommen, daß $\varphi(r,\alpha)$ im wesentlichen eine temperierte Distribution ist, die zu einer lokalintegrierbaren Funktion gehört. Da in der Folge formal Produkte von Distributionen auftreten, ist es geschickt, die Distributionen als Randwerte holomorpher Funktionen aufzufassen. Dazu beachte man, daß für Funktionen aus $\mathfrak{S}'$ die einseitigen Fouriertransformierten erklärt sind.

<u>Bezeichnung</u> für Funktionen aus $\mathfrak{S}'$:

$$\mathfrak{F}_+f(\alpha) := (\mathfrak{F}\, H(t)f(t))(\alpha) \qquad \alpha\in\mathbb{R}; \quad H(t) = \begin{cases} 1 & t\geq 0 \\ 0 & t<0 \end{cases}$$
$$\mathfrak{F}_-f(\alpha) := (\mathfrak{F}\, H(-t)f(t))(\alpha)$$

$\mathfrak{F}_+f$ ist dabei distributioneller Randwert der für $\tau=\operatorname{Im}\alpha>0$ holomorphen Funktion $(\mathfrak{F}_+f)(\sigma+i\tau) := \int_0^\infty f(t)e^{-\tau t}\, e^{i\sigma t}\,dt$ siehe [1], und analog $\mathfrak{F}_-f$ bezüglich der unteren Halbebene. Für $\tau>0$, $\sigma\in\mathbb{R}$ erhält man

$(\mathfrak{F}_-f)(\sigma-i\tau)+(\mathfrak{F}_+f)(\sigma+i\tau) = \int_{-\infty}^{\infty} f(t)e^{-\tau|t|}\, e^{i\sigma t}\,dt$ mit

$\lim_{\tau\to 0} (\mathfrak{F}_-f(\sigma-i\tau)+\mathfrak{F}_+f(\sigma+i\tau)) = \mathfrak{F}f(\sigma)$ im Distributionssinn.

Für $r>\frac{b}{a}$, $r\neq 1$ ergibt sich durch partielle Integration

$$(\mathfrak{F}_\pm \tfrac{\partial}{\partial z}\varphi(r,\cdot))(\alpha) = -i\alpha\,\varphi_\pm(r,\alpha) \mp \varphi(r,z=0), \quad \operatorname{Im}\alpha \gtrless 0$$

$$(\mathfrak{F}_\pm \tfrac{\partial^2}{\partial z^2}\varphi(r,\cdot))(\alpha) = -\alpha^2\varphi_\pm(r,\alpha) \pm i\alpha\varphi(r,z=0) \mp \tfrac{\partial}{\partial z}\varphi(r,z=0), \quad \operatorname{Im}\alpha \gtrless 0$$

wobei $\varphi_\pm(r,\alpha) := (\mathfrak{F}_\pm\varphi(r,\cdot))(\alpha)$.

Die Bedingungen 9. und 1o. zeigen für $r>\frac{b}{a}$, $r\neq 1$:

$$\frac{d}{dr}\varphi_\pm(r,\alpha) = (\mathfrak{F}_\pm\tfrac{\partial}{\partial r}\varphi(r,\cdot))(\alpha), \quad \frac{d^2}{dr^2}\varphi_\pm(r,\alpha) = (\mathfrak{F}_\pm \tfrac{\partial^2}{\partial r^2}\varphi(r,\cdot))(\alpha).$$

Bedingung 2 führt auf

$$\left(\frac{d^2}{dr^2} + \frac{1}{r}\frac{d}{dr} - (\alpha^2-k^2)\right)\varphi_\pm(r,\alpha) = \pm\left(\frac{\partial}{\partial z}\varphi(r,z=0) - i\alpha\varphi(r,z=0)\right) \text{ für}$$

$\operatorname{Im}\alpha \gtrless 0$.

Zur Lösung dieser Differentialgleichungen erkläre man $q = (\alpha^2-k^2)^{\frac{1}{2}}$ in der komplexen Ebene, die von k bis 0 und von 0 bis $i\infty$ aufgeschnitten ist, sowie von $-k$ bis 0 und von 0 bis $-i\infty$. Diese Ebene sei mit E_I bezeichnet. q läßt sich in E_I als holomorphe Funktion erklären mit $\operatorname{Re} q\geq 0$ in E_I und $q^2 = \alpha^2-k^2$. Dabei ist q stetig, wenn man vom 2. Quadranten über den Nullpunkt in den 4. Quadranten gelangt; siehe [6] und [2].

Im Falle $r>1$ zeigen Bedingung 7. und 8. $\lim_{r\to\infty}\varphi_\pm(r,\alpha) = 0$. Als einzige Lösung der Differentialgleichung mit $\lim_{r\to\infty}\varphi_\pm(r,\alpha) = 0$ erhält man für $\alpha\in E_I$, $\operatorname{Im}\alpha \gtrless 0$:

$$\varphi_\pm(r,\alpha) = E_\pm(\alpha) \frac{H_o^{(1)}(iqr)}{H_o^{(1)}(iq)} \pm \mathrm{inh}(r,\alpha) \text{ mit}$$

$$\mathrm{inh}(r,\alpha) = -\frac{i\pi}{4} H_o^{(1)}(iqr) \int_2^r H_o^{(2)}(iq\rho)\rho \left(\frac{\partial}{\partial z} \varphi(\rho,z=0) - i\alpha\varphi(\rho,z=0)\right)d\rho$$

$$-\frac{i\pi}{4} H_o^{(2)}(iqr) \int_r^\infty H_o^{(1)}(iq\rho)\rho \left(\frac{\partial}{\partial z} \varphi(\rho,z=0) - i\alpha\varphi(\rho,z=0)\right)d\rho$$

Die Integrationskonstanten $E_\pm$ lassen sich durch die Randverteilungen für r=1 ersetzen. Nach Bedingung 8. existieren $\lim_{r\to 1+} \varphi_+(r,\alpha) =: \varphi_+(1+,\alpha)$ und $\lim_{r\to 1+} \varphi_-(r,\alpha)$ und sind wieder Fouriertransformierte, also $E_+(\alpha) = \varphi_+(1+,\alpha) - \mathrm{inh}(1,\alpha)$ und $E_-(\alpha) = \varphi_-(1+,\alpha) + \mathrm{inh}(1,\alpha)$.
Nach Bedingung 5. existiert nämlich $\lim_{r\to 1+} \mathrm{inh}(r,\alpha) = \mathrm{inh}(1,\alpha)$. Entsprechend erhält man für $\frac{b}{a} < r < 1$ folgende eindeutig bestimmte Lösung der Differentialgleichung mit $\frac{d}{dr} \varphi_\pm(r,\alpha)\big|_{r=\frac{b}{a}+} = 0$:

$$\varphi_+(r,\alpha)=(\varphi_+(1-,\alpha)-\mathrm{cross}(1,\alpha)) \frac{J_o(iqr)Y_o'(iq\frac{b}{a})-Y_o(iqr)J_o'(iq\frac{b}{a})}{J_o(iq)\,Y_o'(iq\frac{b}{a})-Y_o(iq)\,J_o'(iq\frac{b}{a})} + \mathrm{cross}(r,\alpha)$$

für Im $\alpha>0$ in E_I ,

$$\varphi_-(r,\alpha)=(\varphi_-(1-,\alpha)+\mathrm{cross}(1,\alpha)) \frac{J_o(iqr)Y_o'(iq\frac{b}{a})-Y_o(iqr)J_o'(iq\frac{b}{a})}{J_o(iq)\,Y_o'(iq\frac{b}{a})-Y_o(iq)\,J_o'(iq\frac{b}{a})} - \mathrm{cross}(r,\alpha)$$

für Im $\alpha<0$ in E_I ,

wobei $\mathrm{cross}(r,\alpha):=$

$$-\frac{i\pi}{4} \int_{\frac{b}{a}}^{r} (H_o^{(1)}(iqr)H_o^{(2)}(iq\rho)-H_o^{(2)}(iqr)H_o^{(1)}(iq\rho))\rho\left(\frac{\partial}{\partial z} \varphi(\rho,z=0)-i\alpha\varphi(\rho,z=0)\right)d\rho$$

Offensichtlich ist das Problem der Bestimmung einer Lösung reduziert auf die Bestimmung der Randverteilungen auf Z_l und Z_r. Dazu untersucht man zunächst $\varphi_\pm(1\pm,\alpha)$ und $\frac{d}{dr} \varphi_+(1,\alpha)$ auf Stetigkeitsverhalten auf der reellen Achse. Anschließend gewinnt man eine Wiener-Hopf-Gleichung, die den Drucksprung auf Z_l in Beziehung setzt zur Normalgeschwindigkeitsverteilung auf Z_r. Man betrachte dazu die Normalgeschwindigkeitsverteilung auf Z_l und Z_r durch Differenzieren nach r.

Die Bedingungen 2.,3. und 9. zeigen mit Hilfe von [1] S.89/9o und [5]

1. $\frac{d}{dr}\varphi_+(1+,\alpha)=\varphi_+(1+,\alpha)iq\frac{H_o^{(1)'}(iq)}{H_o^{(1)}(iq)} - inh(1,\alpha)iq\frac{H_o^{(1)'}(iq)}{H_o^{(1)}(iq)}+\frac{d}{dr}inh(1,\alpha)$

für Im α > 0 ,

2. $0=\varphi_-(1+,\alpha)iq\frac{H_o^{(1)'}(iq)}{H_o^{(1)}(iq)} + inh(1,\alpha)iq\frac{H_o^{(1)'}(iq)}{H_o^{(1)}(iq)}-\frac{d}{dr}inh(1,\alpha)$

für Im α < 0 ,

3. $\frac{d}{dr}\varphi_+(1-,\alpha)=(\varphi_+(1-,\alpha)-cross(1,\alpha))iq\frac{J_o'(iq)Y_o'(iq\frac{b}{a})-Y_o'(iq)J_o'(iq\frac{b}{a})}{J_o(iq)Y_o'(iq\frac{b}{a})-Y_o(iq)J_o'(iq\frac{b}{a})}$

$+\frac{d}{dr}cross(1,\alpha)$ für Im α > 0

4. $0=(\varphi_-(1-,\alpha)-cross(1,\alpha))iq\frac{J_o'(iq)Y_o'(iq\frac{b}{a})-Y_o'(iq)J_o'(iq\frac{b}{a})}{J_o(iq)Y_o'(iq\frac{b}{a})-Y_o(iq)J_o'(iq\frac{b}{a})}$

$-\frac{d}{dr}cross(1,\alpha)$ für Im α < 0

Durch Untersuchung von $inh(1,\alpha)$, $\frac{d}{dr}inh(1,\alpha)$, $cross(1,\alpha)$ und $\frac{d}{dr}cross(1,\alpha)$ folgt unter Beachtung von Bedingung 7., daß

$\varphi_-(1+,\alpha)$, $\varphi_+(1+,\alpha)=\varphi_+(1-,\alpha)$, $\prod_{\mu=1}^{\nu_o-1}(\alpha^2-\varkappa_\mu^2)\varphi_-(1-,\alpha)$ und

$\varphi_+'(1,\alpha):=\frac{d}{dr}\varphi_+(1+,\alpha)=\frac{d}{dr}\varphi_+(1-,\alpha)$ für $\alpha\in\mathbb{R}\setminus\{-k,0,k\}$ durch stetige Funktionen repräsentiert werden. Durch Addition der Gleichungen 1. und 2. sowie 3. und 4. erkennt man, daß $\varphi(1+,\alpha)$ und $\varphi(1-,\alpha)$ durch $\varphi_+'(1,\alpha)$ bis auf endlich viele Punkte eindeutig bestimmt sind.

Zur Herleitung der Wiener-Hopf-Gleichung zur Bestimmung von $\varphi_+'(1,\alpha)$ wird der Drucksprung auf der Zylinderfläche r=1 betrachtet:

$\Delta p(z) = \lim_{\varepsilon\to 0}(p(1+\varepsilon,z)-p(1-\varepsilon,z))$ und

$(\mathfrak{F}\Delta p)(\alpha) = -\frac{id_oU}{a\sqrt{1-M^2}}(\alpha-\frac{k}{M})(\varphi(1+,\alpha)-\varphi(1-,\alpha))$. Die Stetigkeit des Druckes für z>0 zeigt $\Delta p(z)=0$ für z>0 .

Also ist $\mathfrak{F}\Delta p$ auf der reellen Achse distributioneller Randwert einer in der unteren Halbebene holomorphen Funktion. Als Bezeichnung sei eingeführt $\Delta p_-(\alpha):=(\mathfrak{F}\Delta p)(\alpha)$.

Die Kantenbedingung 5 fordert $\Delta p(z)=O(|z|^{-1/2})$ für $z\to 0$. Abelsche Asymptotik, siehe [11], zeigt $\Delta p_-(\alpha)=O(|\alpha|^{-1/2})$ für $\alpha\to\infty$ in einem Winkelraum der unteren Halbebene. Drückt man in der Gleichung für die Fouriertransformierte des Drucksprungs $\varphi(1+,\alpha)$ und $\varphi(1-,\alpha)$ durch $\varphi_+'(1,\alpha)$ aus, so erhält man folgende Wiener-Hopf-Gleichung für $\alpha\in\mathbb{R}\setminus\{-k,0,k\}$:

$$i\,\frac{a\sqrt{1-M^2}}{d_0U}\prod_{\mu=0}^{\nu_0-1}(\alpha^2-\varkappa_\mu^2)\;\Delta p_-(\alpha) = -2(\alpha-k)\,\frac{\varphi_+'(1,\alpha)}{{}^1K(\alpha)}\,,\ \text{wobei}$$

$${}^1K(\alpha) = \pi\,\frac{H_0^{(1)'}(iq)}{H_0^{(1)'}(iq\frac{b}{a})}\left(J_0'(iq\tfrac{b}{a})Y_0'(iq)-J_0'(iq)Y_0'(iq\tfrac{b}{a})\right)\Big/\prod_{\mu=1}^{\nu_0-1}(\alpha^2-\varkappa_\mu^2)$$

Dabei gelten folgende Nebenbedingungen:

a) $\Delta p_-(\alpha) = O(|\alpha|^{-1/2})$ in einem Winkelraum der unteren Halbebene,

b) Δp_- enthält an einlaufenden Wellenkomponenten ausschließlich $A_0 u_\nu(r)e^{i\varkappa_\nu z}$ nach Bedingung 6,

c) $\varphi_+'(1,\cdot)$ enthält keine einlaufenden Wellenkomponenten entsprechend Bedingung 7.

Mit Hilfe der Faktorisierung ${}^1K(\alpha)={}^1K_+(\alpha)\,{}^1K_-(\alpha)$ nach [3] kann man leicht eine Lösung der Wiener-Hopf-Gleichung angeben:

$$\varphi_+'(1,\alpha)= \tfrac{i}{2}A_o(-1)^{\nu_o-1}\prod_{\mu=0}^{\nu_o-1}(\varkappa_\nu+\varkappa_\mu)\;{}^1K_-(-\varkappa_\nu)\;{}^1K_+(\alpha)\prod_{\substack{\mu=0\\ \mu\neq\nu}}^{\nu_o-1}(\alpha+\varkappa_\mu)$$

sowie für $\alpha\neq-\varkappa_\nu,\ \varkappa_\mu \quad 0\le\mu\le\nu_0-1$:

$$\Delta p_-(\alpha)= A_0\frac{d_0U}{a\sqrt{1-M^2}}(-1)^{\nu_o}\,\frac{\prod_{\mu=0}^{\nu_o-1}(\varkappa_\nu+\varkappa_\mu)(\alpha-\frac{k}{M})\;{}^1K_-(-\varkappa_\nu)}{\prod_{\mu=0}^{\nu_0-1}(\alpha-\varkappa_\mu)(\alpha+\varkappa_\nu)\;{}^1K_-(\alpha)}$$

Im Falle $\lambda_\nu<k=\lambda_{\nu_0-1}$ erhält man als Lösungsschar mit beliebigem $D_0\in\mathbb{C}$:

$$\varphi_+'(1,\alpha)= \tfrac{i}{2}A_o(-1)^{\nu_o-1} \prod_{\mu=0}^{\nu_o-1} (\varkappa_\nu+\varkappa_\mu)\ {}^1K_-(-\varkappa_\nu)\ {}^1K_+(\alpha) \prod_{\substack{\mu=0\\ \mu\neq\nu}}^{\nu_o-1}(\alpha+\varkappa_\mu)$$

$$+\ D_o\ {}^1K_+(\alpha) \prod_{\mu=0}^{\nu_o-2}(\alpha+\varkappa_\mu)\ ;\ \text{und für } \alpha\neq -\varkappa_\nu,\ \varkappa_\mu\ \ 0\le\mu\le\nu_o-1 :$$

$$i\frac{a\sqrt{1-M^2}}{d_oU}\,\Delta p_-(\alpha)= A_oi(-1)^{\nu_o}\ \frac{\prod_{\mu=0}^{\nu_o-1}(\varkappa_\nu+\varkappa_\mu)(\alpha-\frac{k}{M})\ {}^1K_-(-\varkappa_\nu)}{\prod_{\mu=0}^{\nu_o-1}(\alpha-\varkappa_\mu)(\alpha+\varkappa_\nu)\ {}^1K_-(\alpha)}$$

$$-\ 2D_o\ \frac{(\alpha-\frac{k}{M})}{\alpha^2\prod_{\mu=0}^{\nu_o-2}(\alpha-\varkappa_\mu)\,{}^1K_-(\alpha)}$$

Damit ist eine formale Lösung des gestellten Problems gefunden. Zu zeigen ist die Lösungseigenschaft, und zu klären ist die Eindeutigkeitsfrage.

<u>Eindeutigkeitsproblem</u>: Sei φ_1 eine weitere Lösung. Dann erfüllt $\tilde{\varphi}:=\varphi-\varphi_1$ die Bedingungen 1. bis 1o. mit Zusatz $A_o=0$. Das führt auf folgende Wiener-Hopf-Gleichung für $\alpha\in\mathbb{R}\setminus\{-k,0,k\}$

$$i\frac{a\sqrt{1-M^2}}{d_oU}\prod_{\mu=0}^{\nu_o-1}(\alpha^2-\varkappa_\mu^2)\Delta\tilde{p}_-(\alpha) = -2(\alpha-\frac{k}{M})\ \frac{\tilde{\varphi}_+'(1,\alpha)}{{}^1K(\alpha)}\ .$$

Dabei ist Δp_- einseitige Fouriertransformierte über die negative Halbachse, holomorph für Im $\alpha<0$, stetig auf $\mathbb{R}\setminus\{-k,0,k\}$, im übrigen auf $\mathbb{R}$ distributioneller Randwert von $\Delta p_-(\sigma-i\tau)$ für $\tau\to 0+$. Analoges gilt für $\tilde{\varphi}_+'(1,\cdot)$ auf die obere Halbebene bezogen. Man ziehe die Faktorisierung von ${}^1K$ heran. ${}^1K_-$ ist dabei holomorph in der unteren Halbebene und stetig auf der reellen Achse; außerdem ist $1-{}^1K_-(\alpha)(c+i\alpha)^{\nu_o-\frac{1}{2}}$, $c>0$, klassische Fouriertransformierte einer L^1-Funktion mit Träger in $]-\infty,0[$. Analoges gilt für ${}^1K_+$ und mit Hilfe von [4] auch für $({}^1K_+(\alpha)(c-i\alpha)^{\nu_o-\frac{1}{2}})^{-1}$ auf die obere Halbebene bezogen.

$$\mathfrak{g}(\alpha)=\begin{cases} -2(\alpha-\frac{k}{M})\ \dfrac{\tilde{\varphi}_+'(1,\alpha)}{{}^1K_+(\alpha)} & \text{Im }\alpha\geq 0 \\ i\dfrac{a\sqrt{1-M^2}}{d_0U}\prod_{\mu=0}^{\nu_o-1}(\alpha^2-\varkappa_\mu^2)\,{}^1K_-(\alpha)\Delta\tilde{p}_-(\alpha) & \text{Im }\alpha\leq 0 \end{cases} \qquad \alpha\neq\{-k,0,k\}$$

ist daher holomorph in $\mathbb{C}\setminus\{-k,0,k\}$. Mit einiger Überlegung gelangt man zu der Abschätzung

$$|\mathfrak{g}(\alpha)|\leq M_1(1+|\alpha|^{m+\nu_0+\frac{1}{2}})|\tau|^{\frac{1}{2}-2l} \qquad \text{mit } M_1>0;\ m,l\in\mathbb{N}.$$

Nach [9] S.1o7 folgt, daß $\mathfrak{g}(\alpha)$ eine analytische Darstellung der Distribution $\mathfrak{g}(\sigma+io)-\mathfrak{g}(\sigma-io)$ ist. Das heißt für $\Psi\in\mathfrak{S}$:

$$<\mathfrak{g}(\sigma+io)-\mathfrak{g}(\sigma-io),\Psi> = \lim_{\tau\to 0+}\int_{-\infty}^{\infty}(\mathfrak{g}(\sigma+i\tau)-\mathfrak{g}(\sigma-i\tau))\Psi(\sigma)\,d\sigma\ ,$$

wobei der Grenzwert rechts existiert. Auf jedem Kompaktum der reellen Achse, das $\{-k,0,k\}$ nicht trifft, streben $\mathfrak{g}(\sigma+i\tau)$ und $\mathfrak{g}(\sigma-i\tau)$ gleichmäßig gegen dieselben stetigen Randwerte. Also besteht der Träger von $\mathfrak{g}(\sigma+io)-\mathfrak{g}(\sigma-io)$ nur aus isolierten Punkten. Die Darstellung solcher Distributionen mit Hilfe des δ-Funktionals ist wohlbekannt. Also hat $\mathfrak{g}$ als analytische Darstellung von $\mathfrak{g}(\sigma+io)-\mathfrak{g}(\sigma-io)$ höchstens Pole in -k,0,k. Außerdem ist $\mathfrak{g}$ als analytische Darstellung von $\mathfrak{g}(\sigma+io)-\mathfrak{g}(\sigma-io)$ nur bis auf ein Polynom eindeutig bestimmt. Daher gibt es ein $s\in\mathbb{N}$ und ein Polynom Q mit $(\alpha+k)^s\alpha^s(\alpha-k)^s\mathfrak{g}(\alpha)=Q(\alpha)$. Mit Hilfe der Kantenbedingung 5 folgt: Grad $Q\leq 3s+\nu_0$. So ergibt sich:

$$\tilde{\varphi}_+'(1,\alpha)=-\frac{1}{2}\,{}^1K_+(\alpha)\,\frac{Q_{3s+\nu_0}(\alpha)}{(\alpha-\frac{k}{M})(\alpha^2-k^2)^s\alpha^s} \qquad \text{für Im }\alpha>0 \text{ und}$$

$$\Delta\tilde{p}_-(\alpha)=\frac{d_0U}{ia\sqrt{1-M^2}}\ \frac{Q_{3s+\nu_0}(\alpha)}{{}^1K_-(\alpha)\prod_{\mu=0}^{\nu_o-1}(\alpha^2-\varkappa_\mu^2)(\alpha^2-k^2)^s\alpha^s} \qquad \text{für Im }\alpha<0\ .$$

Sei zunächst $k\neq\lambda_\nu$ für alle $\nu\in\mathbb{N}$. Eine Polstelle in $\frac{k}{M}$ bzw. in k für $\tilde{\varphi}_+'$ würde eine einlaufende Welle bedeuten. Andererseits darf $\Delta\tilde{p}_-$ keine Pole in $-\varkappa_\mu$, $0\leq\mu\leq\nu_0-1$ besitzen, da sonst einlaufende Wellen aufträten. Das zeigt

$$Q_{3s+\nu_0}(\alpha)=(\alpha-\frac{k}{M})(\alpha-k)^s\prod_{\mu=0}^{\nu_o-1}(\alpha+\varkappa_\mu)(\alpha+k)^s\,P_{s-1}(\alpha) \text{ und}$$

$$\tilde{\varphi}_+'(1,\alpha) = -\frac{1}{2}\, {}^1K_+(\alpha) \prod_{\mu=0}^{\nu_0-1} (\alpha+\varkappa_\mu)\, \frac{P_{s-1}(\alpha)}{\alpha^s} \quad \text{für } \operatorname{Im} \alpha>0 ,$$

$$\Delta\tilde{p}_-(\alpha) = \frac{d_0 U}{ia\sqrt{1-M^2}}\; \frac{\alpha-\frac{k}{M}}{{}^1K_-(\alpha) \prod_{\mu=0}^{\nu_0-1} (\alpha-\varkappa_\mu)}\; \frac{P_{s-1}(\alpha)}{\alpha^s} \quad \text{für } \operatorname{Im} \alpha<0 .$$

Dies ist zwar eine Lösung der Wiener-Hopf-Gleichung, führt aber i.a. nicht zu einer Lösung des gestellten Anfangsproblems. In allen Fällen, wo $\varphi_+'(1,\cdot)$ in 0 einen Pol besitzt, wächst $\frac{d}{dr}\varphi(r,z)$ wie ein Polynom für $z\to\infty$. Nähere Untersuchungen zeigen, daß auch $\varphi(1,z)$ für $z\to\infty$ wie ein Polynom wächst im Widerspruch zu Bedingung 7. Also gilt $P_{s-1}=0$. Allein im Falle $k=\lambda_{\nu_0-1}$ erhält man so eine Lösung der Wiener-Hopf-Gleichung, die auch zu einer Lösung des Ausstrahlungsproblems führt, nämlich die schon zitierte Schar mit beliebigem $D_0\in\mathbb{C}$:

$$\tilde{\varphi}_+'(1,\alpha) = D_0\, {}^1K_+(\alpha) \prod_{\mu=0}^{\nu_0-2} (\alpha+\varkappa_\mu) \qquad \text{für } \operatorname{Im} \alpha>0 ,$$

$$\Delta\tilde{p}_-(\alpha) = -i\, \frac{2d_0 U}{a\sqrt{1-M^2}}\, D_0\, \frac{(\alpha-\frac{k}{M})}{\alpha^2\, {}^1K_-(\alpha) \prod_{\mu=0}^{\nu_0-2} (\alpha-\varkappa_\mu)\, {}^1K_-(\alpha)} \qquad \text{für } \operatorname{Im} \alpha<0 .$$

Dabei gilt für den Index der anregenden Welle $\nu<\nu_0-1$, denn im Falle $k=\lambda_{\nu_0-1}$ ist mit den Lösungen $u_{\nu_0-1}(r)\, e^{-i\omega t}$ und $u_{\nu_0-1}(r)\, z\, e^{-i\omega t}$ kein Energietransport verbunden.

Für $\lambda_{\nu_0-1} < k < \lambda_{\nu_0}$ erhält man als Fouriertransformierte von $\varphi(r,z)$ folgenden Ausdruck

$$\varphi(r,\alpha) = A_0 i(-1)^{\nu_0-1} \prod_{\mu=0}^{\nu_0-1} (\varkappa_\mu+\varkappa_\nu)\, {}^1K_-(-\varkappa_\nu)\, \frac{1}{2}\, {}^1K_+(\alpha) \prod_{\substack{\mu=0\\ \mu\neq\nu}}^{\nu_0-1} (\alpha+\varkappa_\mu)\; \frac{1}{iq}\; \frac{H_0^{(1)}(iqr)}{H_0^{(1)\prime}(iq)} ,\ r>1$$

und für $\frac{b}{a}<r<1$, $\alpha\neq-\varkappa_\nu$ sowie $\alpha\neq\varkappa_\mu$ $0\leq\mu\leq\nu_0-1$:

$$\varphi(r,\alpha) = A_0 i(-1)^{\nu_0-1} \prod_{\mu=0}^{\nu_0-1} (\varkappa_\mu+\varkappa_\nu)\, {}^1K_-(-\varkappa_\nu)\, \frac{1}{2}\, {}^1K_+(\alpha) \prod_{\substack{\mu=0\\ \mu\neq\nu}}^{\nu_0-1} (\alpha+\varkappa_\mu) .$$

$$\frac{1}{iq}\;\frac{J_0(iqr)Y_0'(iq\frac{b}{a})-Y_0(iqr)J_0'(iq\frac{b}{a})}{J_0'(iq)\;Y_0'(iq\frac{b}{a})-Y_0'(iq)\;J_0'(iq\frac{b}{a})}$$

Im Falle $k=\lambda_{\nu_0-1}$ kommt jeweils der von $D_0{}^1K_+(\alpha)\prod_{\mu=0}^{\nu_0-2}(\alpha+\varkappa_\mu)$ induzierte Term additiv hinzu, also für $r>1$:

$$\varphi_{D_0}(r,\alpha) = D_0{}^1K_+(\alpha)\prod_{\mu=0}^{\nu_0-2}(\alpha+\varkappa_\mu)\,\frac{1}{iq}\,\frac{H_0^{(1)}(iqr)}{H_0^{(1)\prime}(iq)}\quad \text{und für } \frac{b}{a}<r<1:$$

$$\varphi_{D_0}(r,\alpha) = D_0{}^1K_+(\alpha)\prod_{\mu=0}^{\nu_0-2}(\alpha+\varkappa_\mu)\,\frac{1}{iq}\,\frac{J_0(iqr)Y_0'(iq\frac{b}{a})-Y_0(iqr)J_0'(iq\frac{b}{a})}{J_0'(iq)\;Y_0'(iq\frac{b}{a})-Y_0'(iq)\;J_0'(iq\frac{b}{a})}$$

An den Stellen $-\varkappa_\nu$ und $\varkappa_\mu$ $0\le\mu\le\nu_0-1$ ist $\varphi(r,\alpha)$ für $\frac{b}{a}<r<1$ noch unbestimmt. Dieses Problem löst sich einfach durch geeignete Wahl des Rücktransformationsweges in der Umkehrformel. Dabei geht wesentlich die Sommerfeldsche Ausstrahlungsbedingung ein. Zur Rücktransformation werden folgende Integrationswege angewandt:

$\frac{b}{a}<r<1$:

$1<r$:

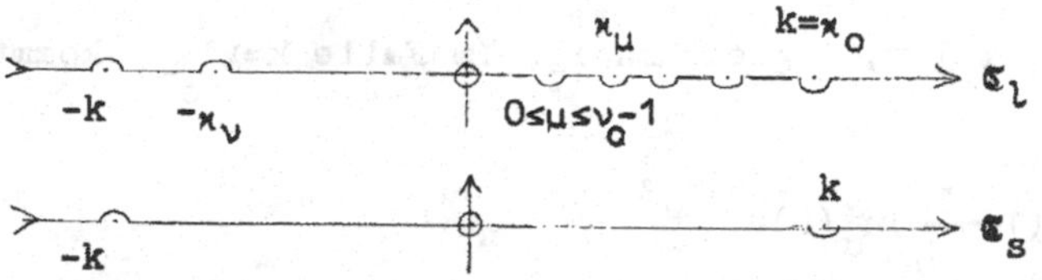

Nach diesen Vorbereitungen läßt sich das Ergebnis formulieren:

<u>Satz 1:</u> Es gibt eine Lösung, die den Bedingungen 1. bis 1o. genügt und zwar:

$$\varphi(r,z):=\frac{1}{2\pi}\int_{\mathfrak{C}_l}\varphi(r,\alpha)e^{-i\alpha z}d\alpha + A_0u_\nu(r)e^{i\varkappa_\nu z}\qquad \text{für } \frac{b}{a}<r<1,$$

$$\varphi(r,z):=\frac{1}{2\pi}\int_{\mathfrak{C}_s}\varphi(r,\alpha)e^{-i\alpha z}d\alpha\qquad \text{für } 1<r,$$

$$\varphi(1,z):=\lim_{r\to 1}\varphi(r,z)\qquad \text{für } z>0\,.$$

Im Falle $\lambda_{\nu_0-1}<k<\lambda_{\nu_0}$ ist die Lösung eindeutig bestimmt.

Auf den Nachweis der Lösungseigenschaften muß hier verzichtet werden. Zur Analyse der Lösung werden noch zwei Ergebnisse angegeben.

Satz 2: Die Lösung φ erfüllt im Ringkanal Bedingung 6 und besitzt folgende Entwicklung nach den Eigenfunktionen des Ringkanals:

$$\varphi(r,z) = A_0\{u_\nu(r)e^{i\varkappa_\nu z} + \sum_{\mu=0}^{\nu_0-1} R_{\nu\mu}\, u_\mu(r)e^{-i\varkappa_\mu z} + \sum_{\mu=\nu_0}^{\infty} b_{\nu\mu}\, u_\mu(r)e^{\varkappa_\mu z}\}$$

Dabei bezeichnen

$$R_{\nu\mu} = - \prod_{\substack{\xi=0\\ \xi\neq\mu}}^{\nu_0-1} \left(\frac{\varkappa_\xi+\varkappa_\nu}{\varkappa_\xi-\varkappa_\mu}\right) \frac{{}^1K_+(\varkappa_\nu)\,{}^1K_+(\varkappa_\mu)}{{}^1K(\varkappa_\mu)} \quad \text{für } 0\leq\mu\leq\nu_0-1$$

die Reflexionskoeffizienten und

$$b_{\nu\mu} = \prod_{\xi=0}^{\nu_0-1} \left(\frac{\varkappa_\xi+\varkappa_\nu}{\varkappa_\xi-i\varkappa_\mu}\right) \frac{1}{\varkappa_\nu+i\varkappa_\mu}\; {}^1K_+(\varkappa_\nu)\; {}^1K_+(i\varkappa_\mu)\; \lim_{\alpha\to i\varkappa_\mu} \frac{\alpha-i\varkappa_\mu}{{}^1K(\alpha)}$$

die übrigen Koeffizienten. Die Konvergenz der Reihe ist für $(r,z) \in [\frac{b}{a},1] \times\ ]-\infty,0]$ gleichmäßig. Im Falle $k=\lambda_{\nu_0-1}$ kommt additiv noch Anteil

$$D\ \{ \sum_{\mu=0}^{\nu_0-2} R_{\nu\mu}(1+\frac{\varkappa_\nu}{\varkappa_\mu})u_\mu(r)e^{-i\varkappa_\mu z}$$

$$+ R_{\nu,\nu_0-1}\left(\sum_{\tau=0}^{\nu_0-2} \frac{\varkappa_\nu}{\varkappa_\tau} + \varkappa_\nu \frac{{}^1K_+'(0)}{{}^1K(0)} - i\varkappa_\nu z\right)u_{\nu_0-1}(r)$$

$$+ \sum_{\mu=\nu_0}^{\infty} b_{\nu\mu}(1+ \frac{\varkappa_\nu}{i\varkappa_\mu})u_\mu(r)e^{\varkappa_\mu z}\}$$ mit beliebigem $D\in\mathbb{C}$ hinzu.

Eine Hälfte von Bedingung 7 ergibt sich aus dem folgenden Satz über die Darstellung des Fernfeldes für $\lambda_{\nu_0-1}< k < \lambda_{\nu_0}$:

Satz 3: In Polarkoordinaten $r=d \sin\Theta$, $z=d\cos\Theta$, $d=\sqrt{z^2+r^2}$ ergibt sich für $-\pi<\Theta<0$ und $0<\Theta<\pi$: $\varphi(d \sin \Theta, d \cos \Theta) =$

$$\frac{A_0}{2\pi}(-1)^{\nu_0+1} \prod_{\mu=0}^{\nu_o-1} (\varkappa_\mu+\varkappa_\nu)\,{}^1K_+(\varkappa_\nu)\,{}^1K_+(-k\cos\Theta) \prod_{\substack{\mu=0\\ \mu\neq\nu}}^{\nu_o-1} (\varkappa_\mu-k\cos\Theta)\ \cdot$$

$$\frac{1}{\sin\Theta\, H_o^{(1)'}(k\sin\Theta)} \quad \frac{e^{ikd}}{kd} + O\left(\frac{1}{kd}\right)^{\frac{3}{2}} \quad \text{für } kd\to\infty \ .$$

Der Beweis erfolgt analog [2] S.173-179 .

LITERATURVERZEICHNIS

[1] Bremermann, H.:
Distributions, Complex Variables, and Fourier Transforms; Addison-Wesley 1965 .

[2] Friedrich, N.:
Untersuchungen zur Schallabstrahlung aus einem in axialer Richtung angeströmten Ringkanal. Deutsche Luft-und Raumfahrt, FB 75-24

[3] Friedrich, N.:
Faktorisierung eines Kerns, der bei Ausstrahlungsproblemen der Elektrodynamik und Aeroelastizität auftritt. Erscheint demnächst.

[4] Krein, M.G.:
Integral equations on a half-line with kernel depending upon the difference of the arguments. Amer.Math.Soc. Transl. (2) (1962) 163-288.

[5] Lavoine, J.:
Transformation de Fourier des Pseudo-Fonctions. CNRS, Paris 1963.

[6] Mittra,R. and Lee,S.W.:
Analytical Techniques in the Theory of Guided Waves. Mac Millan Company New York 1971.

[7] Schwarz, L.:
Theorie des distributions. Hermann Paris 1966.

[8] Söhngen, H.:
Luftkräfte an den schwingenden Schaufeln eines gestaffelten Streckengitters kleiner Teilung. Deutsche Versuchsanstalt für Luft-und Raumfahrt,Mai 1963, Bericht Nr. 245.

[9] Tillmann, H.G.:
Darstellung der Schwartzschen Distributionen durch analytische Funktionen MZ 77, 1o6-124, (1961).

[1o] Watson, G.N.:
A treatise on the theory of Bessel-functions, Cambridge University Press 1922.

[11] Zemanian,A.H.:
Distribution Theory and Transform Analysis, McGraw-Hill Book Company 1965.

BEWEGLICHE SINGULARITÄTEN VON LINEAREN PARTIELLEN DIFFERENTIALGLEICHUNGEN.

Detlef Gronau

I. Einleitung

In der nachfolgenden Arbeit werden formale Existenzfragen einer speziellen Klasse von beweglichen Singularitäten von linearen partiellen Differentialgleichungen mit analytischen Koeffizienten behandelt. Es handelt sich dabei um solche Singularitäten, die eine Darstellung in Form einer verallgemeinerten Laurentreihe in zwei komplexen Variablen zulassen. Es werden notwendige und hinreichende Kriterien in Form von überbestimmten Systemen linearer gewöhnlicher Differentialgleichungen für die formale Existenz angegeben. Im allgemeinen Fall wurden diese Systeme wegen der Schwierigkeit der Theorie noch nicht weiter untersucht. Im speziellen Fall von Differentialgleichungen mit konstanten Koeffizienten ergeben sich jedoch explizite algebraische Bedingungen für das Vorhandensein von solchen beweglichen singulären Lösungen.

Dabei versteht man unter beweglichen Singularitäten in Analogie zur Theorie der gewöhnlichen Differentialgleichungen, singuläre Lösungen von partiellen Differentialgleichungen, wobei grob gesprochen, die Lage der Singularität nicht von jener der Singularitäten der Koeffizienten der partiellen Differentialgleichung abhängt. Oder im Sinne von BIEBERBACH Lösungen, die in solchen Punkten singulär sind, welche innere Punkte der Menge aller singulären Punkte der allgemeinen Lösung der betrachteten partiellen Differentialgleichung sind. Wie bekannt ist, besitzen gewöhnliche lineare Differentialgleichungen keine beweglichen Singularitäten (siehe [3] p.86 f.).

Untersucht werden lineare homogene partielle Differentialgleichungen für die komplexwertige Funktion $w=w(z_1,z_2)$ in den Variablen $z_1,z_2 \in \mathbb{C}$ und zwar:

(1) $w_{z_1 z_2} = F(z_1,z_2)\cdot w_{z_1} + G(z_1,z_2)\cdot w_{z_2} + H(z_1,z_2)\cdot w$

und auch allgemeiner:

$$(2) \quad \sum_{\mu,\nu=0}^{m,n} F^{(\mu\nu)}(z_1,z_2)\cdot\frac{\partial^{\mu+\nu}}{\partial z_1^{\mu}\partial z_2^{\nu}}w = 0$$

mit F, G, H, $F^{(\mu\nu)}$ holomorph in $(z_1,z_2) = (0,0)$ und $F^{(mn)}(0,0) = 1$.

In früheren Arbeiten haben L.REICH und der Verfasser (siehe [9], [6] und [7]) Singularitäten der obigen Differentialgleichungen in der Form

$$w = A(z_1,z_2) + B(z_1,z_2)\cdot\log z_1z_2$$

mit den formalen verallgemeinerten Laurentreihen

$$A, B \in \mathbb{C}\{z_1,z_2\} := \{ \sum_{\substack{\alpha\geq\alpha_o\\ \beta\geq\beta_o}} a_{\alpha\beta}z_1^{\alpha}z_2^{\beta} \mid \alpha_o,\beta_o \in \mathbb{Z},\ a_{\alpha\beta} \in \mathbb{C}\}$$

nachgewiesen.
Hier erhielten wir folgendes Ergebnis (siehe [6] oder [7]):

<u>Satz 1</u>: Eine Lösung von (1) oder (2) der Form $w = A + B\cdot\log z_1z_2 =$
$\sum_{\substack{\alpha\geq\alpha_o\\ \beta\geq\beta_o}} a_{\alpha\beta}z_1^{\alpha}z_2^{\beta} + \sum_{\substack{\alpha\geq\alpha_o\\ \beta\geq\beta_o}} b_{\alpha\beta}z_1^{\alpha}z_2^{\beta}\cdot\log z_1z_2$ mit $\alpha_o<o$, $\beta_o<o$ ist bis auf die frei wählbaren Randkoeffizienten (Anfangsdaten im Goursat'schen Sinne): $a_{\mu j}$ und $a_{i\nu}$ mit $0\leq\mu<m$, $j\geq\beta_o$, $i\geq\alpha_o$ und $0\leq\nu<n$ sowie $b_{\gamma\delta}$ mit $0\leq\gamma<m$, $0\leq\delta<n$ eindeutig bestimmt. Im Fall (1) ist m=n=1 zu setzen. Werden die Randkoeffizienten so gewählt, daß
$\sum_{\substack{i\geq o\\ 0\leq\nu<n}} a_{i\nu}z_1^{i}z_2^{\nu} + \sum_{\substack{j\geq o\\ 0\leq\mu<m}} a_{\mu j}z_1^{\mu}z_2^{j}$ in einer Umgebung von $(z_1,z_2)=(0,0)$ konvergiert, dann konvergiert auch die dadurch bestimmte Lösung $w = A + B\cdot\log z_1z_2$ in einer punktierten (d.h. $z_1\cdot z_2\neq 0$) Umgebung von $(z_1,z_2) = (0,0)$.

Das heißt, daß bewegliche Singularitäten im Gegensatz zur Theorie der gewöhnlichen Differentialgleichungen hier immer existieren. Singularitäten ähnlicher Art ergaben auch Untersuchungen von partiellen Differentialgleichungen in [1], [2], [5] und [10].

Nun erhebt sich die Frage, inwieweit auch logarithmenfreie regulär-singuläre Lösungen von (1) bzw. (2) existieren. Gerade da ergeben sich jedoch größere formale Probleme, über die hier berichtet werden soll. Es stellt sich nämlich heraus, daß nicht in jedem Fall solche singulären Lösungen zulässig sind, sodaß dadurch überhaupt eine Unterscheidung in feste und bewegliche Singularitäten als sinnvoll erscheint.

II. Existenzkriterium in Form von überbestimmten Systemen von gewöhnlichen Differentialgleichungen.

Es sei:

$$F = \sum_{i,j\geq 0} f_{ij} z_1^i z_2^j , \quad G = \sum_{i,j\geq 0} g_{ij} z_1^i z_2^j , \quad H = \sum_{i,j\geq 0} h_{ij} z_1^i z_2^j ,$$

$$F^{(\mu\nu)} = \sum_{i,j\geq 0} f_{ij}^{(\mu\nu)} z_1^i z_2^j \text{ und } (o,\alpha):=1 \; ; \; (\mu,\alpha):=\alpha(\alpha-1)\ldots(\alpha-\mu+1) \text{ für } \mu>o.$$

Setzt man:

$$w=A= \sum_{\alpha\geq\alpha_0,\beta\geq\beta_0} a_{\alpha\beta} z_1^\alpha z_2^\beta \text{ mit } \alpha_o<o,\ \beta_o<o$$

in (1) bzw. (2) ein, so erhält man Rekursionsformeln für die Koeffizienten $a_{\alpha\beta}$:

$$(3) \qquad \alpha\beta a_{\alpha\beta} = \sum_{i,j\geq 0} \{(\alpha-i)a_{\alpha-i,\beta-j-1} f_{ij} + (\beta-j)a_{\alpha-i-1,\beta-j} g_{ij} + a_{\alpha-i-1,\beta-j-1} h_{ij}\}$$

bzw.:

$$(4) \qquad \sum_{\mu,\nu=0}^{m,n} \sum_{i,j\geq 0} a_{\alpha-m+\mu-i,\beta-n+\nu-j} (\mu,\alpha-m+\mu-i)(\nu,\beta-n+\nu-j) f_{ij}^{(\mu\nu)} = o.$$

Rein formale Untersuchungen (siehe [9], [6] oder [7]) ergeben, daß $a_{\alpha\beta} = o$ für $\alpha<o$ und $\beta<o$. Die Reihe A hat also die Gestalt:

$$A = \sum_{\alpha,\beta\geq 0} a_{\alpha\beta} z_1^\alpha z_2^\beta + \sum_{i=1}^{r} z_1^{-i} \sum_{\beta\geq 0} a_{-i,\beta} z_2^\beta + \sum_{j=1}^{s} z_2^{-j} \sum_{\alpha\geq 0} a_{\alpha,-j} z_1^\alpha$$

$$\text{mit } r=-\alpha_0>o \; , \; s=-\beta_0>o.$$

Führt man diesen Ansatz in geschlossener Form in (1) bzw. (2) durch, so ergibt der Koeffizientenvergleich nach den negativen Potenzen von z_1 und z_2 beim Ansatz in (1) je ein System von r+1 gewöhnlichen Differentialgleichungen für die r Reihen $A^{-1\cdot},\ldots,A^{-r\cdot}$ und s+1 gewöhnliche Differentialgleichungen für die s Reihen $A^{\cdot-1},\ldots,A^{\cdot-s}$ wobei:

$$(5) \qquad A^{\alpha\cdot} := \sum_{i\geq 0} a_{\alpha i} z^i \text{ und } A^{\cdot\beta} := \sum_{i\geq 0} a_{i\beta} z^i \; .$$

Diese Systeme lauten:

$$(6) \qquad \alpha\frac{d}{dz}A^{\alpha\cdot} = \sum_{i\geq o} \{(\alpha-i)A^{\alpha-i,\cdot}.F^{i\cdot} + G^{i\cdot}\frac{d}{dz}A^{\alpha-i-1,\cdot} + A^{\alpha-i-1,\cdot}H^{i\cdot}\}$$

mit $\alpha=o,-1,\ldots,\alpha_o=-r$

$$(6') \qquad \beta\frac{d}{dz}A^{\cdot\beta} = \sum_{j\geq o} \{F^{\cdot j}\frac{d}{dz}A^{\cdot\beta-j-1} + (\beta-j)A^{\cdot\beta-j}G^{\cdot j} + A^{\cdot\beta-j-1}H^{\cdot j}\}$$

mit $\beta=o,-1,\ldots,\beta_o=-s$.

Dabei sind die Reihen $F^{i\cdot}$ usw. in Analogie zu (5) gebildet. Beim Ansatz in (2) ergeben sich r+m Differentialgleichungen:

$$(7) \qquad \sum_{i\geq o}\sum_{\mu,\nu=o}^{m,n} [F^{(\mu\nu)}]^{i\cdot} (\mu,\alpha-m+\mu-i)\frac{d^{\nu}}{dz^{\nu}}A^{\alpha-m+\mu-i\cdot} = o$$

mit $\alpha=m-1,m-2,\ldots,\alpha_o$

und s+n Differentialgleichungen:

$$(7') \qquad \sum_{j\geq o}\sum_{\mu,\nu=o}^{m,n} [F^{(\mu\nu)}]^{j\cdot} (\nu,\beta-n+\nu-j)\frac{d^{\mu}}{dz^{\mu}}A^{\cdot\beta-n+\nu-j} = o .$$

mit $\beta=n-1,n-2,\ldots,\beta_o$

Der Hauptteil einer Lösung von (1) bzw. (2) ist durch (6) und (6') bzw. (7) und (7') vollständig bis auf gegebenenfalls frei wählbare Anfangswerte eindeutig bestimmt. Der Potenzreihenanteil der Lösung ist bis auf die in Satz 1 beschriebenen Anfangswerte (Goursat-Daten) ebenfalls durch die Rekursionsformeln (3) bzw. (4) eindeutig bestimmt. Es gilt:

Satz 2: Die Differentialgleichung (1) bzw. (2) besitzt genau dann eine Lösung $w=\sum a_{\alpha\beta}z_1^{\alpha}z_2^{\beta}$ mit nichtverschwindendem Hauptteil, wenn die überbestimmten Systeme(6) oder (6') bzw. (7) oder (7') nichttriviale Lösungen besitzen. Die Konvergenzkriterien für solch eine Lösung sind gleich wie in Satz 1.

Zum Konvergenzbeweis sei bemerkt, daß dieser mittels der Majorantenmethode geführt werden kann. Siehe dazu Abschnitt IV.

III. Partielle Differentialgleichungen mit konstanten Koeffizienten.

Wir betrachten nun die Differentialgleichung:

(1') $w_{z_1 z_2} = f\cdot w_{z_1} + g\cdot w_{z_2} + h\cdot w \qquad f,\ g,\ h \in \mathbb{C}$.

Die Systeme (6) und (6') lauten hier:

(8) $\alpha \frac{d}{dz} A^{\alpha\cdot} = \alpha f A^{\alpha\cdot} + g\frac{d}{dz}A^{\alpha-1\cdot} + hA^{\alpha-1\cdot}$ und

(8') $\beta \frac{d}{dz} A^{\cdot\beta} = f\frac{d}{dz}A^{\cdot\beta-1} + \beta g A^{\cdot\beta} + hA^{\cdot\beta-1}$.

Mit $D := \frac{d}{dz}$ lautet (8) in Matrizenschreibweise:

$$\begin{pmatrix} gD+h & o & o & & & \\ D-f & gD+h & o & & 0 & \\ o & 2(D-f) & gD+h & & & \\ o & o & 3(D-f) & & & \\ & & \cdot & \cdot & \cdot & \\ & & & & (-\alpha_0+1)(D-f) & gD+h \\ & 0 & & & o & -\alpha_0(D-f) \end{pmatrix} \cdot \begin{pmatrix} A^{-1\cdot} \\ \cdot \\ \cdot \\ \cdot \\ A^{\alpha_0\cdot} \end{pmatrix} = 0$$

Dieses System ist äquivalent zu:

$$\begin{pmatrix} i_r & & & \\ & i_{r-1} & 0 & \\ & & \cdot & \\ & 0 & & \cdot \\ & & & \cdot \\ & & & i_1 \\ & 0 & & o \end{pmatrix} \tilde{y} = 0 \qquad \text{mit } i_r | i_{r-1}, \ldots, i_2 | i_1 \text{ (siehe [4], p. 123 f.).}$$

(8) hat also genau dann nur triviale Lösungen, wenn der erste Elementarteiler i_1 der oben angeschriebenen Matrix gleich 1 ist. Dies ist genau dann der Fall, wenn der größte gemeinsame Teiler der Polynome $D-f$ und $gD+h$ gleich 1 ist, also $gf + h \neq o$.
Analoge Untersuchungen ergeben, daß das System (8') ebenfalls nur triviale Lösungen zuläßt, falls $gf+h \neq o$.

Es gilt daher:

Satz 3: Die Differentialgleichung (1') besitzt genau dann regulär-singuläre Lösungen mit nichtverschwindendem Hauptteil beliebig hoher Ordnung bezüglich z_1 und z_2, wenn:

$$f \cdot g + h = o .$$

Ein analoges Ergebnis erhält man auch für Differentialgleichungen höherer Ordnung:

Satz 4: Die Differentialgleichung

$$(2') \qquad \sum_{\mu=o}^{m} \sum_{\nu=o}^{n} f_{\mu\nu} \cdot \frac{\partial^{\mu+\nu}}{\partial z_1^{\mu} \partial z_2^{\nu}} w = o \qquad f_{\mu\nu} \in \mathbb{C},\ f_{mn} = 1$$

besitzt genau dann regulär-singuläre Lösungen der Form $\Sigma a_{\alpha\beta} z_1^{\alpha} z_2^{\beta}$ mit nichtverschwindendem Hauptteil beliebig hoher Ordnung bezüglich z_1 oder z_2, wenn die Polynome $p_\mu(D) := \sum_{\nu=o}^{n} f_{\mu\nu} D^{\nu}$, $\mu=o,\dots,m$ einen nichttrivialen gemeinsamen Teiler haben, oder wenn die Polynome $q_\nu(D) := \sum_{\mu=o}^{m} f_{\mu\nu} D^{\mu}$, $\nu=o,\dots,n$ einen nichttrivialen gemeinsamen Teiler haben.

Beweis: Es geht darum, die Systeme (7) und (7') für die Differentialgleichung (2') zu untersuchen. Diese lauten hier mit den oben eingeführten Polynomen:

$$(9) \qquad \sum_{\mu=o}^{m} (\mu, \alpha-m+\mu) p_\mu(D) A^{\alpha-m+\mu \cdot} = o, \quad \alpha = m-1, m-2, \dots, \alpha_o$$

und

$$(9') \qquad \sum_{\nu=o}^{n} (\nu, \beta-n+\nu) q_\nu(D) A^{\cdot \beta-n+\nu} = o, \quad \beta = n-1, n-2, \dots, \beta_o .$$

(9) besitzt (analog wie vorher System (8)) genau dann nichttriviale Lösungen, wenn die Matrix $S(D)$ mit $r = |\alpha_o|$ Spalten und $r+m$ Zeilen:

$S(D) = (a_{ij})_{i=1,\dots,r+m;\, j=1,\dots,r}$

(mit $a_{ij} = (i-j,-j) p_{i-j}(D)$, wobei $p_i \equiv o$ für $i < o$ oder $i > m$)

einen von 1 verschiedenen ersten Elementarteiler $i_1(D)$ hat. Dies ist aber genau dann der Fall, wenn der größte gemeinsame Teiler $GGT(p_m, p_{m-1}, \dots, p_o) \neq 1$ ist.

Denn, wenn $i_1 \neq 1$ ist und etwa ζ als Nullstelle besitzt, folgt aus der Eigenschaft, daß i_1 jede r-zeilige Unterdeterminante von $S(D)$ teilt:

$$p_m(\zeta) = p_{m-1}(\zeta) = \dots = p_0(\zeta) = o,$$

indem man nacheinander die Unterdeterminanten, welche als Diagonalelemente $p_m, p_{m-1}, \dots$ beziehungsweise p_0 haben, bildet und ζ anstelle von D einsetzt. Also ist $GGT(p_m,\dots,p_0) \neq 1$. Sei nun $GGT(p_m,\dots,p_0) = f(D) \neq 1$. Dann ist mit einer von null verschiedenen Lösung $y(z)$ von $f(D)y = o$ auch $A^{-1\cdot} = y, \dots, A^{\alpha_o \cdot} = y$ eine nichttriviale Lösung von (9). Somit ist $i_1(D) \neq 1$ und (9) hat genau dann nichttriviale Lösungen, wenn $GGT(p_m,\dots,p_0) \neq 1$ ist. Ganz analog zeigt man, daß (9') genau dann nichttriviale Lösungen besitzt, wenn $GGT(q_n, q_{n-1},\dots,q_0) \neq 1$ ist.

IV Konvergenzbeweis.

Zum Abschluß werde noch die Konvergenz einer in Satz 2 beschriebenen formalen Lösung von (2) untersucht. Es gilt:

Satz 5: Es sei $w = \sum a_{\alpha\beta} z_1^{\alpha} z_2^{\beta}$ eine Lösung von (2). Dabei seien die frei wählbaren Randkoeffizienten $a_{\mu j}$ und $a_{i\nu}$ mit $o \leq \mu < m$, $o \leq \nu < n$, $i \geq o$ und $j \geq o$ so angenommen, daß die Reihe

$$\sum_{\substack{i \geq o \\ o \leq \nu < n}} a_{i\nu} z_1^i z_2^{\nu} + \sum_{\substack{o \leq \mu < m \\ j \geq o}} a_{\mu j} z_1^{\mu} z_2^j$$

in einer Umgebung von $(z_1, z_2) = (o,o)$ konvergiert. Dann gibt es eine Umgebung U von (o,o) derart, daß w für alle (z_1, z_2) aus U mit $z_1 \cdot z_2 \neq o$ konvergiert.

Beweis: Wir führen den Konvergenzbeweis mittels der Majorantenmethode. Es sei $\hat{F}^{(m,n)}$ eine Majorante von $F^{(m,n)} - 1$, mit $\hat{F}^{(m,n)}(o,o) = o$. Weiters seien $\hat{F}^{(\mu,\nu)}$ Majoranten von $F^{(\mu,\nu)}$. Zusätzlich setzen wir:

$$\hat{a}_{\alpha\beta} = \begin{cases} o & \text{für } \alpha \geq m \text{ und } \beta \geq n \\ |a_{\alpha\beta}| & \text{für } \alpha < m \text{ oder } \beta < n \end{cases}$$

und bilden damit:

$$\hat{A}_{\mu\nu} = \sum_{\alpha \geq \alpha_0, \beta \geq \beta_0} |(\mu,\alpha)(\nu,\beta)| \hat{a}_{\alpha\beta} z_1^{\alpha-\alpha_o} z_2^{\beta-\beta_o}.$$

Die Reihen $\hat{A}_{\mu\nu}$ konvergieren in einer Umgebung von (o,o), weil sie jeweils aus endlichen Summen von entweder nach Voraussetzung konvergenten Reihen oder von Reihen, die Lösungen von (7) bzw. (7') sind, bestehen.

Durch die Gleichung

$$(10)\quad R = \sum_{\mu,\nu=0}^{m,n} \hat{F}^{(\mu,\nu)} \cdot z_1^{m-\mu} z_2^{n-\nu} (R + \hat{A}_{\mu\nu}) + \hat{A}_{00}$$

wird daher eine in einer Umgebung U von (o,o) holomorphe Funktion R definiert. Setzt man R mit

$$R = \sum_{\alpha\geq\alpha_0, \beta\geq\beta_0} r_{\alpha\beta} z_1^{\alpha-\alpha_0} z_2^{\beta-\beta_0}$$

an, so folgt, wie man leicht aus den resultierenden Rekursionsformeln schließen kann:

$$r_{\alpha\beta} \geq |a_{\alpha\beta}|.$$

Also ist $z_1^{-\alpha_0} z_2^{-\beta_0} \cdot w$ in U konvergent, was Satz 5 beweist.

Literaturverzeichnis:

[1] Bauer,K.W., Peschl,E.: Ein allg.Entwicklungssatz für die Lösungen der Digln $(1+\varepsilon z\bar{z})^2 w + \varepsilon n(n+1)w = o$ in der Nähe isolierter Singularitäten. Sitzungsber.Bayr.Akad.Wiss., Math.-nat.Kl. 1966.

[2] Bergman,S.: Integral Operators in the Theory of Linear Partial Differential Equations, Springer-Verlag, Berlin 1961.

[3] Bieberbach,L.: Theorie der gewöhnl. Differentialgleichungen. 2.Aufl., Springer-Verlag, Berlin 1965.

[4] Gantmacher,F.R.: Matrizenrechnung. Teil I, 2.Aufl. Berlin 1965.

[5] Gilbert,R.P., Hsiao,G.C.: Constructive function theoretic methods for higher order pseudoparabolic equations. Proceedings of the Symposium on Function Theoretic Methods for Partial Differential Equations, Darmstadt 1976.

[6] Gronau,D., Reich,L.: Über lokale regulär-singuläre Lösungen einer Klasse linearer partieller Digln. 2.Ordnung im Komplexen. Sitzungsber.der Österr.Akad.d.Wiss., math.-nat. Kl. Abt.II, 183 (1974), 321-333.

[7] Gronau,D.: Über regulär-sing. logarithmenbehaftete Lösungen einer Klasse linearer part.Dtgln. im Komplexen. Mathematica Balkanica 4 (1974), 229-237.

[8] Gronau,D.: Logarithmenfreie Singularitäten von linearen partiellen Differentialgleichungen. Ber.d.Math.-Statist. Sektion im Forschungszentrum Graz, Nr.51 (1976).

[9] Reich,L.: Über regulär-singuläre Stellen von Lösungen einer Klasse linearer partieller Differentialgleichungen. Berichte der Math.-stat.Sektion im Forschungszentrum Graz, Nr. 3 (1973).

[10] Vekua,I.N.: New Methods for Solving Elliptic Equations, John Wiley, New York, 1967.

Adresse des Autors: II. Math. Institut der
Universität Graz
Steyrergasse 17
A-8010 G r a z

LÖSUNGSDARSTELLUNGEN MITTELS DIFFERENTIALOPERATOREN FÜR DAS DIRICHLET-PROBLEM DER GLEICHUNG $\Delta u+c(x,y)u=0$

Rudolf Heersink

1) Einleitung

In der vorliegenden Arbeit werden gewisse selbstadjungierte elliptische Gleichungen

(1) $\Delta u+c(x,y)u=0$

betrachtet, wobei $c(x,y)$ analytisch in einem Gebiet $D\subset\mathbb{R}^2$ sein soll. Setzt man x und y zu komplexen Werten fort, so erhält man mittels $z:=x+iy, \tilde{z}:=x-iy, \partial/\partial z:=(\partial/\partial x-i\partial/\partial y)/2, \partial/\partial\tilde{z}:=(\partial/\partial x+i\partial/\partial y)/2, w(z,\tilde{z})=u(x,y)$ die (1) zugeordnete Gleichung (2) in den beiden unabhängigen komplexen Veränderlichen z und $\tilde{z}$:

(2) $w_{z\tilde{z}}+C(z,\tilde{z})w=0$ mit $C(z,\tilde{z})=c(\frac{z+\tilde{z}}{2},\frac{z-\tilde{z}}{2i})/4.$

Im folgenden soll D immer ein "Fundamentalgebiet" von (1) sein, d.h.: D sei einfach zusammenhängend und $C(z,\tilde{z}):=c((z+\tilde{z})/2,(z-\tilde{z})/2i)/4$ sei holomorph in $D\times\bar{D}$, wobei $\bar{D}:=\{z|\bar{z}\varepsilon D\}$, $\bar{z}$ konjugiert komplex zu z.

Somit sind (1) und (2) in folgendem Sinne äquivalent (vgl.[5]):

a) jede in $D\times\bar{D}$ holomorphe Lösung w von (2) stellt mit $\tilde{z}=\bar{z}$ eine in D analytische Lösung u von (1) dar;

b) ist $u(x,y)$ eine in D definierte (und somit nach einem Satz von Picard dort analytische) Lösung von (1), so stellt $w(z,\tilde{z}):=u((z+\tilde{z})/2,(z-\tilde{z})/2i)$ eine in $D\times\bar{D}$ holomorphe Lösung von (2) dar.

Mit H_D bzw. $H_{D\times\bar{D}}$ wird im weiteren die Menge der in D bzw. $\bar{D}$ holomorphen Funktionen bezeichnet. Ist $z_0\varepsilon D$ ein beliebiger, aber fester Punkt, so soll $H_D(k,z_0)$, $k\varepsilon\mathbb{N}$, die Menge der Funktionen $g\varepsilon H_D$ sein, für die gilt: $g(z_0)=g'(z_0)=\cdots=g^{(k-1)}(z_0)=0$.

DEFINITION 1: Ein Differentialoperator der Gestalt

$$X_n:=\sum_{k=0}^{n}a_k(z,\tilde{z})\frac{\partial^k}{\partial z^k}\ ,\ n\varepsilon\mathbb{N},\ a_k\varepsilon H_{D\times\bar{D}}\ \text{für}\ k=0,1,\ldots,n; a_n\not\equiv 0\ \text{in}\ D\times\bar{D},$$

soll als "Bauer-Operator" der Ordnung n der Gleichung (2) bezeichnet werden, falls X_ng für alle $g\varepsilon H_D$ eine Lösung $w\varepsilon H_{D\times\bar{D}}$ von (2) darstellt.

Im folgenden werden nur noch Gleichungen (1) betrachtet, für welche Bauer-Operatoren X_n der zugeordneten Gleichung (2) existieren. Diese Klasse enthält einige wichtige Gleichungen (vgl. z.B. [1], [3]). Weiters sollen von nun an nur Gleichungen (1) mit reellwertigem Koeffizienten c(x,y) behandelt werden. Dafür existieren dann reellwertige Lösungen u(x,y) und man kann den folgenden Satz formulieren:

Satz 1

Es sei X_n ein gegebener Bauer-Operator der Ordnung n für eine (1) zugeordnete Gleichung (2), wobei $a_n(z,\tilde{z})\neq 0$ in $D\times\bar{D}$ sei.
a) Zu jeder in D definierten, reellwertigen Lösung u von (1) existiert eine "erzeugende Funktion" $g\varepsilon H_D(n,z_0)$, so daß

$$u(x,y)=\mathrm{Re}\{X_n g|_{\tilde{z}=\bar{z}}\}=\mathrm{Re}\{\sum_{k=0}^{n} a_k(z,\bar{z})g^{(k)}(z)\}.$$

b) Schreibt man $g(z)=(z-z_0)^n g_H(z)$ mit $g_H\varepsilon H_D$, so ist $g_H(z)$ bis auf eine in $g_H(z_0)$ frei wählbare, rein imaginäre Konstante eindeutig durch u(x,y) bestimmt.

Beweis
Lt. [5] stellt

$$(3)\quad w(z,\tilde{z})=\alpha R(z_0,\tilde{z}_0,z,\tilde{z})+\int_{z_0}^{z} R(t,\tilde{z}_0,z,\tilde{z})f(t)dt+\int_{\tilde{z}_0}^{\tilde{z}} R(z_0,\tilde{t},z,\tilde{z})\tilde{f}(\tilde{t})d\tilde{t}$$

mit $\alpha\varepsilon\mathbb{C}$, $f\varepsilon H_D$, $\tilde{f}\varepsilon H_{\bar{D}}$ immer eine Lösung $w\varepsilon H_{D\times\bar{D}}$ von (2) dar, falls $R(z,\tilde{z},t,\tilde{t})$ die Riemannfunktion von (2) ist und $z_0\varepsilon D$, $\tilde{z}_0\varepsilon\bar{D}$ beliebige, aber feste Punkte sind. Umgekehrt kann jede Lösung $w\varepsilon H_{D\times\bar{D}}$ von (2) durch (3) dargestellt werden, wobei α, f und $\tilde{f}$ eindeutig durch

$$(4)\qquad \alpha=w(z_0,\tilde{z}_0),\ f(z)=w_z(z,\tilde{z}_0),\ \tilde{f}(\tilde{z})=w_{\tilde{z}}(z_0,\tilde{z})$$

bestimmt sind.
Betrachtet man nun das Anfangswertproblem

$$(5a)\qquad f(z)=\sum_{k=0}^{n+1} g^{(k)}(z)\{a_{k-1}(z,\tilde{z}_0)+\frac{\partial}{\partial z}a_k(z,\tilde{z}_0)\}\ ,\ a_{-1}:\equiv 0,\ a_{n+1}:\equiv 0,$$

$$(5b)\qquad g(z_0)=g'(z_0)=\cdots\cdots=g^{(n-1)}(z_0)=0,\ g^{(n)}(z_0)=\frac{\alpha}{a_n(z_0,\tilde{z}_0)}$$

mit α und f(z) aus (4), so hat dieses Problem wegen $a_n\neq 0$ in $D\times\bar{D}$ immer genau eine Lösung $g\varepsilon H_D(n,z_0)$ der Form:

(6) $$g(z)=\frac{\alpha}{a_n(z_0,\tilde{z}_0)n!}(z-z_0)^n + g_{n+1}(z),\quad g_{n+1}\varepsilon H_D(n+1,z_0).$$

Bildet man nun $X_n g$ mit diesem so bestimmten $g(z)$, so folgt:

$$w(z,\tilde{z})=X_n g=\alpha R(z_0,\tilde{z}_0,z,\tilde{z})+\int_{z_0}^{z} R(t,\tilde{z}_0,z,\tilde{z})f(t)dt,$$

wobei $f(z)$ durch (5a) gegeben ist. Dies folgt sofort aus (4) und (6) und der Tatsache (vgl. [6]), daß $\partial a_n/\partial\tilde{z}\equiv 0$ für jeden Bauer-Operator. Damit ist gezeigt, daß jede Lösung $w\varepsilon H_{D\times\bar{D}}$ von (2) der Form (3) mit $\tilde{f}(\tilde{z})\equiv 0$ durch $X_n g$ mit genau einem $g\varepsilon H_D(n,z_0)$ dargestellt werden kann, eben mit dem g, welches (5a),(5b) löst.

Die Aussage des Satzes 1 ergibt sich nun unmittelbar (indem man zu $\tilde{z}=\bar{z}$ übergeht) aus dem von Vekua in [5] angegebenen Satz: Zu jeder in D definierten reellwertigen Lösung u von (1) existieren ein $f\varepsilon H_D$ sowie ein $\alpha\varepsilon\mathbb{C}$, so daß

$$u(x,y)= \mathrm{Re}\{\alpha R(z_0,\bar{z}_0,z,\bar{z})+\int_{z_0}^{z} R(t,\bar{z}_0,z,\bar{z})f(t)dt\}.$$

$f(z)$ ist dabei eindeutig, α bis auf eine rein imaginäre Konstante durch $u(x,y)$ bestimmt.

2) Das Dirichlet-Problem für Gleichung (1)

Es sei $K:=\{(x,y)|\ x^2+y^2<R^2,\ R\varepsilon\mathbb{R}^+\}$ die offene Kreisscheibe mit Radius R und Mittelpunkt (0,0), $\tilde{K}$ bzw. ∂K bezeichne die zugehörige abgeschlossene Kreisscheibe bzw. den Rand der Scheibe.

H_K bzw. $H_{K\times K}$ bezeichne die Menge der in K bzw. KxK holomorphen Funktionen, $H_K(l,z_0)$ mit $l\varepsilon\mathbb{N}$ und $z_0\varepsilon K$ sei die Menge der Funktionen $g\varepsilon H_K$ mit $g(z_0)=g'(z_0)=\cdots\cdots=g^{(l-1)}(z_0)=0$.

$C^H(\partial K):=\{f(\xi)|\xi\varepsilon\partial K;\ \bigvee_{\alpha\varepsilon(0,1]}\ \bigwedge_{\xi_1,\xi_2\varepsilon\partial K} |f(\xi_1)-f(\xi_2)|<A|\xi_1-\xi_2|^{\alpha}, A\varepsilon\mathbb{R}^+\}$ sei die Menge der auf ∂K hölderstetigen Funktionen und $C^{l,H}(\partial K)$ sei die Menge der auf ∂K l-mal hölderstetig differenzierbaren Funktionen.

Im folgenden soll $\tilde{K}\subset D$ vorausgesetzt werden, wobei D ein Fundamentalgebiet der betrachteten Gleichung (1) ist. $c(x,y)$ in (1) sei weiterhin als in D reellwertig angenommen. Desweiteren wird in $D\times\bar{D}$ die Existenz eines Bauer-Operators X_n mit der Eigenschaft $a_n\neq 0$ in $D\times\bar{D}$ vorausgesetzt.

Nun soll das folgende Dirichlet-Problem betrachtet werden:

(DR) a) Gesucht ist eine in K definierte reellwertige Lösung u(x,y) von (1), für die gilt:

b) $\lim_{(x,y)\to\xi\varepsilon\partial K} u(x,y) = f(\xi)$, $f\varepsilon C^H(\partial K)$ und reellwertig.

Falls $f(\xi)\equiv 0$ auf ∂K, so soll dies durch (DR_0) angedeutet werden. Wählt man für die im Abschnitt 1 angeführten festen Punkte $z_0=0$ und $\tilde{z}_0=0$, so ergibt sich mit Satz 1:

Ist u(x,y) eine Lösung des Problems (DR) für die vorgegebene Gleichung (1), so existiert eine (bis auf die im Satz 1 erwähnte freie Konstante) eindeutig bestimmte "Erzeugende" $g\varepsilon H_K(n,0)$, mit welcher diese Lösung durch

(7) $$u(x,y)=\mathrm{Re}\{X_n g|_{\tilde{z}=\bar{z}}\}=\mathrm{Re}\{\sum_{k=0}^{n} a_k(z,\bar{z})g^{(k)}(z)\}$$

dargestellt wird.

Für das Weitere benötigt man die folgenden beiden Hilfssätze:

<u>Hilfssatz 1</u>

Ist die vorgegebene Randfunktion $f(\xi),\xi\varepsilon\partial K$, des Problems (DR) hölderstetig: $f\varepsilon C^H(\partial K)$, so existiert $[g^{(n)}(\xi)]^+:=\lim_{z\to\xi\varepsilon\partial K} g^{(n)}(z)$ für die Erzeugende g(z) in (7) und $[g^{(n)}(\xi)]^+\varepsilon C^H(\partial K)$.

<u>Beweisskizze</u>

In [5] wird $\Phi(z):=w(z,0)$ betrachtet, wobei $w(z,\tilde{z})=\alpha R(0,0,z,\tilde{z})+ +\int_0^z R(t,0,z,\tilde{z})f(t)dt$ jene in KxK holomorphe Lösung von (2) ist, die durch $u(x,y)=\mathrm{Re}[w(z,\bar{z})]$ das vorgelegte Problem (DR) löst. Es wird gezeigt, daß $[\Phi(\xi)]^+\varepsilon C^H(\partial K)$. Da $w(z,\tilde{z})$ auch durch $X_n g$ mit $g\varepsilon H_K(n,0)$ darstellbar ist (vgl. Beweis von Satz 1), folgt:

$\Phi(z)=\sum_{k=0}^{n} a_k(z,0)g^{(k)}(z)$. Daraus ergibt sich, daß $[g^{(n)}(\xi)]^+\varepsilon C^H(\partial K)$.

<u>Hilfssatz 2</u>

Für die Erzeugende g(z) in (7) gilt:

a) $[g(\xi)]^+:=\lim_{z\to\xi\varepsilon\partial K} g(z)$ existiert und $[g(\xi)]^+\varepsilon C^{n,H}(\partial K)$.

b) $[g^{(k)}(\xi)]^+:=\lim_{z\to\xi\varepsilon\partial K} g^{(k)}(z)$ existiert und $[g^{(k)}(\xi)]^+\varepsilon C^{n-k,H}(\partial K)$, $k=1,2,\cdots,n$.

c) Schreibt man: $\Phi(\xi):=[g(\xi)]^+$, so gilt: $\Phi^{(k)}(\xi)=[g^{(k)}(\xi)]^+$, $k=1,2,\cdots,n$.

Beweis

Lt. Hilfssatz 1 gilt: $[g^{(n)}(\xi)]^+\varepsilon C^H(\partial K)$. Mit $\alpha(\xi):=[g^{(n)}(\xi)]^+$ folgt (vgl. [2]) $g^{(n)}(z)=\frac{1}{2\pi i}\oint_{\partial K}\frac{\alpha(\xi)}{\xi-z}d\xi$ für alle $z\varepsilon K$. Definiert man:

$\beta(\xi):=\int\alpha(\xi)d\xi$, d.h. $\beta(\xi)\varepsilon C^{1,H}(\partial K)$ und $\beta'(\xi)=\alpha(\xi)$ sowie $F(z):=\frac{1}{2\pi i}\oint_{\partial K}\frac{\beta(\xi)}{\xi-z}d\xi$

für $z\varepsilon K$, so folgt: $F(z)$ ist holomorph in K und $F'(z)=\frac{1}{2\pi i}\oint_{\partial K}\frac{\beta'(\xi)}{\xi-z}d\xi=\frac{1}{2\pi i}\oint_{\partial K}\frac{\alpha(\xi)}{\xi-z}d\xi=$

$=g^{(n)}(z)$. Somit ergibt sich $g^{(n-1)}(z)=F(z)+C$ und $g^{(n-1)}(z)=\frac{1}{2\pi i}\oint_{\partial K}\frac{\beta(\xi)}{\xi-z}d\xi+C$.

Da $\beta\varepsilon C^{1,H}(\partial K)$ ist auch $[g^{(n-1)}(\xi)]^+\varepsilon C^{1,H}(\partial K)$, vgl. [2].
Setzt man $\Phi(\xi):=[g^{(n-1)}(\xi)]^+ \rightarrow \Phi'(\xi)=[g^{(n)}(\xi)]^+$, vgl. [2].
Wiederholt man diesen Beweisgang, so folgen die Aussagen des Hilfssatzes.

Mit obigen beiden Hilfssätzen ergibt sich aus (7):

$$\lim_{(x,y)\to\xi\varepsilon\partial K} u(x,y) = f(\xi) = \mathrm{Re}\sum_{k=0}^{n}\{a_k(\xi)[g^{(k)}(\xi)]^+\}=$$

(8)

$$=\mathrm{Re}\sum_{k=0}^{n}\{a_k(\xi)\cdot([g(\xi)]^+)^{(k)}\} \quad \text{mit } a_k(\xi):=\lim_{z\to\xi\varepsilon\partial K} a_k(z,\bar{z}).$$

Der erste Teil in (8) stellt ein verallgemeinertes Riemann-Hilbert-Poincaré-Problem dar (vgl. [4]). Dieses Problem wird im allgemeinen in eine singuläre Integralgleichung übergeführt (vgl. [4]). Für spezielle Fälle soll in dieser Arbeit das Problem auf ein dazu äquivalentes gewöhnliches Differentialgleichungssystem reduziert werden.

Ist $\hat{\phi}(\xi)=\xi^n\phi(\xi)$ der Randwert von $g(z)=z^n g_H(z)$ in (7): $\hat{\phi}(\xi)=[g(\xi)]^+$, so besitzt diese Erzeugende $g\varepsilon H_K(n,0)$ die Darstellung

(9) $$g(z) = \frac{1}{2\pi i}\oint_{\partial K}\frac{\xi^n\phi(\xi)}{\xi-z}d\xi ,$$

wobei wegen Hilfssatz 2a $\phi\varepsilon C^{n,H}(\partial K)$ gilt. Unter Anwendung der Leibnitz-Regel sowie Umordnen nach aufsteigenden Ableitungen von $\phi(\xi)$ erhält man aus (8):

(10) $$f(\xi) = \mathrm{Re}\{\sum_{j=0}^{n}\phi^{(j)}(\xi)\xi^j S_j(\xi)\} \quad \text{mit}$$

(11) $$S_j(\xi) = \sum_{k=j}^{n}\xi^{n-k}a_k(\xi)\binom{k}{j}\frac{n!}{(n-k+j)!} .$$

Führt man nun den reellen Kurvenparameter $t\varepsilon[0,2\pi]$ durch $\xi=Re^{it}$ ein (und schreibt $f(t)=f[\xi(t)]$ usw.) und spaltet die "Dichtefunktion" $\phi(t)$ in Real- und Imaginärteil auf: $\phi(t)=\phi_1(t)+i\phi_2(t)$, so geht (10) nach einigen Umformungen über in:

$$(12) \qquad f(t) = \sum_{k=0}^{n} A_k(t)\phi_1^{(k)}(t) + \sum_{k=0}^{n} B_k(t)\phi_2^{(k)}(t)$$

mit

$$(13) \quad \begin{cases} A_k(t)=\mathrm{Re}\hat{A}_k, B_k(t)=-\mathrm{Im}\hat{A}_k, \hat{A}_0(t)=S_0(Re^{it}), \hat{A}_j(t)=\sum_{k=j}^{n} \alpha_{k,j}\cdot S_k(Re^{it}) \\ \text{für } j=1,2,\cdots,n, \text{ wobei } \alpha_{k,l}=(-1)^k(i)^l a_{k,l} \text{ mit } a_{0,0}=1, a_{0,1}=0 \\ \text{sowie } a_{k,l}=a_{k-1,l-1}+(k-1)a_{k-1,l} \text{ für } k=1,2,\cdots,n; l=1,2,\cdots,n. \end{cases}$$

Im weiteren benötigen wir die

DEFINITION 2: $\psi_1(\xi)\varepsilon C^H(\partial K)$ und $\psi_2(\xi)\varepsilon C^H(\partial K)$ nennt man zueinander "konjugiert", symbolisch: $\psi_1 \overset{k}{\sim} \psi_2$, wenn eine Funktion $G(z)\varepsilon H_K$ mit $[G(\xi)]^+\varepsilon C^H(\partial K)$ existiert, so daß $\psi_1(\xi)= \lim_{z\to\xi\varepsilon\partial K} [\mathrm{Re}\, G(z)]$ und $\psi_2(\xi)= \lim_{z\to\xi\varepsilon\partial K} [\mathrm{Im}\, G(z)]$.

Sind A_k, B_k durch (13) gegeben und ist $f(\xi)$ die vorgegebene Randfunktion des Problems (DR), so ergibt sich aus dem bisher Gesagten unmittelbar der

Satz 2

a) Ist (ϕ_1,ϕ_2) mit $\phi_1,\phi_2\varepsilon C^{n,H}(\partial K)$ eine Lösung von

$$(14) \quad \begin{cases} f(t) = \sum_{k=0}^{n} A_k(t)\phi_1^{(k)}(t) + \sum_{k=0}^{n} B_k(t)\phi_2^{(k)}(t), \\ \phi_1 \overset{k}{\sim} \phi_2, \end{cases}$$

so stellt (7) mit $g(z)$ nach (9) eine Lösung des Problems (DR) dar (wobei $\phi=\phi_1+i\phi_2$ in (9)).

b) Ist andrerseits $u(x,y)$ eine Lösung von (DR) und (7) sei eine Darstellung von u, so sind (ϕ_1,ϕ_2) in (9) notwendig Lösungen von (14).

(14) stellt also ein zu (DR) äquivalentes Problem dar.

Für die weiteren Untersuchungen wollen wir die folgende (ein - schränkende) Voraussetzung treffen:

Voraussetzung 1: Die vorliegende Gleichung (1) und der zugehörige

Bauer-Operator X_n seien von solcher Gestalt, daß für alle $j=0,1,\cdots,n$ gilt: $S_j(\xi)$ in (11) sei der Randwert einer in K holomorphen Funktion $S_j(z)$, d.h. Re $S_j(\xi) \overset{k}{\sim}$ Im $S_j(\xi)$.

Verwendet man die leicht zu beweisenden Aussagen:

$$(15)\quad \begin{cases} (\phi_1 \overset{k}{\sim} \phi_2) \Leftrightarrow (\phi_2 \overset{k}{\sim} -\phi_1); \\ (\phi_1 \overset{k}{\sim} \phi_2) \wedge \phi_1 \in C^{n,H}(\partial K) \Leftrightarrow \phi_2 \in C^{n,H}(\partial K) \wedge (\phi_1^{(l)} \overset{k}{\sim} \phi_2^{(l)}),\ l=1,2,\cdots,n; \\ (\phi_1 \overset{k}{\sim} \phi_2) \wedge (S_1 \overset{k}{\sim} S_2) \Rightarrow (S_1\phi_1 - S_2\phi_2) \overset{k}{\sim} (S_1\phi_2 + S_2\phi_1), \end{cases}$$

so erhält man den Satz 3:

Satz 3

a) Es sei Voraussetzung 1 erfüllt. Dann ist das Problem (14) und damit (DR) äquivalent zu

$$(16a)\quad \begin{cases} f(t) = \sum\limits_{k=0}^{n} A_k(t)\phi_1^{(k)}(t) + \sum\limits_{k=0}^{n} B_k(t)\phi_2^{(k)}(t) \\ \tilde{f}(t) = \sum\limits_{k=0}^{n} A_k(t)\phi_2^{(k)}(t) - \sum\limits_{k=0}^{n} B_k(t)\phi_1^{(k)}(t) \end{cases}$$

$$(16b)\quad \phi_1 \overset{k}{\sim} \phi_2,$$

wobei $f \overset{k}{\sim} \tilde{f}$.

b) Besitzt das Problem (DR) für jede Randfunktion $f \in C^H(\partial K)$ genau eine Lösung $u(x,y)$, so ist (DR) äquivalent zu (16a) mit der Forderung: $\phi_1^{(k)}(t)$ und $\phi_2^{(k)}(t)$, $k=0,1,\cdots,n$, seien 2π-periodisch, d.h. $\phi_i^{(k)}(t)=\phi_i^{(k)}(t+2\pi)$.

Nach der Floquetschen Theorie gibt es dann genau eine solche Lösung (ϕ_1,ϕ_2) von (16a).

Beweis

a) Daß (16a)$\wedge$(16b)$\Rightarrow$(14) ist offensichtlich. Die Umkehrung folgt, indem man (14) unter Beachtung von Voraussetzung 1 und (15) links und rechts im Sinne von Def.2 "konjugiert".

b) Punkt b folgt mit Hilfe der Floquetschen Theorie für periodische Differentialgleichungssysteme. Bringt man (16a) auf das entsprechende System erster Ordnung von 2n Gleichungen, so sind darin alle Koeffizienten 2π-periodisch. Dieses System besitzt nun entweder für alle $f(\xi)$ genau eine 2π-periodische Lösung (ϕ_1,ϕ_2); da (16a) aber nun genau eine Lösung (ϕ_1,ϕ_2) mit $\phi_1 \overset{k}{\sim} \phi_2$ hat, so ist diese 2π-periodische Lösung die gesuchte "konjugierte" Lösung. Oder aber obiges System besitzt keine bzw. unendlich viele 2π-periodische Lösungen, je nachdem ob $f(\xi)$ gewisse Orthogonalitätsrelationen nicht erfüllt bzw. erfüllt. Da dieses System aber für alle $f(\xi)$ genau eine "konjugierte" und damit mindestens eine 2π-periodische Lösung besitzt, kann dieser Fall unter den in b getroffenen Annahmen nicht eintreten.

Ist also Voraussetzung 1 erfüllt und hat das Problem (DR) für jede Randfunktion $f \varepsilon C^H(\partial K)$ genau eine Lösung u, so ist das Problem, (DR) zu lösen, äquivalent dazu, die eindeutig bestimmte 2π-periodische Lösung (ϕ_1, ϕ_2) von (16a) zu bestimmen.

3) Das Dirichlet-Problem für $\Delta u + [4en(n+1)/(1+e(x^2+y^2))^2] u = 0$

Eine spezielle Gleichungsklasse (1), für die Voraussetzung 1 erfüllt ist, ist

$$(17) \qquad \Delta u + 4e\frac{n(n+1)}{(1+e(x^2+y^2))^2}u = 0, \quad n \varepsilon \mathbb{N}, \; e = \pm 1.$$

Diese Gleichung wurde bereits in einer Reihe von Arbeiten betrachtet, vgl. z.B. [1].

Die (17) zugeordnete Gleichungsklasse der Form (2) ist gegeben durch

$$(18) \qquad w_{z\tilde{z}} + e\frac{n(n+1)}{(1+ez\tilde{z})^2}w = 0.$$

Ein geeigneter Bauer-Operator für (18) ist (vgl. [1])

$$(19) \qquad X_n = \sum_{k=0}^{n} A_k^n \left(\frac{\tilde{z}}{1+ez\tilde{z}}\right)^{n-k} \cdot \frac{\partial^k}{\partial z^k} \quad \text{mit} \quad A_k^n = (-e)^{n-k}\frac{(2n-k)!}{k!(n-k)!}.$$

Damit erhält man für die Funktionen $S_j(\xi)$ in (11):

$$(20) \qquad S_j = \sum_{k=j}^{n} A_k^n \left(\frac{R^2}{1+eR^2}\right)^{n-k} \binom{k}{j} \frac{n!}{(n-k+j)!},$$

wobei R der vorgegebene Radius für das Problem (DR) ist (die Kreisscheibe $\tilde{K}$ sei wieder eine Teilmenge eines Fundamentalgebietes von (17)).

Da alle S_j in (20) konstant sind (d.h. von ξ unabhängig), ist Voraussetzung 1 erfüllt. Mit (20) sowie (13) folgt für das System (16a) im Falle der Gleichung (18):

$$(21) \quad \begin{cases} f(t) = \sum_{k=0}^{[\frac{n}{2}]} A_{2k}\phi_1^{(2k)}(t) + \sum_{k=0}^{[\frac{n+1}{2}]-1} B_{2k+1}\cdot\phi_2^{(2k+1)}(t) \\ \tilde{f}(t) = \sum_{k=0}^{[\frac{n}{2}]} A_{2k}\phi_2^{(2k)}(t) - \sum_{k=0}^{[\frac{n+1}{2}]-1} B_{2k+1}\cdot\phi_1^{(2k+1)}(t), \end{cases}$$

worin alle A_{2k}, B_{2k+1} Konstanten sind, welche sich aus (13) errechnen.

Mit dem bisher Gesagten und Satz 3 ergibt sich somit unmittelbar

Satz 4

a) Das Problem (DR) für Gleichung (17) ist äquivalent dem System (21) mit $\phi_1 \overset{k}{\mathcal{C}} \phi_2$.

b) Sind (ϕ_1,ϕ_2) solche Lösungen von (21), dann stellt $u=\mathrm{Re}[X_n g|_{\tilde{z}=\bar{z}}]$

mit $g(z) = \frac{1}{2\pi i} \oint\limits_{\partial K} \frac{\xi^n[\phi_1(\xi)+i\phi_2(\xi)]}{\xi - z}\, d\xi$ und X_n nach (19) eine Lösung

des Problems (DR) für (17) dar.

Da im Falle von $e=-1$ immer genau eine Lösung von (DR) für (17) existiert (für jeden geeigneten Radius und jede Randfunktion $f\varepsilon C^H(\partial K)$, vgl. z.B. [1]), kann mit Hilfe von Satz 3b gefolgert werden:

Korollar 1

Das Problem (DR) für Gleichung (17) mit $e=-1$ ist äquivalent zu (21) mit der Forderung: (ϕ_1,ϕ_2) sowie alle in (21) auftretenden Ableitungen seien 2π-periodisch. Nach der Floquetschen Theorie existiert genau eine solche Lösung (ϕ_1,ϕ_2), welche mittels der in Satz 4b angegebenen Darstellung die eindeutig bestimmte Lösung u des Problems (DR) liefert.

Im Falle $e=+1$ liegen die Dinge etwas komplizierter. Es wird sich zeigen, daß in diesem Falle auf gewissen - u.zw. endlich vielen- Ausnahmeradien R entweder gar keine oder aber unendlich viele Lösungen von (DR) bzw. von (21) mit $\phi_1 \overset{k}{\mathcal{C}} \phi_2$ existieren.

Dazu nehmen wir für das Weitere an, daß die Randfunktion $f\varepsilon C^H(\partial K)$ in eine absolut und gleichmäßig konvergente Fourierreihe entwickelbar sei:

(22) $$f(t) = \frac{a_0}{2} + \sum_{k=1}^{\infty} a_k \sin kt + \sum_{k=1}^{\infty} b_k \cos kt, \quad a_0, a_k, b_k \varepsilon \mathbb{R},\ k\varepsilon \mathbb{N}.$$

Damit folgt:

(23) $$\tilde{f}(t) = \frac{a_0}{2} + \sum_{k=1}^{\infty} b_k \sin kt - \sum_{k=1}^{\infty} a_k \cos kt.$$

Da $\phi_1,\phi_2 \varepsilon C^{n,H}(\partial K)$, können auch $\phi_1(t)$ und $\phi_2(t)$ in entsprechende Fourierreihen entwickelt werden:

(24) $$\phi_1(t) = \frac{\Phi_0}{2} + \sum_{k=1}^{\infty} \Phi_k \sin kt + \sum_{k=1}^{\infty} \Psi_k \cos kt, \quad \Phi_0, \Phi_k, \Psi_k \varepsilon \mathbb{R},\ k\varepsilon \mathbb{N},$$

(25) $$\phi_2(t) = \frac{\Phi_0}{2} + \sum_{k=1}^{\infty} \Psi_k \sin kt - \sum_{k=1}^{\infty} \Phi_k \cos kt.$$

Wegen $\phi_1^{(k)}, \phi_2^{(k)} \varepsilon C^{n-k,H}(\partial K)$ und wegen (21) können (24) und (25) entsprechend oft gliedweise differenziert werden. Führt man dies aus und setzt in (21) ein, so erhält man durch Koeffizientenvergleich:

(26) $$a_0 = A_0 \Phi_0, \text{ d.h. } \Phi_0 = \frac{a_0}{A_0} ;$$

(27) $$a_k = (\alpha_k - \beta_k)\Phi_k \text{ , d.h. } \Phi_k = \frac{a_k}{\alpha_k - \beta_k} \text{ , } k=1,2,3,\cdots\cdots ;$$

(28) $$b_k = (\alpha_k - \beta_k)\Psi_k \text{ , d.h. } \Psi_k = \frac{b_k}{\alpha_k - \beta_k} \text{ ,}$$

mit

(29) $$\alpha_k := \sum_{j=0}^{\left[\frac{n}{2}\right]} [A_{2j} \cdot k^{2j} (-1)^j] \text{ ; } \beta_k := \sum_{j=0}^{\left[\frac{n+1}{2}\right]-1} [B_{2j+1} \cdot k^{2j+1} (-1)^{j+1}] .$$

Aus Korollar 1 ersieht man für $e=-1$, daß in diesem Falle immer $A_0 \neq 0$ und $(\alpha_k - \beta_k) \neq 0$ für $k=1,2,3,\cdots\cdots$ sein wird. Löst man daher (26), (27) und (28) nach Φ_0, Φ_k, Ψ_k auf, so hat man mit (24) und (25) die gesuchte Lösung des Problems, d.h. es gilt der Satz:

Satz 5

$f(\xi) \varepsilon C^H(\partial K)$ sei eine beliebige vorgegebene Randfunktion des Problems (DR) für Gleichung (17) mit $e=-1$, welche in eine absolut und gleichmäßig konvergente Fourierreihe der Form (22) entwickelbar ist. Dann gilt in diesem Fall:

a) $A_o \neq 0$, $\alpha_k - \beta_k \neq 0$, $k \varepsilon \mathbb{N}$, in (29).

b) Somit existiert immer eine eindeutige Lösung (ϕ_1, ϕ_2) von (21) mit $\phi_1 \overset{k}{\sim} \phi_2$. Stellt man ϕ_1 und ϕ_2 durch die Reihen (24) bzw. (25) dar, so sind die Fourierkoeffizienten eindeutig durch (26) bis (28) bestimmt.

c) Die eindeutig bestimmte Lösung des Problems (DR) ist damit durch

$$u(x,y) = \mathrm{Re}[X_n g|_{\bar{z}=\bar{z}}] \text{ mit } g(z) = \frac{1}{2\pi i} \oint_{\partial K} \frac{\xi^n [\phi_1(\xi) + i\phi_2(\xi)]}{\xi - z} d\xi \text{ gegeben.}$$

Für den Fall $e=+1$ in Gleichung (17) soll abschließend der Beweis des folgenden Satzes kurz skizziert werden.

Satz 6

$f(\xi)$ sei eine vorgegebene Randfunktion des Problems (DR) für Gleichung (17) mit $e=+1$ mit den in Satz 5 angeführten Eigenschaften. Dann gilt:

a_1) Besitzt das homogene Problem (DR_0) nur die triviale Lösung, so besitzt (DR) genau eine Lösung.

a_2) Besitzt dagegen das homogene Problem (DR_0) auch nicht triviale Lösungen, so ist die Dimension dieses Lösungsraumes L_0 endlich. Das inhomogene Problem (DR) hat dann entweder keine Lösung (wenn f im nachfolgenden Sinn nicht orthogonal auf L_0 steht) oder aber unendlich viele Lösungen (wenn f orthogonal auf L_0 steht).

b) Gilt

$$(30) \qquad \sum_{k=0}^{j} A_k^n \left(\frac{R^2}{1+R^2}\right)^{n-k} \cdot \frac{1}{(j-k)!} \neq 0 \text{ für alle } j=1,2,\cdots,n,$$

so liegt der Fall a_1 vor und es gelten wortwörtlich die Aussagen a, b und c aus Satz 5.

c) Liegt jedoch ein "Ausnahmeradius" R vor, d.h. ist die Summe (30)=0 für ein oder mehrere j, so liegt der Fall a_2 vor. In diesem Fall ist immer $A_0=0 \vee (\alpha_k-\beta_k)=0$ für $k\in\hat{\mathbb{N}}_j\subset\mathbb{N}$, wobei $\hat{\mathbb{N}}_j$ endlich viele Elemente enthält. Der Lösungsraum L_0 wird dann durch die Erzeugende g(z) der Form (9) mit $\phi(t)=\phi_1(t)+i\phi_2(t)$ mit

$$(31) \qquad \begin{cases} \phi_1(t) = \dfrac{\Phi_0}{2} + \sum\limits_{k\in\hat{\mathbb{N}}_j} \Phi_k \sin kt + \sum\limits_{k\in\hat{\mathbb{N}}_j} \Psi_k \cos kt , \\ \phi_2(t) = \dfrac{\Phi_0}{2} + \sum\limits_{k\in\hat{\mathbb{N}}_j} \Psi_k \sin kt - \sum\limits_{k\in\hat{\mathbb{N}}_j} \Phi_k \cos kt \end{cases}$$

aufgespannt. Darin sind $\Phi_k,\Psi_k\in\mathbb{R}$, $k\in\hat{\mathbb{N}}_j$ völlig freie Konstanten, $\Phi_0\in\mathbb{R}$ ist völlig frei für $A_0=0$ bzw. $\Phi_0=0$ für $A_0\neq 0$.

Die im Punkt a erwähnte Orthogonalität von f ist nun so zu verstehen, daß man sagt, f steht orthogonal auf L_0, wenn

$$(32) \qquad \int_0^{2\pi} f(t)\phi_1(t)dt = 0.$$

Ist dies erfüllt, dann existieren unendlich viele Lösungen von (DR) der Form $u=\mathrm{Re}[X_n g|_{\tilde{z}=\bar{z}}]$ mit g(z) nach (9), wobei ϕ_1 bzw. ϕ_2 durch (24) bzw. (25) gegeben sind. Die soeben in (31) erwähnten Koeffizienten sind dann völlig frei, die restlichen Koeffizienten sind eindeutig durch (26) bis (28) bestimmt.

Ist dagegen (32) nicht erfüllt, so existiert keine Lösung von (DR).

Beweisskizze

a_1 und a_2 folgen sofort aus (26) bis (29) unter Berücksichtigung der Tatsache, daß für den vorgegebenen Radius R alle A_{2j} und B_{2j+1} in (29) eindeutig festgelegte Zahlen sind; $\alpha_k - \beta_k = 0$ kann damit für höchstens n natürliche Zahlen k erfüllt sein.
b: Dazu vergleiche [1].
c: Diese Aussagen folgen mit Ergebnissen aus [1] sowie dem bisher Gesagten unter Beachtung von (26) bis (29).

Literatur

[1] Bauer, K.W.: Zur Verallgemeinerung des Poissonschen Satzes. Annales Academiae Scientiarum Fennicae. Helsinki (1968).

[2] Gakhov, D.F.: Boundary Value Problems. Pergamon Press (1966).

[3] Heersink, R.: Characterisation of certain Differential Operators in the Solution of Linear Partial Differential Equations. Glasgow Mathematical Journal (im Druck).

[4] Muschelischwili, N.I.: Singuläre Integralgleichungen. Akademie Verlag Berlin (1965).

[5] Vekua, I.N.: New Methods for Solving Elliptic Equations. North-Holland Publishing Company. Amsterdam (1968).

[6] Watzlawek, W.: Über Zusammenhänge zwischen Fundamentalsystemen, Riemann-Funktion und Bergman-Operatoren. Journal f. d. reine u. angew. Mathematik, 251, 200-211 (1971).

PROPERTIES OF A CLASS OF FIRST ORDER ELLIPTIC SYSTEMS

by

Gerald N. Hile

1. Introduction

We consider a special subclass of first order elliptic systems of the form

(1.1) $$\sum_{i=1}^{n} P_i \frac{\partial u}{\partial x_i} = Qu$$

where $u = (u_1,\ldots,u_m)$, and P_i, $i=1,\ldots,n$, and Q are $m \times m$ matrices with entries which are complex valued C^1 functions of $x=(x_1,\ldots,x_n)$. This subclass consists of those systems in which the equations can be differentiated and combined in order to obtain a second order linear system of m equations of the form

(1.2) $$Lu_k = \ldots\ldots\ldots \quad , \; k=1,\ldots,m$$

where the dots represent first and zero order linear terms in $u_1,\ldots,u_m$, and L is a second order elliptic operator with real coefficients,

(1.3) $$L = \sum_{i,j=1}^{n} a_{ij} \frac{\partial^2}{\partial x_i \partial x_j} \quad , \quad a_{ij} = a_{ji} \quad .$$

For example, the system

(1.4) $$p_x - q_y = -2xp + 2yq$$
$$q_x + p_y = -2xp - 2yq$$

is of this type, because upon differentiating and taking linear combinations one obtains the system

(1.5) $$\Delta p = -2(p + q) - 2x(p_x + q_y) + 2y(q_x - q_y)$$
$$\Delta q = -2(p + q) - 2y(q_x + q_y) + 2x(p_y - p_x)$$

In order that one may obtain a system of the form (1.2) in such a way from (1.1), it is necessary that there exist matrices $R_1,\ldots,R_n$ such that

(1.6) $$\left[\sum_{j=1}^{n} R_j \frac{\partial}{\partial x_j} \right] \left[\sum_{i=1}^{n} P_i \frac{\partial}{\partial x_i} \right] = IL + S$$

where I is the $m \times m$ identity matrix, L is given by (1.3), and S is an mxm matrix operator of order less then two. Equating second order terms in (1.6) one obtains the conditions

$$(1.7) \qquad R_jP_i + R_iP_j = 2a_{ij}I \quad , \quad 1\le i,j\le n$$

for some real, symmetric, and positive definite matrix $A = (a_{ij})$.

Note that one solution of (1.4) is $p = q = \exp[-(x^2 + y^2)]$, and that the maximum of p and q for this particular solution occurs at (0,0). Thus an ordinary maximum principle does not hold for solutions of (1.4) nor of (1.5). However in [7] M. H. Protter and the author prove a generalized maximum principle for solutions of (1.2), and they also discuss first order systems from which one may derive second order systems of the type (1.2). These results are described in the next section.

The author in [6] has further considered the case of (1.1) where the P_i's are constant matrices, $Q = 0$, and matrices R_i exist satisfying (1.7). For this case it is seen that one may assume that the P_i's form the basis of a Clifford algebra. A Cauchy integral formula holds along with Taylor series expansions. These results are described in the last section.

For relevant reading regarding maximum principles for systems see [8,9,10,11]. For related results involving function theory in Clifford algebras see [1,2,4,5].

2. A Maximum Principle for Systems

We assume that the operator L in (1.3) is uniformly elliptic in a domain D, i.e., there is a constant $c_0>0$ such that for all x in D and $y = (y_1,\dots,y_n)$ in R^n, the inequality

$$(2.1) \qquad A(x)y\cdot y = \sum_{i,j=1}^{n} a_{ij}(x)y_iy_j \ge c_0 \sum_{i=1}^{n} y_i^2$$

holds. The operator L is the principal part of each equation in the second order linear system

$$(2.2) \quad Lu_k = \sum_{i=1}^{n} \sum_{j=1}^{m} b_{kij} \frac{\partial u_j}{\partial x_i} + \sum_{j=1}^{m} c_{kj} u_j \quad , \quad 1 \leq k \leq m \quad .$$

We suppose that the coefficients b_{kij}, c_{kj}, are bounded complex-valued functions in D. A solution $u = (u_1, \ldots, u_m)$ is a complex-valued C^2 function which satisfies (2.2) in D.

Theorem 2.1 There exists a constant $K \geq 0$ whose size depends only on the coefficients in (2.2) such that if u is a solution of (2.2) and α is a positive C^2 function in D, then the product $\alpha|u|^2 \equiv \alpha \sum_{i=1}^{m} |u_i|^2$ cannot attain a positive maximum at any point where α satisfies

$$(2.3) \quad \alpha^{-1} L\alpha - 2\alpha^{-2}(A\nabla\alpha)\cdot\nabla\alpha > K \quad .$$

Outline of Proof : We set $p = |u|^2 = |u_1|^2 + \ldots + |u_m|^2$, and find

$$(2.4) \quad L(\alpha p) = (L\alpha)p + \alpha(Lp) + 2(A\nabla\alpha)\cdot\nabla p \quad .$$

At a point where αp attains a maximum, we have

$$0 = \nabla(\alpha p) = (\nabla\alpha)p + \alpha(\nabla p)$$

and (2.4) becomes

$$(2.5) \quad L(\alpha p) = p(L\alpha - 2\alpha^{-1}(A\nabla\alpha)\cdot\nabla\alpha) + \alpha(Lp) \quad .$$

A direct computation yields

$$(2.6) \quad Lp = \sum_{k=1}^{m} [2 \operatorname{Re}(\overline{u}_k L u_k) + 2A\nabla u_k \cdot \nabla u_k] \quad .$$

After making several estimates involving the Cauchy-Schwarz inequality and the ellipticity condition (2.1), we obtain from (2.6)

$$(2.7) \quad Lp \geq - Kp$$

where K is a positive constant depending only on the bounds of the coefficients in (2.2) and the constant c_0 in (2.1). Substitution into (2.5) yields

(2.8) $\quad L(\alpha p) \geq \alpha p\,[\alpha^{-1}L\alpha - 2\alpha^{-2}A\nabla\alpha\cdot\nabla\alpha - K]$

which holds at any point in D where αp attains a maximum. Thus αp cannot attain a positive maximum at any point where the quantity in brackets in (2.8) is positive.

Corollary 2.2 (i) There exist positive constants ρ,β depending only on the coefficients in (2.2), such that for any x^0 in R^n and solution u in D, the function $|u(x)|^2 \exp(\beta|x-x^0|^2)$ does not attain a positive maximum in the intersection of D with the sphere $\{x: |x-x^0| \leq \rho\}$. (ii) There exist positive constants σ,τ, depending only on the coefficients in (2.2), such that for any $x^0 = (x_1{}^0,\ldots,x_n{}^0)$ in R^n and solution u in D, the function $|u(x)|^2 \exp(\tau|x_1-x_1{}^0|^2)$ does not attain a positive maximum in the intersection of D with the slab $\{x: |x_1-x_1{}^0| < \sigma\}$.

Outline of Proof : To prove (i), set $\alpha(x) = \exp(\beta|x-x^0|^2)$ in Theorem 2.1. To obtain (2.3), choose $\beta>0$ and sufficiently large, and choose $\rho>0$ and sufficiently small. For (ii), set $\alpha(x) = \exp[\tau(x_1 - x_1{}^0)^2]$. Choose $\tau>0$ and sufficiently large, and choose $\sigma>0$ and sufficiently small.

In [7] it is demonstrated that if matrices R_i exist such that (1.7) holds for some positive definite matrix $A = (a_{ij})$, then (1.1) is necessarily elliptic. In [3] (pg. 529, Theorem 4) it is shown that if the coefficients of (1.1) are C^1 and Hölder-continuous, ellipticity implies that the solutions are C^2. Thus differentiation to obtain the second order system (1.2) is permissible under these conditions.

Ellipticity also implies that each P_i in (1.1) is invertible. Thus we may assume $P_1 = I$. For the case of two independent variables x and y,

(1.1) then takes the form

$$u_x + Pu_y = \ldots\ldots \tag{2.9}$$

where the dots represent zero order linear terms. Conditions (1.7) become

$$\begin{aligned} R_1 &= a_{11}I \\ R_1P + R_2 &= 2a_{12}I \\ R_2P &= a_{22}I \end{aligned} \tag{2.10}$$

which leads to the polynomial equation

$$a_{11}P^2 - 2a_{12}P + a_{22}I = 0 \tag{2.11}$$

Since A is positive definite, we have $a_{12}^2 - a_{11}a_{22} < 0$, and the equation

$$a_{11}\lambda^2 - 2a_{12}\lambda + a_{22} = 0 \tag{2.12}$$

has no real roots. Conversely, suppose P satisfies an equation of the type (2.11), where (2.12) has no real roots. We set $R_1 = a_{11}I$, $R_2 = a_{22}P^{-1}$, and find that (2.10) holds and the 2x2 matrix $A = (a_{ij})$, $a_{21} = a_{12}$, is positive definite (provided we normalize (2.11) by making $a_{11} + a_{22} > 0$). Thus in the case of the system (2.9) in two variables, with coefficients of class C^1 and Hölder-continuous, in order that one may differentiate and obtain a second order system for which our maximum principle holds it is necessary and sufficient that P satisfy a second degree polynomial equation (2.11) which has no real roots. For the special case where P is 2x2 and (2.9) is elliptic, P always satisfies an equation of type, namely its characteristic equation $\det(P - \lambda I) = 0$.

Note that if P satisfies (2.11), then the matrix

$$\tilde{P} = a_{11}\ (a_{11}a_{22} - a_{12}^2)^{-1/2}(P - (a_{12}/a_{11})I)$$

satisfies the equation $(\tilde{P})^2 = -I$. This result has an analogue in more variables, although the analysis is lengthy and will not be presented here. The existence of matrices R_i such that (1.7) holds for some real positive definite matrix $A = (a_{ij})$ implies the existence of a set of mxm matrices $Q_2,\ldots,Q_n$ satisfying

$$Q_iQ_j + Q_jQ_i = -2\,\delta_{ij}\,I \qquad , \quad 2 \le i,j \le n \quad ,$$

and such that each P_i can be expressed as a linear combination of the Q_i's and I involving the a_{ij}'s.

3. Equations with Constant Coefficients

We consider now the homogeneous equation

$$(3.1) \qquad P_1\frac{\partial F}{\partial x_1} + P_2\frac{\partial F}{\partial x_2} + \ldots + P_n\frac{\partial F}{\partial x_n} = 0$$

where each matrix P_i is an sxm matrix of complex constants, and F is a complex matrix valued function with m rows. We assume that there exist mxs complex constant matrices R_i , $i = 1,\ldots,n$, and a real, constant, positive definite nxn matrix $A = (a_{ij})$ such that

$$(3.2) \qquad R_iP_j + R_jP_i = 2a_{ij}I_m \qquad , \quad 1 \le i,j \le n \quad ,$$

where I_m is the mxm identity. As seen in the previous section, these conditions imply that one may differentiate (3.1) to obtain the equation

$$(3.3) \qquad LF = 0 \qquad ,$$

where L is the second order elliptic operator defined by (1.3). We first show that by a change of variables conditions (3.2) may be simplified.

Theorem 3.1 Let R_i, P_i , $i = 1,\ldots,n$, be matrices of dimensions mxs and sxm, respectively, which satisfy (3.2) for some nxn, real, positive definite matrix A. Then there exist matrices $\tilde{R}_i$, $\tilde{P}_i$, $i = 1,\ldots,n$, of dimensions mxs and sxm, respectively, such that

$$(3.4)\qquad \tilde{R}_i\tilde{P}_j + \tilde{R}_j\tilde{P}_i = 2\delta_{ij}\, I_m \quad , 1 \le i,j \le n ,$$

and such that, after a linear change of variables $y = Bx$, any C^1 matrix solution F of (3.1) satisfies

$$(3.5)\qquad \tilde{P}_1 \frac{\partial F}{\partial y_1} + \tilde{P}_2 \frac{\partial F}{\partial y_2} + \ldots + \tilde{P}_n \frac{\partial F}{\partial y_n} = 0$$

Outline of Proof : Since A is positive definite, we may define the inner product on vectors in R^n,

$$\langle x , y \rangle = Ax \cdot y \quad .$$

Let $\xi^1,\ldots,\xi^n$ be vectors in R^n which form an orthonormal basis with respect to this inner product,

$$\langle \xi^i , \xi^j \rangle = \delta_{ij} \quad , \quad 1 \le i,j \le n \quad .$$

Next define

$$\tilde{P}_i = \sum_{j=1}^{n} (\xi^i)_j P_j \quad , \quad \tilde{R}_i = \sum_{j=1}^{n} (\xi^i)_j R_j \quad , \quad 1 \le i \le n \quad ,$$

where $(\xi^i)_j$ is the jth component of ξ^i. Then (3.4) is satisfied. With the change of variables $y = Bx$, where B is the nonsingular nxn matrix $((\xi^i)_j)$, equation (3.1) implies (3.5).

For the case of square matrices, a further simplification of conditions (3.2) can be made.

Theorem 3.2 Let R_i, P_i, $i = 1,\ldots,n$, be mxm matrices which satisfy (3.2) for some nxn, real, positive definite matrix A. Then there exist mxm matrices $M_2,\ldots,M_n$ such that

$$(3.6)\qquad \begin{aligned} M_iM_j &= - M_jM_i \quad , \quad i \ne j \ , \quad 2 \le i,j \le n \\ M_i^2 &= -I_m \quad , \quad 2 \le i \le n \end{aligned}$$

and such that, after a linear change of variables $y = Bx$, any C^1 matrix solution F of (3.1) satisfies

$$(3.7)\qquad \frac{\partial F}{\partial y_1} + M_2 \frac{\partial F}{\partial y_2} + \ldots + M_n \frac{\partial F}{\partial y_n} = 0$$

Outline of Proof : For this special case, the matrices R_i and P_i in the previous theorem are mxm. Since $R_iP_i = I_m$, we have $R_i = (P_i)^{-1}$ for i = 1,...,n. Multiplying (3.5) by R_1, we obtain

$$\frac{\partial F}{\partial y_1} + \tilde{R}_1\tilde{P}_2 \frac{\partial F}{\partial y_2} + \ldots + \tilde{R}_1\tilde{P}_n \frac{\partial F}{\partial y_n} = 0$$

Define $M_i = \tilde{R}_1\tilde{P}_i$, i = 2,...,n . Repeated application of (3.4) gives (3.6).

We assume hereafter that $a_{ij} = \delta_{ij}$ in (3.2). Let D be the operator

$$DF \equiv P_1 \frac{\partial F}{\partial x_1} + P_2 \frac{\partial F}{\partial x_2} + \ldots + P_n \frac{\partial F}{\partial x_n} \tag{3.8}$$

We allow D to operate to the left according to the formula

$$GD \equiv \frac{\partial G}{\partial x_1} P_1 + \frac{\partial G}{\partial x_2} P_2 + \ldots + \frac{\partial G}{\partial x_n} P_n \tag{3.9}$$

Let $\nu = (\nu_1,\ldots,\nu_n)$ denote the outward pointing unit normal on the boundary of a domain Ω , and let N be the sxm matrix

$$N = P_1\nu_1 + P_2\nu_2 + \ldots + P_n\nu_n \quad . \tag{3.10}$$

It is easy to verify a generalized form of the divergence theorem,

$$\int_\Omega [G(DF) + (GD)F]\, dy = \int_{\partial\Omega} GNF \, d\sigma \quad . \tag{3.11}$$

In the case that the P_i's are not constants, we may define

$$DG \equiv \frac{\partial}{\partial x_1}(GP_1) + \frac{\partial}{\partial x_2}(GP_2) + \ldots + \frac{\partial}{\partial x_n}(GP_n)$$

and (3.11) remains true. With the identifications

$$x^* \equiv P_1x_1 + P_2x_2 + \ldots + P_nx_n \tag{3.12}$$

$$\overline{x} \equiv R_1x_1 + R_2x_2 + \ldots + R_nx_n$$

one has

$$\overline{x}\, x^* = (x_1{}^2 + x_2{}^2 + \ldots + x_n{}^2)\, I_m = |x|^2\, I_m \tag{3.13}$$

<u>Theorem 3.3</u> Let Ω be a domain in R^n where the divergence theorem

applies. If F is in $C^1(\overline{\Omega})$ and satisfies (3.1) in Ω, then for x in Ω,

(3.14) $$F(x) = \omega_n^{-1} \int_{\partial\Omega} |y-x|^{-n} \overline{(y-x)}\nu^*(y) F(y)\, d\sigma(y)$$

(ω_n = surface area of unit sphere in R^n).

Outline of Proof : We apply (3.11) where G is the function

(3.15) $$G(y) = |y-x|^{-n} \overline{(y-x)} \quad ,$$

and a small sphere of radius ε about x is deleted from Ω. Then $(G\mathcal{D}_y) = 0$ for $y \neq x$, and (3.14) follows by a standard limiting argument.

The Taylor series expansion obtainable for solutions of (3.1) is in fact valid for solutions of more general systems of equations. We now consider complex matrix solutions of equations of the form

(3.16) $$I\frac{\partial F}{\partial x_1} + C_2\frac{\partial F}{\partial x_2} + \dots + C_n\frac{\partial F}{\partial x_n} = 0$$

where $C_2,\dots,C_n$ are mxm complex constant matrices, and I is the mxm identity. Associated with the matrices $C_2,\dots,C_n$ are the (n-1) matrix variables defined by

(3.17) $$Z_k = x_k I - x_1 C_k \quad , \quad k = 2,\dots,n$$

where $x = (x_1,\dots,x_n)$ is a point in R^n. We let $a = (a_1,\dots,a_n)$ be some fixed point in R^n, and further define

(3.18) $$A_k = a_k I - a_1 C_k \quad , \quad k = 2,\dots,n \quad .$$

Following standard conventions, if $\alpha = (\alpha_1,\dots,\alpha_n)$, where each α_i is a nonnegative integer, we denote $|\alpha| = \alpha_1 + \dots + \alpha_n$, $\alpha! = \alpha_1!\dots\alpha_n!$, $x^\alpha = x_1^{\alpha_1} \dots x_n^{\alpha_n}$, and if u is a function defined in R^n, $u^{(\alpha)} = D^\alpha u = \dfrac{\partial^{|\alpha|}u}{\partial_1^{\alpha_1}\dots\partial_n^{\alpha_n}}$. We let β represent such an n-tuple with $\beta_1 = 0$. Thus $\beta = (0, \beta_2,\dots,\beta_n)$. Associated with the variable x is the (n-1)-tuple of matrices $Z = (Z_2,\dots,Z_n)$ where each

Z_k is given by (3.17). Similarly, set $A = (A_2,\ldots,A_n)$, where A_k is given by (3.18). Define Z^β as the sum of all distinct formal products containing Z_k as a factor β_k times, for $k = 2,\ldots,n$, normalized by dividing by the total number of terms in this sum. For example, $Z^{(0,1,1)} = (Z_2Z_3 + Z_3Z_2)/2$, $Z^{(0,1,0,2)} = (Z_2Z_4{}^2 + Z_4Z_2Z_4 + Z_4{}^2Z_2)/3$. One deduces by elementary counting principles that the normalizing factor is $(|\beta|!)/(\beta!)$. (If $\beta = (0,\ldots,0)$, define $Z^\beta = 1$.) This definition of Z is not necessarily functional, because in general $Z^{\beta+\gamma} \neq Z^\beta Z^\gamma$. However the notation leads to a convenient representation of our Taylor series. (In the special case where the Z_k's all commute with one another, then $Z^\beta = Z_2{}^{\beta_2} \ldots Z_n{}^{\beta_n}$, and the formula $Z^{\beta+\gamma} = Z^\beta Z^\gamma$ is indeed true.) In an obvious and similar manner, we define A^β and $(Z - A)^\beta = (Z_2 - A_2,\ldots,Z_n - A_n)^\beta$.

Following methods in [1] and [4] for functions with values in a Clifford algebra, we obtain:

<u>Theorem 3.4</u> The matrix polynomials $(Z - A)^\beta$, with $\beta_1 = 0$, satisfy equation (3.16), along with the formula

$$D^\beta\,[(Z - A)^\beta\,] = (\beta!)\,I \tag{3.19}$$

Theorem 3.4 can be proved by a direct but somewhat lengthy verification.

We say that a matrix valued function F is <u>real-analytic</u> at a point in R^n if F has a local Taylor series expansion in the real variables $x_1,\ldots,x_n$, valid in some neighborhood of this point.

<u>Theorem 3.5</u> If F has m rows, is real-analytic at a, and satisfies (3.16) in some neighborhood of a, then the power series expansion

$$(3.20) \qquad F(x) = \sum_{\beta_1=0} \frac{(Z - A)^\beta}{\beta\,!} F^{(\beta)}(a)$$

converges uniformly to F in some neighborhood of a. (The sum is taken over all n-tuples β with $\beta_1 = 0$.) Conversely, suppose that the series

$$(3.21) \qquad G(x) = \sum_{\beta_1=0} \frac{(Z - A)^\beta K_\beta}{\beta!}$$

converges uniformly in some neighborhood of the point a in R^n, where each K_β is an $m \times m_1$ complex constant matrix. Then the function G defined by the sum of this series satisfies (3.16) in this neighborhood. Moreover, $K_\beta = G^{(\beta)}(a)$ for each β.

The first half of Theorem 3.5 is proved by first expanding F in a series involving the variables $x_1,\ldots,x_n$, and then rearranging this series to obtain (3.20). The second half follows from Theorem 3.4 and the fact that the series (3.21) may be differentiated termwise. For details see [6].

If equation (3.16) is elliptic, then it is well-known that solutions are real-analytic, and therefore the expansion (3.20) must hold. Also, any polynomial solution of (3.16) has the expansion (3.20), even if the equation is not elliptic.

Returning to our original system (3.1), we note that (3.2) implies that $R_1P_1 = I$ (we still assume $a_{ij} = \delta_{ij}$.) Multiplying (3.1) by R_1, we have the equation

$$(3.22) \qquad I\frac{\partial F}{\partial x_1} + R_1P_2\frac{\partial F}{\partial x_2} + \ldots + R_1P_n\frac{\partial F}{\partial x_n} = 0 \qquad .$$

Since each R_1P_i is square, and the Cauchy integral formula (3.14) implies that solutions F are real-analytic, we have a Taylor series expansion for solutions of (3.1).

References

1. R. Delanghe, On regular-analytic functions with values in a Clifford algebra, Math. Ann., vol. 185 (1970), 91-111.

2. R. Delanghe, On the singularities of functions with values in a Clifford algebra, Math. Ann., vol. 196 (1972), 293-319.

3. A. Douglis and L. Nirenberg, Interior estimates for elliptic systems of partial differential equations, Comm. Pure App. Math., vol. VIII (1955), 503-537.

4. R. Fueter, Über die analytischen darstellungen der regulären funktionen einer quaternionenvariablen, Comment. Math. Helv., vol. 8 (1935-36), 371.

5. H. G. Haefeli, Hyperkomplexe differentiale, Comment. Math. Helv., vol. 20 (1947), 382-420.

6. G. N. Hile, Representations of solutions of a special class of first order systems, to appear.

7. G. N. Hile and M. H. Protter, Maximum principles for a class of first order elliptic systems, Jour. Diff. Eqns., to appear.

8. C. Miranda, Sul teorema del massimo modulo per una classe di sistemi ellitici di equazioni de secondo ordine e per le equazioni a coefficienti complessi, Istituto Lombardo, Ser. A 104 (1970), 736-745.

9. M. H. Protter and H. Weinberger, A maximum principle and gradient bounds for linear elliptic equations, Indiana Univ. Math. Journ. 23 (1973), 239-249.

10. P. Szepticki, Existence theorem for the first boundary value problem for a quasilinear elliptic system, Bull. Acad. Polon. des Sciences, 7 (1959), 419-424.

11. J. Wasowski, Maximum principles for a certain strongly elliptic system of linear equations of second order, Bull. Acad. Polon. des Sciences, 18 (1970), 741-745.

A Neumann Series Representation For Solutions To The Exterior Boundary-value Problems of Elasticity*

by

George C. H s i a o

1. Introduction

The method of the regularized integral equation for treating scattering problems in acoustics has been received much attention lately [1],[6],[7]. The essence of the underlying method is to reformulate the problem as an integral representation analogous to the Green's formula. The integral equation that results is regularized in the sense that the unknown function appears in such a way as to vanish at the weak singularity of the kernel. This regularization then enables one to obtain a Neumann series solution of the problem under consideration. In the present paper, we shall surmarize some of the recent developments concerning this approach to the second fundamental boundary-value problems of elasticity. The materials presented here are based on the results in [2] and in a forthcoming paper [3] by AHNER and HSIAO.

Throughout the paper, we denote by S a closed Lyapunoff surface in R^n, $n = 2$ or 3, and let Ω_i and Ω_e denote the regions interior and exterior to S respectively. The region Ω_e is assumed to be filled with a homogeneous isotropic elastic material with Lamè constants λ and μ.

We consider the second fundamental boundary value problem consisting of the equation

(1) $$(\Delta^* + \omega^2)\underset{\sim}{u}(\underset{\sim}{x}) = \underset{\sim}{0}\,, \quad \underset{\sim}{x} \in \Omega_e$$

together with the boundary condition

(2) $$(T\underset{\sim}{u})(\underset{\sim}{x}) = \underset{\sim}{f}(\underset{\sim}{x})\,, \quad \underset{\sim}{x} \in S$$

and the elastic radiation condition:

(3) $$\frac{\partial}{\partial|\underset{\sim}{x}|}\underset{\sim}{u}_\alpha(\underset{\sim}{x}) - ik_\alpha\,\underset{\sim}{u}_\alpha(\underset{\sim}{x}) = o\,(|\underset{\sim}{x}|^{-(n-1)/2}) \quad \text{as} \quad |\underset{\sim}{x}| \to \infty, \quad i = 1,2\,.$$

---

*This research was supported by the Alexander von Humdoldt Foundation, and in part by the Air Force Office of Scientific Research through AF-AFOSR Grant No. 76-2879

Here $\underset{\sim}{u}(\underset{\sim}{x})$ is the unknown displacement vector field; ω designates the frequency of vibration, and $\underset{\sim}{f}$ is a given stress vector, which is assumed to be sufficiently smooth. (For our purpose, that $\underset{\sim}{f}$ is Hölder continuous will be sufficient). The operators Δ^* and T are respectively defined by

$$\Delta^* \equiv \mu\, \Delta + (\lambda+\mu) \text{ grad div}$$
(4)
$$T \equiv 2\mu \frac{\partial}{\partial \hat{n}} + \lambda\, \hat{n} \text{ div} + \mu(\hat{n} \times \text{curl}) ,$$

where Δ is the Laplacian operator and $\hat{n}$ represents an outward unit normal to S. The constants k_α in (3) are defined by $k_1^2 = \omega^2/(\lambda+2\mu)$, $k_2^2 = \omega^2/\mu$, while the vector fields, $\underset{\sim}{u}_\alpha$, $\alpha = 1,2$, are the corresponding potential and solenoidal parts of $\underset{\sim}{u}$ such that

$$\underset{\sim}{u}_1(\underset{\sim}{x}) = - \frac{1}{k_1^2} \text{ grad div } \underset{\sim}{u}(\underset{\sim}{x}) ,$$

$$\underset{\sim}{u}_2(\underset{\sim}{x}) = \frac{1}{k_1^2} \text{ grad div } \underset{\sim}{u}(\underset{\sim}{x}) + \underset{\sim}{u}(\underset{\sim}{x}) .$$

For simplicity we shall refer to the problem (1) - (3) as the dynamic problem (P); here the harmonic time factor $\exp(-i\omega t)$ has been omitted. We shall also consider the limiting case where $\omega = o$. We then refer to the problem as the static problem (P_o) and denote by $\underset{\sim}{u}_o(\underset{\sim}{x})$ the solution* of (P_o); in this case it is understood that the condition (3) at the infinity should be replaced by the regularity condition:

$(3)_o$
$$\underset{\sim}{u}_o(\underset{\sim}{x}) = O(|\underset{\sim}{x}|^{2-n}) \text{ and } D\underset{\sim}{u}_o(\underset{\sim}{x}) = O(|\underset{\sim}{x}|^{-2}) \text{ as } |x| \to \infty,$$

where D denotes any first order derivative.

2. Regularized Integral Representations

In order to derive the regularized integral representation for the solution $\underset{\sim}{u}$ of (P), we begin with the identity:

(5.a)
$$\int_S \{\Gamma(\underset{\sim}{x},\underset{\sim}{y})\underset{\sim}{f}(\underset{\sim}{y}) - \Gamma_1(\underset{\sim}{x},\underset{\sim}{y})\underset{\sim}{u}(\underset{\sim}{y})\}\, dS_{\underset{\sim}{y}} = \begin{cases} \underset{\sim}{u}(\underset{\sim}{x}) , & \underset{\sim}{x} \in \Omega_e \\ \frac{1}{2}\underset{\sim}{u}(\underset{\sim}{x}), & \underset{\sim}{x} \in S \\ \underset{\sim}{o} , & \underset{\sim}{x} \in \Omega_i , \end{cases}$$

which follows by applying the Betti formula to $\underset{\sim}{u}(\underset{\sim}{x})$ (see e. g. [8]). Here $\Gamma(\underset{\sim}{x},\underset{\sim}{y})$ is the matrix of the fundamental solutions of (1), and $\Gamma_1(\underset{\sim}{x},\underset{\sim}{y})$ is the transpose of

---

*Throughout the paper by a solution to (P) (or (P_o)) we mean a vector field satisfies (1) (2) and (3) (or the corresponding equations for (P_o)) in the classical sense.

$T^{(y)}\Gamma(x,y)$; the entries of the matrices are given by

$$\Gamma_{kj}(x,y) = \frac{1}{\mu}\delta_{kj} E_2(x,y) - \frac{1}{\omega^2}\frac{\partial^2}{\partial y_k \partial y_j}\{E_1(x,y) - E_2(x,y)\}$$

(5.b)
$$\Gamma_{1_{kj}}(x,y) = 2\mu \frac{\partial}{\partial \hat{n}}\Gamma_{kj}(z,y) + \frac{\lambda}{\lambda+2\mu}(\hat{n}\cdot\hat{e}_j)\frac{\partial}{\partial y_k}E_1(x,y)$$
$$+ (\hat{n}\cdot\hat{e}_k)\frac{\partial}{\partial y_j}E_2(x,y) - \delta_{jk}\frac{\partial}{\partial \hat{n}}E_2(x,y),$$

with

$$E_\alpha(x,y) = \begin{cases} \exp(ik_\alpha|x-y|)/4\pi|x-y|, & n=3 \\ -iH_o^{(1)}(k_\alpha|x-y|)/4, & n=2, \end{cases} \qquad \alpha = 1, 2,$$

where $H_o^{(1)}$ is the Hankel function. In (5.b), the vector $\hat{e}_j$ represents the unit vector along the y_j-axis, and δ_{kj} is the Kornecker delta. In the limiting case where $\omega = o$, we have a similar identity for $u_o(x)$:

(5.a)$_o$
$$\int_S\{\overset{o}{\Gamma}(x,y)f(y) - \overset{o}{\Gamma}_1(x,y)u_o(y)\}dS_y = \begin{cases} u_o(x), & x \in \Omega_e, \\ \frac{1}{2}u_o(x), & x \in S, \\ o, & x \in \Omega_i, \end{cases}$$

where the entries of the matrices $\overset{o}{\Gamma}$ and $\overset{o}{\Gamma}_1$ now read

$$\overset{o}{\Gamma}_{kj}(x,y) = \frac{1}{\mu}\delta_{kj} S_n(x,y) - \frac{\lambda+\mu}{4\mu(\lambda+2\mu)}\frac{\partial^2}{\partial y_k \partial y_j}\gamma_n(x,y)$$

(5.b)$_o$
$$\overset{o}{\Gamma}_{1_{kj}}(x,y) = 2\mu\frac{\partial}{\partial \hat{n}}\overset{o}{\Gamma}_{kj}(x,y) + \frac{\lambda+\mu}{\lambda+2\mu}(\hat{n}\cdot\hat{e}_j)\frac{\partial}{\partial y_k}S_n(x,y)$$
$$+ (\hat{n}\cdot\hat{e}_k)\frac{\partial}{\partial y_j}S_n(x,y) - \delta_{jk}\frac{\partial}{\partial \hat{n}}S_n(x,y)$$

with

$$S_n(x,y) = \begin{cases} \frac{1}{2\pi}\log|x-y|, & n=2 \\ \frac{1}{4\pi|x-y|}, & n=3 \end{cases}$$ and

$$\gamma_n(x,y) = \begin{cases} |x-y|^2(\log|x-y| - 1)/2\pi, & n=2, \\ |x-y|/2\pi, & n=3. \end{cases}$$

Since $\Delta^*\hat{e}_k = o$ in Ω_i, an application of the Betti formula then yields

(6)
$$\int_S \overset{\circ}{\Gamma}_1(\underset{\sim}{x},\underset{\sim}{y})\hat{e}_k \, dS_{\underset{\sim}{y}} = \begin{cases} \underset{\sim}{0} , & \underset{\sim}{x} \in \Omega_e \\ \frac{1}{2}\hat{e}_k , & \underset{\sim}{x} \in S , \\ \hat{e}_k , & \underset{\sim}{x} \in \Omega_i . \end{cases}$$

Now from (6), (5.a) and $(5.a)_o$, we arrive at the regularized integral representation for the solutions of (P) and (P_o) respectively,

(7)
$$\underset{\sim}{u}(\underset{\sim}{x}) = \underset{\sim}{F}(\underset{\sim}{x}) - \int_S \overset{\circ}{\Gamma}_1(\underset{\sim}{x},\underset{\sim}{y}) \, [\underset{\sim}{u}(\underset{\sim}{y}) - \underset{\sim}{u}(\underset{\sim}{x})] \, dS_{\underset{\sim}{y}}$$
$$- \int_S [\Gamma_1(\underset{\sim}{x},\underset{\sim}{y}) - \overset{\circ}{\Gamma}_1(\underset{\sim}{x},\underset{\sim}{y})] \, \underset{\sim}{u}(\underset{\sim}{y}) dS_{\underset{\sim}{y}} ,$$

and

$(7)_o$
$$\underset{\sim}{u}_o(\underset{\sim}{x}) = \underset{\sim}{F}_o(\underset{\sim}{x}) - \int_S \overset{\circ}{\Gamma}_1(\underset{\sim}{x},\underset{\sim}{y}) \, [\underset{\sim}{u}_o(\underset{\sim}{y}) - \underset{\sim}{u}_o(\underset{\sim}{x})] \, dS_{\underset{\sim}{y}}$$

for $\underset{\sim}{x} \in S \cup \Omega_e$. Here $\underset{\sim}{F}(\underset{\sim}{x})$ and $\underset{\sim}{F}_o(\underset{\sim}{x})$ are defined by

(8)
$$\underset{\sim}{F}(\underset{\sim}{x}) = \int_S \Gamma(\underset{\sim}{x},\underset{\sim}{y})\underset{\sim}{f}(\underset{\sim}{y}) dS_{\underset{\sim}{y}} \quad \text{and} \quad \underset{\sim}{F}_o(\underset{\sim}{x}) = \int_S \overset{\circ}{\Gamma}(\underset{\sim}{x},\underset{\sim}{y})\underset{\sim}{f}(\underset{\sim}{y}) dS_{\underset{\sim}{y}} ,$$

which are known for the given stress $\underset{\sim}{f}(\underset{\sim}{x})$. We observe that the representations $(7)_o$ and (7) formally resemble those for the solutions of the Laplace equation and the Helmholtz equation in electromagnetic theory. However, it should be noted that here we have the Cauchy type kernel $\overset{\circ}{\Gamma}_1$ and hence it should be understood that the integral involved is only meaningful in the sense of principal value.

3. Neumann Series Solution

From the integral representations (5.a) and $(5.a)_0$, it is clear that solutions to the problems (P) and (P_o) will be determined everywhere in the exterior region Ω_e, if one knows the solutions on the boundary S. Hence it sufficies to construct the solutions of (7) and $(7)_o$ only for points on the boundary. To facilate the presentation, let us first denote by L_o, $M(\omega)$ and L the operators such that

$$(L_o\underset{\sim}{v})(\underset{\sim}{x}) \equiv - \int_S \overset{\circ}{\Gamma}_1(\underset{\sim}{x},\underset{\sim}{y}) \, [\underset{\sim}{v}(\underset{\sim}{y}) - \underset{\sim}{v}(\underset{\sim}{x})] \, dS_{\underset{\sim}{y}}$$
$$(M(\omega)\underset{\sim}{v})(\underset{\sim}{x}) \equiv - \int_S [\Gamma_1(\underset{\sim}{x},\underset{\sim}{y}) - \overset{\circ}{\Gamma}_1(\underset{\sim}{x},\underset{\sim}{y})] \, dS_{\underset{\sim}{y}}$$
$$(L\underset{\sim}{v})(\underset{\sim}{x}) \equiv (L_o\underset{\sim}{v})(\underset{\sim}{x}) + (M(\omega)\underset{\sim}{v})(\underset{\sim}{x})$$

for vector field $\underset{\sim}{v}$ in some suitable function space $B(S)$ (to be specified). Then for $\underset{\sim}{u}, \underset{\sim}{u}_o \in B(S)$, (7) and $(7)_o$ may be rewritten in the form:

(9)
$$(I-L)\underset{\sim}{u}(\underset{\sim}{x}) = \underset{\sim}{F}(\underset{\sim}{x})$$

and

$$(9)_o \qquad (I-L_o)\underset{\sim}{u}_o(\underset{\sim}{x}) = \underset{\sim}{F}_o(\underset{\sim}{x}) ,$$

where I is the identity operator. Our aim here is to establish the fact that both (9) (for small ω) and $(9)_o$ admit the Neumann series solutions in B(S); that is, we shall seek solutions in the form:

$$(10) \qquad \underset{\sim}{u}(\underset{\sim}{x}) = \sum_{n=o}^{\infty} L^n \underset{\sim}{F}(\underset{\sim}{x})$$

and

$$(10)_o \qquad \underset{\sim}{u}_o(\underset{\sim}{x}) = \sum_{n=o}^{\infty} L_o^n \underset{\sim}{F}_o(\underset{\sim}{x}) ,$$

where L_o^n and L^n are the iterated operators.

Let us first specify the function space B(S). It is clear that from the Cauchy type kernel in the integral representations, a naturnal choice of B(S) will be the vector space of vector functions $\underset{\sim}{\phi} \in R^n$, $n = 2,3$, whose components ϕ_i are complex functions defined on S and satisfy the Hölder condition with exponent α, $o < \alpha < 1$.

As usual we equip B(S) with the Hölder norm

$$(11) \qquad ||\phi||_\alpha = \max_i \left\{ \sup_{\underset{\sim}{x} \in S} |\phi_i(\underset{\sim}{x})| + \sup_{\substack{\underset{\sim}{x},\underset{\sim}{y} \in S \\ \underset{\sim}{x} \neq \underset{\sim}{y}}} \frac{|\phi_i(\underset{\sim}{x}) - \phi_i(\underset{\sim}{y})|}{|\underset{\sim}{x} - \underset{\sim}{y}|^\alpha} \right\} .$$

Then clearly B(S) is a Banach space. By following the argument in potential theory, it is easy to sse that both L and L_o map B(S) into itself. Moreover, as we will see, the series (10) and $(10)_o$ indeed converge in B(S). To this end, it is worthwhile to mention that although the inverse $(I-L_o)^{-1}$ is known to exist for the Lyapunoff surface [8] , it is not sufficient to guarantee the convergence of the series in $(10)_o$, since it is not obvious that the norm of L_o, $|||L_o|||$, is indeed less than one. On the other hand, it can be shown that $(I-L_o)^{-1}M(\omega)$ is a bounded linear operator from B(S) into itself; furthermore for ω sufficiently small, one can show that $|||(I-L_o)^{-1}M(\omega)||| < 1$, and hence, the series

$$(12) \qquad \underset{\sim}{u}(\underset{\sim}{x}) = \sum_{n=o}^{\infty} [(I-L_o)^{-1}M(\omega)]^n (I-L_o)^{-1} \underset{\sim}{F}(\underset{\sim}{x})$$

converges in B(S). However, the series representation in (10) has an advantage over that of (12) from the view point of computation, since there is no need to have the explicit form of the inverse $(I-L_o)^{-1}$.

For the convergence of the series in $(10)_o$, we rewrite $(9)_o$ in a slightly more general form by introducing the integral operator K_o ,

$$(K_o \underset{\sim}{u}_o)(\underset{\sim}{x}) \equiv - \int_S 2 \overset{o}{\Gamma}_1(\underset{\sim}{x},\underset{\sim}{y}) \, \underset{\sim}{u}_o(\underset{\sim}{y}) \, dS_{\underset{\sim}{y}} \quad ,$$

and consider the integral equation

$$(13) \qquad (I - \sigma K_o)\underset{\sim}{u}_o(\underset{\sim}{x}) = \underset{\sim}{g}_o(\underset{\sim}{x}) \; , \quad \underset{\sim}{x} \in S$$

where $\underset{\sim}{g}_o(x) = 2\underset{\sim}{F}_o(\underset{\sim}{x})$ and σ is a parameter. We note that from (6), we have $\underset{\sim}{u}_o(\underset{\sim}{x}) = - (\int_S 2 \overset{o}{\Gamma}_1(\underset{\sim}{x},\underset{\sim}{y})dS_{\underset{\sim}{y}}) \, \underset{\sim}{u}_o(x)$ for $\underset{\sim}{x} \in S$. Hence in the special case where $\sigma = 1$, (13) is identical to $(9)_o$. In general it can be shown that the integral equation (13) has no eigenvalue for $|\sigma| < 1$ and for $\sigma = 1$ (see [8 , p. 117]). Moreover, since (13) is a singular integral equation of index zero, the eigenvalues σ of (13) are isolated. Then by using the technique of shifting eigenvalues as for the Laplace equation [5], [9], [10], it follows that even for $\sigma = 1$, (13) may be solved by iteration; more precisely, the sequence $\{\underset{\sim}{u}_o^{(n)}(\underset{\sim}{x})\}$ defined by

$$(14) \qquad \begin{aligned} \underset{\sim}{u}_o^{(o)}(\underset{\sim}{x}) &= \frac{1}{2} \, \underset{\sim}{g}_o(\underset{\sim}{x}) \\ \underset{\sim}{u}_o^{(n)}(\underset{\sim}{x}) &= \frac{1}{2} \{(I+K_o) \, \underset{\sim}{u}_o^{(n-1)}(\underset{\sim}{x}) + \underset{\sim}{g}(\underset{\sim}{x})\} \; , \quad n \geq 1 \; . \end{aligned}$$

converges to the solution $\underset{\sim}{u}_o(\underset{\sim}{x})$ of the integral equation

$$\underset{\sim}{u}_o(\underset{\sim}{x}) = \frac{1}{2} \{(I + K_o) \, \underset{\sim}{u}_o(\underset{\sim}{x}) + \underset{\sim}{g}_o(\underset{\sim}{x})\} \; ,$$

which corresponds to $(9)_o$ for $\underset{\sim}{x} \in S$. Since we may rewrite (14) in the form

$$(15) \qquad \begin{aligned} \underset{\sim}{u}_o^{(n)}(\underset{\sim}{x}) &= L_o \underset{\sim}{u}_o^{(n-1)}(\underset{\sim}{x}) + \underset{\sim}{F}_o(x) \quad , \quad n \geq o \\ &= \sum_{k=o}^{n} L_o^k \, \underset{\sim}{F}_o(\underset{\sim}{x}) \end{aligned}$$

with $\underset{\sim}{u}_o^{(-1)} = \underset{\sim}{o}$, the convergence of the series in $(10)_o$ then follows immediately from that of the sequence $\{u_o^{(n)}\}$.

Similarly, for the series in (10), we consider the integral equation

$$(16) \qquad \{I - \sigma(K_o + 2M(\omega))\} \; \underset{\sim}{u}(\underset{\sim}{x}) = \underset{\sim}{g}(\underset{\sim}{x}), \quad \underset{\sim}{x} \in S$$

with $g(x) = 2F(x)$. Again in the special case where $\sigma = 1$, (16) is identical to (9). For ω sufficiently small, it can be shown that the integral equation (16) has no eigenvalues for $|\sigma| < 1$ and for $\sigma = 1$; in fact, in [3] we will give a sufficient condition for ω to guarantee these results. Now following the same argument as in the static case, we may conclude that the sequence $\{\underset{\sim}{u}^{(n)}(\underset{\sim}{x})\}$,

$$\underset{\sim}{\mu}^{(o)}(\underset{\sim}{x}) = \frac{1}{2}\underset{\sim}{g}(\underset{\sim}{x})$$

$$\underset{\sim}{\mu}^{(n)}(\underset{\sim}{x}) = \frac{1}{2}\{(I+K_o+2M(\omega))\,\underset{\sim}{\mu}^{(n-1)}(\underset{\sim}{x}) + \underset{\sim}{g}(\underset{\sim}{x})\}\ , \quad n \geq 1\ ,$$

converges to the solution of (16), and thus leads to the convergent series (10).

4. Conclusions

In concluding this paper, we present here a simple example* in the static case to illustrate our method. Let S be the unit circle and $\underset{\sim}{f}(\underset{\sim}{x}) = \underset{\sim}{x}$ on S. Then the iteration scheme (15) yields

$$\underset{\sim}{\mu}_o^{(o)}(\underset{\sim}{x}) = -\frac{1}{2(\lambda+2\mu)}\underset{\sim}{x}$$

$$\underset{\sim}{\mu}_o^{(n)}(\underset{\sim}{x}) = \sum_{k=o}^{n}\left(\frac{\lambda+\mu}{\lambda+2\mu}\right)^k \left\{-\frac{1}{2(\lambda+2\mu)}\underset{\sim}{x}\right\}\ ,$$

and hence $\underset{\sim}{u}_o(\underset{\sim}{x}) = -\frac{1}{2\mu}\underset{\sim}{x}$ on S which can be easily verified to be the solution of the integral equation (9) on S.

We remark that the present approach is not limited only to the second fundamental boundary-value problems (the Neumann type problems). The applicability to problems such as mixed boundary value problems as well as the interface problems seems also feasible. Some preliminary developments towards this direction have been recently reported in [4].

---

*The author would like to thank Professor J. F. Ahner for preparing this example.

R E F E R E N C E S

[1] J. F. Ahner and R. E. Kleinman:
The exterior Neumann problem for the Helmholtz equation.
Arch. Rat. Mech. and Anal. 52, 26-43 (1973).

[2] J. F. Ahner and G. C. Hsiao:
A Neumann series representation for solutions to boundary-value problems in dynamic elasticity.
Quart. Appl. Math. 33, 73-80 (1975).

[3] J. F. Ahner and G. C. Hsiao:
On the two-dimensional exterior boundary-value problems of elasticity.
SIAM J. Appl.Math. (in press).

[4] G. C. Hsiao and R. Kittappa:
On an interface problem of elasticity.
Proc. 5th Canadian congress of Appl. Mech. 679-680 (1975).

[5] L. V. Kantorovich and V. I. Krylov:
Approximate Methods fo Higher Analysis
P. Noordhoff, Groningen, 1964.

[6] R. Kittappa and R. E. Kleinman:
Acoustic scattering by penetrable homogeneous objects.
J. Math. Phys. 16, 421-432 (1975).

[7] R. E. Kleinman and W. L. Wendland:
On Neumann's method for the exteriror Neumann problem for the Helmholtz equation.
J. Math. Anal. Appl. (to appear).

[8] V. D. Kupradze:
Potential Methods in the Theory of Elasticity.
Israel Program for Scientific Translations, Jerusalem, 1965.

[9] W. Wendland: Lösung der ersten und zweiten Randwertaufgaben des Innen- und Außengebietes für die Potentialgleichung im R_3 durch Randbelegungen. Dissertation, TU Berlin 1965.

10] W. Wendland:
Die Behandlung von Randwertaufgaben in R_3 mit Hilfe von Einfach- und Doppelschichtpotentialen .
Numerische Mathematik 11, 380-404 (1968).

GREEN'S FUNCTION OF MULTIPLY CONNECTED DOMAIN AND DIRICHLET PROBLEM FOR SYSTEMS OF SECOND ORDER IN THE PLANE

Tadeusz Iwaniec

Introduction Complex methods in the study of Dirichlet problem for systems of two-real 2-nd order equations in the plane have not been as successfull until now as in the case of a single equation. In particular there does not seem to exist a theory capable to handle the uniformly elliptic case with non smooth coefficients. An approach to discuss problems of this type was proposed several years ago by B.Bojarski and the idea was continued in a recent unpublished paper (preprint) by B.Bojarski and T.Iwaniec [5]. It turned out further, that suitably generalized Green's function type kernels, can be used to establish the Fredholm alternative for the Dirichlet problem for general elliptic systems with continuous coefficients, if, roughly speaking, the system can be continuously deformed to the system of two Laplace equations. In this paper we continue the investigation of [3] [4] and give an effective construction of Green's function for circular multiply connected domains. The construction gives explicit expressions for the behaviour of Green's function at the boundary points, the knowledge of which seems to be essential for the construction of the generalized kernels mentioned above. We hope, that the construction method has also an independent interest.

1. The conformal mappings on slit domains

On the basis of Riemann's theorem we known that every multiply connected domain in the plane can be mapped conformally onto certain canonical domain. Here we shall now consider mappings onto Schottky's domains.

DEFINITION 1. A domain Ω which consists of the whole plane except for finite number of disjoint arcs of concentric circles is called Schottky's domain (see figure 1)

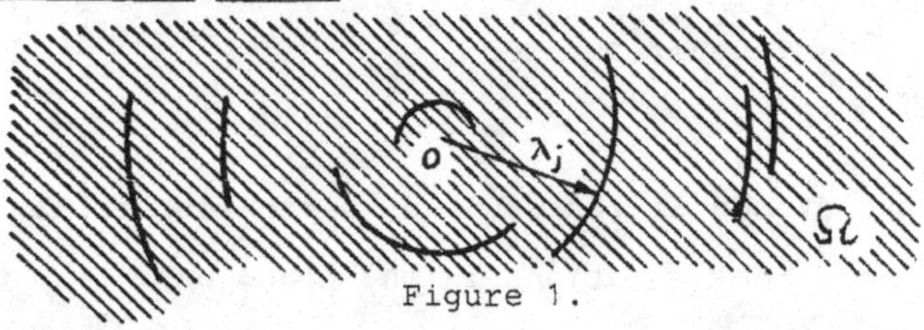

Figure 1.

Let $D \subset \overline{\mathbb{C}}$ be multiply connected domain bounded by Jordan curves $\Gamma_1, \Gamma_2, \ldots, \Gamma_n$ and let $t \neq \infty$ be arbitrary point belonging to D. Then there exists a conformal mapping $\chi(z) = \chi(z,t)$ of D onto Schottky's domain, such that $\chi(t) = \infty$, $\chi(\infty) = 0$. The function $\chi(z,t)$ can be represented in the form $\chi(z,t) = f(z,t)(z-t)^{-1}$, where the function $f(z,t)$ is analytic for $z \in D$, continuous on $D + \Gamma$ $(\Gamma = \Gamma_1 + \Gamma_2 + \ldots + \Gamma_n)$ and never vanishes. The function $f(z,t)$ satisfies the following special boundary conditions

$$(1) \qquad |f(z,t)(z-t)^{-1}| = \lambda_j(t) \quad \text{for} \quad z \in \Gamma_j \quad (j = 1,2,..,n)$$

where the constants $\lambda_j(t)$ are the arcs radii (see Figure 1). It follows from the general conformal mapping theory that one of the λ's may be arbitrarily chosen. The others are then uniquely determined. In this way the construction of the mapping $\chi(z,t)$ reduces to the solution of the following boundary value problem.

PROBLEM 1. Let $t \in D$ be an arbitrary point of D. Find the function $f(z,t)$ analytic for $z \in D$ continuous on $D + \Gamma$ and never vanishing, such that the boundary conditions (1) hold with certain positive constants $\lambda_j(t)$.

This problem is closely related to modified Dirichlet problem discussed by I.N.Muskhelishvili [7]. We shall solve the above problem for circular domains. The function $f(z,t)$ will be effectivel constructed (see also G. Golusin [9] and I. Aleksandrov, A. Sorokin [10]).

2. Circular domains

DEFINITION 2. Let $\bar{D}_1, \bar{D}_2, \ldots, \bar{D}_n$ be disjoint closed discs with centers at the points $z_1, z_2, \ldots, z_n$ and radii $r_1, r_2, \ldots, r_n$ respectively. A set $D = \bar{\mathbb{C}} - \bigcup_{j=1}^{n} \bar{D}_j$ is called a circular domain. The boundary Γ of D consists of n circles $\Gamma_1, \Gamma_2, \ldots, \Gamma_n$

(2) $$\Gamma_j = \{z; \ |z-z_j| = r_j\} \quad \text{for} \quad j = 1,2,\ldots,n.$$

For ε being the smallest distance between discs $D_1, D_2, \ldots, D_n$ we have

(3) $$|z_k - z_j| \geq r_k + r_j + \varepsilon \quad \text{for} \quad k \neq j, \quad k,j = 1,2,\ldots,n$$

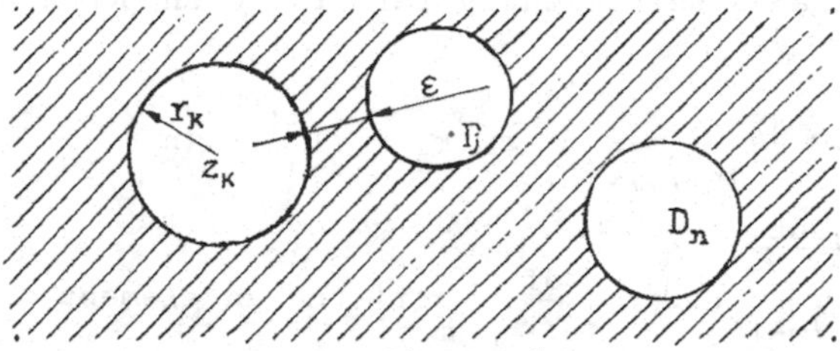

Figure 2.

Let $\tau_j : \bar{\mathbb{C}} \longrightarrow \bar{\mathbb{C}}$ be the reflection in the circle Γ_j (i.e. inversion in the circle Γ_j). It is given by the formula

(4) $$\tau_j(t) - z_j = r_j^2 \, (\bar{t} - \bar{z}_j)^{-1} \qquad \text{for} \quad j = 1,2,\ldots,n$$

The $\tau_j(t)$ is an anti-conformal transformation. It changes the orientation. By $\mathcal{T}$ we denote the group of conformal and anti-conformal mappings of $\bar{\mathbb{C}}$ onto itself generated by reflections $\tau_1, \tau_2, \ldots, \tau_n$. For every j, τ_j^2 = identity. Hence, it follows that each element $\tau \in \mathcal{T}$ can be represented in the form

$\tau = \tau_{j_m} \circ \tau_{j_{m-1}} \circ \ldots \circ \tau_{j_1}$, where $j_1 \neq j_2 \neq \ldots \neq j_m$,

$j_1, j_2, \ldots, j_m \in \{1,2,\ldots,n\}$. Moreover, it is easy to see that $\tau(D) \subset D_{j_m}$. From this it follows that the representation $\tau = \tau_{j_m} \circ \tau_{j_m} \circ \ldots \circ \tau_{j_1}$ is unique and the sets $\tau_1(D)$, $\tau_2(D)$ are disjoint for every $\tau_1 \neq \tau_2 \in \mathcal{T}$. The open set $\bigcup_{\tau \in \mathcal{T}} \tau(D)$ is dense in the plane i.e. $\overline{\bigcup_{\tau\in\mathcal{T}} \tau(D)} = \bar{C}$.

The mapping $\tau = \tau_{j_m} \circ \tau_{j_{m-1}} \circ \ldots \circ \tau_{j_1}$ changes or not the orientation according as to if m is an odd or even integer. We introduce the following notations : $\operatorname{sgn}\tau = (-1)^m$, $|\tau| = m$. For simplicity of calculations we shall assume

$$(5) \qquad \sum_{id \neq \tau \in \mathcal{T}} \operatorname{diam} \tau(D) < \infty$$

This is not always true (as in [9] and [10]). When the discs $D_1, D_2, \ldots, D_n$ are situated sufficiently far from each other we can prove (5).

<u>Proposition 1.</u> <u>For</u> $Q \overset{df}{=} \max_{j=1,2,\ldots,n} r_j (r_j + \varepsilon)^{-1} < (n-1)^{-1/2}$ <u>the series</u> (5) <u>is convergent.</u>

<u>Proof.</u> Let $\tau = \tau_{j_m} \circ \tau_{j_{m-1}} \circ \ldots \circ \tau_{j_1} \overset{df}{=} \tau_{j_m} \circ \tau'$ be an element of the group $\mathcal{T}$. We shall show that $\operatorname{diam} \tau(D) \leq Q^2 \operatorname{diam} \tau'(D)$. We have

$$\operatorname{diam} \tau(D) = \sup_{x,y\in D} |\tau(x) - \tau(y)| = \sup_{p,q\in\tau'(D)} |\tau_{j_m}(p) - \tau_{j_m}(q)|$$

Since $\tau'(D) \subset D_{j_{m-1}}$ it follows that

$$|\tau_{j_m}(p) - \tau_{j_m}(q)| \leq \sup_{t \in D_{j_{m-1}}} |\partial\tau_{j_m}(t)/\partial\bar{t}|\,|p-q| \quad \text{for } p,q \in \tau'(D)$$

We can estimate the derivatives of $\tau_k(t)$ as follows

$$|\partial\tau_k(t)/\partial\bar{t}| = r_k^2 |t - z_k|^{-2} \leq r_k^2 (r_k + \varepsilon)^{-2} \leq Q^2 \quad \text{for } t \in D_l,\ l \neq k$$

and thus the proof of the inequality $\operatorname{diam} \tau(Q) \leq Q^2 \operatorname{diam} \tau'(D)$ is completed. Using it $m-1$ times we get $\operatorname{diam} \tau(D) << Q^{2m}$, so the series (5) is majorized by the power series

$$\sum_{m=1}^{\infty} \sum_{j_1 \neq j_2 \neq \dots \neq j_m} \operatorname{diam} \tau_{j_m j_{m-1} \dots j_1}(D) << \sum_{m=1}^{\infty} (n-1)^m Q^{2m} < \infty .$$

where, for brevity, $\tau_{j_m j_{m-1} \dots j_1}$ stands, for $\tau_{j_m} \circ \tau_{j_{m-1}} \circ \dots \circ \tau_{j_1}$.
This completes the proof.

3. Construction of the function $f(z,t)$ and $\lambda_j(t)$

THEOREM 1. *The solution $f(z,t)$ of Problem 1 is given by the convergent product*

$$(6) \qquad f(z,t) = \prod_{id \neq \tau \in \mathcal{T}} \left(\frac{z-\tau(\infty)}{z-\tau(t)}\right)^{\operatorname{sgn}\tau}, \quad \text{for} \quad (z,t) \in \bar{D} \times D.$$

Proof. The homographic factors in (6) are regular and nonvanishing in D. Product (6) uniformly converges on $\bar{D} \times K$ for every closed subset $K \subset D$ (not necessarily compact) and its value does not depend on the order of factors. This follows from the assumption (5) and the estimate

$$\left|\frac{z-\tau(\infty)}{z-\tau(t)} - 1\right| \leq \frac{|\tau(t)-\tau(\infty)|}{|z-\tau(t)|} \leq C_K \operatorname{diam} \tau(D) \quad \text{for} \quad (z,t) \in \bar{D} \times K$$

where the constant C_K depends only on K.
The function $f(z,t)$ is continuous on $\bar{D} \times D$, analytic with respect to $z \in D$ and nonvanishing on D. Now we shall show that $f(z,t)$ satisfies boundary conditions (1). The λ's will be determined later. For each index $1 \leq j \leq n$ we transform the product (6) by changing its factors into

$$\chi(z,t) = \frac{1}{z-t}\left(\frac{z-\tau_j(t)}{z-\tau_j(\beta)}\right) \times \prod_{j_1 \neq j} \frac{z-\tau_{j_1}(t)}{z-\tau_{j_1}(\infty)} \cdot \frac{z-\tau_{jj_1}(\infty)}{z-\tau_{jj_1}(t)} \times$$

$$\times \prod_{\substack{j_1 \neq j_2 \neq \cdots \neq j_m \\ \neq j}} \left(\frac{z-\tau_{j_m\cdots j_1}(\infty)}{z-\tau_{j_m\cdots j_1}(t)}\right)^{(-1)^m} \cdot \left(\frac{z-\tau_{jj_m\cdots j_1}(\infty)}{z-\tau_{jj_m\cdots j_1}(t)}\right)^{(-1)^{m+1}} \times \cdots$$

In this expression the absolute values of the blocs $\prod\limits_{j_1 \neq j_m \neq j}$ are constant on Γ_j, i.e. for $|z-z_j| = r_j$ we have

$$\left|\frac{1}{z-t} \cdot \frac{z-\tau_j(t)}{z-\tau_j(\infty)}\right| = \frac{1}{|z_j-t|}$$

$$\left|\frac{z-\tau_{j_m\cdots j_1}(\infty)}{z-{}_{j_m\cdots j_1}(t)} \cdot \frac{z-\tau_{jj_m\cdots j_1}(t)}{z-\tau_{jj_m\cdots j_1}(\infty)}\right| = \left|\frac{z_j-\tau_{jj_m\cdots j_1}(t)}{z-\tau_{jj_m\cdots j_1}(\infty)}\right|$$

Hence $|\chi(z,t)|$ does not depend on $z \in \Gamma_j$ and equals to

$$\lambda_j(t) \stackrel{\text{df}}{=} \frac{1}{|z_j-t|} \prod_{m=1}^{\infty} \prod_{j_1 \neq j_2 \neq \cdots \neq j_m \neq j} \left|\frac{z_j-\tau_{j_m\cdots j_1}(t)}{z_j-\tau_{j_m\cdots j_1}(\infty)}\right|^{(-1)^{m+1}}$$

Thus, our theorem is proved.

Remark 1 The product

$$\prod_{|\tau| \geq 2} \left(\frac{z-\tau(\infty)}{z-\tau(t)}\right)^{\operatorname{sgn}\tau}$$

uniformly converges on a certain neighborhood of $\bar{D} \times \bar{D}$. The proof of this fact is similar to the proof of Theorem 1. In particular it represents an analytic function in $z \in \bar{D}$. Therefore, interior singular points of $\chi(z,t)$ as well as boundary ones, appear only in the term

$$\frac{1}{z-t} \prod_{j=1}^{n} (z-\tau_j(t))$$

We shall make use of this remark for the study of singular points of Green's function in multiply connected domains.

4. Singularity of the Green's function

Let $h_j(z)$ be harmonic measures of Γ_j's components of Γ, $j = 1,2,\ldots,n$. The harmonic function $h_j(z)$ has the boundary value 0 on all Γ, except on Γ_j, where it has the value 1. The linear combination $\sum_{k=1}^{n} h_k(z)\ln\lambda_k(t)$ has the boundary value $\ln\lambda_j(t)$ on Γ_j, $j = 1,2,\ldots,n$. So, the Green's function of D can be written in the form

$$(7) \qquad \mathcal{G}_D(z,t) = \ln|\chi(z,t)| - \sum_{k=1}^{n} h_k(z)\ln\lambda_k(t)$$

where $\chi(z,t) = f(z,t)(z-t)^{-1}$ is a solution of Problem 1. From (7) by conformal or quasiconformal transformation we can obtain the Green's function for arbitrary circular domains. For instance if $\tau \in \mathcal{T}$, then $\mathcal{G}_{\tau(D)}(\zeta,\xi) = \mathcal{G}_D(\tau^{-1}(\zeta),\tau^{-1}(\xi))$. Taking into acount Remark 1 we get

$$\mathcal{G}_D(z,t) = -\ln|z-t| + \sum_{j=1}^{n} \ln|z-\tau_j(t)| + \text{regular terms}$$

The first and second order derivatives of Green's function can be written as follows

$$(8) \qquad \begin{cases} \mathcal{G}_z(z,t) = -\dfrac{1}{2(z-t)} + \dfrac{1}{2}\displaystyle\sum_{j=1}^{n} \dfrac{\bar{z}_j-\bar{t}}{r_j^2-(z-z_j)(\bar{t}-\bar{z}_j)} + \text{regular terms} \\ \mathcal{G}_{zz}(z,t) = \dfrac{1}{2(z-t)^2} - \dfrac{1}{2}\displaystyle\sum_{j=1}^{n} \dfrac{(\bar{z}_j-\bar{t})^2}{[r_j^2-(z-z_j)(\bar{t}-\bar{z}_j)]^2} + \text{regular terms} \end{cases}$$

The above formulae hold for every circular domain (not necessarily satisfying (5) or unbounded), the proof can be obtained by applying a quasiconformal transformation. The formulae (8) may be derived also in an other manner.

5. Systems of equations of second order in the plane

DEFINITION 3. The system of two equations of second order for two unknown functions u,v

(9) $$Aw_{xx} + 2Bw_{xy} + Cw_{yy} + aw_x + bw_y + cw = g$$

where $w = (u,v)$, A,B,C, a,b,c are the square matrices of type 2×2 is called elliptic if

(10) $$\det(A\xi^2 + 2B\xi\zeta + C\zeta^2) \neq 0 \quad \text{for} \quad \xi^2 + \zeta^2 = 1.$$

Using the complex variables $z = x+iy$, $w = u+iv$, we can write this system in the form of one equation with complex coefficients

$$\alpha w_{zz} + \beta w_{\bar{z}\bar{z}} + \gamma\overline{w_{zz}} + \delta\overline{w_{\bar{z}\bar{z}}} + \mu w_{z\bar{z}} + \nu\overline{w_{z\bar{z}}} + T(w) = h$$

where T is a linear (over $\mathbb{R}$) differential operator of order ≤ 1 and h is a given complex function. Such system determines the point $(\alpha,\beta,\gamma,\delta,\mu,\nu)$ of the complex space $\mathbb{C}^6$ and the condition of ellipticity defines an open subset $\mathcal{E}\subset\mathbb{C}^6$. As it was proved by B.Bojarski (see [3][4]) the set $\mathcal{E}$ has precisely six components. We shall deal with only one component $\mathcal{E}_0$, namely this which contains the Laplace system $w_{z\bar{z}} = 0$. B.Bojarski has proved that the systems from $\mathcal{E}_0$ can be uniquely expressed as follows

(11) $$\mathcal{H}(w) \overset{\text{def}}{=\!=} w_{z\bar{z}} + \alpha w_{zz} + \beta w_{\bar{z}\bar{z}} + \gamma\overline{w_{zz}} + \delta\overline{w_{\bar{z}\bar{z}}} = -Tw + h$$

where the condition of ellipticity (10) may be described in terms of coefficients $\alpha,\beta,\gamma,\delta$, by the following condition

(12) $$|1 + \alpha\xi + \beta\bar{\xi}| > |\gamma\bar{\xi} + \delta\xi| \quad \text{for} \quad |\xi| = 1$$

6. The Dirichlet Problem

The Fredholm alternative for the Dirichlet problem for general elliptic systems is not true. As it was demonstrated by A.V.Bitsadze in [6] the system $w_{zz} = 0$, which belongs to $\mathcal{E}$ has infinitely many linearly independent solutions on the unit disc vanishing identically on its boundary. We shall examine the Dirichlet problem for the systems of the class $\mathcal{E}_o$. Without loss of generality we can consider circular domains only.

Let Γ be the boundary of the circular domain D. We find the solution of the Dirichlet problem

(13) $$\begin{cases} w_{z\bar{z}} + \alpha w_{zz} + \beta w_{\bar{z}\bar{z}} + \gamma \overline{w_{zz}} + \delta \overline{w_{\bar{z}\bar{z}}} = -Tw + h \\ w|_{\Gamma} = 0 \end{cases}$$

in the form

(14) $$w(z) = \frac{2}{\pi} \iint_D \mathcal{G}(z,t)\omega(t)d\sigma_t$$

where $\mathcal{G}(z,t)$ is the Green's function of the domain D and $\omega(t)$ is the unknown function. If we substitute this into the equation (13) using (8) we shall obtain the singular integral equation

(15) $$\mathcal{L}(\omega) = \omega + \alpha\mathcal{S}\omega + \beta\mathcal{R}\omega + \gamma\overline{\mathcal{S}\omega} \quad \delta\overline{\mathcal{R}\omega} = \mathcal{K}\omega + h$$

where the operators $\mathcal{S}$ and $\mathcal{R}$ are given by the formulae

$$\mathcal{S}\omega = \frac{1}{\pi} \iint_D \frac{\omega(t)d\sigma_t}{(z-t)^2} - \sum_{j=1}^{n} \frac{1}{\pi} \iint_D \frac{(\bar{t}-\bar{z}_j)^2 \omega(t)d\sigma_t}{[r_j^2-(\bar{t}-\bar{z}_j)(z-z_j)]^2}$$

$$\mathcal{R}\omega = \overline{\mathcal{S}\bar{\omega}}$$

On the basis of the Calderon Zygmund theory [8] we can prove that the operators $\mathcal{S}$ and $\mathcal{R}$ act in the space $L_p(D)$ for all $p > 1$. We are able to evaluate the norms $\|\mathcal{S}\|_p = \|\mathcal{R}\|_p \overset{\text{df}}{=} \Lambda_p$ only for $p=2$. Then, by the Riesz-Thorins theorem they can be estimated for all p near to 2.

The operator $\mathcal{K}$ in equation (15) has a weak singular kernel and so it is compact in the space $L_p(D)$. Now we assume for the moment that the coefficients of equation (13) are measurable functions and satisfy the following condition

$$(16) \qquad \sup_D(|\alpha(z)| + |\gamma(z)|) + \sup_D(|\beta(z)| + |\delta(z)|) \overset{\text{df}}{=} q_0 < \Lambda_2^{-1} < 1 \; .$$

The assumption (16) ensures invertibility of $\mathcal{L}$ in $L_p(D)$ for sufficiently small $|p-2|$. Therefore, Fredholm alternative for equations (13) holds in the Sobolev class $W_2^p(D)$.

7. <u>The Dirichlet problem in the general case</u>

Let D be the circular domain introduced in section 2. Let us consider the solution $w(z) \in W_2^p(D)$ of the Dirichlet problem (13) such that the coefficients $\alpha, \beta, \gamma, \delta$ satisfy ellipticity condition (12) and the coefficients of the operator T belong to $L_p(D)$ for any $p > 2$. It is clear that the function $w(z)$ is of class $C^1(D)$. We extend this function onto the whole plane in the following manner

$$w^*(z) = \begin{cases} w(z) & \text{for} \quad z \in \bar{D} \\ -w(\tau^{-1}(z)) & \text{for} \quad z \in \tau(\bar{D}) \quad id \neq \tau \in \mathcal{T} \\ 0 & \text{for} \quad z \in \mathcal{C} \stackrel{df}{=} \bar{\mathbb{C}} - \bigcup_{\tau \in \mathcal{T}} \tau(\bar{D}) \end{cases} \quad {}^{*)}$$

The function $w^*(z)$ is continuous on the closed plane $\bar{\mathbb{C}}$ because of the condition $w_{|\Gamma} = 0$. Now, we shall examine the first derivatives of $w^*(z)$. For $z \in \tau_j(\bar{D})$ we have

$$w^*_z(z) = r_j^2(z-z_j)^{-2} w_{\bar{\tau}_j}(\tau_j(z)) \;,\; w^*_{\bar{z}}(z) = r_j^2(\bar{z}-\bar{z}_j)^{-2} w_{\tau_j}(\tau_j(z))$$

The boundary condition $w_{|\Gamma_j} = 0$ implies $(z-z_j)w_z = (\bar{z}-\bar{z}_j)w_{\bar{z}}$ for $z \in \Gamma_j$ and in consequence $w^*_z(z)$ and $w^*_{\bar{z}}(z)$ are continuous on $\bar{D} \cup \tau_j(\bar{D})$. Hence w^* is of class $C^1(\bigcup_{\tau \in \mathcal{T}} \tau(\bar{D}))$. From the theory of the generalized derivatives in the sense of Sobolev it follows that the function w^* has the second order generalized derivatives.

<u>Remark 2</u> <u>We can prøove that if the function</u> w <u>is of class</u> $C^2(\bar{D})$ <u>and</u> $w_{z\bar{z}}{}_{|\Gamma} = 0$ (<u>for example if</u> $\alpha, \beta, \gamma, \delta,$ h <u>and coefficients of</u> T <u>in</u> (13) <u>belong to</u> $C_0(D)$) <u>then</u> w <u>is class</u> $C^2(\bigcup_{\tau \in \mathcal{T}} \tau(\bar{D}))$ <u>also</u>.

The coefficients of (13) can be extended onto the whole plane in such a way that $w^*(z)$ will satisfy equation (13) with new coefficients. For simplicity, we shall discuss this for simply connected domains, $n = 1$

8. <u>The Dirichlet problem for the unit disc</u>

Let D be the unit disc and $w(z) \in W_2^p(D)$ be the solution of the problem (13). Let

$$w^*(z) = \begin{cases} w(z) & \text{for} \quad |z| \leq 1 \\ -w(\frac{1}{\bar{z}}) & \text{for} \quad |z| \leq 1 \end{cases}$$

*) It follows from condition (5) that the one-dimensional Hausdorff measure of $\mathcal{C}$ equals zero

As it was proved earlier the function $w^*(z)$ belongs to $C^1(\bar{\mathbb{C}})$ and has generalized derivatives of second order on the whole plane. We shall show now, that $w^* \in W_2^p(\bar{\mathbb{C}})$. We have

$$(17)\quad \begin{cases} w^*_{z\bar{z}}(z) = -|z|^{-4} w_{z\bar{z}}(\frac{1}{\bar{z}}) & \text{for } |z| \geq 1 \\ w^*_{\bar{z}\bar{z}}(z) = -\bar{z}^{-4} w_{zz}(\frac{1}{\bar{z}}) - 2\bar{z}^{-3} w_z(\frac{1}{\bar{z}}) & \text{for } |z| \geq 1 \\ w^*_{zz}(z) = -z^{-4} w_{\bar{z}\bar{z}}(\frac{1}{\bar{z}}) - 2z^{-3} w_{\bar{z}}(\frac{1}{\bar{z}}) & \text{for } |z| \geq 1 \end{cases}$$

The functions $w_z(\bar{z}^{-1})$, $w_{\bar{z}}(\bar{z}^{-1})$ are bounded in $|z| \geq 1$, thus $2z^{-3} w_{\bar{z}}(\bar{z}^{-1})$ belongs to $L_p(|z|\geq 1)$ and $2\bar{z}^{-3} w_z(\bar{z}^{-1})$ belongs to $L_p(|z| \geq 1)$. Changing the variables we obtain

$$\iint_{|z|\geq 1} |z^{-4} w_{z\bar{z}}(\bar{z}^{-1})|^p d\sigma_z = \iint_{|z|\leq 1} |w_{z\bar{z}}(z)|^p |z|^{4p} |z|^{-4} d\sigma_z < \infty$$

Analogously we prove p-integratability of the two other second derivatives of w^*. This completes the proof of the fact $w^* \in W_2^p(\bar{\mathbb{C}})$. The extended function $w^*(z)$ satisfies the following elliptic equation on the whole plane

$$(18)\quad w^*_{z\bar{z}} + \alpha^*(z) w^*_{zz} + \beta^*(z) w^*_{\bar{z}\bar{z}} + \gamma^*(z)\overline{w^*_{zz}} + \delta^*(z)\overline{w^*_{\bar{z}\bar{z}}} + Tw^* = h^*(z)$$

where

$$\alpha^*(z) = \begin{cases} \alpha(z) & |z|\leq 1 \\ \bar{z}^{-2} z^2 \beta(\bar{z}^{-1}), & |z|>1 \end{cases}, \quad \beta^*(z) = \begin{cases} \beta(z) & |z|\leq 1 \\ \bar{z}^2 z^{-2} \alpha(\bar{z}^{-1}), & |z|>1 \end{cases}$$

$$\gamma^*(z) = \begin{cases} \gamma(z) & |z| \le 1 \\ z^{-2}\bar{z}^{2}\delta(\bar{z}^{-1}), & |z|>1 \end{cases}, \quad \delta^*(z) = \begin{cases} \delta(z) & |z| \le 1 \\ z^{2}\bar{z}^{-2}\gamma(\bar{z}^{-1}), & |z|>1 \end{cases}$$

$$h^*(z) = \begin{cases} h(z) & , \ |z| \le 1 \\ -|z|^{-4}h(\bar{z}^{-1}), & |z|>1 \end{cases}$$

and T^* is the linear (over $\mathbb{R}$) differential operator of order ≤ 1. If we assume the coefficients $(\alpha,\beta,\gamma,\delta)$ to be measurable functions of z then $(\alpha^*,\beta^*,\gamma^*,\delta^*)$ in (18) will be also measurable. However, when the coefficients in (13) are smooth the coefficients in (18) can be said only, in general, to be piecewise continuous. If $\alpha,\beta,\gamma,\delta$ vanish on Γ then $\alpha^*,\beta^*,\gamma^*,\delta^*$ are continuous. The coefficients $\alpha^*,\beta^*,\gamma^*,\delta^*$ in (18) always satisfy the ellipticity condition (12), so the extended equation (18) is of the class $\mathcal{E}_0$.

The solution $w^* \in W_2^p(\bar{\mathbb{C}})$ of equation (18) can be expressed in the form

$$w^*(z) = \frac{2}{\pi} \iint_{\mathbb{C}} \ln|z-t|\,\omega(t)\,d\sigma_t$$

where $\omega(t) = w^*_{z\bar{z}}(t) = -|t|^{-4}w_{z\bar{z}}(\bar{t}^{-1})$ for $|t| \ge 1$. Substituting this into the equation (18) we get the strongly singular integral equation on the whole plane.

$$(19) \qquad L\omega \overset{df}{=} \omega + \alpha^* S\omega + \beta^* R\omega + \gamma^* S\omega + \delta^* R\omega = K\omega + h$$

where S and R are well known fundament al singular operators

$$S\omega = -\frac{1}{\pi}\iint_{\mathbb{C}} \frac{\omega(t)d\sigma_t}{(z-t)^2} \quad , \qquad R\omega = -\frac{1}{\pi}\iint_{\mathbb{C}} \frac{\omega(t)d\sigma_t}{(\bar{z}-\bar{t})^2} \quad .$$

They have the following properties (see I.N.Vekua [2])

a) S and R are isometries of $L_2(\bar{\mathbb{C}})$ and may be extended to continuous operators on $L_p(\mathbb{C})$ for $p > 1$.

b) $SR = RS = I$ - identity, $S\bar{\omega} = \overline{R\omega}$, $R\bar{\omega} = \overline{S\omega}$

c) The commutators

$$qS\omega - Sq\omega = -\frac{1}{\pi}\iint_{\mathbb{C}} \frac{q(z)-q(t)}{(z-t)^2}\omega(t)d\sigma_t$$

$$qR\omega - Rq\omega = -\frac{1}{\pi}\iint_{\mathbb{C}} \frac{q(z)-q(t)}{(\bar{z}-\bar{t})^2}\omega(t)d\sigma_t$$

are compact if $q(z)$ is continuous and bounded.

In the next section we shall show that under some weak conditions on the smoothness of coefficients $\alpha^*, \beta^*, \gamma^*, \delta^*$ the operator L is of Fredholm type.

8. <u>The operator L</u> With every integral operator (19) we associate another operator

$$\hat{L} = I + \bar{\beta}^* S + \bar{\alpha}^* R - \gamma^* \overline{S^*} - \delta^* \overline{R^*}$$

We have the following identity $\hat{\hat{L}} = L$. It will be clear later that operators L and $\hat{L}$ are connected with the polynomial

$$Q(\lambda) = (\bar{\beta}^*\alpha^* - \delta^*\bar{\gamma}^*)\lambda^4 + (\alpha^* + \bar{\beta}^*)\lambda^3 + (1 + |\alpha^*|^2 + |\beta^*|^2 - |\gamma^*|^2 - |\delta^*|^2)\lambda^2 +$$

$$+ (\beta^* + \bar{\alpha}^*)\lambda + (\beta^*\bar{\alpha}^* - \bar{\delta}^*\gamma^*)$$

Now ellipticity condition (12) implies that the roots of Q have the form $\bar{q}_1$, $\bar{q}_2$, q_1^{-1}, q_2^{-1}, where $|q_1| < 1$, $|q_2| < 1$.

Hence we have

$$Q(\lambda) = (\bar{\beta}^*\alpha^*-\delta^*\bar{\gamma}^*)q_1^{-1}q_2^{-1}(1-\lambda q_1)(1-\lambda q_2)(\lambda-\bar{q}_1)(\lambda-\bar{q}_2) \qquad ^{*)}$$

Suppose that $|q_i(z)| \le q_0 < 1$ $(i=1,2)$. In this case the corresponding system (13) is called uniformly elliptic. If $q_i(z)$ are continuous $(i=1,2)$ then the commutators Sq_i-q_iS, Rq_i-q_iR, $S\bar{q}_i-\bar{q}_iS$, $R\bar{q}_i-\bar{q}_iR$ $(i=1,2)$ are compact. Taking into account (a), (b), (c) the composition $\hat{L}L$ has a form

$$\hat{L}L = (\bar{\beta}^*\alpha^*-\delta^*\bar{\gamma}^*)q_1^{-1}q_2^{-1}(I-q_1S)(I-q_2S)(I-\bar{q}_1R)(I-\bar{q}_2R) + \text{compact operator}$$

The operators $I-q_iS$ and $I-\bar{q}_iR$ for $i=1,2$ are invertible in $L_p(\mathbb{C})$ for sufficiently small $|p-2|$. Hence it follows that $\hat{L}L$ is Fredholm operator in the space $L_p(\mathbb{C})$. The index of $\hat{L}L$ equals zero. Repeating the above consideration for the operator $\hat{L}$ we obtain that $L\hat{L}$ is also Fredholm operator.

It follows from the general theory of Fredholm operators that L and $\hat{L}$ are Fredholm operators, too. Moreover index L = -index $\hat{L}$.

I wish to thank Mr. H. Löffler and Professor W. Wendland for drawing my attention to the papers: G. M. Golusin [9] and I. A. Aleksandrov, A. S. Sorokin [10], which contain a lot of similar results.

---

*) It can be proved by allpying Rouché's theorem. The polynomial Q has the following form:

$$Q(\lambda) = (\alpha^*\lambda^2+\lambda+\beta^*)(\bar{\beta}^*\lambda^2+\lambda+\bar{\alpha}^*)-(\delta^*\lambda^2+\gamma^*)(\bar{\gamma}^*\lambda^2+\bar{\delta}^*)$$

It follows from the ellipticity condition (12) that the plynomial Q has no roots on the unit circle.

References

[1] I.N.Vekua, A method for a solution of boundary value problems for partial differential equations. DAN SSSR 101, No 4, 1955

[2] I.N.Vekua, Generalized analytic functions.Pergamon Press 1962

[3] B.Bojarski, On the first boundary value problem for elliptic system of second order in the plane.Bull.Acad.Pol.Sci. VII, No 9, 1959

[4] B.Bojarski, Investigations on elliptic systems. Doctoral dissertation, Moscow, 1960

[5] B.Bojarski and T.Iwaniec, On systems of two second order elliptic equations with non regular coefficients(unpublished preprint)

[6] A.V.Bitsadze, Boundary value problems for elliptic equations of second order "Nauka" Moscow 1966

[7] I.N.Muskhelishvili, Singular integral equations, Groningen-Holland, 1953

[8] A.P.Calderon A.Zygmund, On the existence of certain singular integrals. Acta Math.88, 1952.

[9] G. M. Golusin, Auflösung einiger ebenen Grundaufgaben der mathematischen Physik im Falle der Laplaceschen Gleichung und mehrfach zusammenhängender Gebiete, die durch Kreise begrenzt sind (Rec. math. Moscow 41, S. 246-276, 1934, Russian)

[10] I. A. Aleksandrov, A. S. Sorokin, The problem of Schwarz for multiply connected circular domains. Sibirskii Matematicheskii Jhurnal, Vol. 13, No. 5, pp. 971-1001, 1972.

Institute of Mathematics
University of Warsaw
PKiN. 9p. Warsaw
Poland

AUTOMORPHE LÖSUNGEN DER EULER-DARBOUX GLEICHUNG

G. Jank
Technische Universität Graz

1. Einleitung

In der vorliegenden Arbeit betrachten wir die Differentialgleichung

$$(1) \qquad w_{z\bar{z}} - \frac{m+1}{z-\bar{z}} w_z + \frac{n+1}{z-\bar{z}} w_{\bar{z}} = 0, \qquad m,n \in \mathbb{N}_0 .$$

Diese geht durch die Transformation

$$v(z) = (z-\bar{z})^{m+1} w(z) \qquad \text{bzw.} \qquad v(z) = (z-\bar{z})^{n+1} w(z)$$

aus der Differentialgleichung

$$(2) \qquad v_{z\bar{z}} + \frac{n-m}{z-\bar{z}} v_{\bar{z}} + \frac{(m+1)n}{(z-\bar{z})^2} v = 0$$

bzw.

$$(3) \qquad v_{z\bar{z}} + \frac{n-m}{z-\bar{z}} v_z + \frac{(n+1)m}{(z-\bar{z})^2} v = 0$$

hervor. Für den Fall $n = m$ sind (2) und (3) identisch und wurden u.a. von K.W. Bauer und E. Peschl z.B. in [1] untersucht.
Betrachtet man einen beliebigen Automorphismus der oberen Halbebene H, definiert durch eine Matrix $A \in SL(2,\mathbb{R})$, so ist im Fall $n = m$ (2) invariant gegenüber derartigen Transformationen. Dies führt auf die Frage nach der Darstellung von Lösungen , die gegenüber einer ganzen Gruppe $\Gamma \subset SL(2,\mathbb{R})$ invariant sind. Ergebnisse in dieser Richtung gelangen K.W. Bauer und E. Peschl durch die Betrachtung gewisser Differentialinvarianten. Beschränkt man sich auf endlich erzeugte Fuchs'sche Gruppen erster Art, was wir in Hinkunft immer tun wollen, und auf Lösungen von (2) mit $n = m$ und höchstens endlich vielen isolierten logarithmenfreien Singularitäten in einem Fundamentalbereich, die gegenüber den Transformationen einer derartigen Gruppe invariant sind, so konnten I. Haeseler und St. Ruscheweyh in [3] alle derartigen Funktionen darstellen. Verwendet wurden dabei wesentlich die von L. Bers in [2] eingeführten singulären Eichlerintegrale. Nachdem $(z-\bar{z})^2 \frac{\partial^2}{\partial z \partial \bar{z}}$ im wesentlichen der einzige mit allen Substitutionen aus $SL(2,\mathbb{R})$ vertauschbare lineare Operator ist, kann man für die Diffe-

rentialgleichung (1) keine invarianten Lösungen für $n \neq m$ erwarten. Jedoch besitzt jede Lösung w von (1) das bekannte Transformationsverhalten für $A \in SL(2,\mathbb{R})$, mit $A = \begin{pmatrix} \cdot & \cdot \\ c & d \end{pmatrix}$,

$$w_A(z) := \frac{w(Az)}{(cz+d)^{n+1}(c\bar{z}+d)^{m+1}},$$

wobei $w_A(z)$ wieder eine Lösung von (1) bezeichnet. Gilt für eine Gruppe $\Gamma \subset SL(2,\mathbb{R})$ $w_A(z) = w(z)$ für alle $A \in \Gamma$, so könnte man derartige Funktionen als verallgemeinerte automorphe Formen ansehen. In [5] wurde die Existenz derartiger automorpher Formen gezeigt. Weiters ist es möglich, durch eine Erweiterung des Begriffs des singulären Eichlerintegrals die in [3] angegebenen Darstellungen auf diesen Fall zu übertragen und damit ein zu den dort erzielten Ergebnissen analoges Resultat zu erhalten. Das Anliegen dieses Aufsatzes ist, in Ergänzung zu [3] automorphe Formen, die Lösungen von (1) sind und eine logarithmische Singularität besitzen, durch geeignete Poincaré'sche Reihen darzustellen.

2. Allgemeine Darstellungssätze

Aus [4] übernehmen wir folgende Sätze zur Darstellung von Lösungen von (1) mittels Differentialoperatoren, soweit sie für unsere Betrachtungen von Interesse sind.

Satz 1

(i) Zu jeder in einem einfach zusammenhängenden Gebiet G der oberen Halbebene H definierten Lösung w von (1) gibt es zwei in G holomorphe Funktionen f, g, so daß

$$w = \frac{\partial^{n+m}}{\partial z^n \partial \bar{z}^m} \frac{f(z)+\overline{g(z)}}{z-\bar{z}}. \tag{4}$$

(ii) Umgekehrt stellt (4) für jedes Paar von in G holomorphen Funktionen eine Lösung von (1) dar.

Satz 2

Bezeichnet w eine in $\dot{U}(z_0) = \{z \mid 0 < |z-z_0| < r\} \subset H$ definierte Lösung von (1), so besitzt w eine Darstellung gemäß (4), nun aber mit den mehrdeutigen Erzeugenden

$$f(z) = f_1(z) + S(z)\log(z-z_0)$$
$$g(z) = g_1(z) + \overline{S(\bar{z})}\log(z-z_0),$$

wobei f_1 und g_1 in $\dot{U}(z_o)$ holomorphe Funktionen bezeichnen, und $S(z)$ aus der Menge π_{n+m} der Polynome vom Grad $\leq n+m$ ist.

Anmerkung: Die Polynome S sind bei Vorgabe einer Lösung w eindeutig bestimmt. Satz 1(ii) gilt analog für Satz 2.

3. Definition der automorphen Formen und Poincaré-Reihen

Wie bereits erwähnt, bezeichnet $\Gamma \subset SL(2,\mathbb{R})$ eine endlich erzeugte Fuchs'sche Gruppe erster Art. Damit geben wir die folgende

Definition

Die Funktion w heißt automorphe (m,n)-Form bezüglich der Gruppe Γ, wenn gilt:

(i) w(z) ist reell analytisch in H mit Ausnahme isolierter Singularitäten und erfüllt die Differentialgleichung (1).

(ii) $\dfrac{w(Az)}{(cz+d)^{n+1}(c\bar{z}+d)^{m+1}} = w(z)$ für alle $A \in \Gamma$, mit $A = \begin{pmatrix} \cdot & \cdot \\ c & d \end{pmatrix}$,

$m,n \in \mathbb{N}_o$ und alle $z \in H$.

Den Vektorraum dieser automorphen (m,n)-Formen bezeichnen wir mit $\mathfrak{A}_{m,n}(\Gamma)$.

Ist $w \in \mathfrak{A}_{m,n}(\Gamma)$, so erkennt man sofort, daß $(z-\bar{z})^{n+m+1}w \in \mathfrak{A}_{-n,-m}(\Gamma)$.

Wir wollen hier (m,n)-Formen mit logarithmischen Singularitäten durch geeignete Poincaré'sche Reihen darstellen.

Sei $S \in \pi_{n+m}$ und $\zeta \in H$, so definieren wir mit der Erzeugenden $S(z)\log\frac{z-\zeta}{z-\bar{\zeta}}$ die Funktion

$$(5) \qquad w(z;S;\zeta) = \frac{\partial^{n+m}}{\partial z^n \partial \bar{z}^m}\left[\frac{S(z)\log\frac{z-\zeta}{z-\bar{\zeta}} - S(\bar{z})\log\frac{\bar{z}-\zeta}{\bar{z}-\bar{\zeta}}}{z-\bar{z}}\right],$$

welche nach Satz 1 und 2 eine in $H-\{\zeta\}$ definierte Lösung von (1) ist. Man beachte, daß (5) eine eindeutige Funktion bezeichnet, obwohl die Erzeugenden mehrdeutige Funktionen sind.

Wie man leicht nachrechnet, ist diese Lösung auf der reellen Achse gleichmäßig beschränkt, und für $z \to \infty$ gilt $w(z;S;\zeta) = O(z^{-n-m-2})$. Mit Hilfe dieser Grundlösung (5) könnte man bereits durch Anwendung der Gruppenelemente und unter Beachtung des Transformationsverhaltens einer Lösung von (1) formal eine Poincaré-Reihe konstruieren. Bei der Untersuchung auf Konvergenz stößt man aber vorerst auf Schwierigkeiten. Daher ist es vorteilhafter, für den Fall $n \geq m$ auf Lösungen von

(2) bzw. für $n \leq m$ auf Lösungen der Differentialgleichung (3) zurückzugreifen. Diese ergeben sich aus (5) zu

$$(6) \qquad \begin{aligned} w_1(z;S;\zeta) &= (z-\bar{z})^{m+1} w(z;S;\zeta), \quad n \geq m \\ &\text{bzw.} \\ w_2(z;S;\zeta) &= (z-\bar{z})^{n+1} w(z;S;\zeta), \quad n \leq m. \end{aligned}$$

Die so erhaltenen Lösungen von (2) bzw. (3) haben die Eigenschaft, daß für alle $A \in SL(2,\mathbb{R})$

$$w_{1,A}(z;S;\zeta) := \frac{w_1(Az;S;\zeta)}{(cz+d)^{n-m}}$$

bzw.

$$w_{2,A}(z;S;\zeta) := \frac{w_2(Az;S;\zeta)}{(c\bar{z}+d)^{m-n}}$$

wieder eine Lösung von (2) bzw. (3) darstellt.
Bildet man nun mittels der Transformation $\frac{z-\zeta}{z-\bar{\zeta}} = u$ H konform auf den Einheitskreis ab und setzt

$$w_1(z;S;\zeta) = W_1(u)$$

bzw.

$$w_2(z;S;\zeta) = W_2(u),$$

so ergibt sich aus dem "Randverhalten" von $w(z;S;\zeta)$

$$W_1(u) = O(1-|u|^2)^{m+1} \quad \text{für} \quad |u| \to 1, \quad n \geq m,$$

bzw.

$$W_2(u) = O(1-|u|^2)^{n+1} \quad \text{für} \quad |u| \to 1, \quad n \leq m,$$

unabhängig von arg u.
Bilden wir nun die Poincaré-Reihen

$$\sum_{A\in\Gamma} w_{1,A}(z;S;\zeta) \quad \text{für} \quad n \geq m$$

bzw.

$$\sum_{A\in\Gamma} w_{2,A}(z;S;\zeta) \quad \text{für} \quad n \leq m,$$

so konvergieren diese absolut und auf jenen Kompakta gleichmäßig, die keinen zu ζ Γ-äquivalenten Punkt enthalten, wenn für jedes $\varepsilon > 0$ und $i = 1, 2$ gilt

$$W_i(u) = O(1-|u|^2)^{1+\varepsilon} \quad \text{für } |u| \to 1 \text{ unabh. von arg u.}$$

Diese Aussage ergibt sich in Analogie zu [6] S. 266-267, wenn man zu-

sätzlich beachtet, daß die Menge $\{x \in \mathbb{R} \mid x = |cz+d|^{-1},\ A = \begin{pmatrix} \cdot & \cdot \\ c & d \end{pmatrix} \in \Gamma\}$ auf kompakten Teilmengen von H beschränkt bleibt.
Wir können damit folgende Aussage beweisen.

Satz 3
Die Poincaré-Reihe zum Punkt $\zeta \in H$

$$P(z;S;\zeta) := \sum_{A\in\Gamma} \frac{w(Az;S;\zeta)}{(cz+d)^{n+1}(c\bar{z}+d)^{m+1}},$$

mit $w(z;S;\zeta)$ gemäß (5) und mit $m \geq 1$, $n \geq 1$, stellt eine Funktion aus $\mathfrak{U}_{m,n}(\Gamma)$ dar mit einer logarithmischen Singularität in ζ, sofern nicht $P(z;S;\zeta) \equiv 0$.

Daß P das richtige Transformationsverhalten hat, ist evident. Zur Konvergenzbetrachtung sei vorerst $n \geq m \geq 1$ angenommen. Wie wir gesehen haben, konvergiert die Reihe

$$\sum_{A\in\Gamma} w_{1,A}(z;S;\zeta).$$

Durch Multiplikation mit $(z-\bar{z})^{-m-1}$ ergibt sich die Aussage in diesem Fall. Ist andrerseits $m\geq n\geq 1$, so ergibt sich die entsprechende Aussage unter Benutzung der zweiten Reihe.
Anmerkung: Unter Verwendung dieser Reihen und der in [3] erzielten Ergebnisse bzw. deren Verallgemeinerung auf den Fall $n \neq m$ kann eine Darstellung aller Funktionen aus $\mathfrak{U}_{m,n}(\Gamma)$ erzielt werden.

Literatur

[1] Bauer, K. W. und E. Peschl
Ein allgemeiner Entwicklungssatz für die Lösungen der Differentialgleichungen $(1+\varepsilon z\bar{z})^2 w_{z\bar{z}} + \varepsilon n(n+1)w = 0$ in der Nähe isolierter Singularitäten. S.-ber. d. Bayer. Akad. d. Wiss., math.-naturw. Klasse, S. 113-146 (1965).

[2] Bers, L.
Eichler integrals with singularities. Acta Math. 127, 11-22 (1971).

[3] Haeseler, I. und St. Ruscheweyh
Singuläre Eichlerintegrale und verallgemeinerte Eisensteinreihen. Math. Ann. 203, 251-259 (1973).

[4] Jank, G.
Funktionentheoretische Untersuchungen von Lösungen gewisser elliptischer Differentialgleichungen. Topics on Differential Equations, Colloquia Mathematica Societatis Janos Bolyai, Vol. 13.

[5] Maaß, H.
Lectures on modular functions of one complex variable. Bombay, Tata Institute 1964.

[6] Roelcke, W.
Das Eigenwertproblem der automorphen Formen in der hyperbolischen Ebene II. Math. Ann. 168, 261-324 (1967).

Wie dem Verfasser erst nachträglich bekannt wurde, hat E. Meister (Dipl. Arbeit Univ. Heidelberg, 1956) nach einer Methode, wie sie in [5] dargestellt wird, allgemeine nicht analytische automorphe Formen durch Poincaré-Reihen dargestellt.

Gerhard Jank
Institut für Mathematik I
Technische Universität
Kopernikusgasse 24
A-8010 Graz

HOLOMORPHICITY OF SEMIGROUPS OF OPERATORS GENERATED BY SUBLAPLACIANS ON LIE GROUPS

Jan Kisyński

Institute of Mathematics, University of Warsaw, Poland

Let $\mathcal{G}$ be a connected Lie group with the neutral element e and with other elements denoted by x, y, z. Let $\hat{C}$ be the space of all real functions continuous on $\mathcal{G}$ with limit at infinity equal zero. The bounded Borel signed measures on $\mathcal{G}$ are linear functionals on $\hat{C}$ and, if μ and ν are two such measures, then their convolution $\mu * \nu$ is the measure such that $\int_{\mathcal{G}} \varphi(z)(\mu*\nu)(dz) = \int_{\mathcal{G}}\int_{\mathcal{G}} \varphi(xy)\mu(dx)\nu(dy)$ for every $\varphi \in \hat{C}$. The convolution semigroup of probability measures on $\mathcal{G}$ is a family $\{p_t : 0 \leqslant t < \infty\}$ of Borel probability measures on $\mathcal{G}$ such that p_0 = the unit mass concentrated at e, $p_{t+s} = p_t * p_s$ for every $t \geqslant 0$ and $s \geqslant 0$, and that p_t depends on t $*$-weakly continuously.

If p_t, $t \geqslant 0$, is a convolution semigroup of probability measures on $\mathcal{G}$ and if R is a strongly continuous representation of $\mathcal{G}$ by linear isometric automorphisms of a Banach space E, then the formula

$$(1) \qquad S(t)u = \int_{\mathcal{G}} R(x)u \, p_t(dx) \, , \quad u \in E \, ,$$

defines a strongly continuous one parameter semigroup $S(t)$, $t \geqslant 0$, of linear endomorphisms of E (see [2]). In particular, in the case of the representation of $\mathcal{G}$ by left translations in $\hat{C}$, we obtain the important semigroup $T(t)$, $t \geqslant 0$, of endomorphisms of $\hat{C}$, which act onto elements $u \in \hat{C}$ according to the rule

$$(2) \qquad (T(t)u)(x) = \int_{\mathcal{G}} u(y^{-1}x)\, p_t(dy) .$$

Evidently, the functional $u \to (T(t)u)(e)$ determines p_t .

It follows from the results of G.A.Hunt [4], and it was proved later by an other method by Palle E.T.Jørgensen [5], that if $X_0, X_1, \dots, X_n$ are right invariant vector fields on $\mathcal{G}$, then the differential operator

$$(3) \qquad P = \sum_{\nu=1}^{n} X_\nu^2 + X_0 ,$$

defined a priori on $C_0^\infty(\mathcal{G})$, has the closure in $\hat{C}$ which is the <u>infinitesimal generator</u> of a strongly continuous semigroup $T(t)$, $t \geqslant 0$, of endomorphisms of $\hat{C}$, connected with a convolution semigroup p_t , $t \geqslant 0$, of probability measures by means of the formula (2) . Let us mention that if $X_0 = 0$ then P is called a (right invariant) <u>sublaplacian</u> and, if $X_0 = 0$, $n = \dim \mathcal{G}$ and the fields $X_1, \dots, X_n$ are linearly independent, then P is called a <u>laplacian.</u>

Let R be a strongly continuous representation of $\mathcal{G}$ in a Banach space E and let E_∞ denote the set of all the C^∞ -vectors of this representation. The image of the differential operator P by means of the differential dR of the representation R is the linear endomorphism $dR(P)$ of E_∞ defined by the formula

$$dR(P)u = PR(x)u \Big|_{x=e} , \quad u \in E \quad .$$

It was proved by Jørgensen [5] that, <u>for any P of the form (3) and for any strongly continuous representation R of $\mathcal{G}$ in a Banach space E , the operator $dR(P)$ has the closure in E which is the infinitesimal generator of a strongly continuous semigroup $S(t)$, $t \geqslant 0$, of endomorphisms of E . Moreover the endomorphisms $S(t)$ have the form (1) , where p_t , $t \geqslant 0$, is the convolution semigroup</u>

<u>of probability measures such that the $\hat{C}$-closure of $P|C_0^\infty(\mathcal{G})$ is the infinitesimal generator of the corresponding semigroup $T(t)$, $t \geqslant 0$.</u>

Following L.Gårding [1], denote by $\tau_0(x)$ the geodesic distance from e to x in the sense of an arbitrarily fixed right invariant riemannian metric on $\mathcal{G}$. Then $\tau_0(x^{-1}) = \tau_0(x)$ and $\tau_0(xy) \leqslant \tau_0(x) + \tau_0(y)$. Moreover, for any strongly continuous representation R of $\mathcal{G}$ in a Banach space, there is a finite $\lambda \geqslant 0$ such that

$$(4) \qquad \sup\left\{e^{-\lambda\tau_0(x)}\|R(x)\| \; : \; x \in \mathcal{G}\right\} < \infty .$$

Let dx denote the element of a left invariant Haar measure on $\mathcal{G}$ and, for a fixed $s \geqslant 0$, denote by H the Hilbert space of all complex functions on $\mathcal{G}$ square integrable with respect to the measure $e^{-s\tau_0(x)}dx$. As Gårding proved ([1], p.75), there is a non negative constant λ_0 such that

$$(5) \qquad \int e^{-\lambda_0\tau_0(x)}\,dx < \infty ,$$

so that, if $\lambda \leqslant \frac{1}{2}(s - \lambda_0)$, then $e^{\lambda\tau_0} \in H$. Let $\mathcal{L}$ denote the representation of $\mathcal{G}$ by left translations in H. If $\lambda \geqslant 0$ and if s is choosen so large that $\lambda \leqslant \frac{1}{2}(s - \lambda_0)$, then, since $\lambda\tau_0(y) \leqslant \lambda\tau_0(x) + \lambda\tau_0(y^{-1}x)$, the Jørgensen theorem applied to $\mathcal{L}$ and H implies that $\left(\int e^{\lambda\tau_0(y)}\,p_t(dy)\right)^2 \int e^{-(s+2\lambda)\tau_0(x)}\,dx \leqslant \int\left(\int e^{\lambda\tau_0(y^{-1}x)}p_t(dy)\right)^2 e^{-s\tau_0(x)}dx = \|S(t)e^{\lambda\tau_0}\|_H^2 < \infty$. Thus, for a convolution semigroup p_t, $t \geqslant 0$, such that the corresponding semigroup $T(t)$ has the infinitesimal generator $\overline{P|C_0^\infty(\mathcal{G})}$ with P of the form (3), the integral $\int_{\mathcal{G}} e^{\lambda\tau_0(y)}\,p_t(dy)$ is finite for every $\lambda \geqslant 0$ and every $t \geqslant 0$. This last statement essentially goes back to E.Nelson ([9], lemma 8.1) and it is the crucial point in the proof of the mentioned theorem of Jørgensen.

In the present paper, mainly thanks to a study of the semigroup (1) for the case of the representation $\mathcal{L}$ in the space H, we shall complete the results of Jørgensen by the following

THEOREM 1. Let R be a strongly continuous representation of the Lie group $\mathcal{G}$ in a complex Banach space E. Let $X_0, X_1, \ldots, X_n$ be right--invariant vector fields on $\mathcal{G}$ and let P be the differential operator of the form (3). Then the operator $dR(P)$, defined on the set E_∞ of all C^∞-vectors of R, has the closure in E which, as we already know from the Jørgensen theorem, is the infinitesimal generator of a strongly continuous semigroup $S(t)$, $t \geqslant 0$, of endomorphisms of E. We claim that if

(6) $\quad X_0 \in \mathrm{lin}\,(X_1, \ldots, X_n)$

then the semigroup $S(t)$, $t \geqslant 0$, can be extended holomorphically into the open right half-plane $\mathrm{Re}\, t > 0$. Moreover, if E is a Hilbert space, then the holomorphic extension, which we still denote by $S(t)$, has the following property of sectorial strong continuity at $t = 0$:

(7) $\quad \lim\limits_{S_\alpha \ni t \to 0} \| S(t)u - u \| = 0$

for every $u \in E$ and every $\alpha \in (0, \frac{\pi}{2})$, where

$$S_\alpha = \{ t : t \in \mathbb{C},\ t = 0 \text{ or } |\mathrm{Arg}\, t| \leqslant \alpha \} .$$

Before we come up to the proof let us perform a preparation. Firstly, the weight function $e^{-s\tau_0}$ may not be sufficiently smooth for our purposes and therefore, following Hulanicki ([3], p.274--275) we shall replace τ_0 by $\tau = \varphi * \tau_0$, where $\varphi \in C_0^\infty(\mathcal{G})$

is non-negative and such that $\int \varphi(y)dy = 1$. Then, as easy to be seen,

(8) $\sup\{|\tau(x) - \tau_0(x)| : x \in \mathcal{G}\} < \infty$.

Moreover, for any set $X_1,\dots,X_m$ of right invariant vector fields on $\mathcal{G}$, we have

(9) $\sup\{|X_m \dots X_1 \tau(x)| : x \in \mathcal{G}\} < \infty$.

In order to prove (9), observe that $|\tau_0(xy) - \tau_0(y)| \leq \tau_0(x)$, from which it follows at once that τ_0 is a Lipchitz function and that, for any right-invariant vector field X, the derivative $X\tau_0$ exists almost everywhere on $\mathcal{G}$ and is a function essentially bounded on $\mathcal{G}$. Recall the theorem of Rademacher [10] (see [11], chapter IX, § 14 or [8], chapter VII, § 1) which states that a Lipschitz function on an open subset Ω of R^n has the total differential at almost every point of Ω. Let $Y_1,\dots,Y_d$ be a linear basis in the set of all right-invariant vector fields on $\mathcal{G}$. Obviously, it is sufficient to prove (9) for $X_1,\dots,X_m = Y_{\nu_1},\dots,Y_{\nu_m}$, where $\nu_k = 1,\dots,d$. Following Hulanicki ([3], p.275) we write

$$(Y_{\nu_1}\tau)(x) = Y_{\nu_1}\int \varphi(y)\tau_0(y^{-1}x)dx = \lim_{t\to 0}\int \varphi(y)\tfrac{1}{t}\left(\tau_0(y^{-1}e^{tY_{\nu_1}}x) - \tau_0(y^{-1}x)\right)dy$$
$$= \lim_{t\to 0}\int \varphi(y)\tfrac{1}{t}\left(\tau_0(e^{tAd(y^{-1})Y_{\nu_1}}y^{-1}x) - \tau_0(y^{-1}x)\right)dy,$$

from which, applying the theorem of Rademacher and the Lebesgue bounded convergence theorem, we conclude that

$$Y_{\nu_1}\tau = \sum_{\mu=1}^{d}\varphi^{\mu}_{\nu_1} * Y_\mu \tau_0$$

where $\varphi^{\mu}_{\nu_1}(y) = a^{\mu}_{\nu_1}(y^{-1})\varphi(y)$, a^{μ}_{ν} being the matrix elements of the adjoint representation defined by $Ad(y)Y_\nu = \sum_{\mu=1}^{d} a^{\mu}_{\nu}(y)Y_\mu$. Since a^{μ}_{ν} are real-analytic functions on $\mathcal{G}$, it follows that $\varphi^{\mu}_{\nu_1} \in C^{\infty}_{0}(\mathcal{G})$, so that the subsequent derivatives $Y_{\nu_2},\dots,Y_{\nu_m}$ of $Y_{\nu_1}\tau$ can be calculated by means of the $\varphi^{\mu}_{\nu_1}$, namely $Y_{\nu_m}\dots Y_{\nu_1}\tau = \sum_{\mu=1}^{d}(Y_{\nu_m}\dots Y_{\nu_2}\varphi^{\mu}_{\nu_1}) * Y_\mu \tau_0$.

Consequently $\sup|Y_{\nu_m}\cdots Y_{\nu_1}\tau| \leqslant \sum_{\mu}\int|Y_{\nu_m}\cdots Y_{\nu_2}\varphi^{\mu}_{\nu_1}(y)|dy\cdot \operatorname{ess\,sup}|Y_{\mu}\tau_0|$

and so (9) is proved.

The second point of our preparation for the proof of theorem 1 is to recall how the infinitesimal generators of some holomorphic semigroups in Hilbert spaces can be defined by means of bilinear forms. Following Lions [6], consider a pair V, H of Hilbert spaces such that $V \subset H$, V is dense in H and the embedding is continuous. If u and v are in V , then denote their scalar product in V by $((u,v))$ and put $|||u||| = ((u,u))^{\frac{1}{2}}$. If u and v are in H , then denote their scalar product in H by (u,v) and put $\|u\| = (u,u)^{\frac{1}{2}}$. Let $a(u,v)$ be a bilinear (i.e. linear in u and antilinear in v) complex form continuous on $V \times V$ and define the linear operator A with domain $\mathcal{D}(A)$ by the conditions

(10) $\mathcal{D}(A) = \{u : u \in V$, the functional $V \ni v \longrightarrow a(u,v) \in \mathbb{C}$ is continuous on V with respect to the topology of H $\}$,

(11) $(Au,v) = a(u,v)$ for every $u \in \mathcal{D}(A)$ and every $v \in V$.

For the operators A defined in such a way we have the following

THEOREM 2. If there are finite constants $\alpha > 0$, λ_0 and $K \geqslant 0$, such that

(12) $$-\operatorname{Re} a(u,u) + \lambda_0 \|u\|^2 \geqslant \alpha |||u|||^2$$

and

(13) $$|\operatorname{Re} a(u,v) - \operatorname{Re} a(v,u)| \leqslant K |||u||| \cdot \|v\|$$

for every $u \in V$ and every $v \in V$, then the operator A is the infinitesimal generator of a strongly continuous semigroup $S(t)$, $t \geqslant 0$, of endomorphisms of H . This semigroup can be extended holomorphically into the open right half-plane $\mathrm{Re}\, t>0$ and the extension, still denoted by $S(t)$, has the property (7) of the sectorial strong continuity at $t = 0$.

The theorem 2 was presented by the author in his lectures at Autumn Mathematical Course on Control Theory and Topics in Functional Analysis, 1974, in International Centre for Theoretical Physics, Trieste, Italy. The proof runs as follows. Consider the direct sum $V \oplus H$ and write its elements as columns $\binom{v}{u}$, $v \in V$, $u \in H$. Then it is essentially the result of Lions [7] , that under assumptions of theorem 2 the operator $\mathcal{A} = \begin{pmatrix} 0 & 1 \\ A & 0 \end{pmatrix}$ with the domain $\mathcal{D}(\mathcal{A}) = \mathcal{D}(A) \oplus \oplus V$ is the infinitesimal generator of an one-parameter strongly continuous group $G(t) = \begin{pmatrix} G_{11}(t) & G_{12}(t) \\ G_{21}(t) & G_{22}(t) \end{pmatrix}$, $t \in R^1$, of automorphisms of $V \oplus H$. Afterwards it is proved that the operators $S(0) = \mathrm{id}$ and $S(t) = \frac{1}{2\sqrt{\pi t}} \int_{-\infty}^{\infty} e^{-\frac{\sigma^2}{4t}} G_{22}(\sigma) d\sigma$, $t>0$, constitute the one-parameter semigroup with the infinitesimal generator equal to A. The holomorphic extendability is deduced directly from the formula expressing $S(t)$ by $G_{22}(\sigma)$.

PROOF OF THE THEOREM 1. THE CASE OF LEFT TRANSLATIONS IN THE SPACE H . Let, as before, $\mathcal{L}$ denote the representation of $\mathcal{G}$ by left translations in the Hilbert space H of complex functions square integrable on $\mathcal{G}$ with respect to the measure $e^{-s\,\tau_0(x)}\,dx$. Choose the "left" version of the distribution theory on $\mathcal{G}$, i.e. such that the embedding of the set of locally integrable functions into the set of distributions is realized by associating to any locally integrable f the measure $f(x)dx$. In this formalism the set H_∞ of all the

C^∞ -vectors of the representation $\mathcal{L}$ consists of all the elements u of H such that the distribution Qu belongs to H for every right invariant differential operator Q on $\mathcal{G}$. By the Jørgensen theorem the operator $\mathcal{A} = \overline{P|_{H_\infty}}$, where the closure is taken in H , is the infinitesimal generator of a strongly continuous semigroup $T(t)$, $t \geqslant 0$, of endomorphisms of H of the form

$$(14) \qquad (T(t)u)(x) = \int_{\mathcal{G}} u(y^{-1}x)p_t(dy) \ , \quad u \in H \ ,$$

where p_t , $t \geqslant 0$, is a convolution semigroup of probability measures on $\mathcal{G}$.

Denote by V the set of all the functions u in H such that their distributional derivatives $X_1u,\dots,X_2u$ again belong to H . Let τ be a C^∞ -function on $\mathcal{G}$ satisfying the conditions (8) and (9) and introduce in H and V the scalar products

$$(u,v) = \int_{\mathcal{G}} u(x)\overline{v(x)}\, e^{-s\tau(x)}dx \quad \text{for} \quad u,v \in H$$

and

$$((u,v)) = \int_{\mathcal{G}} \Big(u(x)\overline{v(x)} + \sum_{\nu=1}^{n} X_\nu u(x)\, X_\nu \overline{v(x)}\Big) e^{-s\tau(x)}dx \ \text{for} \ u,v \in V \ .$$

Consider the bilinear form $a(u,v)$ on $V \times V$ defined by the formula

$$(15) \qquad a(u,v) = -\sum_{\nu=1}^{n} \int_{\mathcal{G}} X_\nu u \, X_\nu(\overline{v}e^{-s\tau})dx + \int_{\mathcal{G}} (X_0 u)\overline{v}e^{-s\tau}\, dx$$

and let A be the corresponding operator defined by the conditions (10) and (11). From the definition of the distributional derivation, by some elementary reasonings, it follows that

$$(16) \qquad \mathcal{D}(A) = \Big\{ u : u \in V, \text{ the distribution } \sum_{\nu=1}^{n} X_\nu^2 u + X_0 u \text{ belongs to } H \Big\},$$

$$Au = \sum_{\nu=1}^{n} X_\nu^2 u + X_0 u \quad \text{for} \quad u \in \mathcal{D}(A) \ .$$

We shall show that the form (15) satisfies the conditions (12) and (13). If $u \in C_0^\infty(\mathcal{G})$, then $2\operatorname{Re} a(u,u) = -2\sum_{\nu=1}^{n}\int_{\mathcal{G}} |X_\nu u|^2 e^{-s\tau}dx - \sum_{\nu=1}^{n}\int_{\mathcal{G}} X_\nu |u|^2 X_\nu e^{-s\tau}dx$

$+\int_{\mathcal{G}}(X_0|u|^2)e^{-s\tau}dx = -2((u,u)) + \int_{\mathcal{G}}|u|^2\left(\sum_{\nu=1}^{n}X_\nu^2 - X_0 + 2\right)e^{-s\tau}dx$, so that

(17) $\quad -\operatorname{Re} a(u,u) + \lambda_s \|u\|^2 \geq |||u|||^2$

with $\lambda_s = \frac{1}{2}\sup\left\{ |e^{s\tau}\left(\sum_{\nu=1}^{n} X_\nu^2 - X_0 + 2\right)e^{-s\tau}| : x \in \mathcal{G}\right\}$. Since $C_0^\infty(\mathcal{G})$ is a densesubset of the Hilbert space V and since the bilinear form (15) is continuous on $V \times V$, the inequality (17) is still valid for every $u \in V$. Thus the condition (12) is verified. It remains to verify (13). If $u \in C_0^\infty(\mathcal{G})$ and $v \in C_0^\infty(\mathcal{G})$ then $a(u,v) = -((u,v)) + (u,v) + s\sum_{\nu=1}^{n}\int_{\mathcal{G}}(X_\nu u)(X_\nu \tau)\bar{v}e^{-s\tau}dx + \int_{\mathcal{G}}(X_0 u)\bar{v}e^{-s\tau}dx$
$= -((u,v)) + (u,v) - s\sum_{\nu=1}^{n}\int_{\mathcal{G}} u(X_\nu \bar{v})(X_\nu \tau)e^{-s\tau}dx - \int_{\mathcal{G}} u X_0\bar{v}e^{-s\tau}dx +$
$\int_{\mathcal{G}} u\bar{v}\left(\sum_{\nu=1}^{n}X_\nu^2 - X_0\right)e^{-s\tau}dx$, so that $\operatorname{Re} a(u,v) - \operatorname{Re} a(v,u) =$
$2\operatorname{Re}\int_{\mathcal{G}}\left(X_0 u + s\sum_{\nu=1}^{n}X_\nu\tau X_\nu u\right)\bar{v}e^{-s\tau}dx - \int_{\mathcal{G}}\bar{u}v\left(\sum_{\nu=1}^{n}X_\nu^2 - X_0\right)e^{-s\tau}dx$
and consequently

(18) $\quad |\operatorname{Re} a(u,v) - \operatorname{Re} a(v,u)| \leq \left(M_s\|u\| + 2\|X_0 u\| + 2sN_s\sum_{\nu=1}^{n}\|X_\nu u\|\right)\|v\|$,

where $M_s = \sup|e^{s\tau}\left(\sum_{\nu=1}^{n}X_\nu^2 - X_0\right)e^{-s\tau}|$ and $N_s = \sup\{|X_\nu\tau| : x \in \mathcal{G}$, $\nu = 1,\dots,n\}$. Since $X_0 \in \operatorname{lin}(X_1,\dots,X_n)$, both the parts of (18) are functions of (u,v) continuous on $V \times V$ with respect to the topology of V , so that, since $C_0^\infty(\mathcal{G})$ is dense in V , the inequality (18) is still valid for arbitrary $u \in V$ and $v \in V$. Thus we see that the condition (13) is satisfied. Consequently, by the theorem 2 the operator A is the infinitesimal generator of a strongly continuous semigroup of endomorphisms of H , holomorphic on $\operatorname{Re} t > 0$ and sectorially continuous at $t = 0$. It follows from (16) that $P|_{H_\infty} \subset A$ and so, since A as an infinitesimal generator is closed, $\mathcal{A} \subset A$. But we already know that both the operators, $\mathcal{A}$ and A , are infinitesimal generators of strongly continuous semigroups and so the inclusion $\mathcal{A} \subset A$ implies the equality $\mathcal{A} = A$. This completes the proof of theorem 1 for the case of the representation $\mathcal{L}$ in the space H .

THE CASE OF AN ARBITRARY REPRESENTATION IN A HILBERT SPACE. Let R be a strongly continuous representation of $\mathcal{G}$ in a Hilbert space E and let p_t , $t \geq 0$, be the convolution semigroup corresponding to the operator $P = \sum_{\nu=1}^{n} X_\nu^2 + X_0$. According to the Jørgensen theorem, the formula (1) defines a strongly continuous semigroup $S(t)$, $t \geq 0$, of endomorphisms of E , the infinitesimal generator of which is $\overline{dR(P)}$. Let λ be so large that (4) holds and fix an $s \geq \lambda_0 + 2\lambda$. Then, by (5), $(v,R^{-1}(\cdot)u) \in H$ for every $u,v \in E$. Moreover, from (1) and (14) we see that, for any $u,v \in E$,

$$(19) \qquad (v,R(x^{-1})S(t)u) = \left[T(t)(v,R^{-1}(\cdot)u)\right](x)$$

for almost every $x \in \mathcal{G}$. The tensor product $E \otimes H$ is the Hilbert space of all the E-valued functions on $\mathcal{G}$ Bochner measurable and square integrable with respect to the measure $e^{-s\tau_0(x)}\,dx$. Consider the operators $\mathcal{J} : E \ni u \longrightarrow R^{-1}(\cdot)u \in E \otimes H$ and $\mathcal{A} : E \otimes H \ni f \longrightarrow (\int_{\mathcal{G}} e^{-s\tau_0(x)}\,dx)^{-1} \int_{\mathcal{G}} e^{-s\tau_0(x)} R(x)f(x)dx \in E$. We then have $\mathcal{A}\mathcal{J} = id_E$. Let $\{e_i : i \in I\}$ be an orthonormal basis in E and for any $i \in I$ define the operator $P_i \in \mathcal{L}(E \otimes H; H)$ by the condition that $(P_i f)(x) = (e_i, f(x))$ almost everywhere on $\mathcal{G}$, for every $f \in E \otimes H$. Substituting e_i in the place of v in (21) we obtain that $P_i \mathcal{J} S(t)u = T(t)P_i \mathcal{J} u$, so that

$$(20) \qquad S(t)u = \mathcal{A}\left[\bigoplus_{i \in I_u} e_i \otimes T(t)P_i \mathcal{J} u\right] ,$$

where I_u is a countable subset of I such that $\{R(x)u : x \in \mathcal{G}\} \subset \operatorname{lin}\{e_i : i \in I_u\}$. As we already proved, the semigroup $T(t), t \geq 0$, can be extended holomorphically into $\operatorname{Re} t > 0$ and the extension, which we still denote by $T(t)$, is sectorially strongly continuous at $t = 0$. The former implies that for any $\alpha \in (0,\frac{\pi}{2})$ there are constants $K_\alpha \geqslant 1$ and c_α such that $\|T(t)\| \leqslant K_\alpha e^{c_\alpha |t|}$ for $t \in S_\alpha$. Since, in $E \otimes H$, $\mathcal{J}u = \bigoplus_{i \in I_u} e_i \otimes P_i \mathcal{J}u$ and so $\|\mathcal{J}u\|^2_{E\otimes H} = \sum_{i\in I_u} \|P_i \mathcal{J}u\|^2_H$, we conclude that $\sum_{i\in I_u} \|T(t)P_i\mathcal{J}u\|^2_H \leqslant K_\alpha^2 e^{2c_\alpha|t|}\|\mathcal{J}u\|^2_{E\otimes H}$ for $t \in S_\alpha$. Consequently

$$\bigoplus_{i\in I_u} e_i \otimes T(t)P_i \mathcal{J}u \tag{21}$$

is a series of $E\otimes H$-valued functions of t defined on $\{0\} \cup \{\mathrm{Re}\ t > 0\}$ strongly convergent almost uniformly in every sector S_α, $\alpha \in (0,\frac{\pi}{2})$. Therefore, since the functions $e_i \otimes T(t)P_i\mathcal{J}u$ are holomorphic in $\mathrm{Re}\ t > 0$ and are strongly continuous in every sector S_α, $\alpha \in (0,\frac{\pi}{2})$, it is the same with the sum of the series (21). Now, it is evident that the right member of (20) represents the holomorphic extension of $S(t)u$, $t \geqslant 0$, into $\mathrm{Re}\ t > 0$. It is also evident that the extension is sectorially strongly continuous at $t = 0$.

THE CASE OF A REPRESENTATION IN A BANACH SPACE. The linear set $\mathfrak{g}$ of all right invariant vector fields on $\mathcal{G}$ with the Lie bracket defined as the commutator of fields is the Lie algebra of $\mathcal{G}$. Let $\mathfrak{g}_0$ be its Lie subalgebra generated by the fields $X_0, X_1, \ldots, X_n$ and let $\mathcal{G}_0$ be the corresponding Lie subgroup of $\mathcal{G}$. Let p_t, $t \geqslant 0$, be the convolution semigroup corresponding to the differential operator $p = \sum_{\nu=1}^{n} X_\nu^2 + X_0$. Then, as proved by Jørgensen ([5], theorem 3.1), the probabilistic measures p_t have their supports in $\mathcal{G}_0$. Consider the representation $\mathcal{L}_0$ of $\mathcal{G}_0$ by left translations in the space H^0 of all complex functions on $\mathcal{G}_0$ square integrable on $\mathcal{G}_0$ with respect to the measure $e^{-s\,\tau_0(x)}dx$. The nonnegative con -

stant s is choosen similarly as in the preceding part of the proof. By the first case of the theorem 1 (applied to $\mathcal{G}_0$, H^0 and $\mathcal{L}_0$ instead of $\mathcal{G}$, H and $\mathcal{L}$) already proved we conclude that the semigroup $T_0(t)$, $t \geqslant 0$, of endomorphisms of H^0 defined by the formula

$$(22) \qquad (T_0(t)u)(x) = \int_{\mathcal{G}_0} u(y^{-1}x)p_t(dx), \; u \in H^0, \; x \in \mathcal{G}_0 ,$$

can be extended holomorphically into $\mathrm{Re}\, t > 0$. We denote the extension still by $T_0(t)$. The infinitesimal generator G of the semigroup (22) is the closure in H^0 of $P/_{H^0_\infty}$, where H^0_∞ denotes the set of all C^∞-vectors of the representation $\mathcal{L}_0$. The holomorphicity of the semigroup (22) implies that for any complex t with $\mathrm{Re}\, t > 0$ and for any $u \in H^0$ we have $T_0(t)u \in \bigcap_{m=1}^{\infty} \mathcal{D}(G^m)$. Moreover, if $\bigcap_{m=1}^{\infty} \mathcal{D}(G^m)$ is treated as a Fréchet space with the topology given by the system of norms $\|u\|_k = \|u\|_E + \|Gu\|_E + \ldots + \|G^k u\|_E$, $k = 1,2,\ldots$, then $T(t)$ appears as a holomorphic function of t on $\mathrm{Re}\, t > 0$ with its values in $\mathcal{L}(E; \bigcap_{m=1}^{\infty} \mathcal{D}(G^m))$. But, as follows immediately from a theorem by Jørgensen ([5], theorem 2.1), $\bigcap_{m=1}^{\infty} \mathcal{D}(G^m) = H^0_\infty$ and the natural Fréchet space topologies in both these sets coincide. Since H^0_∞ consist of all the functions $u \in H^0$ such that the distribution Qu is also in H^0 for every right invariant differential operator Q, the Sobolev embedding lemma implies that any $u \in H^0_\infty$ is (equivalent to) a C^∞-function on $\mathcal{G}_0$. Thus, for any $u \in H^0_\infty$, the mapping $\{\mathrm{Re}\, t > 0\} \ni t \longrightarrow T_0(t)u \in C^\infty(\mathcal{G}_0)$ is well defined and holomorphic. Similarly to (19), for the semigroup (1) and for any $u \in E$ and $f \in E'$ we have $\langle f, R^{-1}(\,\cdot\,)u \rangle \in H^0$ and

$$(23) \quad \langle f, R(x^{-1})S(t)u \rangle = \int_{\mathcal{G}_0} \langle f, R^{-1}(y^{-1}x)u \rangle p_t(dy) = [T_0(t)\langle f, R^{-1}(\cdot)u \rangle](x), \; x \in \mathcal{G}_0 .$$

Let $Ev_e : C^\infty(\mathcal{G}_0) \to \mathbb{C}$ be the operator of evaluation at the neutral element e of $\mathcal{G}_0$. We see from the formula (23) that, for $t>0$ and for any $u\in E$ and any $f\in E'$,

$$(24)\qquad \langle f, S(t)u\rangle = Ev_e\, T_0(t)\langle f, R^{-1}(\,\cdot\,)u\rangle \ .$$

But $\langle f, R^{-1}(\,\cdot\,)u\rangle \in H^0$ and so, the mapping $\{\mathrm{Re}\, t>0\}\ni t \longrightarrow T_0(t)\langle f, R^{-1}(\,\cdot\,)u\rangle \in C^\infty(\mathcal{G}_0)$ is strongly holomorphic. Thus it is clear that the right member of (24) represents the holomorphic extension, say $h(t;f,u)$ of the function $[0,\infty)\ni t\longrightarrow\langle f, S(t)u\rangle\in\mathbb{C}$ into $\mathrm{Re}\, t>0$.

Let now $u\in E_\infty$, $t_0>0$, and consider the Taylor series

$$(25)\qquad S(t_0)u + \sum_{k=1}^{\infty} \frac{h^k}{k!} \frac{d^k}{dt^k}\Big|_{t=t_0} S(t)u \ .$$

Since $E_\infty\subset\bigcap_{m=1}^{\infty}\mathcal{D}(\overline{dR(P)/_{E_\infty}}^{\,m})$, the coefficients $\frac{d^k}{dt^k}\Big|_{t=t_0} S(t)u$ are well defined elements of E. So, if $u\in E_\infty$, $t_0>0$ and $f\in E'$, then, by the established above holomorphic extendability of $t\rightarrow\langle f, S(t)u\rangle$, the power series

$$\langle f, S_0(t)u\rangle + \sum_{k=1}^{\infty}\frac{h^k}{k!}\Big\langle f, \frac{d^k}{dt^k}\Big|_{t=t_0} S(t)u\Big\rangle$$

has the convergence radius not less then t_0. Consequently, if $0<\tilde{h}<t_0$ and $u\in E_\infty$, then $\lim_{k\to\infty}\frac{\tilde{h}^k}{k!}\langle f, \frac{d^k}{dt^k}\big|_{t=t_0} S(t)u\rangle = 0$ for every $f\in E'$. But this implies that $\frac{\tilde{h}^k}{k!}\frac{d^k}{dt^k}\Big|_{t=t_0} S(t)u$, $k=1,2,\ldots$, is a bounded sequence of elements of E and so, if $|h|<\tilde{h}$, then the series (25) is strongly convergent in E. Consequently, if $u\in E_\infty$, then, for any $t_0>0$ the series (25) is strongly convergent in the open disc $|h|<t_0$, so that the mapping $[0,\infty)\ni t\rightarrow S(t)u\in E$ has strongly holomorphic extension into $\mathrm{Re}\, t>0$. Let us denote this extension still by $S(t)u$.

It remains to consider the case of an arbitrary $u\in E$. Before doing that, observe that the above discussed holomorphicity properties of the semigroup $T_0(t)$ imply the following. For any compact subset $\mathcal{K}$ of $\operatorname{Re} t>0$ there is a constant $C(\mathcal{K})$ such that $|Ev_e\, T_0(t)\varphi| \leqslant C(\mathcal{K})\, \|\varphi\|_{H^0}$ for every $\varphi \in H^0$ and $t \in \mathcal{K}$. Consequently, for $t \in \mathcal{K}$,

$$(26) \quad |h(t;f,u)| \leqslant C(\mathcal{K}) \| \langle f, \bar{R}^{1}(\cdot)u\rangle \|_{H^0} \leqslant D_s^{\frac{1}{2}} C(\mathcal{K}) \|f\|_{E'} \|u\|_E ,$$

where $D_s = \int_{\mathcal{G}_0} e^{(2\lambda - s)\tau_0(x)}dx < \infty$.

Let now u be an arbitrary element of E and take a sequence u_p, $p = 1,2,\ldots$, of elements of E_∞ such that $\|u_p - u\|_E \to 0$. For any compact subset $\mathcal{K}$ of $\operatorname{Re} t>0$ we have, by (26),

$$\sup_{t\in\mathcal{K}} \|S(t)u_p - S(t)u_q\|_E = \sup\{|h(t;f,u_p-u_q)| : f \in E', \|f\|=1,\ t\in\mathcal{K}\}$$
$$\leqslant D_s^{\frac{1}{2}} C(\mathcal{K})\, \|u_p - u_q\|_E .$$

Therefore the sequence $S(t)u_p$, $p = 1,2,\ldots$, of E-valued functions of t, holomorphic in $\operatorname{Re} t>0$, is strongly convergent almost uniformly in $\operatorname{Re} t>0$. Clearly, the limit of this sequence is an E-valued function holomorphic in $\operatorname{Re} t>0$ and equal $S(t)u$ for real $t>0$. This completes the proof.

Bibliography

[1] L.Gårding, Vecteurs analytiques dans les représentations des groupes de Lie, Bull.Soc.Math.France 88(1960), p.73-93.

[2] U.Grenander, Probabilities on algebraic structures, 1968.

[3] A.Hulanicki, Subalgebra of $L_1(G)$ associated with laplacian on a Lie group, Colloquium Math. XXX, 2(1974), p.259-287.

[4] G.A.Hunt, Semi-groups of measures on Lie groups, Trans.Amer.Math. Soc., 81(1956), p.264-293.

[5] Palle E.T.Jørgensen, Representations of differential operators on

a Lie group, Journal of Functional Analysis 20(1975), p.105-135.

[6] J.L.Lions, Équations différentielles opérationnelles et problèmes aux limites, 1961.

[7] J.L.Lions, Une remarque sur les applications du théorème de Hille-Yosida, J.Math.Soc.Japan 9(1957), p.62-70.

[8] S.Łojasiewicz, An introduction to theory of real functions (in polish), Warszawa 1973.

[9] E.Nelson, Analytic vectors, Annals of Mathematatis, Vol.70, No 3, November 1959, p.572-615.

[10] H.Rademacher, Über partielle und totale Differenzierbarkeit I and II, Math.Annalen 79(1919), p.370-359 and 81)1920), p.52-63.

[11] S.Saks, Theory of the Integral .

ITERATIVE SOLUTIONS OF BOUNDARY VALUE PROBLEMS*

R. E. Kleinman

University of Delaware

ABSTRACT

Recent work has shown that Neumann's method in potential theory could be extended to solve boundary value and transition problems for the Helmholtz equation. The procedure consists of formulating the problem as a boundary integral equation which is then rewritten, with the use of a homographic transformation of the associated eigenvalue equation, so that the spectral radius of the resulting integral operator is less than one for small perturbations of the corresponding potential operator. The present paper describes two extensions of this work. The transformation which maximizes the distance to the spectrum of the resolvent point of interest in potential theory (wave number equal to zero) is shown to be not optimal for non zero wave numbers. Thus while this transformation optimizes the rate of convergence of the Neumann series for zero wave number, this is not true in general. Some analytic and numerical examples are presented.

INTRODUCTION

C. Neumann's classical iterative method for solving boundary value problems in potential theory has recently been shown to apply to problems for the Helmholtz equation by Ahner and Kleinman [1973]. Deficiencies in that analysis were corrected by Kleinman and Wendland [1976] who proved that the exterior Neumann problem in R^3 could be solved using Neumann's method for a very general class of boundaries including piecewise Lyapunoff surfaces with no convexity restrictions.

---

* Research supported under AFOSR Grant 74-2634

This work generalized Wendland's [1968] treatment of the potential problem (see also Král [1967]) in which the problem is cast as a boundary integral equation and Plemelj's Theorem on the location and nature of the eigenvalues of the integral operator is used to transform the integral operator into one with spectral radius less than one. This approach is well known in potential theory for smooth convex boundaries [e.g. Goursat (1964, p. 195) Kantorovich and Krylov (1964, p. 118)]. The transformation of the integral operator in that case was observed by Goursat to be equivalent to a Möbius transformation of the eigenvalue parameter of the operator and that Neumann's method corresponded to one of an infinite number of such transformations.

The iteration scheme analysed by Kleinman and Wendland is exactly analogous to Neumann's original method however Kress and Roach [1976] have found that the rate of convergence is increased if a different iteration method is employed.

In the present paper it is shown that the Kress and Roach method corresponds to a different Möbius transformation than the standard Neumann iteration and that both are special cases of the generalized over relaxation iterative method of Petryshyn [1962] (see also Patterson [1974, p. 54]). Furthermore the Möbius transformation that minimizes the spectral radius of the transformed operator is derived and shown to correspond to Kress and Roach's method if all of the eigenvalues of the original operator are of the same sign. Actually the results are shown to apply even when the operator does not have a pure point spectrum.

Wendland and Kleinman pointed out that the method could be applied to other integral equations provided the analogue of Plemelj's theorem was available. Such extensions have been made in acoustic scattering by Kittappa and Kleinman [1975] and elasticity by Ahner and Hsiao [1975]. It is noted in the present paper that the method may also be extended to the exterior boundary value problem for Maxwell's equations with perfectly conducting boundaries.

PROBLEM STATEMENT

The essentials of the extended Neumann method are as follows. Let B be a Banach space with norm $\|\ \|$ and K_k an operator valued function of the complex parameter k which for fixed k is a bounded linear operator, mapping B into itself. One desires the solution $u \in B$ of the equation

(1) $$u = K_k u + g$$

where $g \in B$ is specified, or equivalently,

(2) $$u = \lambda K_k u + g$$

for $\lambda = 1$. Because of the way in which λ is introduced in (2) the spectral radius of K_k is

(3) $$r_\sigma(K_k) = \sup_{\lambda \varepsilon \sigma(K_k)} \left\{\frac{1}{|\lambda|}\right\}$$

where the spectrum of K_k, $\sigma(K_k)$, is the set of all points λ for which either $(I-\lambda K_k)^{-1}$ does not exist, is unbounded, or the range of $(I-\lambda K_k)$ is not dense in B. If $r_\sigma(K_k) < 1$, then the convergent Neumann series solution of (1) is

(4) $$u = \sum_{n=0}^{\infty} (K_k)^n g .$$

Furthermore, if

(5) $$r_\sigma(K_0) \leq \rho < 1$$

and

(6) $$\|K_k - K_0\| \to 0 \quad \text{as} \quad |k| \to 0$$

then it follows, (Kleinman and Wendland [1976], Taylor [1958, p. 256]), that for $|k|$ sufficiently small the Neumann series (4) still converges. The argument breaks down if $r_\sigma(K_k) \geq 1$ but one may attempt to salvage the method by introducing a new operator through a Möbius transformation of the eigenvalue parameter in (2), i.e., define

(7) $$\mu := \frac{a\lambda+b}{c\lambda+d} \Rightarrow \lambda = \frac{-d\mu+b}{c\mu-a}$$

substitute in (2) and simplify, obtaining,

(8) $$(aI+bK_k)u = \mu(cI+dK_k)u + g(a-c\mu) \ .$$

Choosing $b = 0$, or equivalently letting 0 be a fixed point of the transformation (6), has the effect of trivializing the otherwise appreciable task of constructing $(aI+bK_k)^{-1}$. With this choice (8) may be written

(9) $$u = \mu\left(\frac{c}{a} I + \frac{d}{a} K_k\right)u + \left(1 - \frac{c}{a}\mu\right) g \ .$$

Of interest is the solution of equation (9) when

(10) $$\mu = \mu(1) = \frac{a}{c+d}$$

which is critically dependent on the spectral radius of the operator

(11) $$L_k = \mu(1)\left(\frac{c}{a} I + \frac{d}{a} K_k\right) = \frac{1}{c+d}(cI+dK_k) \ .$$

Since L_k is independent of a, provided $a \neq 0$, no loss of generality results by choosing $a = 1$. Furthermore, aside from k, there is really only one other independent parameter in L_k (e.g. $\frac{c}{d}$) since, with $p := \frac{c}{c+d}$,

(11a) $$L_k = pI + (1-p)K_k$$

Equivalently one could have specified 1 as another fixed point of the Möbius transformation since

(12) $$\mu(1) = 1 \Rightarrow \frac{1}{c+d} = 1 \Rightarrow d = 1-c$$

and equation (11) becomes

(13) $$L_k = cI + (1-c)K_k.$$

Thus the question at issue is whether it is possible to choose the third coefficient of the Möbius transformation, (7), in such a way that the $r_\sigma(L_k) < 1$

even though $r_\sigma(K_k) \geq 1$.

FURTHER PROPERTIES OF THE TRANSFORMATION

First we observe that, with equation (13),

$$I-\lambda K_k = \left(1 + \frac{\lambda c}{1-c}\right)\left(I - \frac{\lambda}{\lambda c+1-c} L_k\right) \tag{14}$$

hence it follows that, for $c \neq 1$,

$$\lambda \in \sigma(K_k) \Leftrightarrow \frac{\lambda}{\lambda c+1-c} \in \sigma(L_k) \tag{15}$$

and thus

$$r_\sigma(L_k) = \sup_{\mu\in\sigma(L_k)} \left\{\frac{1}{|\mu|}\right\} = \sup_{\lambda\in\sigma(K_k)} \left\{\left|\frac{\lambda c+1-c}{\lambda}\right|\right\} . \tag{16}$$

Since the spectral radius is a function of the complex parameter c, define

$$\begin{aligned} \overline{r}(L_k) &= \inf_{c\in C} r_\sigma(L_k) = \inf_{c\in C} \sup_{\lambda\in\sigma(K_k)} \left\{\left|\frac{(1-c)}{\lambda} + c\right|\right\} \\ &= \inf_{c\in C} |1-c| \sup_{\lambda\in\sigma(K_k)} \left\{\left|\frac{1}{\lambda} - \frac{c}{c-1}\right|\right\} \end{aligned} \tag{17}$$

where C is the set of complex numbers. Thus $\overline{r}(L_k)$ is the smallest spectral radius that may be achieved through a Möbius transformation with the origin as a fixed point. It has been shown that 1 may also be a fixed point while the image of a third point, which would completely determine the transformation, involves finding that value of c for which $r_\sigma(L_k) = \overline{r}(L_k)$. This requires more information on the spectrum of K_k. In the important case where $\sigma(K_k)$ is real, the following results obtain. First define positive and negative bounds on the spectrum

$$\lambda^+ := \inf_{\substack{\lambda>0 \\ \lambda\in\sigma(K_k)}} \{\lambda\} , \quad \text{and} \quad \lambda^- := \sup_{\substack{\lambda<0 \\ \lambda\in\sigma(K_k)}} \{\lambda\} \tag{18}$$

Lemma 1: If $\sigma(K_k)$ is real then

$$\sup_{\lambda\in\sigma(K_k)} \left|\frac{1}{\lambda} - \frac{c}{c-1}\right| = \max\left\{ \left|\frac{1}{\lambda^+} - \frac{c}{c-1}\right| , \left|\frac{1}{\lambda^-} - \frac{c}{c-1}\right|\right\} \tag{19}$$

Proof.

$$\text{a)}\quad \lambda \geq \lambda^+ \geq 0 \Rightarrow 0 \leq \frac{1}{\lambda} \leq \frac{1}{\lambda^+} \Rightarrow \frac{1}{\lambda^-} \leq \frac{1}{\lambda} \leq \frac{1}{\lambda^+}$$

(20)

$$\text{b)}\quad \lambda \leq \lambda^- \leq 0 \Rightarrow \frac{1}{\lambda^-} \leq \frac{1}{\lambda} \leq 0 \Rightarrow \frac{1}{\lambda^-} \leq \frac{1}{\lambda} \leq \frac{1}{\lambda^+}$$

Thus

(21) $$\lambda \in \sigma(K_k) \Rightarrow \frac{1}{\lambda^-} \leq \frac{1}{\lambda} \leq \frac{1}{\lambda^+}$$

or, subtracting $\operatorname{Re} \frac{c}{c-1}$ from each term,

(22) $$\frac{1}{\lambda^-} - \operatorname{Re} \frac{c}{c-1} \leq \frac{1}{\lambda} - \operatorname{Re} \frac{c}{c-1} \leq \frac{1}{\lambda^+} - \operatorname{Re} \frac{c}{c-1} \quad .$$

Since for x, y, z real, $x \leq y \leq z \Rightarrow y^2 \leq \max(x^2, z^2)$ it follows that

(23) $$\left(\frac{1}{\lambda} - \operatorname{Re} \frac{c}{c-1}\right)^2 \leq \max\left\{\left(\frac{1}{\lambda^-} - \operatorname{Re} \frac{c}{c-1}\right)^2, \quad \left(\frac{1}{\lambda^+} - \operatorname{Re} \frac{c}{c-1}\right)^2\right\}$$

Adding $\left(\operatorname{Im} \frac{c}{c-1}\right)^2$ to both sides establishes the lemma.

Lemma 2:

(24) $$\text{a)}\quad \operatorname{Re} \frac{c}{c-1} \geq \frac{1}{2}\left(\frac{1}{\lambda^+} + \frac{1}{\lambda^-}\right) \Rightarrow \sup_{\lambda \in \sigma(K_k)} \left|\frac{1}{\lambda} - \frac{c}{c-1}\right| = \left|\frac{1}{\lambda^-} - \frac{c}{c-1}\right|$$

$$\text{b)}\quad \operatorname{Re} \frac{c}{c-1} \leq \frac{1}{2}\left(\frac{1}{\lambda^+} + \frac{1}{\lambda^-}\right) \Rightarrow \sup_{\lambda \in \sigma(K_k)} \left|\frac{1}{\lambda} - \frac{c}{c-1}\right| = \left|\frac{1}{\lambda^+} - \frac{c}{c-1}\right|$$

Proof.

(25) $$\operatorname{Re} \frac{c}{c-1} \geq \frac{1}{2}\left[\frac{1}{\lambda^+} + \frac{1}{\lambda^-}\right] \Rightarrow \frac{1}{\lambda^-} + \frac{1}{\lambda^+} - 2\operatorname{Re} \frac{c}{c-1} \leq 0$$

Since $\frac{1}{\lambda^-} - \frac{1}{\lambda^+} \leq 0$ it follows that

(26) $$\left(\frac{1}{\lambda^-} + \frac{1}{\lambda^+} - 2 \operatorname{Re} \frac{c}{c-1}\right)\left(\frac{1}{\lambda^-} - \frac{1}{\lambda^+}\right) \geq 0$$

or

(27) $$\left(\frac{1}{\lambda^-} - \operatorname{Re} \frac{c}{c-1}\right)^2 - \left(\frac{1}{\lambda^+} - \operatorname{Re} \frac{c}{c-1}\right)^2 \geq 0$$

Therefore

(28) $$\left(\frac{1}{\lambda^+} - \operatorname{Re} \frac{c}{c-1}\right)^2 \leq \left(\frac{1}{\lambda^-} - \operatorname{Re} \frac{c}{c-1}\right)^2$$

from which it follows that

(29) $$\left|\frac{1}{\lambda^+} - \frac{c}{c-1}\right| \leq \left|\frac{1}{\lambda^-} - \frac{c}{c-1}\right| .$$

With Lemma 1, a) follows. The proof of b) is exactly the same with the inequalities reversed. Although c thus far is a complex parameter, it is next shown that for present purposes it may be taken to be real.

Lemma 3: If $\sigma(K_k)$ is real then

(30) $$\inf_{c\varepsilon C}\ \sup_{\lambda\varepsilon\sigma(K_k)} \left\{\left|\frac{1-c}{\lambda} + c\right|\right\} = \inf_{c\varepsilon R}\ \sup_{\lambda\varepsilon\sigma(K_k)} \left\{\left|\frac{1-c}{\lambda} + c\right|\right\}$$

<u>Proof:</u> Assume $c = x+iy$. With Lemma 1, it follows that

(31) $$\sup_{\lambda\varepsilon\sigma(K_k)} \left\{\left|\frac{1-c}{\lambda} + c\right|\right\} = \max\left\{\left|\frac{1-c}{\lambda^+} + c\right|,\ \left|\frac{1-c}{\lambda^-} + c\right|\right\}$$

$$= \max\left\{\left[\left(\frac{1-x}{\lambda^+} + x\right)^2 + y^2\left(1 - \frac{1}{\lambda^+}\right)^2\right]^{1/2},\ \left[\left(\frac{1-x}{\lambda^-} + x\right)^2 + y^2\left(1 - \frac{1}{\lambda^-}\right)^2\right]^{\frac{1}{2}}\right\}$$

$$\geq \max\left\{\left|\frac{1-x}{\lambda^+} + x\right|,\ \left|\frac{1-x}{\lambda^-} + x\right|\right\} = \sup_{\lambda\varepsilon\sigma(K_k)} \left\{\left(\frac{1-x}{\lambda} + x\right)\right\}$$

from which the Lemma follows.

Define two real valued functions of the real variable c

(32) $$f^{\pm}(c) = \left|\frac{1-c}{\lambda^{\pm}} + c\right| = \left|c\left(1 - \frac{1}{\lambda^{\pm}}\right) + \frac{1}{\lambda^{\pm}}\right| .$$

Their graphs are shown in Figure 1. Observe that $f^{\pm}(c)$ is monotone increasing for $c > c^{\pm}$ and monotone decreasing for $c < c^{\pm}$ where

(33) $$f^+(c^+) = 0, \quad c^+ = \frac{1}{1-\lambda^+} \quad \begin{array}{ll} < 0 & \text{if } \lambda^+ > 1 \\ > 1 & \text{if } 0 < \lambda^+ \leq 1 \end{array}$$

and

(34) $$f^-(c^-) = 0, \quad c^- = \frac{1}{1-\lambda^-} \leq 1 .$$

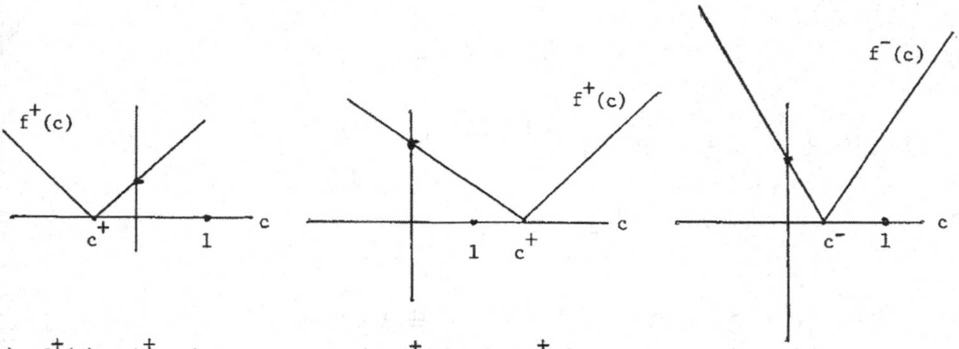

a) $f^+(c)$, $\lambda^+ > 1$ b) $f^+(c)$, $0 < \lambda^+ \leq 1$ c) $f^-(c)$

FIGURE 1

Introducing the shorthand

(35)
$$s^- = \left\{ c \,\middle|\, \frac{c}{c-1} \geq \frac{1}{2}\left(\frac{1}{\lambda^+} + \frac{1}{\lambda^-}\right)\right\}$$

and

(36)
$$s^+ = \left\{ c \,\middle|\, \frac{c}{c-1} \leq \frac{1}{2}\left(\frac{1}{\lambda^+} + \frac{1}{\lambda^-}\right)\right\}$$

the previous Lemmas may be condensed with equation (17) to yield

Lemma 4: If $\sigma(K_k)$ is real

(37)
$$\overline{r}(L_k) = \max\left\{ \inf_{c \varepsilon s^-} f^-(c), \quad \inf_{c \varepsilon s^+} f^+(c) \right\} .$$

Next it is necessary to establish some technical inequalities.

Lemma 5: Define

(38)
$$\xi := \frac{\frac{1}{\lambda^+} + \frac{1}{\lambda^-}}{\frac{1}{\lambda^+} + \frac{1}{\lambda^-} - 2}$$

a) If $\lambda^+ > 1$ <u>then</u>

(39)
$$\frac{1}{\lambda^+} + \frac{1}{\lambda^-} < 2$$

<u>and</u>

(40)
$$1 \geq \frac{1}{1-\lambda^-} > \xi > \frac{1}{1-\lambda^+}$$

b) If $0 < \lambda^+ \leq 1$ <u>and</u> $\frac{1}{\lambda^+} + \frac{1}{\lambda^-} < 2$ <u>then</u>

(41)
$$\frac{1}{1-\lambda^+} > 1 \geq \frac{1}{1-\lambda^-} > \xi$$

c) If $0 < \lambda^+ < 1$ and $\frac{1}{\lambda^+} + \frac{1}{\lambda^-} > 2$ <u>then</u>

(42)
$$\xi > \frac{1}{1-\lambda^+} > 1 \geq \frac{1}{1-\lambda^-}$$

<u>Proof:</u> To establish (39) note that

$$\lambda^+ > 1 \Rightarrow \frac{1}{\lambda^+} - 1 < 0 \quad \text{and} \quad \lambda^- \leq 0 \Rightarrow \frac{1}{\lambda^-} - 1 \leq 0$$

Adding these two inequalities yields the desired result. Next observe that

$$\lambda^- < \lambda^+ \Rightarrow 1 > \frac{\lambda^+}{\lambda^-} \Rightarrow 2 > 1 + \frac{\lambda^+}{\lambda^-} = \lambda^+\left(\frac{1}{\lambda^+} + \frac{1}{\lambda^-}\right).$$

Hence $-\lambda^+\left(\frac{1}{\lambda^+} + \frac{1}{\lambda^-}\right) > -2$

Adding $\frac{1}{\lambda^+} + \frac{1}{\lambda^-}$ to both sides yields

(43)
$$(1-\lambda^+)\left(\frac{1}{\lambda^+} + \frac{1}{\lambda^-}\right) > \frac{1}{\lambda^+} + \frac{1}{\lambda^-} - 2.$$

In a similar way it may be established that

(44)
$$(1-\lambda^-)\left(\frac{1}{\lambda^+} + \frac{1}{\lambda^-}\right) > \frac{1}{\lambda^+} + \frac{1}{\lambda^-} - 2 \quad .$$

If $\lambda^+ > 1$ then $1 - \lambda^+ < 0$, and with (39) it follows from (43) that

$$\xi > \frac{1}{1-\lambda^+} .$$

On the other hand $1 - \lambda^- > 0$ thus again using (39) one obtains with (44)

$$\xi < \frac{1}{1-\lambda^-} .$$

Furthermore $\lambda^- \leq 0 \Rightarrow 1 - \lambda^- \geq 1 \Rightarrow \frac{1}{1-\lambda^-} \leq 1$ which completes the proof of (40).

The previous inequality, $\frac{1}{1-\lambda^-} \leq 1$, is equally valid in b) and c). Similarly $0 < \lambda^+ \leq 1 \Rightarrow 1 - \lambda^+ < 1 \Rightarrow \frac{1}{1-\lambda^+} > 1$ which also holds for b) and c). To complete b), use the fact that $\frac{1}{\lambda^+} + \frac{1}{\lambda^-} - 2 < 0$ in (44) to yield $\xi < \frac{1}{1-\lambda^-}$. (Note that if $\lambda^+ = 1$ then $\frac{1}{\lambda^+} + \frac{1}{\lambda^-} - 2 < 0$). The proof of c) is concluded using the assumptions $0 < \lambda^+ < 1$ ($\Rightarrow 1 - \lambda^+ > 0$) and $\frac{1}{\lambda^+} + \frac{1}{\lambda^-} > 2$ in (39) to yield $\xi > \frac{1}{1-\lambda^+}$.

MAIN RESULTS

With these preliminary results it is possible to establish

Theorem 1. If $0 < \lambda^+ \leq 1$ there is no choice of c (no Möbius transformation of λ with 0 as a fixed point) for which $r_\sigma(L_k) < 1$.

Proof. Since $0 < \lambda^+ \leq 1$, if $\frac{1}{\lambda^+} + \frac{1}{\lambda^-} < 2$ (this must obtain if $\lambda^+ = 1$), Lemma 5b) applies and, with (35) and (36), it can be shown that

(45) $$s^- = \left\{c \mid 1 \geq \frac{1}{1-\lambda^-} > \xi > c\right\} \cup \left\{c \mid c > 1 \geq \frac{1}{1-\lambda^-} > \xi\right\}$$

whereas

(46) $$s^+ = \left\{c \mid \frac{1}{1-\lambda^+} > 1 > c \geq \xi\right\} .$$

As is clear from Figure 1

(47) $$\inf_{c \in s^-} f^-(c) = \min\left\{\inf_{1 \geq \frac{1}{1-\lambda^-} > \xi > c} f^-(c),\ \inf_{c > 1 \geq \frac{1}{1-\lambda^-} > \xi} f^-(c)\right\} = \min\{f^-(\xi), f^-(1)\}$$

while

(48) $$\inf_{c \in s^+} f^+(c) = \inf_{\frac{1}{1-\lambda^+} > 1 > c \geq \xi} f^+(c) = f^+(1) .$$

Since

(49) $$f^+(1) = f^-(1) = 1$$

it follows that

(50) $$\bar{r}(L_k) = \max\left\{\inf_{c\varepsilon s^-} f^-(c),\ \inf_{c\varepsilon s^+} f^+(c)\right\}$$

$$= \max\{\min(f^-(\xi),1),\ 1\} = 1 .$$

Similarly if $\frac{1}{\lambda^+} + \frac{1}{\lambda^-} > 2$, Lemma 5c applies hence

(51) $$s^- = \left\{c \mid \xi \geq c > 1 \geq \frac{1}{1-\lambda^-}\right\}$$

(52) $$s^+ = \left\{c \mid \xi > \frac{1}{1-\lambda^+} > 1 > c\right\} \cup \left\{c \mid c > \xi > \frac{1}{1-\lambda^+} > 1\right\}$$

Again using the monotonic character of f^+ and f^- it follows that

(53) $$\inf_{c\varepsilon s^-} f^-(c) = f^-(1)$$

and

(54) $$\inf_{c\varepsilon s^+} f^+(c) = \min(f^+(1),\ f^+(\xi))$$

and, as before,

(55) $$\bar{r}(L_k) = 1 .$$

If $\frac{1}{\lambda^+} + \frac{1}{\lambda^-} = 2$ then $s^- = \{c \mid c > 1\}$, $s^+ = \{c \mid c < 1\}$ and again $\bar{r}(L_k) = 1$.

Since $\bar{r}(L_k) \leq r_\sigma(L_k)$ the theorem is proven.

It is easily seen that the transformation (7) with $a = 1$, $b = 0$, $d = 1-c$ maps the point $\lambda = \frac{c-1}{c+1}$ into -1. Furthermore if $\lambda^+ > 0$ and

(56) $$\lambda^- < \frac{c-1}{c+1} < 0$$

the line segment $(\frac{c-1}{c+1}, 1)$ in the λ-plane is mapped into $(-1,1)$ in the μ-plane and these line segments contain no points of the spectrum. This is sufficient to establish

Theorem 2. If $\sigma(K_k)$ is real, $\lambda^+ > 1$ and $\frac{1+\lambda^-}{1-\lambda^-} < c < 1$ (this is equivalent to (56)) then $r_\sigma(L_k) < 1$, where L_k is defined in (13).

Corollary. If $\frac{1+\lambda^-}{1-\lambda^-} < c < 1$ and $u = K_k u + g$ then (with 14) $u = (1-c) \sum_{n=0}^{\infty} L_k^n g$.

Moreover an optimal choice of c is given in

Theorem 3. If $\sigma(K_k)$ is real, $\lambda^+ > 1$, and

(57) $$c = \xi = \frac{\frac{1}{\lambda^+} + \frac{1}{\lambda^-}}{\frac{1}{\lambda^+} + \frac{1}{\lambda^-} - 2}$$

then

(58) $$r_\sigma(L_k) = \overline{r}(L_k) = \frac{\frac{1}{\lambda^+} - \frac{1}{\lambda^-}}{2 - \frac{1}{\lambda^+} - \frac{1}{\lambda^-}} < 1 .$$

Proof. $\lambda^+ > 1$ implies Lemma 5a) from which, with (35) and (36), it may be shown that

(59) $$s^- = \left\{c \mid c > 1 \geq \frac{1}{1-\lambda^-} > \xi\right\} \cup \left\{c \mid 1 \geq \frac{1}{1-\lambda^-} > \xi > c\right\}$$

and

(60) $$s^+ = \left\{c \mid 1 > c > \xi > \frac{1}{1-\lambda^+}\right\}.$$

With the monotonic behavior of f^+ and f^- illustrated in Figure 1, it follows that

(61) $$\inf_{c \in s^-} f^-(c) = \min(f^-(1), f^-(\xi))$$

and

(62) $$\inf_{c \in s^+} f^+(c) = f^+(\xi) .$$

It is easily seen from (32) that

(63) $$f^+(\xi) = f^-(\xi) = \left|\frac{\frac{1}{\lambda^+} - \frac{1}{\lambda^-}}{\frac{1}{\lambda^+} + \frac{1}{\lambda^-} - 2}\right| .$$

Hence, with Lemma 4,

(64) $$\overline{r}(L_k) = \max\{\min(f^-(1), f^-(\xi)), f^+(\xi)\} = \left|\frac{\frac{1}{\lambda^+} - \frac{1}{\lambda^-}}{\frac{1}{\lambda^+} + \frac{1}{\lambda^-} - 2}\right|$$

which is assumed when $c = \xi$.

Next observe that $\lambda^+ > 1 \Rightarrow 1 - \frac{1}{\lambda^+} > 0$ and the inequality is strengthened by adding a positive quantity on the left. Hence

$$2 - \frac{1}{\lambda^+} - \frac{1}{\lambda^-} > 0 . \tag{65}$$

Furthermore $1 - \frac{1}{\lambda^+} > 0 \Rightarrow 2 - \frac{1}{\lambda^+} > \frac{1}{\lambda^+}$ and subtracting $\frac{1}{\lambda^-}$ from both sides yields

$$2 - \frac{1}{\lambda^+} - \frac{1}{\lambda^-} > \frac{1}{\lambda^+} - \frac{1}{\lambda^-} \tag{66}$$

which, with (65), establishes

$$\left| \frac{\frac{1}{\lambda^+} - \frac{1}{\lambda^-}}{\frac{1}{\lambda^+} + \frac{1}{\lambda^-} - 2} \right| = \frac{\frac{1}{\lambda^+} - \frac{1}{\lambda^-}}{2 - \frac{1}{\lambda^+} - \frac{1}{\lambda^-}} < 1 . \tag{67}$$

This completes the proof.

Corollary 1. If $\sigma(K_k)$ is real and non-positive $(\lambda^+ \to \infty)$ then $\bar{r}\,(L_k) = \frac{1}{2-\lambda^-} < 1$ and this is achieved when $c = \frac{1}{2-\lambda^-}$.

Corollary 2. If $\lambda^- = -1$ and $\lambda^+ > 1$ then $\bar{r}\,(L_k) = \frac{1+\lambda^+}{3\lambda^+ - 1}$ and this is achieved when $c = \frac{1-\lambda^+}{1-3\lambda^+}$. This applies in the particular case when K_k satisfies Plemelj's theorem.

Corollary 3. If $\sigma(K_k)$ is real and non positive and $\lambda^- = -1$ then $\bar{r}\,(L_k) = \frac{1}{3}$ and this is achieved when $c = \frac{1}{3}$.

Corollary 4. If $\sigma(K_0)$ is real and $\lambda^+_0 > 1$ then for $c = \xi_0$, $\bar{r}\,(L_0) < 1$ where λ^+_0, λ^-_0 are bounds on the spectrum of K_0 and ξ_0 is similarly defined in terms of these bounds. Then, as shown by Kleinman and Wendland, $r_\sigma(L_k) < 1$, if $||L_k - L_0||$ is sufficiently small, where L_k is defined by (13) with $c = \xi_0$.

DISCUSSION

In the problem treated by Kleinman and Wendland, K_k was a boundary integral operator arising from the exterior Neumann problem for the Helmholtz equation. In that case the operator was compact hence had a pure point spectrum. Moreover the operator K_0 satisfied Plemelj's theorem (real eigenvalues, $\lambda^- = -1$ and all other eigenvalues greater than 1 in absolute value.) Hence the corollary to theorem 2 applied. That work, modelled after Neumann's original proof corresponds to a choice of $c = \frac{1}{2}$. Kress and Roach (1976) used $c = \frac{1}{3}$ which, from corollaries 2 and 3 of Theorem 3, is seen to be the optimal choice if there are no positive points in the spectrum and if $k = 0$. If $k \neq 0$ the spectrum is no longer real. A numerical example is given Chertock (1968). If one has no information about λ^+, except that $\lambda^+ > 1$ then corollary 2 is not determining. However since $c = \frac{1-\lambda}{1-3\lambda}$ is an increasing function of λ for $\lambda > 1$ it may be inferred that the optimal choice of c is bounded by

$$0 < \frac{1-\lambda^+}{1-3\lambda^+} \leq \frac{1}{3} .$$

The results described here completely specify the optimal Möbius transformation when the spectrum of K_k is real and have been shown to apply even when $\sigma(K_k)$ is complex but $\sigma(K_0)$ is real and $||K_k - K_0||$ is sufficiently small. It should be noted that this transformation and the subsequent iterative solution of the operator equation is identical with Petryshyn's over relaxation method [see Patterson, 1974, p. 56, equation (9) with D, S, Q and w replaced by I, 0, K_k, and 1-c]. More complicated analytic transformations of the eigenvalue parameter are discussed briefly by Kantorovich and Krylov (1964) but this appears to be an area worthy of further study.

In closing it is noted that the general methods of establishing convergence of the Neumann series for operators with $\lambda^+ > 1$ has been extended to the boundary integral operator arising in electromagnetic scattering from a perfectly conducting surface by Gray (1976), by generalizing the results of Müller and Niemeyer (1961) and Werner (1966) on the spectrum of the operator.

REFERENCES

Ahner, J. F. and G. C. Hsiao (1975), A Neumann Series Representation for Solutions to Boundary Value Problems in Dynamic Elasticity, Quart. Appl. Math, XXXIII, 73-80.

Ahner, J. F. and R. E. Kleinman (1973), The Exterior Neumann Problem for the Helmholtz Equation, Archive for Rat. Mech. and Anal., 52, 26-43.

Chertock, G. (1968), Convergence of Iterative Solutions to Integral Equations for Sound Radiation, Quart. Appl. Math., XXVI, 268-272.

Goursat, E. (1964), A Course in Mathematical Analysis, Vol. III, Part 2, Dover Publications, New York (translation).

Gray, G. (1976), Ph.D. Dissertation, University of Delaware, Newark, Delaware.

Kantorovich, L. V. and V. I. Krylov (1964), Approximate Methods of Higher Analysis, P. Noordhoff, Groningen (translation).

Kittappa, R. and R. E. Kleinman (1975), Acoustic Scattering by Penetrable Homogeneous Objects, J. Math. Phys., 16, 421-432.

Kleinman, R. E. and W. Wendland (1976), On Neumann's Method For the Exterior Neumann Problem For the Helmholtz Equation, J. Math Anal. and Appl., to appear.

Král, J. (1966), The Fredholm Method in Potential Theory, Trans. Amer. Math Soc., 125, 511-547.

Kress, R. and G. F. Roach (1976), On the Convergence of Successive Approximations for an Integral Equation in a Green's Function Approach to the Dirichlet Problem, To Appear.

Müller, C. and H. Niemeyer (1961), Greensche Tensoren und asymptotische Gesetze der elektromagnetischen Hohlraumschwingungen, Arch. Rat. Mech. and Anal., 7 , 305-348.

Patterson, W. M. (1974), Iterative Methods for the Solution of a Linear Operator Equation in Hilbert Space - A Survey, Lecture Notes in Mathematics No. 394, Springer-Verlag, Berlin, Heidelberg, New York.

Petryshyn, W. V. (1962), The Generalized Overrelaxation Method For the Approximate Solution of Operator Equations in Hilbert Space, SIAM J. Appl. Math, 10, 675-690.

Taylor, A. (1958), Introduction to Functional Analysis, John Wiley, New York.

Wendland, W. (1968), Die Behandlung von Randwertaufgaben im R_3 mit Hilfe von Einfach-und Doppelschicht-potentialen, Numerische Mathematik, 11, 380-404.

Werner, P. (1966), On an Integral Equation in Electromagnetic Diffraction Theory, J. Math. Anal. and Appl., 14, 445-456.

NEUMANN PROBLEM ON A SYMMETRIC BRELOT'S HARMONIC SPACE

Tosiaki KORI

1. Introduction

The classical Neumann problem is posed as follows:

Given a function f on the circle $B = \{ z ; |z| = 1 \}$, find a solution u of the following boundary problem

$$(1.1) \quad \begin{cases} \Delta u = 0 & \text{in } \Omega = \{ z ; |z| < 1 \} \\ \dfrac{\partial u}{\partial n} = f & \text{on } B \quad , \end{cases}$$

where Δ is the Laplacian operator and $\dfrac{\partial}{\partial n}$ is the operator of inner normal derivative.

The condition

$$\int_B f \, d\sigma = 0$$

is necessary for the existence of such a solution.

What kind of generalization of the above problem is possible ? We can replace the Laplacian by a more general elliptic differential operator. From the point of view of potential theory this would not be difficult. In fact we can even treat such an operator with discontinuous coefficients. Next we consider the generalization of the domain Ω : for example Ω is an open Riemann surface. In this case the formulation of the problem itself is one of the essential difficulties because we lack the concepts " the boundary" and "the normal derivative". Our purpose in this report is to (pose and) solve the Neumann problem in the case of Brelot's axiomatic theory of harmonic functions.

Here we will briefly review the method of solving the problem (1.1). This method is a modification of the one given by J. L. Doob and seems to be new.

The Dirichlet quadratic form (or the Dirichlet integral) is defined as follows;

$$D(u, u) = \int_{\Omega} | \text{grad. } u |^2 \, dV .$$

By Green's formula:

$$D(u, v) = \int_{\Omega} \Delta u \, v \, dV - \frac{1}{2\pi} \int_{B} \frac{\partial u}{\partial n} v \, d\sigma ,$$

the problem (1.1) reduces to find a harmonic functionusatisfying

$$D(u, v) = \int_B f \, v \, d\sigma' , \quad \sigma' = \frac{1}{2\pi} \sigma$$

for all smooth functions v on $\overline{\Omega}$.

Let $\mathcal{E}$ be the completion of the set of smooth functions on $\overline{\Omega}$ by the quadratic form $D(u, v)$, and $\mathfrak{h}$ be the elements of $\mathcal{E}$ that are harmonic in Ω. $\mathfrak{h}$ is a closed subspace of $(\mathcal{E} ,D)$. Therefore we have $D(u, v) = D(u, v_h)$ for any $u \in \mathfrak{h}$ and any $v \in \mathcal{E}$, where v_h is the projection of v on $\mathfrak{h}$. So the above problem becomes to;

(1.2) find a harmonic function $u \in \mathfrak{h}$ such that

$$D(u, v) = \int_B f \, v \, d\sigma'$$

for all $v \in \mathfrak{h}$ (under the condition $\int_B f \, d\sigma = 0$).

The last problem can be solved by the method of Hilbert space if we can show the estimate:

$$(1.3)' \quad | \int_B f \, v \, d\sigma' | \leq C(f) \, D(v, v)^{1/2}$$

for all $v \in \mathfrak{h}$, where $C(f)$ is a constant depending on f.
Or, if we note the fact that the integral of f on B vanishes, the following inequality will be more plausible;

$$(1.3) \quad | \int_B (f - \int_B f \, d\sigma') \, v \, d\sigma' | \leq C(f) \, D(v, v)^{1/2} .$$

For this purpose we shall introduce the Ⓗ-kernel, that

is nothing but the normal derivative of Poisson kernel of Ω, which is also the normal derivative of Green's funtion of Ω.

Here is a formula due to Doob - Osborn:

$$(1.4)\quad D(v,v) = \iint_{B \times B} Ⓗ(x,y)(v(x) - v(y))^2 \sigma (dx\, dy) ,$$

where

$$Ⓗ(x , y) = \frac{1}{2\pi} \frac{1}{1 - \cos((x,y))} , \quad |x| = |y| = 1 ,$$

and $((x,y))$ denotes the angle between the two vectors x and y. The importance of the above formula is that we can calculate the Dirichlet integral (which is an integral over Ω) by an integral on the boundary.

(1.4) yields the following inequality;

$$(1.5)\quad \int_B (v - \int_B v\, d\sigma')^2 d\sigma'$$

$$= \frac{1}{4\pi^2} \iint_{B \times B} (v(x) - v(y))^2 \sigma (dx\, dy)$$

$$\leqq D(v , v)$$

for all $v \in \mathfrak{h}$.

From this inequality we have

$$\left| \int_B (f - \int_B f\, d\sigma')\, v\, d\sigma' \right| = \left| \int_B f\, (v - \int_B v\, d\sigma')\, d\sigma' \right|$$

$$\leqq (\int_B f^2\, d\sigma')^{1/2} (\int_B (v - \int_B v\, d\sigma')^2 d\sigma')^{1/2}$$

$$\leqq (\int_B f^2\, d\sigma')^{1/2} D(v , v)^{1/2} .$$

Therefore, for $f \in L^2(B, d\sigma)$ with $\int_B f\, d\sigma = 0$, the inequality (1.3)' is satisfied and there exists a $u \in \mathfrak{h}$ which satisfies (1.2). The Neumann problem is solved.

We remark that the inner product space $(\mathfrak{h} , D)$ is not separated, in fact

$$R^1 = \{ v \in \mathfrak{h} \ ; \ D(u, v) = 0 \text{ for all } u \in \mathfrak{h} \} .$$

This means that the solution of Neumann's problem is unique up to a constant.

2. Neumann's problem on a symmetric harmonic space

Let $(X, \mathcal{H})$ be a harmonic space with the sheaf $\mathcal{H}$ of harmonic functions satisfying the axioms 1, 2, 3 of Brelot. We suppose that the constant function 1 is harmonic on X and that there exists a symmetric Green function $p_y(x)$, that is, for each $y \in X$, p_y is a potential on X which is harmonic out of the point y, and the function

$$(x, y) \rightsquigarrow p_y(x)$$

is lower semi-continuous and continuous out of the diagonal $\{x=y\}$.

Let $\overline{X}$ be the Martin compactification of $(X, \mathcal{H})$.

For each $x \in X$, the function $y \rightsquigarrow (p_y(x_0))^{-1} p_y(x)$ has a continuous extension over $\overline{X} - x_0$, such that the extended functions $\xi \rightsquigarrow k_\xi(x)$ separate the points of $\Delta = \overline{X} - X$, and the function $k_\xi(\cdot)$ is harmonic on X for each $\xi \in \Delta$. We shall call this function the Poisson-Martin kernel. Every harmonic function $h \geq 0$ in X has a unique integral representation

$$h = \int_\Delta k_\xi \, \mu(d\xi)$$

with the aid of a measure $\mu \geq 0$ on Δ, that is concentrated on the set Δ_1 of all minimal points of Δ. We note that

$$k_\xi(x_0) = 1 .$$

In the sequel σ denotes the measure representing the constant 1 ; $1 = \int_\Delta k_\xi(\cdot) \, \sigma(d\xi)$.

<u>Theorem</u> 2.1. There exists a kernel-function $Ⓗ(\xi, \eta)$ defined on $\overline{X} \times \overline{X}$ with the following properties:

(1) $Ⓗ(x, \xi) = (p_x(x_0))^{-1} k_\xi(x)$, $x \in X$, $\xi \in \overline{X}$.

(2) $Ⓗ(\xi, \eta) = Ⓗ(\eta, \xi)$

(3) For each $\eta \in \overline{X}$, the function $\xi \rightsquigarrow Ⓗ(\xi, \eta)$ is lower semi-continuous on $\overline{X}$.

(4) $Ⓗ(\xi, \eta) = \liminf_{X \ni y \to \xi} Ⓗ(\eta, y)$, so Ⓗ is determined by its values on $X \times X$.

(5) There is a constant $c > 0$ such that $Ⓗ(\xi, \eta) \geq c$ for all $\xi, \eta \in \overline{X}$.

We set

$$\mathfrak{h}^\circ = \left\{ u \in \mathcal{H}(X);\ u = \int_\Delta k_\xi \overline{u}(\xi)\, \sigma(d\xi) \text{ with its boundary values } \overline{u} \in L^2(\Delta, d\sigma) \right\}$$

and we define, for $u, v \in \mathfrak{h}^\circ$,

$$D(u, v) = \frac{1}{2} \iint_{\Delta \times \Delta} Ⓗ(\xi, \eta)(\overline{u}(\xi) - \overline{u}(\eta))(\overline{v}(\xi) - \overline{v}(\eta))\sigma(d\xi\, d\eta).$$

Let

$$\mathfrak{h} = \left\{ u \in \mathfrak{h}^\circ;\ D(u, u) < \infty \right\}.$$

By Theorem 2.1 (5) we have the following inequality;

$$(2.1)\quad D(u, u) \geq \frac{c}{2} \iint_{\Delta \times \Delta} (\overline{u}(\xi) - \overline{u}(\eta))^2\, \sigma(d\xi\, d\eta)$$

$$= c \int_\Delta \left(\overline{u}(\xi) - \int_\Delta \overline{u}\, d\sigma\right)^2 \sigma(d\xi)$$

for all $u \in \mathfrak{h}^\circ$.

From this inequality we can prove without difficulty the

<u>Theorem</u> 2.2. $(\mathfrak{h}, D(\ ,\ ))$ is a complete inner-product space with

$$\left\{ u \in \mathfrak{h};\ D(u, v) = 0 \text{ for all } v \in \mathfrak{h} \right\} = R^1.$$

Now we shall introduce the concept of normal derivative of a harmonic function.

Let

$$\mathcal{D} = \left\{ u \in \mathfrak{h} \;;\; \begin{array}{l} \text{there is a constant } a \geqq 0 \text{ such that} \\ D(u, v) \leqq a \left(\int_\Delta \overline{v}(\xi)^2 d\sigma(\xi) \right)^{1/2} \\ \text{for all } v \in \mathfrak{h} \end{array} \right\},$$

and

$$\mathcal{N} = \left\{ f \in L^2(\Delta, d\sigma);\; \begin{array}{l} \int_\Delta f(\xi)\overline{u}(\xi) d\sigma(\xi) = 0 \\ \text{for all } u = \int_\Delta k_\xi \overline{u}(\xi) d\sigma(\xi) \in \mathfrak{h} \end{array} \right\}.$$

For any function $u \in \mathcal{D}$ we can find a $f \in L^2(\Delta, d\sigma)$ which satisfies

$$D(u, v) = \int_\Delta f(\xi)\, \overline{v}(\xi)\, d\sigma(\xi) \quad \text{for all } v \in \mathfrak{h}.$$

Such a function f is called a normal derivative of u. In this case any function of the form $f + g$, where $g \in \mathcal{N}$, is also a normal derivative of u, so the normal derivative of $u \in \mathfrak{h}$ is defined up to a function of $\mathcal{N}$.

By the same argument as we have explained in section 1 for the classical Neumann problem we get the following result:

<u>Theorem</u> 2.3. For a function $f \in L^2(\Delta, d\sigma)$ to be a normal derivative of a harmonic function $u \in \mathcal{D}$, it is necessary and sufficient that $\int_\Delta f(\xi)\, d\sigma(\xi) = 0$.

Now we give the estimate of the above solution u.

Let $u \in \mathfrak{h}$ be such that $D(u, v) = \int_\Delta f(\xi)\, \overline{v}(\xi)\, d\sigma(\xi)$ for a $f \in L^2(\Delta, d\sigma)$ satisfying $\int_\Delta f(\xi)\, d\sigma(\xi) = 0$.

From the inequality (2.1) we have

$$c \int_\Delta \left(\overline{u} - \int_\Delta \overline{u}\, d\sigma \right)^2 d\sigma \leqq D(u, u) = \int_\Delta f\, \overline{u}\, d\sigma$$

$$= \int_\Delta (\bar{u} - \int_\Delta \bar{u}\, d\sigma)\, f\, d\sigma \leqq \left(\int_\Delta (\bar{u} - \int_\Delta \bar{u}\, d\sigma)^2 d\sigma \right)^{1/2} \times \left(\int_\Delta f^2\, d\sigma \right)^{1/2}$$

hence

$$(2.2) \quad \left(\int_\Delta (\bar{u} - \int_\Delta \bar{u}\, d\sigma)^2 d\sigma \right)^{1/2} \leqq c^{-1} \left(\int_\Delta f^2\, d\sigma \right)^{1/2} .$$

In particular if we take the solution u with the property

$$u(x_0) = \int_\Delta k_\xi(x_0) \bar{u}(\xi)\, d\sigma(\xi) = \int_\Delta \bar{u}(\xi) d\sigma(\xi) = 0 ,$$

we have the estimate

$$(2.3) \quad |u(x)| = \left| \int_\Delta k_\xi(x) \bar{u}(\xi) d\sigma(\xi) \right| \leqq c^{-1} \sup_\xi k_\xi(x) \, \| f \|_{L^2(\Delta, d\sigma)} .$$

We can also introduce the precise normal derivative of a harmonic function

$$- \text{fine limit}_{y \to \xi} \, (p_y(x_0))^{-1} (u(y) - \bar{u}(\xi)) = - \int_\Delta Ⓗ(\eta, \xi)(\bar{u}(\eta) - \bar{u}(\xi))\, d\sigma(\eta) .$$

The normal derivative that we have defined previously coinsides with this normal derivative almost everywhere on Δ_1 if the functions in $\mathcal{N}$ vanish almost everywhere on Δ_1 .

We terminate this section with the following remark.

Let $u \in \mathfrak{h}^\circ$. The superharmonic function $-u^2$ can be decomposed in the following way

$$-u^2 = -\int_\Delta k_\xi \, \bar{u}(\xi)^2 d\sigma(\xi) + p$$

where p is a potential on X.

We define the Dirichlet integral of $u \in \mathfrak{h}^\circ$ by

$$D'(u, u) = \frac{1}{2}\nu(1)$$

where ν is the measure representing the potential

$$p = \int_X p_y \, \nu(dy) \quad ,$$

(Definition due to Doob). This is an abstract version of the classical formula

$$\int_{\Omega} \Delta(u^2)\, dV = 2\int_{\Omega} (\operatorname{grad} u)^2\, dV .$$

We can prove the Doob- Osborn formula generalized on symmetric Brelot's harmonic space:

$$D'(u, u) = D(u, u) .$$

3. Reproducing kernel of $\mathfrak{h}$

We consider the following function φ_y on Δ

$$\varphi_y(\xi) = k_\xi(y) - 1 ,$$

for each $y \in X$. Since it satisfies the condition

$$\int_\Delta \varphi_y(\xi)\, d\sigma(\xi) = 0 ,$$

there is a unique harmonic function b_y on X that satisfies

$$(3.1) \quad D(b_y, v) = \int_\Delta \varphi_y(\xi)\, \overline{v}(\xi)\, d\sigma(\xi) , \quad \forall v \in \mathfrak{h} ,$$

and

$$b_y(x_0) = 0 .$$

From the definition of φ_y , we have

$$D(b_y, u) = u(y) - u(x_0) \quad \text{for all } u \in \mathfrak{h} .$$

In particular,

$$D(b_y, b_z) = b_y(z) = b_z(y) .$$

This means that the functions b_y form a reproducing kernel of the complete inner-product space $(\mathfrak{h}, D)$

Lemma 3.1. We have the estimate

$$0 \leq \sup_K b_y(y) \leq c^{-1} M_K^2$$

for any compact set K, where M_K is the Harnack's constant: a constant $M_K > 1$ such that

$$\sup_K h \leq M_K h(x_0)$$

for all non-negative harmonic functions on X.

Proof Let $\overline{b_y}$ be the boundary value of b_y;

$$b_y(x) = \int_\Delta k_\xi(x)\,\overline{b_y}(\xi)\,d\sigma(\xi).$$

From (2.3) we have

$$b_y(y) \leq c^{-1} \sup_\xi k_\xi(y) \;\|\varphi_y\|_{L^2(\Delta,\, d\sigma)}$$

$$\leq c^{-1} M_K \|\varphi_y\|_{L^2(\Delta,\, d\sigma)} .$$

On the other hand

$$\|\varphi_y\|_{L^2(\Delta,\, d\sigma)} = \left(\int_\Delta (k_\xi(y)-1)^2 \sigma(d\xi)\right)^{1/2} \leq (M_K - 1)^{1/2} .$$

These inequalities yield the lemma.

Theorem 3.2.

$$\sup_K |u(y) - u(x_0)| \leq c^{-1/2} M_K D(u, u)^{1/2}$$

for all $u \in \mathfrak{h}$.

Proof The above lemma implies the inequality:

$$|u(y) - u(x_0)| = |D(b_y, u)| \leq (b_y(y))^{1/2} D(u, u)^{1/2}$$

$$\leq c^{-1/2} M_K D(u, u)^{1/2} ,$$

for all $y \in K$ and all $u \in \mathfrak{h}$.

From this theorem we see that, on the set

$$\{u \in \mathfrak{h}^0 \;;\; D(u, u) < \infty ,\; u(x_0) = 0\}$$

the topology induced by the norm $D(,)$ is stronger than the topology of compact uniform convergence, the latter is known to be the natural topology on harmonic functions.

Remark: If $X = R^n$, $n \geq 3$, and $\mathcal{H}$ is the sheaf of ordinary harmonic functions on R^n. We have $\mathfrak{h} = R^1$ and $D(u, u) = 0$ for all $u \in \mathfrak{h}$. The above theorem becomes to assert only $0 = 0$.

4. Neumann kernel-function

In this section we consider the general version (on a harmonic space of Brelot) of the following classical problem:

$$\begin{cases} \Delta u = f & \text{in } \{ z ; \ |z| < 1 \} \\ \dfrac{\partial u}{\partial n} = 0 & \text{on } \{ z ; \ |z| = 1 \} . \end{cases}$$

Again the condition

$$\int_{\Omega} f \, dV = 0$$

is necessary for the existence of solution. We shall give a kernel-function $n_y(x)$ which gives the solution:

$$u(x) = \int_{\Omega} n_y(x) f(y) \, dV(y) .$$

Let $(X, \mathcal{H})$ be a symmetric harmonic space and $p_y(x) = p_x(y)$ its Green function. Using the reproducing kernel b_y for $(\mathfrak{h}, D)$ that we have introduced in section 3 we set

$$n_y(x) = p_y(x) + b_y(x).$$

We have $n_y(x) = n_x(y)$. n_y is a superharmonic function on X which is harmonic out of the point y, and $(x, y) \rightsquigarrow n_y(x)$ is lower semi-continuous and continuous out of the diagonal $x = y$.

In the sequel m denotes a bounded measure on X which satisfies the condition

$$(4.1) \qquad \xi \rightsquigarrow \int_X k_\xi(y) \, dm(y) \in L^2(\Delta, d\sigma).$$

This condition is in particular satisfied by a bounded measure of compact support.

Since the function

$$\varphi(\xi) = \int_X (k_\xi(y) - 1) \, m(dy)$$

satisfies the condition

$$\int_{\Delta} \varphi \, d\sigma = 0$$

there is a unique function $U^m \in \mathfrak{H}$ such that

$$(4.2) \quad \begin{cases} D(U^m, v) = \int_{\Delta} \varphi(\xi)\, \bar{v}(\xi)\, d\sigma(\xi), \\ \qquad\qquad\qquad\qquad \text{for all } v \in \mathfrak{H}, \\ U^m(x_0) = 0 \end{cases}$$

From this and the reproducing property of b_y we have

$$(4.3) \quad U^m(y) = D(U^m, b_y) = \int_{\Delta} \varphi(\xi)\, \bar{b}_y(\xi) d\sigma(\xi)$$

$$= \int_X \int_{\Delta} (k_{\xi}(z) - 1)\, \bar{b}_y(\xi) d\sigma(\xi)\, m(dz)$$

$$= \int_X b_y(z)\, m(dz) .$$

Here we introduce the normal derivative of a function of potential type

$$p = \int_X p_y \, \mu(dy)$$

as follows:

$$\frac{\partial p}{\partial n}(\xi) = - \int_X k_{\xi}(y) \mu(dy) .$$

This definition is justified because we can show that

$$\operatorname*{fine\ limit}_{x \to \xi} \frac{1}{p_x(x_0)}\, p(x) = \int_X k_{\xi}(y)\, \mu(dy),$$

when they are well defined.

Now if we set

$$Gm(x) = \int_X p_y(x)\, m(dy),$$

we have

$$\int_{\Delta} \left(\frac{\partial}{\partial n} Gm\right)(\xi)\, \bar{v}(\xi) d\sigma(\xi) + \int_X v(y)\, m(dy) = 0$$

for all $v \in \mathfrak{H}$, especially we have

$$\int_{\Delta} \left(\frac{\partial}{\partial n} Gm\right)(\xi) d\sigma(\xi) + m(1) = 0 .$$

Hence, from (4.2),

$$D(U^m , v) = - \int_{\Delta} (\frac{\partial}{\partial n} Gm - m(1)) \bar{v} \, d\sigma ,$$

for all $v \in \mathfrak{h}$, that is, the normal derivative of the <u>harmonic function</u> U^m is equal to

$$- \frac{\partial}{\partial n} Gm + m(1) .$$

So we have the

<u>Theorem</u> 4.1. Let m be a measure satisfying the condition (4.1). Then the function

$$\int_X n_y(\cdot) \, m(dy)$$

has its normal derivative equal to m(1).

In particular, for a given constant a there is a superharmonic function whose normal derivative is equal to a .

5. What we have developped in the above seems somewhat to be a formal theory except the basic inequality

$$Ⓗ(\xi , \eta) \geq c > 0 , \quad \xi , \eta \in \bar{X} ,$$

on which all our arguments have depended. (Ⓗ-kernel on a harmonic space was introduced earlier (1969) by the author and the first proof of the above inequality was given by an easy but long standing calculation depending on the very definition of it, recently G. Mokobodzki suggested me a simple and beautiful proof of it.)

As for the further developpment of our theory we give a general version of the following classical problem:

$$\begin{cases} \Delta u = 0 & \text{in } \{ a < |z| < 1 \} \\ u = f & \text{on } \{ |z| = a \} \\ \frac{\partial u}{\partial n} = 0 & \text{on } \{ |z| = 1 \} . \end{cases}$$

To solve this problem on a Brelot's harmonic space we should introduce a Dirichlet space localized at the point at infinity. This is no longer a formal theory but a rather hard theory. Even to define the necessary concepts we must explain before the duality relation of harmonic sheaf which itself is a sufficiently long theory. Here we omit this theory but give a one of its consequences.

<u>Minimum Principle</u> Let $(X, \mathcal{H})$ be a symmetric harmonic space of Brelot and K be a compact subset of X. Let s be a superharmonic function on the domain $\omega = X - K$ such that:

(1) $\liminf_{y \to x} s(y) \geq 0$ at every boundary point $x \in \partial\omega$,

(2) The normal derivative of s is non-negative on Δ .

Then $s \geq 0$.

References

1. H. Bauer, Harmonische Räume und ihre Potentialtheorie, Lecture Notes in Mathematics 22.

2. J. L. Doob, Boundary properties of functions with finite Dirichlet integrals, Ann. Inst. Fourier, 12

3. T. Kori, Problème de Neumann sur les espaces harmoniques, (to be published in Math. Annalen)

4 T. Kori, Théorie des espaces fonctionnels à nullite 1 et le problème de Neumann sur les espaces harmoniques (to appear)

SPEZIELLE BERGMAN-OPERATOREN NEBST ANWENDUNGEN

Manfred W. Kracht, Universität Düsseldorf

0. Übersicht

Neben der Gewinnung von Existenz- und Eindeutigkeitsaussagen und von expliziten Lösungsdarstellungen ist bei S. Bergmans Integraloperatoren ([3a-c, 7a-b]) von besonderem Interesse die Übertragung von Ergebnissen aus der Funktionentheorie in einer komplexen Variablen auf die Lösungen linearer partieller Differentialgleichungen zweiter Ordnung in zwei Variablen. Für diese Zwecke sind Bergman-Operatoren mit "einfachen" Kernen, wie z. B. mit Kernen 1. Art ([3c]), Polynomkernen ([11c]), Polynomkernen 1. Art ([10]), Exponentialkernen ([11a]) sowie Kernen 2. Art ([3c, 14]), besonders geeignet.

Wir wollen hier Operatoren mit Kernen 1. Art, Polynomkernen und Polynomkernen 1. Art behandeln: Nach Bereitstellung der notwendigen Definitionen und Hilfsmittel in Abschn. 1 diskutieren wir in Abschn. 2 Existenz, Eindeutigkeit und Darstellung der Kerne 1. Art sowie der hierdurch erzeugten Lösungen von (Goursatschen) Anfangswertproblemen. Zudem gehen wir auf die Zusammenhänge zwischen den Operatoren von Bergman, LeRoux und Riemann ein. Abschn. 3 beschäftigt sich mit der Existenz von Operatoren mit Polynomkernen bzw. Polynomkernen 1. Art und zeigt, daß die hiermit gewonnenen Lösungen sich integralfrei darstellen und als durch einen (in Spezialfällen auf K. W. Bauer ([1a-d]) zurückgehenden) Differentialoperator erzeugt auffassen lassen ([9a]). Existenzsätze und Konstruktionsverfahren von M. Kracht und E. Kreyszig ([9c-d]) für Polynomkerne und für die Charakterisierung der zugelassenen Klassen von Gleichungen sowie eine Aussage bzgl. der Eindeutigkeit der Lösungsdarstellung mittels Differentialoperatoren folgen in Abschn. 4. Letztlich führen wir in Abschn. 5 als Beispiele für das erwähnte Übertragungsprinzip für Lösungen aus obigen Klassen von Gleichungen Ergebnisse an bzgl. Werteverteilung (vgl. [3b, 4]) und bzgl. Singularitäten (vgl. [11b]).

1. Bergman-Operatoren mit speziellen Kernen

1.1. Sei die homogene lineare Differentialgleichung

$$L_o w := D_1 D_2 w + a_1 D_1 w + a_2 D_2 w + a_3 w = 0$$

in zwei unabhängigen komplexen Variablen mit $D_j := \partial/\partial z_j$ $(j = 1, 2)$ und $a_1, a_2, a_3 \in C^\omega(G)$ ($G \subset \mathbb{C}^2$ ein einfach-zusammenhängendes Gebiet) vorgelegt. (Falls a_1, a_2, a_3 zusätzlich die Bedingungen $a_1(z_1,\bar{z}_1) = \bar{a}_2(\bar{z}_1,z_1)$ und $a_3(z_1,\bar{z}_1) = \bar{a}_3(\bar{z}_1,z_1)$ für $z \in G$ $(z = (z_1,z_2))$ erfüllen, kann man diese Gleichung auch als Fortsetzung zu komplexen Koordinaten einer in Normalform gegebenen elliptischen Gleichung in zwei reellen Variablen x, y auffassen, auf die die Transformation $z_1 = x+iy$, $z_2 = x-iy$, $\partial/\partial z_j = \frac{1}{2}(\partial/\partial x + (-1)^j i\partial/\partial y)$ $(j = 1, 2)$ angewandt worden ist. In diesem Falle erhält man aus den im folgenden hergeleiteten Lösungsdarstellungen für $z_2 = \bar{z}_1$ und Realteilbildung reellwertige Lösungen der Gleichung $L_o w = 0$ und auch der Ausgangsgleichung.)

1.2. Sei $s = (s_1,s_2) \in G$; j habe stets die Werte 1, 2; $K_j(s_j;r_j) \subset \mathbb{C}$ sei eine offene Kreisscheibe in der z_j-Ebene mit Zentrum s_j und Radius r_j (> 0) derart, daß $H := K_1(s_1;r_1) \times K_2(s_2;r_2) \subset G$ ist. K_t bzw. $\bar{K}_t$ bezeichnen die offene bzw. abgeschlossene Einheitskreisscheibe in der komplexen t-Ebene und $\gamma_j \subset \bar{K}_t$ eine rektifizierbare Kurve von -1 nach +1. Unter Verwendung einer Funktion $\tilde{g}_j \in C^\omega(H \times K_t)$ definieren wir einen Operator $B_j : C^\omega(K_j(s_j;\rho_j)) \to C^\omega(H_j^*)$ durch

$$(B_j f_j)(z) = \int_{\gamma_j} \tilde{g}_j(z,t)\, f_j(s_j+\tfrac{1}{2}(z_j-s_j)(1-t^2))\,(1-t^2)^{-1/2}\, dt .$$

Hierbei bedeutet H_j^* den Dizylinder

$$H_j^* = K_1(s_1;r_1^*) \times K_2(s_2;r_2^*) \Big| \; r_{3-j}^* = r_{3-j},\; r_j^* = \min(r_j, 2\rho_j) .$$

B_j heißt Bergman-Operator (mit Bezugspunkt s) mit Bergman-Kern (bzw. -Erzeugender) $\tilde{g}_j$ in H, wenn durch $w_j = B_j f_j$ $(f_j \in C^\omega(K_j(s_j;\rho_j)))$ zur Differentialgleichung $L_o w = 0$ eine Lösung $\in C^\omega(H_j^*)$ gegeben wird. Erhält man durch den eingliedrigen Lösungsansatz, d.h.: durch $w_j = B_j f_j$, eine Lösung in H_j^*, so natürlich auch durch den zweigliedrigen Lösungsansatz $w = \alpha_1 w_1 + \alpha_2 w_2$ $(\alpha_1 \neq 0 \neq \alpha_2)$ in $H_1^* \cap H_2^*$.

Ein Bergman-Kern in H liegt aber vor, wenn die durch $g_j = e_j^{-1}\tilde{g}_j$ mit $e_j(z) = \exp[\eta_j(z_j) - \int_{s_{3-j}}^{z_{3-j}} a_j(z^*)\big|_{z_j^* = z_j} dz_{3-j}^*]$ $(\eta_j \in C^\omega(G)$ beliebig) definierte Funktion g_j den folgenden drei Bedingungen genügt:

$$M_j g_j := [(1-t^2)D_{3-j}D_t - \tfrac{1}{t}D_{3-j} + 2(z_j-s_j)t\, L^{(j)}]\, g_j = 0$$
für $(z,t) \in H \times K_t$,

$$[(z_j-s_j)t]^{-1} D_{3-j} g_j \in C^0(H \times \gamma_j) ,$$

$$(1-t^2)^{1/2} D_{3-j} g_j(z,t) \to 0 \text{ für } t \to \pm 1 \text{ gleichmäßig für } z \in H.$$

Hierin bedeutet $D_t = \partial/\partial t$, und $L^{(j)} := D_1 D_2 + b_j D_{3-j} + c_j$ hat die Koeffizienten

$$b_j(z) = D_j \eta_j(z_j) + a_{3-j}(z) - \int_{s_{3-j}}^{z_{3-j}} D_j(a_j(z^*)\big|_{z_j^* = z_j})\, dz_{3-j}^* ,$$

$$c_j(z) = a_3(z) - a_1(z)a_2(z) - D_j a_j(z) .$$

(Dabei ist die erste Bedingung, $M_j g_j = 0$, die wesentlichste, da z.B. im Falle der unten definierten Bergman-Kerne 1. Art die anderen Bedingungen automatisch erfüllt sind.)

1.3. Drei Klassen von Bergman-Kernen haben aufgrund von Anwendungen bisher besonderes Interesse gefunden: Kerne 1. Art, Polynomkerne (bzw. minimale Polynomkerne) und Polynomkerne 1. Art:

Ein Bergman-Kern $\tilde{g}_j$ in $H \subset G$ heißt Kern 1. Art (mit Bezugspunkt $s \in H$), wenn $g_j(z_1,s_2,t) = 1 = g_j(s_1,z_2,t)$ für (z_1,s_2,t), $(s_1,z_2,t) \in H \times \bar{K}_t$. Ein Polynomkern $\tilde{g}_j$ vom Grade n (in t^2) liegt vor, wenn

$$g_j(z,t) = \sum_{\mu=0}^{n} q_{j,2\mu}(z)((z_j-s_j)t^2)^\mu , \quad q_{j,2n} \neq 0 \quad (n \in \mathbb{N}_0) \text{ gilt;}$$

er heißt minimal, wenn es zur betrachteten Gleichung in H keinen Polynomkern kleineren Grades gibt. Letztlich, ein Polynomkern 1. Art ist ein Kern 1. Art, der zugleich Polynomkern ist.

2. Eigenschaften der Bergman-Kerne erster Art

2.1. Zu jeder Gleichung $L_o w = 0$ kann man die Existenz von Kernen $\tilde{g}_1$ und $\tilde{g}_2$ 1. Art nachweisen, und zwar beweist man mittels Reihenansatzes und der Majorantenmethode, daß für L_o-Koeffizienten $\in C^{\omega}(H)$ diese Kerne zumindest in $\tilde{H}_1 \times K_t$ bzw. $\tilde{H}_2 \times K_t$ existieren (siehe dazu [3c]; vgl. [8b]); hierbei ist

$$\tilde{H}_j := K_1(s_1;\tilde{r}_1) \times K_2(s_2;\tilde{r}_2) \Big|_{\tilde{r}_{3-j} = r_{3-j},\ \tilde{r}_j = r_j/2} .$$

Wie wir in [10] gezeigt haben, sind Kerne 1. Art eindeutig bestimmt bis auf eine in t ungerade Funktion (die allerdings wegen der Form der Lösungsdarstellung keinen Beitrag leistet). Beispielsweise besitzt die gemäß 1.1 auf die Gestalt

$$L_o^{(a)} w := D_1 D_2 w + k^2 w = 0 \qquad (k \in \mathbb{C})$$

transformierte zeitfreie Helmholtzgleichung in zwei Ortsvariablen die folgenden Kerne 1. Art mit Bezugspunkt $s = (0,0)$ $(j = 1, 2)$:

$$\tilde{g}_{j1}(z,t) = \cos\,(2k\sqrt{z_1 z_2}\; t) \text{ und } \tilde{g}_{j2}(z,t) = \exp\,(i2k\sqrt{z_1 z_2}\; t).$$

Gemäß Definition eines Polynomkernes ist aber ein Polynomkern 1. Art eindeutig bestimmt.

Mittels des zweigliedrigen Lösungsansatzes (siehe 1.2) mit Bergman-Operatoren mit Kernen 1. Art lassen sich alle $C^{\omega}(\tilde{H})$-Lösungen ($\tilde{H} = \tilde{H}_1 \cap \tilde{H}_2$) gewinnen. Der Nachweis hängt zusammen mit der Lösung von Goursatschen Anfangswertproblemen auf den charakteristischen Ebenen $z_1 = s_1$ bzw. $z_2 = s_2$, wodurch man auf eine Abelsche Integralgleichung geführt wird. Als Darstellung der sog. B_j-Zugeordneten f_j von w durch die Anfangsdaten von w erhält man in $K_j(s_j;\frac{1}{2}r_j)$ (siehe [8c]):

$$f_j(\zeta_j) = \frac{1}{\pi} w(s) + \frac{4}{\pi}\, \zeta_j \int_0^{\pi/2} u_j'(2\zeta_j \sin^2\vartheta)\, \sin\vartheta\, d\vartheta$$
$$- e_{3-j}(z)\Big|_{z_{3-j} = s_{3-j}}\, f_{3-j}(s_{3-j}) ,$$

wobei $u_j(z_j) = w(z)\Big|_{z_{3-j} = s_{3-j}}$ gesetzt wurde. Offenbar sind die beiden

B_j-Zugeordneten f_j eindeutig festgelegt, sobald man für eine der beiden den Wert im Bezugspunkt vorschreibt.

2.2. Die Kerne 1. Art hängen eng zusammen mit speziellen Kernen von LeRouxschen Integraloperatoren (siehe dazu [8c, 16a]). Da die Riemannsche Funktion im Falle selbstadjungierter Gleichungen (d.h.: $a_1 = 0 = a_2$) ein solcher LeRouxscher Kern ist, kann man für diese Gleichungen auch Kerne 1. Art aus der komplexen Riemann-Funktion gewinnen, und zwar besteht die Beziehung (wenn man o.B.d.A. als Bezugspunkt $s = (0,0)$ wählt)

$$\tilde{g}_j(z,t) = \sum_{\mu=0}^{\infty} \frac{\mu!}{(2\mu)!} (-4z_j t^2)^\mu r_{j\mu}(z)$$

mit

$$r_{j\mu}(z) = \left[\frac{\partial^\mu}{\partial \zeta_j^{\,\mu}} \{R(\zeta;z)\big|_{\zeta_{3-j}=0}\}\right]\Big|_{\zeta_j = z_j}, \qquad \zeta = (\zeta_1,\zeta_2),$$

wobei R die Riemannsche Funktion zu L_o bezeichne.

Als Beispiel betrachten wir die Gleichung

$$L_o^{(b)}w := D_1D_2w + k(D_1D_2 \ln(\varphi_1 + \varphi_2))w = 0 ,$$

wobei $k=\text{const}\in\mathbb{C}$ und φ_1 bzw. φ_2 beliebige Funktionen in einer Variablen z_1 bzw. z_2 sind, derart daß der Koeffizient von w in G $(\ni(0,0))$ holomorph (ergänzbar) ist. Zur Bestimmung der Riemann-Funktion machen wir den Ansatz $R(\zeta;z) = \tilde{R}(\sigma(\zeta;z))$ mit der Hilfsvariablen

$$\sigma(\zeta;z) = -\frac{[\varphi_1(z_1) - \varphi_1(\zeta_1)][\varphi_2(z_2) - \varphi_2(\zeta_2)]}{[\varphi_1(z_1) + \varphi_2(z_2)][\varphi_1(\zeta_1) + \varphi_2(\zeta_2)]} ;$$

dann erhalten wir als Bedingungsgleichung für $\tilde{R}$ die gewöhnliche hypergeometrische Differentialgleichung

$$\sigma(1-\sigma)\tilde{R}'' + [c-(a+b+1)\sigma]\tilde{R}' - ab\tilde{R} = 0$$

mit den Faktoriellen $a = \frac{1}{2}(1+\sqrt{4k+1})$, $b = \frac{1}{2}(1-\sqrt{4k+1})$, $c = 1$ und der Nebenbedingung $\tilde{R}(0) = 1$. Die im Nullpunkt reguläre Lösung dieses Problems ist aber die hypergeometrische Funktion $\tilde{R}(\sigma) = {}_2F_1(a,b;c;\sigma)$, so daß sich die $r_{j\mu}$ in obiger Darstellung von $\tilde{g}_j$ zu

$$r_{j\mu} = \left[\frac{\partial^\mu}{\partial \zeta_j^{\,\mu}} \{{}_2F_1(a,b;1;\sigma(\zeta;z))\big|_{\zeta_{3-j}=0}\}\right]\Big|_{\zeta_j = z_j}$$

ergeben. Ist speziell $k = n(n+1)$, $n \in \mathbb{N}$, so bricht die Reihenentwicklung von ${}_2F_1$ um den Nullpunkt wegen $a = n+1$, $b = -n$ ab, und

$$\tilde{g}_j(z,t) = \sum_{\mu=0}^{n} \frac{\mu!}{(2\mu)!} (-4z_j t^2)^\mu \Big[\frac{\partial^\mu}{\partial \zeta_j{}^\mu} \{P_n(1-2\sigma(\zeta;z))\Big|_{\zeta_{3-j}=0}\}\Big]\Big|_{\zeta_j = z_j}$$

ist Kern 1. Art mit Bezugspunkt (0,0) zur Gleichung $L_o^{(b)} w = 0$ mit $k = n(n+1)$, $n \in \mathbb{N}$. (P_n bedeutet hierin das Legendre-Polynom vom Grade n.)

3. Eigenschaften der Polynomkerne

3.1. Im Unterschied zu den Kernen 1. Art existieren nicht zu jeder Gleichung $L_o w = 0$ Polynomkerne, da sich hierfür aus der Gleichung $M_j g_j = 0$ das folgende überbestimmte Gleichungssystem ergibt:

$$D_{3-j} q_{j,2\mu} + \frac{2}{2\mu-1} L^{(j)} q_{j,2\mu-2} = 0 \ , \quad \mu = 0,\dots,n+1 \quad ,$$

wobei $q_{j,o}$ nur von z_j abhängt und $q_{j,2n+2} = 0$ zu setzen ist. Deshalb wollen wir folgende Klassen von Differentialoperatoren einführen: P_{jn} bezeichne die Menge aller L_o, für die es zu $L_o w = 0$ in G einen Polynomkern $\tilde{g}_j$ vom Grade n gibt. Ist der Grad n minimal, so schreiben wir P^o_{jn}; es ist also

$$P^o_{jn} = \{L_o \in P_{jn} \,|\, \nexists\, m < n \ (m \in \mathbb{N}_o) : L_o \in P_{jm}\} \ .$$

Im Falle, daß ein Polynomkern $\tilde{g}_j$ 1. Art in G existiert, definieren wir analog:

$$EP_{jn} = \{L_o \in P_{jn} \,|\, \exists \tilde{g}_j : \tilde{g}_j \text{ Polynomkern 1. Art vom Grade n in G}\} \ ,$$

$EP^o_{jn} = EP_{jn} \cap P^o_{jn}$. (Vorstehend definierte Klassen von Operatoren hängen offensichtlich vom Gebiet G ab. Der Übersichtlichkeit halber wollen wir jedoch, falls nichts anderes gesagt ist, stets voraussetzen, daß $(0,0) \in G$ ist, und dann - wie in der Literatur meistens üblich, insbesondere bei Kernen 1. Art - den Nullpunkt als Bezugspunkt wählen.)

Die Zusammenhänge zwischen den soeben eingeführten Klassen ersieht man u.a. durch die folgenden Beispiele $L_o^{(c)}$, $L_o^{(d)}$, $L_o^{(e)}$:
Sei $L_o^{(c)} := D_1 D_2 + \alpha n(n+1)(1+\alpha z_1 z_2)^{-2}$ $(\alpha \in \mathbb{C}\setminus\{0\},\ n \in \mathbb{N})$, so folgt $L_o^{(c)} \in EP^o_{1n}$, $L_o^{(c)} \in EP^o_{2n}$, $L_o^{(c)} \in EP^o_{1n} \cap EP^o_{2n}$; als minimale Polynomkerne 1. Art mit Bezugs-

punkt (0,0) ergeben sich bei Lösung obigen Systems für die $q_{j,2\mu}$ unter den Nebenbedingungen, die ein Kern 1. Art zu erfüllen hat, oder auch als Spezialfall des in 2.2 aus der Riemann-Funktion hergeleiteten Kerns 1. Art (mit $\varphi_1(z_1) = (\alpha z_1)^{-1}$, $\varphi_2(z_2) = z_2$, $k = n(n+1)$) (vgl. [9a-c, 10]):

$$\tilde{g}_1(z,t) = \sum_{\mu=0}^{n} \binom{n+\mu}{2\mu} \left(\frac{-4\alpha z_1 z_2 t^2}{1+\alpha z_1 z_2} \right)^{\mu} = \tilde{g}_2(z,t) .$$

Für $L_0^{(d)} := D_1 D_2 + 4z_1(1+z_1{}^2 z_2)^{-2}$ gilt $L_0^{(d)} \in EP_{12}$, $L_0^{(d)} \notin EP_{12}^0$, $L_0^{(d)} \in P_{11}^0$; denn der eindeutig bestimmte Polynomkern $\tilde{g}_1$ 1. Art mit $s = (0,0)$ wird durch $\tilde{g}_1(z,t) = 1+8z_1{}^2 z_2(1+z_1{}^2 z_2)^{-1}(-t^2+\frac{2}{3}t^4)$ gegeben, während aber mit $\tilde{g}_1^*(z,t) = 1+(2-8z_1{}^2 z_2(1+z_1{}^2 z_2)^{-1})t^2$ ein (minimaler) Polynomkern kleineren Grades existiert.

$$L_0^{(e)} := D_1 D_2 + \alpha z_2 D_2 - \alpha, \quad \alpha \in \mathbb{R} \wedge \alpha > 0 ,$$

hat einerseits als Polynomkern $\tilde{g}_1$ 1. Art mit Bezugspunkt (0,0)

$$\tilde{g}_1(z,t) = 1+2\alpha z_1 z_2 t^2 ,$$

aber

$$\tilde{g}_2(z,t) = \exp(-\alpha z_1 z_2)\ {}_1F_1(2;\tfrac{1}{2};\alpha z_1 z_2 t^2)$$

ist der zweite Kern 1. Art zum zweigliedrigen Lösungsansatz. Da die Reihenentwicklung der Kummerschen Funktion ${}_1F_1$ um den Nullpunkt nicht abbricht, gilt also wohl $L_0^{(e)} \in EP_{11}^0$, aber andererseits $\forall m \in \mathbb{N}_0 : L_0^{(e)} \notin EP_{2m}$.

Sind j und n beliebig aber fest ($j=1$ oder $j=2$; $n \in \mathbb{N}_0$), so läßt sich insgesamt zeigen (vgl. [9c-d, 10]):

$$L_0 \in P_{jn} \Rightarrow \exists m \in \{0,\dots,n\} : L_0 \in P_{jm}^0 ,$$

$$L_0 \in P_{jn}^0 \Rightarrow \forall m \in \mathbb{N} \setminus \{0,\dots,n\} : L_0 \in P_{jm} ,$$

$$L_0 \in EP_{jn} \Rightarrow L_0 \in P_{jn} ;$$

$$L_0 \in P_{jn} \underset{\text{i.a.}}{\not\Rightarrow} \exists m \in \mathbb{N}_0 : L_0 \in EP_{jm} ,$$

$$L_0 \in P_{jn} \underset{\text{i.a.}}{\not\Rightarrow} \exists m \in \mathbb{N}_0 : L_0 \in P_{3-j,m} ,$$

$$L_0 \in EP_{jn} \underset{\text{i.a.}}{\not\Rightarrow} L_0 \in EP_{jn}^0 ,$$

$$L_0 \in EP_{jn} \underset{\text{i.a.}}{\not\Rightarrow} \exists m \in \mathbb{N}_0 : L_0 \in EP_{3-j,m} .$$

(Obige Beispiele $L_0^{(d)}$, $L_0^{(e)}$ bestätigen die beiden letzten der vorstehenden

Aussagen im Falle n=2, j=1 bzw. n=1, j=1.)

3.2. Bevor wir auf die Charakterisierung der Klassen P_{jn} und die Gewinnung von Polynomkernen näher eingehen, wollen wir noch auf eine weitere Möglichkeit der Lösungsdarstellung zu Gleichungen $L_o w = 0$ hinweisen, falls m, $n \in \mathbb{N}_o$ existieren, so daß $L_o \in P_{jn}$ bzw. $L_o \in P_{1n} \cap P_{2m}$ gilt. Dann lassen sich nämlich die Lösungsdarstellungen $w_j = B_j f_j$ mit Polynomkernen integralfrei ausdrücken und auffassen als durch einen Differentialoperator $\tilde{B}_j$ erzeugt. Insbesondere erhält man auf diesem Wege die von K.W. Bauer zu speziellen Gleichungen mit $L_o \in P_{jn}$ eingeführten Differentialoperatoren ([1a-d]). Es gilt nämlich (siehe [9a, c]):
Existiert in H ein Polynomkern $\tilde{g}_j$, $g_j(z,t) = \sum_{\mu=0}^{n} q_{j,2\mu}(z)((z_j-s_j)t^2)^\mu$, und ist $w_j = B_j f_j$ zu $L_o w = 0$ eine Lösungsdarstellung in H mittels des zugehörigen Bergman-Operators B_j und der B_j-Zugeordneten f_j $\in C^\omega(K_j(s_j;\frac{1}{2}r_j))$, $f_j(z_j) = \sum_{\varkappa=0}^{\infty} \gamma_{j\varkappa}(z_j-s_j)^\varkappa$, so gilt $w_j = \tilde{w}_j$, wenn

$$\tilde{w}_j(z) = (\tilde{B}_j\tilde{f}_j)(z), \quad \tilde{B}_j := e_j(z) \sum_{\mu=0}^{n} \frac{(2\mu)!}{4^\mu \mu!} q_{j,2\mu}(z)\, D_j^{\,n-\mu} \quad \text{und}$$

$$\tilde{f}_j(z_j) = \sum_{\nu=n}^{\infty} \Big[\frac{(2(\nu-n))!\pi}{8^{\nu-n}(\nu-n)!\nu!} \sum_{\lambda=0}^{\infty} \binom{\lambda+\nu-n}{\nu-n} s_j^{\lambda} \gamma_{j,\lambda+\nu-n}\Big] (z_j-s_j)^\nu .$$

Hiermit ergibt sich zu unserem Beispiel $L_o^{(c)}$ der folgende Differentialoperator $\tilde{B}_j^c$ von K.W. Bauer und E. Peschl:

$$\tilde{B}_j^c = \sum_{\mu=0}^{n} \frac{(2n-\mu)!}{(n-\mu)!\mu!} \left(\frac{-\alpha z_{3-j}}{1+\alpha z_1 z_2} \right)^{n-\mu} D_j^{\,\mu} .$$

4. Existenz und Darstellung von Polynomkernen

4.1. Die Klassen P_{jn} lassen sich außer durch das System in 3.1 auch durch Integralgleichungen zwischen b_j und c_j von $L^{(j)}$ (deren Ordnung aber leider vom Grad der Polynomkerne abhängt) festlegen ([8b]). Darüber hinaus haben spezielle Ansätze für g_j, z.B. als Polynom in einer gewissen Hilfsfunktion ([11d] , vgl. auch [2, 6, 8b, 10, 11e, 16b-c]), zu expliziten Ergebnissen für Teilklassen von P_{jn} geführt.

4.2. Unter Ausnutzung der Theorie der Laplace-Invarianten lassen sich jedoch allgemeine Konstruktionsmöglichkeiten für Gleichungen $L_0 w = 0$, zu denen Polynomkerne existieren, und auch für explizite Darstellungen solcher Kerne aufzeigen ([9c-d]). Auf diesem Wege erhält man die folgende Charakterisierung der Klassen P_{jn} : Sei $h_{j\rho} = D_j a_{j\rho} + a_{1\rho} a_{2\rho} - a_{3\rho}$ $(\rho \in \mathbb{N}_0)$; hierin seien die Funktionen $a_{1\rho}$, $a_{2\rho}$, $a_{3\rho}$ die Koeffizienten in $L_\rho := D_1 D_2 + a_{1\rho} D_1 + a_{2\rho} D_2 + a_{3\rho}$ $(a_{10}=a_1,\ a_{20}=a_2,\ a_{30}=a_3)$, wobei die Operatoren $L_\rho(\rho \in \mathbb{N})$ gewonnen werden aus $L_{\rho-1}$ durch

$$L_\rho T_{j\rho} = h_{j,\rho-1} T_{j\rho} h_{j,\rho-1}^{-1} L_{\rho-1} \quad \text{im Falle } h_{j,\rho-1} \neq 0 ,$$

bzw.

$$L_\rho T_{j\rho} = T_{j\rho} L_{\rho-1} \qquad \text{im Falle } h_{j,\rho-1} = 0$$

und $T_{j\rho}$ durch $T_{j\rho} := D_{3-j} + a_{j,\rho-1}$ definiert ist.

Dann ist für ein $n \in \mathbb{N}$ genau dann $L_0 \in P^o_{jn}$, wenn mit $L_0 w = 0$ für $w_{jn} := T_{jn} \cdots T_{j1} w$ die Gleichung $L_n w_{jn} = 0$ mit $h_{jn} = 0$, aber $h_{j\rho} \neq 0$ für $0 \leq \rho < n$ gilt. Ein minimaler Polynomkern zu $L_0 w = 0$ ist dann

$$\tilde{g}_j = e_{3-j} \sum_{\mu=0}^{n} \frac{\mu! 4^\mu}{(2\mu)!} \Big(\sum_{|\varkappa|=\mu} \prod_{\lambda=0}^{n-1} h_{j\lambda}^{-1} D_j^{\varkappa_{\lambda+1}} \frac{e_{jn}}{e_{3-j}} \Big) \ ((z_j - s_j) t^2)^\mu .$$

Dabei ist e_{jn} wie e_j definiert (jedoch mit a_{jn} statt $a_{jo}=a_j$), und es ist $\varkappa = (\varkappa_1,\ldots,\varkappa_n)$, $|\varkappa| = \varkappa_1+\ldots+\varkappa_n$, $\varkappa_\lambda = 0,\ 1$ für $\lambda = 1,\ldots,n$.

4.3. Für den zweigliedrigen Lösungsansatz läßt sich das Problem der Existenz von (minimalen) Polynomkernen $\tilde{g}_1$ und $\tilde{g}_2$ der Gerade n und m reduzieren auf die Betrachtung einer homogenen linearen gewöhnlichen Differentialgleichung der Ordnung n+m+1 mit holomorphen Koeffizienten. Aus den Lösungen einer solchen Gleichung lassen sich dann Operatoren $L_n \in P^o_{jo} \cap P_{3-j,n+m}$ und $L_0 \in P^o_{jn} \cap P_{3-j,m}$ konstruieren. Insbesondere existieren für alle n, $m \in \mathbb{N}_0$ Operatoren $L_n \in P^o_{jo} \cap P^o_{3-j,n+m}$ und $L_0 \in P^o_{jn} \cap P^o_{3-j,m}$. Bei dieser Konstruktion ergeben sich zugleich explizit die Funktionen $h_{j\rho}$ $(\rho=0,\ldots,n-1)$, e_{jn}, e_{3-j} (j=1, 2), so daß damit gemäß 4.2 Polynomkerne zu Bergmanschen Integraloperatoren bekannt sind und damit auch Differentialoperatoren $\tilde{B}_1$ und $\tilde{B}_2$ gemäß 3.2. Wegen der genauen Details hinsichtlich des Konstruktionsverfahrens sei auf [9d] verwiesen. Als Beispiel sei aber hierzu die folgende Gleichung

$$L_o^{(f)}w := D_1D_2w+(n-m)(D_1 \ln \omega)D_2w+n(m+1)(D_1D_2 \ln \omega)w = 0$$

mit $\omega(z) = \varphi_1(z_1)+\varphi_2(z_2)$($\varphi_1$, φ_2 wie in $L_o^{(b)}$) betrachtet.

Nach 4.2 erhält man (vgl. [9c, e]) als minimalen Polynomkern zu $L_o^{(f)}$:

$$\tilde{g}_j(z,t) = e_j(z) \sum_{\mu=o}^{k(j)} q_{j,2\mu}(z)((z_j-s_j)t^2)^\mu, \quad j = 1, 2 ,$$

mit

$$e_j = \omega^{k(j)}, \; k(1) = n, \; k(2) = m ,$$

$$q_{j,2\mu}(z) = \sum_{\nu=o}^{\mu} \omega(z)^{-\nu} \rho_{j\mu\nu}(z_j) ,$$

$$\rho_{j\mu\nu} = \frac{(-1)^\nu(k(3-j)+\nu)!}{\nu!\,(k(j)-\nu)!} \sum_{|\varkappa(j,\nu)|=\mu-\nu} \left[\prod_{\lambda=1}^{k(j)-\nu} ((D_j\varphi_j)^{-1}D_j^{\varkappa_\lambda})\right] \frac{4^\mu \mu!}{(2\mu)!} ,$$

$$\varkappa(j,\nu) = (\varkappa_1,\dots,\varkappa_{k(j)-\nu}), \; |\varkappa(j,\nu)| = \varkappa_1+\dots+\varkappa_{k(j)-\nu} ,$$

$$\varkappa_\lambda = 0, 1 \text{ für } \lambda = 1,\dots,k(j)-\nu.$$

Hieraus folgt nach 3.2 unmittelbar

$$\tilde{B}_j = \omega^{k(j)-n} \sum_{\mu=o}^{k(j)} \sum_{\nu=o}^{\mu} \omega^{-\nu} \rho_{j\mu\nu} D_j^{k(j)-\mu} ,$$

und es ergibt sich aus der Lösungsdarstellung nach 1.2, $w = B_1f_1+B_2f_2$, nach geeigneter Umformung die Darstellung

$$w = \tilde{B}_1\tilde{f}_1+\tilde{B}_2\tilde{f}_2$$
$$= \sum_{j=1}^{2} \sum_{\sigma=o}^{k(j)} \frac{(-1)^{k(j)-\sigma}(n+m-\sigma)!}{\sigma!\,(k(j)-\sigma)!} \omega^{\sigma-n}((D_j\varphi_j)^{-1}D_j)^\sigma \tilde{f}_j ,$$

d.h.: Die Lösungsdarstellung vermittels Differentialoperatoren nach Bauer [1d].

4.4. Im Falle des zweigliedrigen Lösungsansatzes mit Differentialoperatoren $\tilde{B}_1$, $\tilde{B}_2$ nach 3.2 sind im Unterschied zum Ansatz mit Bergman-Operatoren (vgl. 2.1) die $\tilde{B}_j$-Zugeordneten $\tilde{f}_1$ und $\tilde{f}_2$ noch nicht eindeutig festgelegt, wenn ein Funktionswert einer der beiden Funktionen vorgegeben ist. (Dies ist auch intuitiv nicht zu erwarten, da ja die Anfangsdaten der Lösung, die in f_1 und f_2 enthalten sind, gewissermaßen n-fach bzw. m-fach zur Gewinnung von $\tilde{f}_1$ und $\tilde{f}_2$ aufintegriert worden sind.) Hinsichtlich der Eindeutigkeit der integralfreien zweigliedrigen Lösungsdarstellung gilt vielmehr im Falle $L_0 \in P^o_{jn} \cap P^o_{3-j,m}$:

$\widetilde{w} = \widetilde{B}_1\widetilde{f}_1+\widetilde{B}_2\widetilde{f}_2$ ist eindeutig bis auf einen Summanden der Form

$$w_0 = \sum_{\mu=o}^{n} A_{j\mu} \sum_{\lambda=1}^{n+m+1} k_\lambda D_j^{\mu}\xi_{j\lambda} + \sum_{\rho=o}^{m} A_{3-j,\rho} \sum_{\lambda=1}^{n+m+1} k_\lambda D_{3-j}^{\rho}\eta_{j\lambda} \qquad (k_\lambda \in \mathbb{C}) .$$

Hierin sind $\xi_{j\lambda} \in C^{\omega}(G)$ $(\lambda = 1,\dots,n+m+1)$ linear unabhängige Funktionen, die nur von z_j abhängen, und $\eta_{j\lambda} \in C^{\omega}(G)$ $(\lambda = 1,\dots,n+m+1)$ ist eine Basis von Lösungen einer gewöhnlichen Differentialgleichung der in 4.3 erwähnten Art. Die $A_{j\mu}$ $(\mu = 0,\dots,n)$ und $A_{3-j,\rho}$ $(\rho = 0,\dots,m)$ sind gewisse Determinanten, die gemäß [9d, Satz 14] aus den vorstehenden Funktionen $\xi_{j\lambda}$ und $\eta_{j\lambda}$ und ihren Ableitungen gebildet werden und mit deren Hilfe sich die Koeffizienten von L_0 darstellen lassen. Hiermit ergibt sich insbesondere zur Gleichung

$$L_0^{(g)}w := D_1D_2w-n(n+1)(z_1+z_2)^{-2}w = 0 \qquad (n \in \mathbb{N}_o)$$

zu einer vorgegebenen Lösung $\widetilde{w} = \widetilde{B}_1\widetilde{f}_1+\widetilde{B}_2\widetilde{f}_2$ als allgemeinstes Zugeordnetenpaar $(\widetilde{f}_1(z_1)+ \sum_{\lambda=o}^{2n} k_\lambda(-1)^\lambda z_1^\lambda,\ \widetilde{f}_2(z_2)- \sum_{\lambda=o}^{2n} k_\lambda z_2^\lambda)$, d.h.: das Ergebnis stimmt überein mit dem von Bauer in [1b] behandelten Spezialfall.

5. Funktionentheoretische Anwendungen

5.1. Unter den mannigfachen Anwendungsmöglichkeiten der beschriebenen Integral- und Differentialoperatoren aus den oben charakterisierten Teilklassen wollen wir nur zwei Ergebnisse herausgreifen, nämlich hinsichtlich Werteverteilung und Singularitäten der erzeugten Lösungen.

Für unser Beispiel $L_0^{(c)}w = 0$ mit $\alpha = -1$ wurde von S. Ruscheweyh [15b] ein Analogon zum Picardschen Satz aufgestellt. Ergebnisse im Hinblick auf den ersten Fundamentalsatz von R. Nevanlinna finden sich bei Bergman [3b, insbesondere § 8]. Analoga zum zweiten Fundamentalsatz von Nevanlinna finden sich bei Chernoff [4].

Sei nun $u = w_1$. (Entsprechende Sätze wie hier in 5.1 und 5.2 lassen sich selbstverständlich auch für w_2 formulieren, indem man die Variable z_2 anstelle z_1 in den Vordergrund rückt.) Es sei $G = \mathbb{C}^2$, so daß also der Kern 1. Art gemäß 2.1 eine ganze Funktion in z_1, z_2 ist, und seien $z_2 = \bar{z}_1$, $s = (0,0)$. Ferner lasse sich $\tilde{g}_1$ darstellen als Funktion in den Variablen z_1, $r = (z_1\bar{z}_1)^{1/2} = |z_1|$ und t. Die B_1-Zugeordnete f_1 werde als ganze Funktion gewählt. Dann ist die Funktion u eine ganze Funktion für jeden Wert $r > 0$, d.h.: sie nimmt auf jedem Kreisrand $\partial K_1(0;r)$ die Werte einer ganzen Funktion u_r an, $u(z_1,\bar{z}_1)\big|_{\partial K_1(0;r)} = u_r(z_1)$. Damit läßt sich die klassische Funktionentheorie einer komplexen Variablen anwenden. Man definiert deshalb:

$$n[r,(u(z_1,\bar{z}_1)-a)^{-1}] := \frac{1}{2\pi i}\int_{|z_1|=r} d\log[u(z_1,\bar{z}_1)-a].$$

Wenn u keine a-Stellen auf $|z_1| = r$ hat und dort nicht konstant ist, stimmt dieser Ausdruck überein mit der Anzahl der a-Stellen (einschließlich ihrer Ordnung) $\tilde{n}[r,(u_r(z_1)-a)^{-1}]$ der Funktion u_r in $|z_1| \leq r$. In Analogie zu $\tilde{m}[r,u_r(z_1)] = \frac{1}{2\pi}\int_0^{2\pi} \log^+|u_r(re^{i\varphi})|d\varphi$ für meromorphe Funktionen führt man jetzt weiter ein

$$m[r,u(z_1,\bar{z}_1)] := \frac{1}{2\pi}\int_0^{2\pi} \log^+|u(re^{i\varphi},re^{-i\varphi})|\, d\varphi .$$

Da $u(z_1,\bar{z}_1) = u_r(z_1)$ auf $|z_1| = r$ (u_r ganz) ist, existiert der vorstehende Ausdruck und, falls $u(z_1,\bar{z}_1) \neq a$ auf $|z_1| = r$, ebenfalls $m[r,(u(z_1,\bar{z}_1)-a)^{-1}] = \tilde{m}[r,(u_r(z_1)-a)^{-1}]$. (Da $u(z_1,\bar{z}_1)$ stets endlich ist, kann $m[r,u(z_1,\bar{z}_1)]$ als Größe für das Wachstum der Funktion u aufgefaßt werden.)

H. Chernoff beweist nun durch Zurückgehen auf die klassische Darstellung von Nevanlinna im Appendix zu [4] eine modifizierte Form des zweiten Fundamentalsatzes und überträgt diesen auf Lösungen $u = w_1 = B_1f_1$; so erhält er obere Schranken für $m[r,u(z_1,\bar{z}_1)]$ in Termen von Ausdrücken $n[r,(u(z_1,\bar{z}_1) -a_\nu)^{-1}]$ ([4], Theorems 3, 3a, 4). Hiermit oder auch mittels eines Korollars (u_r holomorph für $|z_1| \leq r \wedge u_r(0) = 0 \wedge |u_r(z_1)| > |a|$ für $|z_1| = r$ $\Rightarrow$ $\exists z_1 \in K_1(0;r)$: $u_r(z_1) = a$) zum Satz von Rouché beweist man letztlich folgende Form des Picardschen Satzes für Lösungen u :

Wenn Konstante r_0, $\tilde{r}_0$ und q_0 existieren, so daß für $|z_1| \leq r_0$, $r > \tilde{r}_0$ $\tilde{g}_1(z_1,r,t) = \tilde{\tilde{g}}_1(r,t)(1+\varepsilon e^{i\delta})$ gilt, wobei $|\varepsilon| \leq q_0 < 1$ und $\tilde{\tilde{g}}_1(r,t)$ reell für reelle $t \in [-1, +1]$ und nach unten beschränkt für $r > 0$ ist, wenn $f_1(z_1) = \sum_{\kappa=\lambda}^{\infty} \gamma_{1\kappa} z_1^{\kappa}$ mit $\gamma_{1\lambda} \neq 0$, $\lambda > 0$, ist und ferner

$$\rho_\lambda(r) := 2^{-\lambda} \int_{-1}^{+1} \tilde{\tilde{g}}_1(r,t)(1-t^2)^{\lambda-(1/2)} dt \to \infty \quad \text{für} \quad r \to \infty ,$$

dann folgt, daß $u = w_1 = B_1 f_1$ alle endlichen Werte annimmt.

Die Voraussetzungen dieses Satzes werden beispielsweise durch die folgenden Kerne erfüllt:

$$\tilde{g}_1(z_1,r,t) = \tilde{\tilde{g}}_1(r,t) = \cosh (rt)$$

zu $L_o^{(a)} w = 0$ mit $k = i/2$ (also zu einer Gleichung, welche gemäß 1.1 aus $\Delta w - w = 0$ $(\Delta = \partial^2/\partial x^2 + \partial^2/\partial y^2)$ gewonnen werden kann) und $\tilde{g}_1^*(z_1,r,t) = \tilde{\tilde{g}}_1^*(r,t) = 1+2\alpha r^2 t^2$ $(\alpha > 0)$ zu $L_o^{(e)} w = 0$.

Denn im ersten Falle gilt für $\tilde{\tilde{g}}_1$

$$\rho_\lambda(r) = \frac{\pi(2\lambda)!}{8^\lambda \ \lambda!} \sum_{\mu=0}^{\infty} (\mu!(\mu+\lambda)!)^{-1} \left(\frac{r}{2}\right)^{2\mu}$$

$$\geq \frac{\pi(2\lambda)!}{8^\lambda \ \lambda!} \left(\frac{\exp (r/2)}{2 \ (r/2)^{2\lambda}} - \sum_{\mu=0}^{\lambda-1} \frac{1}{(2\mu)!} \left(\frac{2}{r}\right)^{2(\lambda-\mu)} \right) \to \infty \quad \text{für } r \to \infty ,$$

und im zweiten Fall gilt für $\tilde{\tilde{g}}_1^*$

$$\rho_\lambda^*(r) = \frac{\pi(2\lambda)!}{8^\lambda(\lambda!)^2} \left(1 + \frac{1}{\lambda+1} \alpha r^2\right) \to \infty \quad \text{für} \quad r \to \infty .$$

5.2. Unter den vielen Möglichkeiten, Sätze über Singularitäten aus der Theorie der analytischen Funktionen einer komplexen Variablen auf Lösungen $u = w_1 = B_1 f_1$ zu übertragen, wollen wir hier nur eine aufzeigen, die von einem auf Borel und Hadamard zurückgehenden Satz Gebrauch macht (vgl. auch [3b], [11b]).

Seien die Reihenentwicklungen von $e_1 z_1^{\mu} q_{1,2\mu}$ $(\mu = 0,\dots,n)$ im Bezugspunkt $s = (0,0)$ gegeben durch $\sum_{\sigma,\tau=0}^{\infty} a_{\sigma\tau}^{(\mu)} z_1^{\sigma} z_2^{\tau}$ und von f_1 durch $\sum_{\kappa=0}^{\infty} \gamma_{1\kappa} z_1^{\kappa}$.

Durch Einsetzen in die Integraldarstellung ergibt sich dann $u(z_1,0) = \sum_{\rho=0}^{\infty} \beta_\rho z_1^{\rho}$ mit

$$\beta_\rho := \sum_{\kappa=0}^{\rho} \sum_{\mu=0}^{n} a_{\rho-\kappa,0}^{(\mu)} \ \gamma_{1\kappa} \ \frac{\pi(2\mu)! \ (2\kappa)!}{2^{2\mu+3\kappa} \mu!\kappa!(\mu+\kappa)!} .$$

Seien $\Delta(p,q) := \det((\varepsilon_{\sigma\tau}^{(p,q)})_{\sigma,\tau})$ mit

$$\varepsilon_{\sigma\tau}^{(p,q)} = \beta_{p+\sigma+\tau-2} \quad \text{und} \quad \sigma,\tau = 1,\ldots,q+1; \quad p,q\in\mathbb{N} .$$

Mittels obigen Satzes ergibt sich somit das folgende Ergebnis:

Eine Lösung $u = w_1 = B_1 f_1$ mit Polynomkern $\tilde{g}_1$ vom Grade n hat höchstens endlich viele Pole der Ordnung $\leq q$ ($q\in\mathbb{N}$) auf der Ebene $z_2 = 0$, wenn höchstens endlich viele der (q+1)-reihigen Determinanten $\Delta(p,q)$, $p\in\mathbb{N}$, nicht verschwinden.

Ist $\tilde{g}_1$ sogar Polynomkern 1. Art und ist L_o nicht der Laplace-Operator, so läßt sich eine weitere Darstellungsformel aus [3c, § I.3] anwenden, und die Pole auf der Ebene $z_2 = 0$, d.h. Schnitte von Singularitätenebenen des z_1,z_2-Raumes mit der Ebene $z_2 = 0$, erweisen sich als auf Singularitätenebenen liegend, die Polebenen der Ordnung $\leq q$ und für $z_2 \neq 0$ zugleich logarithmische Verzweigungsebenen sind ("polähnliche Singularitäten").

6. Literatur

[1] BAUER, K.W.: (a) Über eine der Differentialgleichung $(1\pm z\bar{z})^2 w_{z\bar{z}} \pm n(n+1)w = 0$ zugeordnete Funktionentheorie. Bonner math. Schriften 23, 98 pp. (1965). - (b) Über eine partielle Differentialgleichung 2. Ordnung mit zwei unabhängigen komplexen Variablen. Monatsh. Math. 70, 385-418 (1966). - (c) Differentialoperatoren bei partiellen Differentialgleichungen. Ber. Ges. Math. Datenverarb., Bonn 77, 7-17 (1973). - (d) Polynomoperatoren bei Differentialgleichungen der Form $w_{z\bar{z}}+Aw_{\bar{z}}+Bw = 0$. J. reine angew. Math. (im Druck).

[2] BAUER, K.W., FLORIAN, H.: Bergman-Operatoren mit Polynomerzeugenden. Appl. Analysis (im Druck).

[3] BERGMAN, S.: (a) Linear operators in the theory of partial differential equations. Trans. Amer. Math. Soc. 53, 130-155 (1943). - (b) Certain classes of analytic functions of two real variables and their properties. Trans. Amer. Math. Soc. 57, 299-331 (1945). - (c) Integral operators in the theory of linear partial differential equations. Ergebn. Math. Grenzgeb. 23, 3rd pr., Berlin-Heidelberg-New York: Springer 1971.

[4] CHERNOFF, H.: Complex solutions of partial differential equations. Amer. J. Math. 68, 455-478 (1946).

[5] COLTON, D.: Complete families of solutions for parabolic equations with analytic coefficients. SIAM J. Math. Anal. 6, 937-947 (1975).

[6] FLORIAN, H., JANK, G.: Polynomerzeugende bei einer Klasse von Differentialgleichungen mit zwei unabhängigen Variablen. Monatsh. Math. 75, 31-37 (1971).

[7] GILBERT, R.P.: (a) Function theoretic methods in partial differential equations. New York - London: Academic Press: 1969. - (b) Constructive methods for elliptic equations. Lecture Notes Math. 365, Berlin -

Heidelberg-New York: Springer 1974.

[8] KRACHT, M.: (a) Integraloperatoren für die Helmholtzgleichung. Z. angew. Math. Mech. 50, 389-396 (1970). - (b) Zur Existenz und Charakterisierung von Bergman-Operatoren. I: Der eingliedrige Lösungsansatz. J. reine angew. Math. 265, 202-220 (1974). - (c) Zur Existenz und Charakterisierung von Bergman-Operatoren. II: Paare von Zugeordneten. J. reine angew. Math. 266, 140-158 (1974).

[9] KRACHT, M., KREYSZIG, E.: (a) Bergman-Operatoren mit Polynomen als Erzeugenden. Manuscripta math. 1, 369-376 (1969). - (b) Zur Darstellung von Lösungen der Gleichung $\Delta\psi + c(1 + x^2 + y^2)^{-2}\psi = 0$. Z. angew. Math. Mech. 50, 375-380 (1970). - (c) Zur Konstruktion gewisser Integraloperatoren für partielle Differentialgleichungen. Teil I. Manuscripta math. 17, 79-103 (1975). - (d) Zur Konstruktion gewisser Integraloperatoren für partielle Differentialgleichungen. Teil II. Manuscripta math. 17, 171-186 (1975). - (e) Konstruktive Methoden bei gewissen Bergman-Operatoren. (In Vorbereitung).

[10] KRACHT, M., SCHRÖDER, G.: Bergmansche Polynom-Erzeugende erster Art. Manuscripta math. 9, 333-355 (1973).

[11] KREYSZIG, E.: (a) On a class of partial differential equations. J. Rational Mech. Anal. 4, 907-923 (1955). - (b) Relations between properties of solutions of partial differential equations and the coefficients of their power series development. J. Math. Mech. 6, 361-381 (1957). - (c) Über zwei Klassen Bergmanscher Operatoren. Math. Nachr. 37, 197-202 (1968). - (d) On Bergman operators for partial differential equations in two variables. Pacific J. Math. 36, 201-208 (1971). - (e) Representations of solutions of certain partial differential equations related to Liouville's equation. Abh. math. Sem. Univ. Hamburg (im Druck).

[12] LANCKAU, E.: Zur Lösung gewisser partieller Differentialgleichungen mittels parameterabhängiger Bergman-Operatoren. Nova Acta Leopoldina N. F. 201 / Bd. 36, Leipzig: Barth 1971.

[13] MITCHELL, J.: Approximation to the solutions of linear partial differential equations given by Bergman integral operators. Ber. Ges. Math. Datenverarb., Bonn 77, 97-107 (1973).

[14] ROSENTHAL, P.: On the location of the singularities of the function generated by the Bergman operator of the second kind. Proc. Amer. Math. Soc. 44, 157-162 (1974).

[15] RUSCHEWEYH, S.: (a) Gewisse Klassen verallgemeinerter analytischer Funktionen. Bonner math. Schriften 39, 79 pp. (1969). - (b) Geometrische Eigenschaften der Lösungen der Differentialgleichung $(1 - z\bar{z})^2 w_{z\bar{z}} - n(n + 1)w = 0$. J. reine angew. Math. 270, 143-157 (1974).

[16] WATZLAWEK, W.: (a) Über Zusammenhänge zwischen Fundamentalsystemen, Riemann-Funktion und Bergman-Operatoren. J. reine angew. Math. 251, 200-211 (1971). - (b) Über lineare partielle Differentialgleichungen zweiter Ordnung mit Bergman-Operatoren der Klasse P. Monatsh. Math. 76, 356-369 (1972). - (c) Hyperbolische und parabolische Differentialgleichungen der Klasse P. Ber. Ges. Math. Datenverarb., Bonn 77, 147-179 (1973).

Anschrift des Autors:

Mathematisches Institut, Universität Düsseldorf
Universitätsstr. 1, D - 4000 Düsseldorf

EINE FUNKTIONALGLEICHUNG ZUR SCHALLBEUGUNG

N. Latz
Technische Universität Berlin

1. Formulierung der Aufgabe

Die Ausbreitung stationärer akustischer Schwingungen in stückweise homogenen und isotropen Medien wurde von Kupradse in |3| mit Hilfe der Integralgleichungsmethode untersucht. Es handelt sich dabei um die Lösung eines speziellen Transmissionsproblems elastischer Schwingungen. Diese Aufgabe wird im folgenden zugrundegelegt zur Herleitung einer Funktionalgleichung für holomorphe Abbildungen, nämlich die Fourier-Laplacetransformierten des elastischen Verschiebungsvektors und der Hauptachsenkomponente des Spannungstensors. Das Schallbeugungsproblem lautet folgendermaßen:

Es sei G die Vereinigungsmenge endlich vieler offener und zusammenhängender Gebiete G_l aus R^3, $l\varepsilon T(r) = \{1,...,r\geq 2, r\varepsilon N\}$. Sie seien punktfremd, und es gelte $\bar{\bar{G}} = R^3$. Jedes Gebiet sei Träger eines homogenen und isotropen akustischen Mediums, das gekennzeichnet ist durch die Lamékonstante $\lambda_l > 0$, die Dichtekonstante $Q_l > 0$ und den Dämpfungsfaktor $R_l > 0$ (|6|). Der zeitharmonische akustische Schwingungszustand der Medien werde erzeugt durch eine stetige Volumenkraft $F:G \to C^3$. Gesucht sind der elastische Verschiebungsvektor $U:G \to C^3$ und die Komponente $\sigma:G \to C$ in der Hauptachse des Spannungstensors; sie genügen folgenden Bedingungen:

i) U und σ sind stetig differenzierbar und erfüllen die Elastizitätsgleichung und das Hookesche Gesetz mit der Lamékonstanten $\mu_l = 0$, $l\varepsilon T(r)$:

$$\operatorname{grad} \sigma + \kappa U + F = 0$$
$$\sigma - \lambda \operatorname{div} U = 0$$

Dabei bezeichnet λ bzw. κ eine Treppenfunktion auf G, die auf G_l erklärt ist durch λ_l bzw. $\kappa_l = \omega^2 Q_l + i\omega R_l$, $i^2 = -1$, $l\varepsilon T(r)$. Die Konstante $\omega > 0$ ist die Frequenz der zeitharmonischen akustischen Schwingung.

ii) Längs des glatten Randes δG mit der Normalen n gelten die Transmissionsbedingungen:

$$\langle U_+, n\rangle = \langle U_-, n\rangle, \quad \sigma_+ = \sigma_- .$$

Die untere Indizierung bezieht sich dabei auf die Grenzwertbildung der Ortsvariablen gegen gemeinsame Randpunkte aus verschiedenen Gebieten der Anordnung. Die Grenzwerte sollen existieren.

iii) Die Abbildungen F, U und σ klingen im Unendlichen gleichmäßig für alle Richtungen exponentiell ab mit dem Exponentialfaktor $\gamma>0$.

Vorausgesetzt sei, daß G so beschaffen ist, daß die formulierten Bedingungen ausreichen, um aus Greenschen Umformungen für beschränkte Gebiete solche für unbeschränkte zu gewinnen.

2. Herleitung der Funktionalgleichung

Es seien $p = (p_1,p_2,p_3) = \rho + i\tau \in C^3$ mit $|\rho|<\gamma$ und $\langle p,x\rangle$ das Skalarprodukt aus p und $x\in R^3$. Dann gilt unter den gemachten Voraussetzungen folgende Greensche Umformung (|5|):

$$\int_{G_l} \exp-\langle p,x\rangle \operatorname{grad} \sigma(x)\, dx =$$

$$\int_{\delta G_l} n(s)\exp-\langle p,s\rangle\sigma(s)\, ds$$

$$+\ p\int_{G_l} \exp-\langle p,x\rangle\, \sigma(x)\, dx, \qquad l\in T(r).$$

Summieren wir die linke und die rechte Seite dieser Gleichung über $l\in T(r)$, so ergibt sich unter Ausnutzung der Elastizitätsgleichung, der zweiten Transmissionsbedingung und der Ausstrahlungsbedingung:

$$p\,\hat{\sigma}(p) + \sum \kappa_l \hat{U}_l(p) + \hat{F}(p) = 0, \quad l\in T(r).$$

Dabei bezeichnet $\hat{\sigma}$ bzw. $\hat{U}_l$ die Fourier-Laplacetransformierte von σ über R^3 bzw. von U über G_l. Ganz analog lassen sich über den Satz von Gauss das Hookesche Gesetz und die erste Transmissionsbedingung transformieren; man erhält:

$$\sum \lambda_l^{-1}\hat{\sigma}_l(p) - \langle p,\hat{U}(p)\rangle = 0, \qquad l\in T(r).$$

Die beiden voranstehenden Gleichungen können folgendermaßen zusammengefaßt werden:

$$(1) \quad \sum m_l(p)\ \phi_l(p) = \omega(p), \quad l\in T(r), \text{für}$$

$$p = (p_1,p_2,p_3) = \rho + i\tau \in C^3 : |\rho|<\gamma,$$

$$\text{mit } \phi_l(p)' = (\hat{\sigma}_l(p),\ \hat{U}_l(p)) = \int_{G_l} \exp-\langle p,x\rangle(\sigma(x),\ U(x))\, dx,$$

$$\omega(p)' = (0,-\hat{F}(p)),$$

$$m_l(p) = \begin{pmatrix} \lambda_l^{-1}, & -p_1, & -p_2, & -p_3 \\ p_1, & \kappa_l, & 0, & 0 \\ p_2, & 0, & \kappa_l, & 0 \\ p_3, & 0, & 0, & \kappa_l \end{pmatrix}.$$

In (1) repräsentieren die Matrizen die Differentiationseigenschaften der Feldgrößen. Die Transmissionsbedingungen der Aufgabe werden in (1) durch die Summenbildung eingebracht. Wir können folgendes Ergebnis notieren:

Folgerung 1

Aus den Bedingungen des formulierten Problems ergibt sich über Greensche Umformungen die Funktionalgleichung (1). Sie bezieht sich auf die holomorphen Fourier-Laplacetransformierten der gesuchten Feldgrößen.

3. Lösung der Funktionalgleichung

Wir beschäftigen uns im folgenden mit einer Lösungsaussage für die Gleichung (1) zu $p = i\tau$, $\tau \varepsilon R^3$. In diesem Falle sind die Fourier-Laplacetransformierten Elemente des Lebesgue-Raumes $L_2 = L_2(R^3)$ aus 4-dimensionalen meßbaren Abbildungen auf R^3 mit endlicher Hilbertnorm. Dementsprechend interpretieren wir (1) als lineare Gleichung auf L_2. Es sei dazu:

$$\kappa_0 \varepsilon C : |\kappa_0| = 1, \ 0 < \arg \kappa_o < \min \{\arg \kappa_l, l \varepsilon T(r)\}.$$

Diese Festlegung ist sinnvoll, da die Vorgaben κ_l, $l \varepsilon T(r)$, im offenen ersten Quadranten C^+ der komplexen Ebene liegen sollen. Dann gilt:

$$\alpha_l := \kappa_o \ \lambda_l^{-1} \varepsilon \ C^+ \ \text{und} \ \beta_l := \kappa_o^{-1} \kappa_l \ \varepsilon \ C^+, \ l \varepsilon T(r).$$

Es sei außerdem für $\tau = (\tau_1, \tau_2, \tau_3) \varepsilon R^3$ und $l \varepsilon T(r)$:

$$f_l(\tau)' = (\kappa_o^{-1} \hat{\sigma}(i\tau), \ \hat{U}(i\tau)),$$

$$g\ (\tau)' = (0, -\kappa_0^{-1} \hat{F}(i\tau)),$$

$$n_l(\tau) = \begin{pmatrix} \alpha_l, & -i\tau_1, & -i\tau_2, & -i\tau_3 \\ i\tau_1, & \beta_l, & 0, & 0 \\ i\tau_2, & 0, & \beta_l, & 0 \\ i\tau_3, & 0, & 0, & \beta_l \end{pmatrix}.$$

Dann ergibt sich aus (1):

$$(2) \ \sum n_l(\tau) f_l(\tau) = g\ (\tau), \ \tau \varepsilon R^3, \ l \varepsilon T(r).$$

Folgerung 2

Die Spezialisierung von (1) auf $p = i\tau$, $\tau \varepsilon R^3$, kann äquivalent umgeformt werden in (2).

Mit (2) kommen wir folgendermaßen zu einer linearen Gleichung auf L_2: es ist $f \varepsilon L_2$ mit $f(\tau) = \sum f_l(\tau)$, $\tau \varepsilon R^3$, $l \varepsilon T(r)$; die rechte Seite von (2) erklärt ein $g \varepsilon L_2$. Aus (2) ergibt sich dann folgende lineare Gleichung für $f \varepsilon L_2$ bei gegebenem $g \varepsilon L_2$:

$$(3) \quad Af = g \text{ mit } A = \sum N_l P_l, \; l \varepsilon T(r).$$

Die Operatoren N_l und P_l sind erklärt durch:

$$(N_l f)(\tau) = n_l(\tau) f(\tau),$$

$$(P_l f)(\tau) = (F_2 \Gamma_l F_2^{-1}) f(\tau), \tau \varepsilon R^3.$$

Dabei bezeichnen F_2 die Fouriertransformation auf L_2 und Γ_l, $l \varepsilon T(r)$, die charakteristische Funktion von G_l. Der Operator A ist erklärt auf dem Teilraum $D(A) = \{f \varepsilon L_2 : Af \varepsilon L_2\}$. Mit A ist auf D(A) auch folgende Schar B(t), t>0, von Multiplikatoren definiert:

$$B(t)f(\tau) = b(\tau,t)f(\tau) \text{ mit}$$

$$b(\tau,t) = \begin{pmatrix} it, & -i\tau_1, & -i\tau_2, & -i\tau_3 \\ i\tau_1, & it, & 0, & 0 \\ i\tau_2, & 0, & it, & 0 \\ i\tau_3, & 0, & 0, & it \end{pmatrix}$$

für $\tau = (\tau_1, \tau_2, \tau_3) \varepsilon R^3$ und $t>0$.
Die Schar B(t) operiert auf D(A), da $C(t) = A - B(t)$ auf L_2 beschränkt ist. Durch Rechnung beweist man für diese Operatoren:

Folgerung 3

Die Scharen $B^{-1}(t)$ und $C(t) = A - B(t)$, $t>0$, sind auf L_2 beschränkt mit

$$||B^{-1}(t)|| \leqslant t^{-1} \text{ und}$$

$$||C(t)|| \leqslant \max\{|\alpha_l - \eta|, |\beta_l - \eta|, \eta = it, t>0, l \varepsilon T(r)\}.$$

Zu bemerken ist, daß zum Beweis der Normabschätzung für $B^{-1}(t)$ die zugehörige Matrix $b^{-1}(\tau,t)$ nicht benötigt wird. Es genügt, von $b(\tau,t)$ auszugehen und dafür die Spektralnorm zu berechnen (|1|). Unter Ausnutzung von Folgerung 3 zeigt man nun über eine einfache geometrische Konstruktion:

Folgerung 4:
Es existiert ein $t_o>0$, so daß für alle $t\geq t_o$ gilt: $||B^{-1}(t)||\cdot||C(t)||<1$.

Dabei ist wesentlich zu beachten, daß die Zahlen α_1 und β_1 im offenen ersten Quadranten der komplexen Ebene liegen.
Für die lineare Gleichung (3) bedeuten die voranstehenden Ergebnisse: Der Operator A ist beschränkt invertierbar (|1|). Damit besitzt (3) genau eine Lösung auf L_2. Die Gleichung (3) kann außerdem in folgende iterierbare überführt werden:

$$(4)\quad (I-K(t))f = h(t) \quad \text{mit } K(t) = B^{-1}(t)C(t),\ h(t) = B^{-1}(t)g.$$

Zusammengefaßt erhalten wir folgendes Ergebnis:

Satz

Die lineare Gleichung (3) besitzt genau eine Lösung $f\epsilon L_2$. Diese kann gewonnen werden durch Iteration der Gleichung (4) für $t\geq t_o$ mit einem gegeeigneten $t_o>0$. Die Lösung f erfüllt für fast alle $\tau\epsilon R^3$, die Funktionalgleichung (2). Für fast alle $p = i\tau, \tau\epsilon R^3$, ergibt sich daraus eine Lösung der Funktionalgleichung (1).

4. Bemerkungen

Das vorliegende Ergebnis ist im folgenden Sinne unbefriedigend: Die Funktionalgleichung (1) gilt für holomorphe Abbildungen, die Lösungsaussage bezieht sich jedoch nur auf eine Spezialisierung. Legt man zur Untersuchung der Gleichung statt des L_2 Hardyräume zugrunde (|2|), so gewinnt man nach dem beschriebenen Verfahren eine holomorphe Lösung.
Ganz analog zum Problem der Schallbeugung kann die von Kupradse in |3| gelöste Aufgabe der Beugung einer linear polarisierten elastischen Querwelle durch eine Funktionalgleichung der Form (1) gekennzeichnet werden. Dasselbe gilt für das allgemeine elastische Transmissionsproblem, das in |4| gelöst wurde. Die sich ergebende Funktionalgleichung bezieht sich in diesem Falle auf holomorphe Abbildungen mit mehr als 4 Komponenten.
Wir diskutieren zum Schluß eine Spezialisierung des Schallbeugungsproblems, die vorliegt, wenn gilt: $\kappa_1 = \kappa \ \epsilon\ C^+$, $1\epsilon T(r)$. Setzen wir dann $k_1^2 = \lambda_1^{-1}\kappa^2$, so ergibt sich aus den gekoppelten Funktionalgleichungen des Abschnitts 2 eine skalare für $\hat{\sigma}(p)$:

$$(5)\quad \sum(\langle p, p\rangle + k_1^2)\hat{\sigma}_1(p) + \langle p,\hat{F}(p)\rangle = 0,\ 1\epsilon T(r).$$

Diese Gleichung kann zu $p = i\tau$, $\tau\epsilon R^3$, umgeformt werden in:

$$(6)\quad \hat{\sigma}(p) - \sum(k^2-k_1^2)(\langle p,p\rangle + k^2)^{-1}\hat{\sigma}_1(p) + \langle p,\hat{F}(p)\rangle (\langle p, p\rangle + k^2)^{-1} = 0,\ 1\epsilon T(r) \text{ mit } k\epsilon C,\ \operatorname{Im} k>0.$$

Die Umkehrung der Fouriertransformation liefert dann anhand von (6) folgende Schar von Integralgleichungen für $\sigma : G \to C$:

$$(7) \quad \sigma(x) - \int_{R^3} S_k(|x-y|)\Gamma(y)\sigma(y)\,dy = \Psi(x),\ x \in R^3,$$

$$\text{mit } S_k(r) = (4\pi r)^{-1} \exp ikr\ ,\ r>0,$$

$$\Gamma(y) = (k^2 - k_1^2) \text{ für } y \in G_1,\ 1 \in T(r).$$

Die rechte Seite von (7) ist bekannt und definiert ein Element des skalaren L_2. Der Parameter $k \in C$ kann so gewählt werden, daß die Gleichung (7) auf L_2 durch Iteration lösbar wird. Dieser Lösungsansatz für die spezialisierte Schallbeugungsaufgabe ist verschieden von demjenigen in |3|. Zur Lösung von (7) braucht der Integraloperator nicht kompakt zu sein.

Literatur

|1| L. COLLATZ, Funktionalanalysis und numerische Mathematik, 79, 140, Berlin 1964, Springer-Verlag.

|2| E. HILLE, Analytic function theory II, 429, New-York 1962, Ginn and Company.

|3| W. D. KUPRADSE, Randwertaufgaben der Schwingungstheorie und Integralgleichungen, 83-85, Berlin 1956, VEB Deutscher Verlag der Wissenschaften.

|4| V. D. KUPRADZE, Dynamical Problems in Elasticity, 64-73. In: I. N. SNEDDON and R. HILL (Eds.), Progress in Solid Mechanics III, Amsterdam 1963, North-Holland Publishing Company.

|5| R. LEIS, Vorlesungen über partielle Differentialgleichungen zweiter Ordnung, 21-22, Mannheim 1967, Hochschultaschenbücher-Verlag.

|6| R. LEIS, Zur Theorie elastischer Schwingungen, GMD-Bericht Nr. 72, Bonn 1973, Gesellschaft für Mathematik und Datenverarbeitung.

Properties of Solutions of Linear Partial Differential Equations given by Integral Operators

by

Josephine Mitchell

§1. Introduction.

Let n be a positive integer, $x = (x_1,\ldots,x_n)$ a point in Euclidean space R^n. Let

$$(1) \quad \Delta_n u + \sum_{j=1}^{n} a_j(x) u_{x_j} + b(x)u = 0 ,$$

where Δ_n is the Laplace operator in n real variables and $a_j(x)$ and $b(x)$ are real-valued analytic functions of x in a domain in R^n. Bergman [1,2] and Vekua [14] constructed integral operators, which map analytic functions of one complex variable onto C^2 solutions of (1) when $n = 2$ and obtained many properties of such solutions. Bergman also considered the Laplace equation and other special equations for $n = 3$ [2]. Colton extended the integral operator solutions to $n = 3$ [4] and Colton and Gilbert to $n = 4$ [5], whileas Kukral proved that this method, which uses a variable of a particular form, cannot be extended to $n \geq 5$ variables [9].

Consider the case $n = 3$. Continue a_j and b $(j = 1,2,3)$ to complex functions A, B, C and D respectively and introduce the new variables $x_1 = X$, $Z = \frac{1}{2}(x_2 + ix_3)$, $Z^* = \frac{1}{2}(-x_2 + ix_3)$. Then (1) becomes in complex form

$$(2) \quad L[U] = U_{XX} - U_{ZZ^*} + AU_X + BU_Z + CU_{Z^*} + DU = 0 .$$

Let $\chi = (X,Z,Z^*) \in \mathbb{C}^3_\chi$. By a further substitution Colton reduces $L[U]$ to standard form and then constructs an integral operator solution for (2), viz $U(\chi) = P_3\{f\}$, where the "associate" f is an arbitrary analytic function of two complex variables in the Cartesian product of the polydisc $\Delta^1_\mu(0;\rho)$ and $B = \{\zeta\colon 1 - \epsilon < |\zeta| < 1 + \epsilon,\ 0 < \epsilon < \frac{1}{2}\}$.

To simplify the notation and calculations we assume that the coefficients

$A = B = C = 0$ and $D = Q$ is an entire function of χ, although most of our results hold for the more general equation (2). Then (1) becomes

(3) $\Delta_3 u + q(x)u = 0$

and in complex form

(4) $U_{XX} - U_{ZZ^*} + QU = 0$

with the integral operator solution

(5) $U(\chi) = P_3\{f\} = \frac{1}{2\pi i} \int_{|\zeta|=1} \int_{\gamma} E(\chi;\zeta,t) f(\mu(1 - t^2),\zeta) \frac{dt}{\sqrt{1 - t^2}} \frac{d\zeta}{\zeta}$,

where γ is a path in $T = \{t: |t| \leq 1\}$ joining $t = -1$ to $t = +1$ and $U(\chi)$ is regular in a neighborhood of the origin in C^3_χ.

Following Bergman for the case $n = 2$ [2] it is convenient to replace (5) by a series expansion, which is done in §2. In §3 we obtain approximation theorems for $U(\chi)$, using approximation theorems for the associate $g(\mu,\zeta)$ obtained in the series representation such as the Oka-Weil theorem [15] and Mel'nikov's theorem on analytic capacity [11,15]. In §4 the subordination principle is applied to the associate using results of MacGregor [10] for $n = 2$ and Cima [3] for $n = 3$. This gives approximation theorems for the corresponding integral operator solutions. In §5 we prove that if the associate belongs to the Hardy class H^p, then the integral operator solution does also for $p \geq 1$.

§2. Series representation for $U(\chi)$.

Let $\xi = (\xi_1,\xi_2,\xi_3)$, where

(1) $\xi_1 = 2\zeta Z$, $\xi_2 = X + 2\zeta Z$, $\xi_3 = X + 2\zeta^{-1}Z^*$, $\mu = X + \zeta Z + \zeta^{-1}Z^*$.

The generating function in (1.5) $E(\chi;\zeta,t) = E^*(\xi;\zeta,t)$ has the series expansion

(2) $E^*(\xi;\zeta,t) = 1 + \sum_{n=1}^{\infty} t^{2n}\mu^n p^{(n)}(\xi;\zeta)$,

where $p^{(n)}(\xi;\zeta)$ given by the recursion relation (2.14) in [4a], is regular in $\bar{\Delta}^3_\xi(0;r) \times B$ and E satisfies the auxiliary partial differential equation (2.11) in

[4a]. The series expansion for E^* converges absolutely and uniformly in any compact subset of the region $\Delta^3_\zeta(0;r) \times B \times T$.

Following the method of Bergman [2, p. 15] we replace the integral representation (1.5) by a series representation. Substitute the series (2) for E^* into (1.5) and use the uniform convergence of series (2) to interchange $\int$ and Σ. We get

$$(3)\quad U(\chi) = \frac{1}{2\pi i} \int_{|\zeta|=1} \int_\gamma f(\mu(1-t^2),\zeta) \frac{dt}{\sqrt{1-t^2}} \frac{d\zeta}{\zeta}$$

$$+ \sum_{n=1}^{\infty} \mu^n \frac{1}{2\pi i} \int_{|\zeta|=1} p^{(n)}(\xi;\zeta) \frac{d\zeta}{\zeta} \int_\gamma t^{2n} f(\mu(1-t^2),\zeta) \frac{dt}{\sqrt{1-t^2}}.$$

Set

$$(4)\quad g(\mu,\zeta) = \int_\gamma f(\mu(1-t^2),\zeta) \frac{dt}{\sqrt{1-t^2}},$$

where $g(\mu,\zeta)$ is an analytic function of μ and ζ in $\Delta^1_\mu(0;\rho) \times B$. Let $f(\mu,\zeta) = \Sigma^{\infty}_{m,n=0} a_{mn}\mu^m\zeta^n$. By the same computation as in Bergman [2, p. 16] we get

$$\int_\gamma t^{2n} f(\mu(1-t^2),\zeta) \frac{dt}{\sqrt{1-t^2}} = \frac{\Gamma(2n+1)}{\Gamma(n+1)2^{2n}} \mu^{-n} G(\mu,\zeta)_n,$$

where

$$(5)\quad G(\mu,\zeta)_n = \int_0^\mu \int_0^{\mu_1} \cdots \int_0^{\mu_{n-1}} g(\mu_n,\zeta)\, d\mu_n \cdots d\mu_1 .$$

Thus (3) becomes

$$(6)\quad U(\chi) = \frac{1}{2\pi i} \int_{|\zeta|=1} g(\mu,\zeta) \frac{d\zeta}{\zeta} + \sum_{n=1}^{\infty} \frac{\Gamma(2n+1)}{\Gamma(n+1)2^{2n}} \frac{1}{2\pi i} \int_{|\zeta|=1} p^{(n)}(\xi;\zeta) G_n(\mu,\zeta) \frac{d\zeta}{\zeta}$$

$$\equiv P_3^{(1)}\{g\} + P_3^{(2)}\{g\},$$

where the series converges uniformly in a neighborhood of the origin in C^3_χ.

Proof. For $|\mu_n| \le |\mu| < \rho$ and $|\zeta| = 1$, $g(\mu_n,\zeta)$ is holomorphic; thus $|g(\mu_n,\zeta)| \le B$. By (2.34) in [4a]

(7) $|\mu^n p^{(n)}(\xi;\zeta)| \le M(r,\alpha)(2n)^{-1}(2n-1)^{-1}$,

where $M(r,\alpha)$ is a constant depending on the arbitrary positive number r and α which satisfies inequality (2.35) and $r = \alpha R$. Thus by (5)

(8) $$|G(\mu,\zeta)_n| \le B \int_0^{\mu}\int_0^{\mu_1}\cdots\int_0^{\mu_{n-1}} |d\mu_n \cdots d\mu_1| = \frac{B|\mu|^n}{\Gamma(n+1)} ,$$

so that in (6) using (7) and (8)

$$|U(\chi)| \le B + M(r,\alpha)BS ,$$

where S is the sum of the convergent series

$$S = \sum_1^\infty \frac{\Gamma(2n+1)}{2^{2n}\Gamma^2(n+1)} \frac{1}{2n(2n-1)} \le \sum_1^\infty \frac{1}{2n(2n-1)} < \infty .$$

§3. Approximation theorems for $n = 2$ and 3.

Bergman for $n = 2$ [2, p. 22] and Colton for $n = 3$ [4a, Theorem 3.2] proved that the real equations have the Runge approximation property and obtained a complete system of real solutions for bounded simply-connected domains.

1. Application of the Oka-Weil theorem.

Let K be a compact set in $\mathbb{C}^n$. The polynomially convex hull of K is

$$\hat{K} = \{z \in \mathbb{C}^n : |p(z)| \le \|p\|_K = \sup_{z\in K} |p(z)| \text{ for all polynomials } p\} .$$

If $K = \hat{K}$, then K is polynomially convex. [7, p. 38].

We apply the following theorem:

Theorem (Oka and Weil). Let K be compact and polynomially convex in $\mathbb{C}^n$. If f is holomorphic in a neighborhood of K , then there exists a sequence $\{p_j\}$ of polynomials with p_j converging to f uniformly on K [15].

From this theorem follows easily.

Theorem 1. Let $U(\chi)$ be a complex solution of (1.4). Let $g(\mu,\zeta)$ be a holomorphic function on $D = \Delta^1_\mu(0;1) \times B$. Let K be a compact polynomially convex set in D , containing $|\zeta| = 1$ and $(0,\zeta)$ in its interior. Then $U(\chi)$ can be uniformly approximated on the inverse image of K in C^3_χ by solutions $\{U_j\}$ of (1.4) with polynomial associates.

Proof. Since $g(\mu,\zeta)$ satisfies the hypothesis of the Oka-Weil theorem there exists a sequence of polynomials $\{p_j\}$ in (μ,ζ) such that p_j converge to g uniformly on K. Call $U_j(\chi) = P_3\{p_j\}$. By formula (2.6) and the linearity of the operator P_3

$$|U(\chi) - U_j(\chi)| = |P_3\{g - p_j\}| \le |P_3^{(1)}\{g - p_j\}| + |P_3^{(2)}\{g - p_j\}|$$

But

$$|P_3^{(1)}\{g - p_j\}| \le \frac{1}{2\pi} \int_{|\zeta|=1} |g(\mu,\zeta) - p_j(\mu,\zeta)| \left|\frac{d\zeta}{\zeta}\right| < \frac{\epsilon}{2}$$

for j sufficiently large, and from (2.6)-(2.8)

$$|P_3^{(2)}\{g - p_j\}| \le \frac{\epsilon}{2} M(r,\alpha) B \Sigma_{n=1}^{\infty} \frac{1}{2n(2n-1)},$$

for j sufficiently large, where the line $\overline{0\mu}$ lies in K since K is polynomially convex. Also the inverse image of the compact set K in D is closed by (2.1) and bounded and hence compact.

The hypotheses may be put on $U(\chi)$ instead of on $g(\mu,\zeta)$. We get

Theorem 2. Let $U(\chi)$ be a complex solution of (1.4) in a polydisc neighborhood $\Delta^3_\chi(0,R_0)$ and K a compact set in $\Delta^3_\chi(0,R_0)$. Let $K_0 = \mu(K, |\zeta| = 1)$, $0 \in K_0$, and $K_1 = K_0 \times \{|\zeta| = 1\}$ be polynomially convex. Then U can be uniformly approximated in K by solutions $\{U_j\}$ of (1.4), whose associates are polynomials in μ and ζ.

Proof. Since μ is a continuous function of χ and ζ the set K_0 is compact in the μ-plane and K_1 is compact in the $\mu\zeta$-plane and polynomially convex. By Theorem 2.2 of [4a] there exists an analytic function $f(\mu,\zeta)$, holomorphic in $\Delta^1_\mu(0;\rho) \times B$ and by (2.4) $g(\mu,\zeta)$ is holomorphic in the same domain. Then in some $\Delta^3_\chi(0;R_0)$, $U(\chi) = P_3\{g\}$. By the Oka-Weil theorem $g(\mu,\zeta)$ can be uniformly approximated on the polynomially convex set K_1 by a sequence of polynomials

$\{p_j\}$ in (μ,ζ), converging uniformly to g on K_1. The rest of the proof follows as in Theorem 1.

2. Analytic C-capacity and Mel'nikov's theorem for $n = 2$ and 3.

Let G be a bounded subset of $\mathbb{C}'$ and $F(G) = \{f: f$ continuous on $\mathbb{C}^1$ and holomorphic outside some closed subset of G, $|f(z)| \le 1$ and $f(\infty) = 0\}$.

Then

$$\alpha(G) = \sup_{f \in F(G)} \lim_{z\to\infty} |zf(z)|$$

is the *analytic C-capacity of the set* G. For example, the analytic C-capacity of the circle equals its radius [15, p. 20].

We apply

Theorem (Mel'nikov). *Let $f(z)$ be continuous on the closed disk $|z| \le 1$ and holomorphic at points of the complement, $\mathbb{C}(G)$, of G lying in the open disk. Then*

$$\left|\int_{|z|=1} f(z)dz\right| \le C \max_{|z|\le 1} |f(z)|\alpha(G),$$

where C *is an absolute constant and* $\alpha(G)$ *is the analytic C-capacity of* G [11,15].

a. If $n = 2$ the complex form of (1.1) is

(1) $L[u] = U_{zz^*} + AU_z + BU_{z^*} + CU = 0,$

A, B and C holomorphic functions of z and z^* in $\Delta^2(0;R)$ $(R > 2)$. Bergman proved that if $f(z)$ is analytic in a neighborhood of the origin a solution of (1) is $U(z,z^*) = \int_\gamma E(z,z^*,t)f(\tfrac{1}{2}z(1-t^2))dt/\sqrt{1-t^2}$. The function $E(z,z^*,t)$ is regular in $\Delta^2(0;\tfrac{1}{2}R) \times \{|t| \le 1\}$ [2, pp. 12-13, 8]. A series representation for $U(z,z^*)$ is

(2) $U(z,z^*) = G(z,z^*)[g(z) + \Sigma_{n=1}^{\infty} c_n Q^{(n)}(z,z^*)G(z)_n] \equiv U_1(z,z^*) + U_2(z,z^*),$

[2, p. 16] where $G(z,z^*)$ is holomorphic in $\Delta^2(0;R)$ and depends on the coefficient A in (1), c_n is a constant, $G(z)$ is given by (2.5) with $z = \mu$ and ζ omitted, $Q^{(n)}(z,z^*)$ is holomorphic in $\Delta^2(0;\frac{1}{2}R)$ and the analytic function g is connected with f by [2, p. 15 (7)]. Note that (2) is still defined for continuous $g(z)$, although U need not satisfy (1).

Theorem 3. Let A, B, C of (1) be holomorphic in $\Delta^2_{zz^*}(0;R)\,(R > 2)$. Let G be a closed set of $\bar{\Delta}^1_z(0;1)$, not containing the origin, and such that $\Delta^1_z(0;1) \cap \complement(G)$ is star-shaped with respect to the origin. Let g be analytic in $\Delta^1_z(0;1) \cap \complement(G)$ and continuous on $\bar{\Delta}^1_z(0;1)$. Then

$$\Big| \iint_{\substack{|z|=1 \\ |z^*|=1}} U(z,z^*)\,dz\,dz^* \Big| \le CB \max_{|z|\le 1} |g(z)|\,\alpha(G),$$

where C is an absolute constant, and B depends only on the coefficients of (1).

Proof. The function $g(z)G_0(z)$, $G_0(z) = \int_{|z^*|=1} G(z,z^*)\,dz^*$ satisfies the hypotheses of Mel'nikov's theorem. Hence

$$(3)\quad \Big| \int_{|z|=1} [g(z)G_0(z)\,dz \Big| \le C \max|G_0(z)g(z)|\,\alpha(G) \le CB_1 \max_{|z|\le 1} |g(z)|\,\alpha(G),$$

where C is the absolute constant of Mel'nikov's theorem and $B_1 = B_1(A)$ is an upper bound of $G_0(z)$. Integrate $U_2(z,z^*)$ over $(|z| = 1) \times (|z^*| = 1)$. By [2] the series in (2) converges absolutely and uniformly on compact subsets of $\Delta^2(0;1)$. Use Fubini's theorem. Then the integrand $\sum_{n=1}^{\infty} c_n G(z) \int_{|z^*|=1} G(z,z^*)Q^{(n)}(z,z^*)\,dz^*$ satisfies the hypotheses of Mel'nikov's theorem. Thus

$$(4)\quad \Big| \iint_{\substack{|z|=1 \\ |z^*|=1}} U_2(z,z^*)\,dz\,dz^* \Big| \le CB_2 \max_{|z|\le 1} |g(z)|\,\alpha(G),$$

where B_2 depends on the coefficients of (1). The theorem follows from (3) and (4).

b. If $n = 3$ we have

Theorem 4. Let G be a closed set in the polydisc $\bar{\Delta}^1_{\mu}(0;1), 0 \notin G$. Let $g(\mu,\zeta)$

be continuous on $\overline{\Delta}^2_{\mu\zeta}(0;1)$ and holomorphic in $[\Delta^1_{\mu}(0;1) \cap \mathfrak{C}(G)] \times B$ where $\Delta^1_{\mu} \cap \mathfrak{C}(G)$ is star-shaped with respect to 0. Let $U(\chi)$ be a solution of (1.4) given by (1.5) and $\mathfrak{L}$ a closed curve of length ℓ in $\mathbb{C}^3_{\chi}$, in the domain of definition of $U(\chi)$, not enclosing the origin. Then

$$(5) \quad \left|\int_{\mathfrak{L}} U(\chi)ds\right| \le 2\pi C\ell B' \max_{\substack{|\zeta|\le 1\\ |\mu|\le 1}} |g(\mu,\zeta)|\alpha(G),$$

where C is an absolute constant.

Proof. If we choose X, Z, Z^*, μ as independent variables, the Jacobian $J = \partial(\xi_1,\xi_2,\xi_3,\zeta)/\partial(X,Z,Z^*,\mu) \ne 0$. As remarked in §2 $p^{(n)}(\xi;\zeta)$ is holomorphic in $\Delta^3_{\xi}(0;r) \times B$. By the change of variables (2.1), ζ can be obtained from the last equation as a holomorphic function of χ and μ. Then $\tilde{p}^{(n)}(\chi;\mu) = p^{(n)}(\xi;\zeta)$ is a holomorphic function of χ and μ in a certain domain of $\mathbb{C}^4_{\chi\mu}$ if $J \ne 0$. Formula (2.6) gives a solution of equation (1.4) where $g(\mu,\zeta)$ and $\tilde{p}^{(n)}(\chi;\mu)$ are holomorphic and is continuous at other points in the domain of definition. Now integrate over $(|\mu| = 1) \times \mathfrak{L}$. This gives

$$(6) \quad \int_{|\mu|=1} d\mu \int_{\mathfrak{L}} U(\chi)ds = \int_{|\mu|=1} d\mu \int_{\mathfrak{L}} P_3^{(1)}\{g\}ds + \int_{|\mu|=1} d\mu \int_{\mathfrak{L}} P_3^{(2)}\{g\}ds,$$

where ds is arc length on $\mathfrak{L}$. Use Fubini's theorem on the terms on the right of (6). This gives for the first term by Mel'nikov's theorem

$$(7) \quad \frac{\ell}{2\pi} \int_{|\zeta|=1} \left|\frac{d\zeta}{\zeta}\right| \left|\int_{|\mu|=1} g(\mu,\zeta)d\mu\right| \le \ell C \max_{\substack{|\mu|\le 1\\ |\zeta|\le 1}} |g(\mu,\zeta)|\alpha(G)$$

In the second integral on the right of (6) use Fubini's theorem to interchange $\int d\mu$ and $\int ds$ and Mel'nikov's theorem on the integrand

$$\frac{1}{2\pi i} \int_{|\zeta|=1} \sum_{n=1}^{\infty} \frac{\Gamma(2n+1)}{\Gamma(n+1)2^{2n}} p^{(n)}(\xi;\zeta)G(\mu,\zeta)_n \frac{d\zeta}{\zeta},$$

which satisfies the hypotheses of the theorem on $|\mu| \le 1$. This gives

$$\int_{\mathfrak{L}} ds \left|\int_{|\mu|=1} P_3^{(2)}\{g\}d\mu\right| \le C \int_{\mathfrak{L}} ds \max_{|\mu|\le 1} \left|\frac{1}{2\pi i} \int_{|\zeta|=1} \sum_{n=1}^{\infty} \frac{\Gamma(2n+1)}{\Gamma(n+1)2^{2n}} |p^{(n)}(\xi;\zeta)G(\mu,\zeta)_n \frac{d\zeta}{\zeta}\right|\alpha(G).$$

Then use (2.7) on $p^{(n)}(\xi;\zeta)$ and (2.5) and (2.8) on $G(\mu,\zeta)_n$ to get

$$(8)\quad \int_{\mathcal{L}} ds \Big| \int_{|\mu|=1} P_3^{(2)}\{g\}\, d\mu \Big| \le C\ell M(r,\alpha) S \max_{\substack{|\mu|\le 1\\ |\zeta|\le 1}} |g(\mu,\zeta)|\,\alpha(G) \; .$$

(7) and (8) give (5) in Theorem 4.

A different type of approximation theorem for real equations is found in [13, p. 236].

§4. Subordination applied to integral operator solutions of linear partial differential equations.

Let F and G be two biholomorphic mappings on the polydisc $\Delta^n = \Delta^n(0;1)$ with $F(0) = G(0)$, then F is subordinate to G if $F(\Delta^n) \subset G(\Delta^n)$. This is written $F < G$.

1. Case $n = 2$. For the unit disc we use

Theorem (MacGregor). *Let $g(z)$ be analytic and univalent in $\Delta^1 = \Delta^1(0;1)$. Then there exists a sequence of polynomials $\{p_j(z)\}$ $(j = 1,2,\ldots)$, $p_j(z)$ of degree j and univalent in Δ^1 such that $p_1(z) < p_2(z) < \ldots$ in Δ^1 and $p_j(z)$ converges to $g(z)$ uniformly on compact subsets of Δ^1* [10].

A complex solution $\mathcal{U}(z,z^*)$ of (3.1) is real if it becomes real when $z^* = \bar{z}$: $\mathcal{U}(z,z^*) = \frac{1}{2}[U(z,z^*) + \bar{U}(z^*,z)]$. It is convenient to choose $A = 0$ in (3.1). An application of MacGregor's theorem to $\mathcal{U}(z,z^*)$ gives

Theorem 5. *Let $\mathcal{U}(z,z^*)$ be a real solution of (3.1) on $\Delta^2 = \Delta^2(0;1)$, where $A = 0$ in (3.1), with univalent associate g on $\Delta^1 = \Delta^1(0;1)$. Then there exists a sequence $\{\mathcal{U}_j(z,z^*)\}$ of real solutions of (3.1) with polynomial associates $\{p_j(z)\}$, $p_j(z)$ of degree j, $p_j(0) = 0$ and univalent in Δ^1 such that $\mathcal{U}_j(z,z^*)$ converges uniformly on compact subsets of Δ^2. If $p_j + U_{2,j} < p_{j+1} + U_{2,j+1}$ on Δ^2, then $\mathcal{U}_j < \mathcal{U}_{j+1}$ on Δ^2.*

Proof. Let K be a compact subset of Δ^2. The projection K_1 of K on the z-plane is compact. Let $\{p_j\}$ be the sequence of polynomials satisfying

MacGregor's theorem for $g(z)$. Set $U_j(z,z^*) = P_2\{p_j\}$, where $P_2\{p_j\}$ is the integral operator solution of U_j for $n = 2$ given in Paragraph 2a. By (3.2)

$$(2) \quad |U_1(z,z^*) - U_{1,j}(z,z^*)| = |g(z) - p_j(z)| < \frac{\epsilon}{2}$$

for j sufficiently large. Use the analogous inequality to (2.8) on $G(z)_n$ in (3.2). Let T be the sum of the resulting series of absolute values. Then

$$(3) \quad |U_2(z,z^*) - U_{2,j}(z,z^*)| \le \frac{\epsilon}{2} T$$

for j sufficiently large. By (2) and (3) $U_j(z,z^*)$ converges to $U(z,z^*)$ absolutely and uniformly on compact subsets of Δ^2 and similarly for $\{\overline{U}_j(z^*,z)\}$ and hence for $\{\mathcal{U}_j(z,z^*)\}$.

By hypothesis $p_j(0) = 0$ $(j = 1,2,\ldots)$. Let $G_j(z)_n = G(z)_n$ with g replaced by p_j. By definition $G_j(0)_n = 0$ and by (3.2) $U_j(0,0) = 0$. Since $U_j(z,z^*) = p_j(z) + U_{2,j}(z,z^*)$, $U_j < U_{j+1}$. (Note that $U_{2,j}(z,z^*)$ is a polynomial in z of degree j with holomorphic coefficients in (z,z^*).) Similarly for $\overline{U}_j(z^*,z)$. Thus $\mathcal{U}_j(z,z^*) < \mathcal{U}_{j+1}(z,z^*)$.

2. <u>Case $n = 3$</u>. We use a theorem of Cima [3] on approximation theorems for biholomorphic mappings.

Theorem (Cima). <u>Let</u> F <u>be a biholomorphic mapping of</u> Δ^n <u>into</u> C^n <u>with</u> $F(0) = 0$, $F(z) = (f_1(z),\ldots,f_n(z))$. <u>Then there exist polynomial mappings</u> $P_j(z)$, <u>which are biholomorphic on</u> Δ^n <u>and of degree</u> j <u>such that</u> $P_j < P_{j+1} < \ldots (P_j < F)$ $(j = 1,2,\ldots)$ <u>and</u> P_j <u>converges uniformly to</u> $F(z)$ <u>on compact subsets of</u> Δ^n [3].

We prove

Theorem 6. <u>Let</u> $U(\chi)$ <u>be a mapping of a neighborhood</u> D <u>of the origin containing</u> $\Delta^3 = \Delta^3_\chi(0;R_0)$ $(R_0 > 0)$, <u>whose components</u> U_1,U_2,U_3 <u>are complex solutions of</u> (1.4) <u>given by</u> (2.6) <u>with</u> $U(0) = 0$. <u>The associate</u>, $g(\mu,\zeta) = (g_1,g_2,g_3)$, <u>of</u> U <u>is biholomorphic on</u> $\Delta^2 = \Delta^1_\mu(0;1) \times \Delta^1_\zeta(0;1+\epsilon)$. <u>Then there exists a sequence</u> $\{U_j(\chi)\}$ <u>of solutions of</u> (1.4) <u>with polynomial</u>

<u>associates</u> $\{P_j(\mu,\zeta)\}$, P_j <u>of degree</u> j <u>and a biholomorphic mapping in</u> Δ^2 <u>and</u> $P_j < P_{j+1}$ $(j = 1,2,\ldots)$. <u>The sequence</u> $\{U_j(\chi)\}$ <u>converges uniformly on compact subsets of</u> Δ^3 <u>to</u> $U(\chi)$.

Proof. Let K be a compact set in Δ^3 . Take $n = 3$ in Cima's theorem. By the hypothesis of the theorem and (2.6) $|P_3^{(1)}\{g\} - P_3^{(1)}\{P_j\}| < \frac{\epsilon}{2}$ for j sufficiently large. Again by the hypotheses of Theorem 6, (2.7), (2.5), (2.8) and the fact that $\Delta_\mu^1(0;1)$ is star-shaped with respect to the origin so that the lines $\overline{0\mu}_{n-1},\ldots,\overline{0\mu}$ lie in $\Delta_\mu^1(0;1)$, $|P_3^{(2)}\{g\} - P_3^{(2)}\{P_j\}| < \frac{\epsilon}{2}$ for j sufficiently large. Combining these two inequalities proves that $\{U_j(\chi)\}$ converges uniformly to $U(\chi)$ on K .

§5. <u>Hardy spaces of solutions of (1.4).</u>

A function f , holomorphic on the polydisc $\Delta^n(0;1)$, belongs to the <u>Hardy class</u> H^p, $0 < p < \infty$, if $\sup_{0<r<1} M_p(r,f) < \infty$, where

$$M_p(r,f) = \{(\tfrac{1}{2\pi})^n \int_0^{2\pi} \cdots \int_0^{2\pi} |f(re^{i\theta})|^p d\theta\}^{\frac{1}{p}}, \ \theta = (\theta_1,\ldots,\theta_n)$$

$\sup_{0<r<1} M_p(r,f)$ is called the pth norm, $\|f\|_p$, of f .

Assuming that the associates g in (3.2) and (2.6) in cases $n = 2$ and $n = 3$ respectively belong to H^p on a suitable polydisc we can prove that U belongs to H^p also if $p \geq 1$. Then many of the properties for g in H^p [6] could be carried over to U .

Let $n = 3$. Then

Theorem 7. <u>Let</u> $g(\mu,\zeta) \in H^p(D)$ <u>where</u> $D = \Delta_\mu^1(0;3) \times \Delta_\zeta^1(0;1 + \epsilon)$. <u>Then</u> $U(\chi) = P_3\{g\} \in H^p(D_1)$ <u>where</u> $D_1 = \Delta_\chi^3(0;1)$.

Proof. From (2.1) if $|\zeta| = |X| = |Z| = |Z^*| = 1$, then $|\mu| \leq 3$.

Take absolute values in (2.6), raise both sides to the pth power $(p > 1)$ and integrate over $0 \le \psi, \theta_j (j = 1,2,3) < 2\pi$ where $\mu = \rho e^{i\psi}$, $X = |X| e^{i\theta_1}$, $Z = |Z| e^{i\theta_2}$, $Z^* = |Z^*| e^{i\theta_3}$. Using Minkowski's inequality on the right we have

$$(1) \quad \{ (\tfrac{1}{2\pi})^4 \int_C \int_0^{2\pi} |U(\chi)|^p d\psi d\theta \}^{\frac{1}{p}} \le \{ (\tfrac{1}{2\pi})^4 \int_C \int_0^{2\pi} |P_3^{(1)}\{g\}|^p d\psi d\theta \}^{\frac{1}{p}}$$

$$+ \{ (\tfrac{1}{2\pi})^4 \int_C \int_0^{2\pi} |P_3^{(2)}\{g\}|^p d\psi d\theta \}^{\frac{1}{p}}$$

where $\int_C = \int_0^{2\pi} \int_0^{2\pi} \int_0^{2\pi}$. In the first term on the right replace $P_3^{(1)}\{g\}$ by its value in (2.6) and use Hölder's inequality on the inner integral. Let p' be the conjugate exponent to p and $\zeta = re^{i\varphi}$. Then

$$\{ (\tfrac{1}{2\pi})^3 \int_C d\theta \, \frac{1}{2\pi} \int_0^{2\pi} d\psi \{ \frac{1}{2\pi} \int_0^{2\pi} |g(\rho e^{i\psi}, re^{i\varphi})|^p d\varphi \}^{\frac{p}{p}} (\frac{1}{2\pi} \int_0^{2\pi} d\varphi)^{\frac{p}{p'}} \}^{\frac{1}{p}}$$

$$(2) \quad \le \{ (\tfrac{1}{2\pi})^3 \int_C M_p(\rho, r; g) d\theta \}^{\frac{1}{p}}$$

But $M_p(\rho,r;g) \le M_p(\rho',\rho';g)$ where $\rho' = \max(\rho,r)$ by the monotonicity of $M_p(r;g)$. Thus (2) is

$$\le \|g\|_p \{ (\tfrac{1}{2\pi})^3 \int_C d\theta \}^{\frac{1}{p}} = \|g\|_p$$

In the second term on the right of (1), replace $P_3^{(2)}\{g\}$ by its value in (2.6). Here we first use Hölder's inequality on the inner integral, then Minkowski's inequality in infinite form [16, p. 19, (9.12)] to interchange $\int_C d\theta \int_0^{2\pi} \int_0^{2\pi} d\varphi \, d\psi$ and $\Sigma_{n=1}^{\infty}$. Now use inequality (2.7) on the term $p^{(n)}(\xi;\zeta)$ and remove it from under the integral signs. We have left under the integral sign

$$\{ (\tfrac{1}{2\pi})^3 \int_C d\theta (\tfrac{1}{2\pi})^2 \int_0^{2\pi} \int_0^{2\pi} |G(\mu,\zeta)_n|^p d\psi \, d\varphi \}^{\frac{1}{p}} = \{ (\tfrac{1}{2\pi})^2 \int_0^{2\pi} \int_0^{2\pi} |G(\mu,\zeta)_n|^p d\psi d\varphi \}^{\frac{1}{p}}$$

Replace $G(\mu,\zeta)_n$ by its value (2.5) and use Hölder's inequality on $g(\rho_n e^{i\psi}, re^{i\varphi})$ to give

$$(3) \quad \{ \frac{\rho^n}{\Gamma(n+1)} \}^{\frac{1}{p'}} \{ \frac{1}{2\pi^2} \int_0^{2\pi} \int_0^{2\pi} \int_0^{\rho} \int_0^{\rho_1} \cdots \int_0^{\rho_{n-1}} |g(\rho_n e^{i\psi}, re^{i\varphi})|^p d\rho_n \cdots d\rho_1 d\varphi \, d\psi \}^{\frac{1}{p}}$$

By Fubini the integral in (3) equals

$$(4)\quad \{\frac{\rho^n}{\Gamma(n+1)}\}^{\frac{1}{p'}}\{(\frac{1}{2\pi})^2\int_0^{\rho}\int_0^{\rho_1}\cdots\int_0^{\rho_n} d\rho_n\cdots d\rho_1 M_p^p(\rho_n,r;g)\}^{\frac{1}{p}}$$

Using the same procedure as for (2) this is

$$\leq \|g\|_p \left\{\frac{\rho^n}{\Gamma(n+1)}\right\}^{\frac{1}{p'}+\frac{1}{p}} = \|g\|_p \frac{\rho^n}{\Gamma(n+1)}$$

and as in §2 the resulting series converges. Thus

$$((\frac{1}{2\pi})^3\{\int_C |U(X)|^p d\Theta)^{\frac{1}{p}} < \infty \quad \text{so that} \quad U \in H^p .$$

References

[1] S. Bergman, Zur Theorie der Funktionen, die eine lineare partielle Differentialgleichung befriedigen. I. Math. Sbornik, N.S. (2) (1937), 1169-1198.

[2] S. Bergman, Integral Operators in the Theory of Linear Partial Differential Equations, Ergeb. der Math. vol. 23 (1969), 2nd. ed.

[3] Joseph A. Cima, An approximation theorem for biholomorphic functions on D^n, Trans. Amer. Math. Soc. 175 (1973), 491-7.

[4] David Colton, (a) Integral operators for elliptic equations in several independent variables. I, Applicable Analysis, 4(1974), 77-95.
(b) Integral operators for elliptic equations in three independent variables, II. Applicable Analysis 4 (1975), 283-295.

[5] David Colton and Robert P. Gilbert, An integral operator approach to Cauchy's problem for $\Delta_{p+2}u(X) + F(X)u(X) = 0$, Siam J. Math. Anal. 2 (1971), 113-132.

[6] Peter L. Duren, The Theory of H^p Spaces, Academic Press, New York and London (1970).

[7] R. C. Gunning and Hugo Rossi, Analytic Functions of Several Complex Variables, Prentice-Hall, Inc. Englewood Cliffs, N. J. (1965).

[8] Edwin T. Hoefer, Properties of solutions of linear partial differential equations given by Bergman integral operators, Ph.D. Dissertation, State University of New York at Buffalo (1974) (Unpublished).

[9] Dean K. Kukral, On a Bergman-Whittaker type operator in five or more variables, Proc. Amer. Math. Soc. 39 (1973), 122-124.

[10] Thomas H. MacGregor, Approximation by polynomials subordinate to a univalent function, Trans. Amer. Math. Soc. 148 (1970), 199-209.

[11] M. S. Mel'nikov, Analytic capacity and the Cauchy integral, Soviet Math. Dokl. 8 (1967), 20-23.

[12] Josephine Mitchell, Approximation to the solution of linear partial differential equations given by Bergman integral operators, Berichte der Gesellschaft f. Math. u. Datenverarbeitung mbH Bonn 77 (1973), 97-107.

[13] R. Narasimhan, Analysis on Real and Complex Manifolds, North-Holland Publishing Co., Amsterdam (1968).

[14] I. N. Vekua, New Methods for Solving Elliptic Equations, John Wiley, New York (1967).

[15] A. G. Vitushkin, Uniform approximations by holomorphic functions, J. Functional Anal. 20 (1975), 149-157.

[16] A. Zygmund, Trigonometric Series, vol. 1, Cambridge University Press (1959).

State University of New York at Buffalo

ÜBER DIE LINEAREN PARTIELLEN QUASIELLIPTISCHEN DIFFERENTIALOPERATOREN MIT KONSTANTEN KOEFFIZIENTEN

E. Pehkonen
Universität Helsinki
Mathematisches Institut
Hallituskatu 15
SF-00100 Helsinki 10
Finnland

1. Hier betrachten wir die linearen partiellen quasielliptischen Differentialgleichungen mit konstanten komplexen Koeffizienten auf einem Gebiet des euklidischen Raumes R^n. Die quasielliptischen Operatoren, die auch "semielliptisch" oder "regulär pseudoparabolisch" genannt werden bilden eine Teilklasse der hypoelliptischen Operatoren, und andererseits enthalten sie die elliptischen und parabolischen Operatoren.

Wir benutzen die folgenden üblichen Bezeichnungen: Für zwei Multiindizes $\sigma = (\sigma_1, \dots, \sigma_n)$ und $\tau = (\tau_1, \dots, \tau_n)$ aus N_0^n sei $s\sigma := (s\sigma_1, \dots, s\sigma_n)$, $\sigma + \tau := (\sigma_1+\tau_1, \dots, \sigma_n+\tau_n)$ und

$$\binom{\sigma}{\tau} := \binom{\sigma_1}{\tau_1} \cdots \binom{\sigma_n}{\tau_n} .$$

Ferner sei $\sigma \leqq \tau$ genau dann, wenn $\sigma_j \leqq \tau_j$ für alle $j = 1, \dots, n$ gilt; wir schreiben $\sigma < \tau$, wenn $\sigma \leqq \tau$ ist und wenn $\sigma_j < \tau_j$ für mindestens ein $j = 1, \dots, n$ gilt. Für $x = (x_1, \dots, x_n)$ aus R^n sei $x^\sigma := x_1^{\sigma_1} \cdots x_n^{\sigma_n}$; den Differentialoperator D^σ erklären wir durch $D^\sigma := D_1^{\sigma_1} \cdots D_n^{\sigma_n}$ mit $D_j := -i\,\partial/\partial x_j$ und $i^2 = -1$.

Es sei $\rho = (\rho_1, \dots, \rho_n)$ ein fester Multiindex mit positiven Komponenten ρ_j. Wir betrachten die Differentialoperatoren

$$(1) \qquad L_\rho(D) := \sum_{|\sigma:\rho| \leqq 1} a_\sigma D^\sigma$$

mit konstanten komplexen Koeffizienten a_σ, wobei $|\sigma:\rho| := \sum_{j=1}^n \sigma_j/\rho_j$ ist. Mit

$$L_\rho(y) := \sum_{|\sigma:\rho| \leqq 1} a_\sigma y^\sigma$$

sei das durch die Substitution $D_j \mapsto y_j$ dem Operator $L_\rho(D)$ zugeordnete Polynom (der Veränderlichen $y = (y_1, \dots, y_n)$ aus R^n) bezeichnet. Man nennt Operator (1) <u>quasielliptisch</u>, falls

$$\sum_{|\sigma:\rho| = 1} a_\sigma y^\sigma \neq 0$$

für alle $y \in R^n - \{0\}$ gilt; entsprechend heisst (1) stark quasielliptisch, falls $\mathrm{Re} \sum a_\sigma y^\sigma > 0$ für alle $y \in R^n - \{0\}$ ist. Zum Beispiel ist der Operator $i D_1^3 + D_2^2 + D_3^2$ quasielliptisch, aber nicht stark quasielliptisch.

Dann führen wir den dem Operator $L_\rho(D)$ entsprechenden Sobolevraum analog wie im elliptischen Fall ein. Sei G ein Gebiet in R^n. Für jedes $s \in N_0$ definieren wir den Raum $C^{s\rho}(G)$ als die Menge der in G erklärten komplexwertigen Funktionen u, die in G stetige Ableitungen $D^\sigma u$ für alle σ mit $|\sigma:\rho| \leqq s$ besitzen. Für Elemente $u \in C^{s\rho}(G)$ sei

$$\|u\|_{s\rho,G} := \Big(\sum_{|\sigma:\rho| \leqq s} \int_G |D^\sigma u(x)|^2 dx \Big)^{1/2} .$$

Es sei $C_*^{s\rho}(G)$ die Menge der Funktionen u aus $C^{s\rho}(G)$ mit $\|u\|_{s\rho} < \infty$. Den Sobolevraum $H^{s\rho}(G)$ definieren wir als Vervollständigung des Raumes $C_*^{s\rho}(G)$ bezüglich Norm $\|.\|_{s\rho}$.

Sei u eine Funktion aus $C^\rho(G)$. Durch partielle Integration ergibt sich $(\phi , L_\rho(D)u)_0 = (L_\rho(D)^* \phi , u)_0$ für alle $\phi \in C_0^\infty(G)$, wobei $(\phi , u)_0 := \int_G \phi(x)\overline{u(x)}dx$ ist; hierbei ist

$$L_\rho(D)^* := \sum_{|\sigma:\rho| \leqq 1} \overline{a}_\sigma D^\sigma$$

der dem Operator $L_\rho(D)$ formal adjungierte Differentialoperator. Sei dann f ein Element aus $L^2(G)$. Eine Funktion u aus $L^2(G)$, die der Relation $(L_\rho(D)^* \phi , u)_0 = (\phi , f)_0$ für alle $\phi \in C_0^\infty(G)$ genügt, nennen wir eine <u>schwache Lösung</u> der Differentialgleichung $L_\rho(D)u = f$ in G.

<u>2.</u> Für schwache Lösungen stark quasielliptischer Differentialgleichungen mit variablen Koeffizienten gab Giusti [2] einen Regularitätsbeweis. Seine Resultate, die man wohl auch aus den Ergebnissen von Hörmander [3] herleiten kann, sind analog denen für stark elliptische Operatoren. Wir geben jetzt einen elementaren Beweis mit Hilfe der Fouriertransformierten. Grob gesprochen bilden wir die Fouriertransformation der Differentialgleichung, multiplizieren sie mit einer gewissen Gewichtsfunktion, die einem bestimmten Ableitungsoperator entspricht, und schätzen das Resultat ab.

<u>Regularitätssatz</u>. <u>Sei</u> $L_\rho(D)$ <u>ein (nicht notwendig stark) quasielliptischer Differentialoperator, und sei</u> f <u>ein Element aus</u> $H^{s\rho}(G)$. <u>Falls</u> $u \in L^2(G)$ <u>eine schwache Lösung der Differentialgleichung</u> $L_\rho(D)u = f$ <u>in</u> G <u>ist, so bestehen für alle Teilgebiete</u> G' <u>von</u> G <u>mit</u> $\overline{G}' \subset G$ <u>die Relationen</u> $u \in H^{(s+1)\rho}(G')$ <u>und</u>

$$\|u\|_{(s+1)\rho,G'} \leqq c\,(\|f\|_{s\rho,G} + \|u\|_{0,G}) .$$

Im folgenden bezeichnet c jeweils eine positive Konstante.

Beweisskizze. Nach dem Heine - Borelschen Satz genügt es die Behauptungen für eine Kugel zu beweisen. Sei $x_0 \in G'$ ein beliebiger fester Punkt, und sei $K_0, K_1, \dots, K_{R+1}$ ein System offener Kugeln mit x_0 als Mittelpunkt und mit $\overline{K}_{R+1} \subset K_R, \dots, \overline{K}_1 \subset K_0, \overline{K}_0 \subset G$, wobei $R := (s+1) \max \rho_j$ ist. Für jedes $j = 0, 1, \dots, R$ sei $\eta_j \in C_0^\infty(K_j)$ eine reellwertige Abschneidefunktion mit $\eta_j = 1$ in K_{j+1}. Wir setzen die Funktion η_j als Null ausserhalb K_j fort.

Für jedes $k \in Z$ sei p_k die durch

(2) $$p_k(y) := \left(1 + \sum_{j=1}^{n} |y_j|^{\rho_j}\right)^{k/r}, \qquad y \in R^n,$$

erklärte Gewichtsfunktion, wobei $r := \max \rho_j$ ist. Sei ferner Fu die Fouriertransformierte von u. Nachdem wir gezeigt haben, dass in G die Ungleichung

(3) $$\|p_k\, F(\eta_k u)\|_0 \leqq c\,(\|f\|_{s\rho} + \|u\|_0)$$

für alle $k = 0, 1, \dots, R$ besteht, folgern wir hieraus für $k = R$ mittels der Friedrichsschen Glättung (siehe [1]) die Behauptungen.

Wir beweisen (3) durch Induktion: Nach der Leibnizschen Regel erhalten wir

$$(L_\rho(D)^* (\eta_k \psi), u)_0$$

$$= (L_\rho(D)^* \psi, \eta_k u)_0 + \sum_{|\sigma:\rho| \leqq 1} \overline{a}_\sigma \sum_{0<\beta\leqq\sigma} \binom{\sigma}{\beta} (D^\beta \eta_k\, D^{\sigma-\beta} \psi, u)_0$$

für alle $\psi \in C^\infty(K_0)$. Zur Vereinfachung der Schreibweise setzen wir $u_k := \eta_k u$, $f_k := \eta_k f$, $u_k^\beta := u\, D^\beta \eta_k$, die alle in $L^2(K_0)$ liegen und einen kompakten Träger in K_0 haben. Weil u eine schwache Lösung der Gleichung $L_\rho(D)\, u = f$ in K_0 ist und weil $\eta_k \psi$ in $C_0^\infty(K_0)$ liegt, erhalten wir

$$(L_\rho(D)^* \psi, u_k)_0 = (\psi, f_k)_0 - \sum_{|\sigma:\rho| \leqq 1} \overline{a}_\sigma \sum_{0<\beta\leqq\sigma} \binom{\sigma}{\beta} (D^{\sigma-\beta} \psi, u_k^\beta)_0 .$$

Wir wählen speziell $\psi(x) := e^{i(y,x)}$, $x \in R^n$, wobei $y \in R^n$ beliebig ist und $(y, x) := y_1 x_1 + \dots + y_n x_n$ bedeutet; dann erhalten wir mittels der Fouriertransformierten

$$L_\rho(y)\, F u_k(y) = F f_k(y) - \sum_{|\sigma:\rho| \leqq 1} a_\sigma \sum_{0<\beta\leqq\sigma} \binom{\sigma}{\beta} y^{\sigma-\beta}\, F u_k^\beta(y)$$

für alle $y \in R^n$. Weil $1 + |L_\rho(y)| \neq 0$ ist, ergibt sich

$$(4) \qquad |F\,u_k(y)|\;p_k(y) \;\leqq\; \frac{|F\,f_k(y)| + |F\,u_k(y)|}{1+|L_\rho(y)|}\,p_k(y)$$

$$+ \sum_{|\sigma:\rho|\leqq 1} |a_\sigma| \sum_{0<\beta\leqq\sigma} \binom{\sigma}{\beta}\,|F\,u_k^\beta(y)|\;\frac{|y^{\sigma-\beta}|}{1+|L_\rho(y)|}\,p_k(y)\,.$$

Gemäss der Quasielliptizität ersieht man

$$|y^{\sigma-\beta}|\;p_1(y) \;\leqq\; c\,(1+|L_\rho(y)|)\,,$$

und wegen Definition (2) erhält man

$$p_k(y) \;\leqq\; c\,(1+|L_\rho(y)|)\;p_{k-r}(y)\;;$$

also schliessen wir aus (4)

$$\|p_k\;F\,u_k\|_0 \;\leqq$$

$$c\,(\|p_{sr}\;F\,f_k\|_0 + \|p_{k-1}\;F\,u_k\|_0 + \sum_{|\sigma:\rho|\leqq 1} |a_\sigma| \sum_{0<\beta\leqq\sigma} \binom{\sigma}{\beta}\,\|p_{k-1}\;F\,u_k^\beta\|_0)\,.$$

Nachdem wir gezeigt haben, dass in G die Abschätzung

$$(5) \qquad \|p_{k-1}\;F\,u_k^\beta\|_0 \;\leqq\; c\,(\|f\|_{s\rho} + \|u\|_0)$$

besteht, ergibt sich Behauptung (3).

Beweis für (5). Weil aus $\eta_k(x) \neq 0$ die Beziehung $\eta_{k-1}(x) = 1$ folgt, haben wir $u_k^\beta = u_{k-1}\,D^\beta\,\eta_k$ (weil $u_{k-1} = \eta_{k-1}\,u$ ist). Bekanntlich gilt

$$(6) \qquad F\,(u_{k-1}\,D^\beta\,\eta_k)_y = (2\,\pi)^{-n} \int F\,u_{k-1}(y-z)\;F\,(D^\beta\,\eta_k)(z)\,dz$$

für alle $y \in R^n$. Weil $D^\beta\,\eta_k$ in $C_0^\infty(R^n)$ liegt, gibt es zu jedem $t \in N$ eine positive Konstante c derart, dass $|F\,(D^\beta\,\eta_k)(z)|^2 \leqq c\,(1+|z|)^{-t}$ für alle $z \in R^n$ besteht. Jetzt können wir aus (6) folgern:

$$\|p_{k-1}\;F\,u_k^\beta\|_0^2 \;\leqq\; c \iint |F\,u_{k-1}(y-z)|^2\,(1+|z|)^{-t}\;p_{k-1}(y)^2\,dz\,dy\,.$$

Dann ersetzen wir auf der rechten Seite $y-z$ durch x und erhalten

$$\|p_{k-1}\;F\,u_k^\beta\|_0 \;\leqq\; c\;\|p_{k-1}\;F\,u_{k-1}\|_0\;;$$

hieraus folgt Behauptung (5).

3. Mit ähnlichen Methoden kann man die Regularität schwacher Lösungen des stark quasielliptischen Dirichletproblems im Falle eines Quaders bis zum Rande beweisen (siehe [5]). Die Beweisidee (die ich Herrn C. G. Simader verdanke) besteht darin, dass man zuerst die schwache

Lösung u der stark quasielliptischen Differentialgleichung nach der Methode von Friedrichs [1] zu einer C^∞-Funktion u_ε $(\varepsilon > 0)$ glättet, für diese das Bestehen einer stark quasielliptischen Differentialgleichung nachweist und a-priori-Abschätzungen für u_ε beweist. Schliesslich folgen hieraus die gewünschten Regularitätsresultate für die ursprünglich betrachtete schwache Lösung $\varepsilon \to 0$, was durch Fouriertransformierten geschieht.

Die hier benötigte Beweismethode kann man wohl mit kleinen Modifikationen auch im Falle variabler Koeffizienten benutzen, und man erhält entsprechende Resultate. Man darf vielleicht auch vermuten, dass man die Regularität für die stark 2t-koerzitiven (siehe [4]), hypoelliptischen Operatoren mit dieselben elementaren Methoden verifizieren könnte.

Literatur

[1] Friedrichs, K. O.: The identity of weak and strong extensions of differential operators. - Trans. Amer. Math. Soc. 55, 1944, S.132-151.

[2] Giusti, E.: Equazioni quasi-ellittiche e spazi $W^{p,\theta}(\Omega,\delta)$ (II). - Ann. Scuola Norm. Sup. Pisa (3) 21, 1967, S.353-372.

[3] Hörmander, L.: Linear partial differential operators. - [Third revised printing.] Die Grundlehren der mathematischen Wissenschaften in Einzeldarstellungen 116. Springer-Verlag, Berlin / Heidelberg / New York, 1969.

[4] Louhivaara, I. S., und C. G. Simader: Über nichtelliptische lineare partielle Differentialoperatoren mit konstanten Koeffizienten. - Ann. Acad. Sci. Fenn. Ser. A. I. 513, 1972.

[5] Pehkonen, E.: Regularität der schwachen Lösungen linearer quasielliptischer Dirichletprobleme. - Univ. Jyväskylä Math. Inst. Bericht 16, 1976.

ÜBER DIE LÖSUNG EINER NICHT-LINEAREN ANFANGSWERTAUFGABE IN DER THERMOELASTIZITÄTSTHEORIE

Adam PISKOREK
(Warszawa)

1. EINLEITUNG. Unter Verwendung der neuen Ergebnisse der Kontinuumsmechanik (siehe [8], [9]) lässt sich das Differentialgleichungssystem für die Bewegungen und thermische Effekte des thermoelastischen, inhomogenen, anisotropen Mediums verallgemeinern. Dieses verallgemeinerte System hat (vgl. z.B. [8], S.919 und [9], S.189) folgende Form[1)]

$$(1) \qquad \rho\partial_t^2 u_j - \partial_h(a_{jhkl}(x,t)\partial_l u_k) + \partial_h(b_{jh}(x,t)T) = F_j(x,t)$$

$$(2) \qquad c\partial_t T - \partial_h(a_{lh}(x,t)\partial_l T) + Tb_{lh}(x,t)\partial_t(\partial_l u_h + \partial_h u_l) = Q(x,t)$$

wo $u=(u_1,u_2,u_3)$ die Verschiebungsvektorfunktion des Mediums, T die Temperatur des Mediums, ρ die Dichte des Mediums, $F = (F_1,F_2,F_3)$ die Massenkräfte, Q die Dichte der Wärmequelle sind.

---

1) Lateinische Indizes gehen (ohne t !) von 1 bis 3. Über doppelte vorkommende Indizes ist stets über den entsprechenden Bereich zu summieren (EINSTEINsche Summenkonvention). Wir benutzen folgende Bezeichnung: $\partial_\alpha w$ ist die partielle Ableitung von w nach der α-Koordinate für $\alpha=1,2,3,t$; $\partial_\alpha\partial_\beta w$, $\partial_\alpha\partial_\beta\partial_\chi w$ - die höheren partiellen Ableitungen von w. Der LAPLACEsche Operator Δ ist erklärt durch $\Delta = \partial_j\partial_j$.

Die Koeffizienten a_{jhkl}, a_{lh}, b_{lh}, c sind beschränkt, beliebig oft differenzierbar und genügen folgenden (vgl. [4], S.103, (2.5) und [9], S.188, (2.5.14)) Bedingungen

$$a_{jhkl} = a_{kljh} = a_{hjkl} \tag{3}$$

$$a_{lh} = a_{hl} \qquad \text{und} \qquad c > 0 \tag{4}$$

Wir setzen voraus, dass die quadratischen Formen $a_{jhkl} z_{jh} z_{kl}$ und $a_{lh} z_l z_h$ positiv definit für jedes x und $t > 0$ sind.

Unter diesen Voraussetzungen betrachten wir folgende Anfangswertaufgabe für das Differentialgleichungssystem (1), (2): Gesucht ist eine Lösung u, T des Systemes (1), (2) im EUKLIDischen Raum $\mathbb{R}_3$ für $t > 0$, welche den Anfangsbedingungen

$$u(x,+0) = u^0(x) \quad , \quad (\partial_t u)(x,+0) = u^1(x) \; , \; T(x,+0) = T_0(x) \tag{5}$$

genügt. Nach V.M.SCHALOV (siehe [8], S.918, Definition 3) setzen wir noch voraus, dass die Anfangswerte u^0, u^1, T_0 zu SOBOLEVschen Räumen[1] $H^s(\mathbb{R}_3, \mathbb{R}_3)$, $H^{s-1}(\mathbb{R}_3, \mathbb{R}_3)$, $H^s(\mathbb{R}_3, \mathbb{R})$ gehören, und der Einfachheit halber $F = 0 = Q$. In dieser Arbeit werden wir die Anfangswertaufgabe für (1), (2) mit (5) im Funktionenraum[2] $C([0,\vartheta], H^s)$ untersuchen. Zuerst betrachten wir den Fall des homogenen und isotropen Mediums und dann - den Fall des inhomogenen und anisotropen Mediums.

---

1) Mit $H^s(D, \mathbb{R}_m)$ bezeichnen wir die Menge aller Abbildungen vom Gebiet D im $\mathbb{R}_3$ in $\mathbb{R}_m$ der Klasse H^s (= B_{2k} nach Kap.II, Ab. 2.2 im [7]). Mit $\|.\|_s$ bezeichnen wir die Norm im H^s.

2) Mit $C(I,E)$ bezeichnen wir den Raum aller stetigen Funktionen auf dem Intervall I im $\mathbb{R}$ mit Werten im BANACHraum E. Jedes Element w von $C(I,E)$ nennt man die stetige Kurve in E. Mit Norm $\|w\| = \sup \|w(t)\|_E$ ist $C(I,E)$ ein BANACHraum.

2. HOMOGENES UND ISOTROPES MEDIUM. In diesem Falle haben die Gleichungen (1),(2) konstante Koeffizienten (siehe [4],S.103,(2.9) und [8],S.920) und nehmen sehr einfache (vgl. [3],(8),(9)) Form

$$(6) \qquad \rho\partial_t^2 u - (\lambda + \mu)\mathrm{grad\,div}u - \mu\Delta u + \gamma\,\mathrm{grad}T = 0$$

$$(7) \qquad \varkappa^{-1}\partial_t T - \Delta T + \eta T\partial_t \mathrm{div}u = 0$$

an, wo $\lambda, \mu, \varkappa, \gamma, \eta$ physikalische Konstante sind.

Wir setzen voraus, dass die Anfangswerte u^0, u^1 und T_0 von SOBOLEVschen Räumen $H^s(\mathbb{R}_3,\mathbb{R}_3)$, $H^{s-1}(\mathbb{R}_3,\mathbb{R}_3)$ und $H^s(\mathbb{R}_3,\mathbb{R})$ für

$$(8) \qquad s > r + 3/2 \quad , \quad r \geqq 4$$

sind.

Mit Hilfe des HELMHOLTZschen Satzes (siehe z.B. [5],S.122,Cor. 7.3), nach welchem jedes H^s-Vektorfeld u die Darstellung

$$u = v - \mathrm{grad}\phi$$

(v - divergenzfreies H^s-Vektorfeld und ϕ - skalares H^{s+1}-Potential) hat, reduziert sich nicht lineare Anfangswertaufgabe (siehe [3],(11),(12),(13)) für (6),(7) mit Anfangswerten (5) von SOBOLEVschen Räumen mit dem Index s, der (8) genügt, auf folgende Anfangswertaufgaben

$$(9) \qquad L_a v = 0 \quad , \quad v(x,+0) = v^0(x) \quad , \quad (\partial_t v)(x,+0) = v^1(x)$$

$$(10) \qquad L_b\phi = \frac{\gamma}{\rho}T \quad , \quad \phi(x,+0) = \phi_0(x) \quad , \quad (\partial_t\phi)(x,+0) = \phi_1(x)$$

$$(11) \qquad \varkappa^{-1}\partial_t T = \Delta T - \eta T\Delta\partial_t\phi \quad , \quad T(x,+0) = T_0(x) \quad ,$$

worin L_j für $j=a,b$ den D'ALEMBERTschen Operator bedeutet, $a = (\mu/\rho)^{1/2}$, $b = ((\lambda + 2\mu)/\rho)^{1/2}$ und die Anfangswerte v^k, ϕ_k für $k = 0,1$ durch die einfachen Beziehungen und den HELMHOLTZschen Satz aufgrund von (5) und (8) gegeben (siehe [3],Remark 1) sind. Hieraus folgt dass v^0 von $H^s(\mathbb{R}_3,\mathbb{R}_3)$, v^1 von

$H^{s-1}(\mathbb{R}_3,\mathbb{R}_3)$, ϕ_0 von $H^{s+1}(\mathbb{R}_3,\mathbb{R})$, ϕ_1 von $H^s(\mathbb{R}_3,\mathbb{R})$ sind. Mit Hilfe der Fundamentallösung G_a der Wellengleichung $L_a w = 0$ erhält man die Lösung der Anfangswertaufgabe (9) in der expliziten Gestalt[1)]

$$v = G_a *_3 v^1 + (\partial_t G_a) *_3 v^o \tag{12}$$

Um die nicht lineare Anfangswertaufgabe (10),(11) zu lösen , nehmen wir, ohne Beschränkung der Allgemeinheit, b = 1 an und nun reduzieren wir diese Anfangswertaufgabe auf die äquivalente (s. [2], Kap.VI, §3, Ab.8,9) Anfangswertaufgabe

$$\partial_t U = A_j \partial_j U + \Theta \tag{13}$$

$$\varkappa^{-1} \partial_t T = \triangle T - \eta T \triangle U_4 \tag{14}$$

$$U(x,+0) = U^0(x) \quad , \quad T(x,+0) = T_0(x) \tag{15}$$

worin (13) das symmetrische, hyperbolische Differentialgleichungssystem erster Ordnung ist, und

$$U = \begin{bmatrix} U_1 \\ U_2 \\ U_3 \\ U_4 \end{bmatrix} , \quad \Theta = \begin{bmatrix} 0 \\ 0 \\ 0 \\ \frac{\gamma}{\varrho} T \end{bmatrix} , \quad U^0 = \begin{bmatrix} \partial_1 \phi_0 \\ \partial_2 \phi_0 \\ \partial_3 \phi_0 \\ \phi_1 \end{bmatrix} .$$

Sei X die Menge aller Kurven W von $C([0,\vartheta],H^s(\mathbb{R}_3,\mathbb{R}_4))$ derart, dass $W(0)=U^0$ und $\|W(t) - U^0\|_s \leq M$ für t von $[0,\vartheta]$. Wir definieren die Abbildung S der Menge X in den BANACHraum $C([0,\vartheta],H^s(\mathbb{R}_3,\mathbb{R}_4))$

$$X \ni W \longrightarrow (SW)(t) = U^0 + \int_0^t A_j \partial_j (SW)(\sigma) d\sigma + \int_0^t \frac{\gamma}{\varrho} \begin{bmatrix} 0 \\ 0 \\ 0 \\ T_0 *_3 G_\eta \triangle W_4 \end{bmatrix} (\sigma) d\sigma \tag{16}$$

---

1) Mit $*_3$ bezeichnen wir die Faltung im $\mathbb{R}_3$.

wo $G_{\eta\Delta W_4}$ die Fundamentallösung der verallgemeinerten Wärmegleichung $\varkappa^{-1}\partial_t T - \Delta T + T\eta\Delta W_4 = 0$ ist.

Diese Abbildung ist wohldefiniert aufgrund von der Theorie der symmetrische hyperbolischenSysteme erster Ordnung (siehe z.B.[2], Kap.VI,§8,Ab.1,2) und vom verallgemeinerten ARSENIEVschen Lemma, welches (vgl.[1],Lemma 1.1 und Lemma 1.2;siehe [3],Remark 2,die Formeln (i),(ii),(iii)) lautet:

<u>LEMMA</u> 1. Es seien w_1 und w_2 stetige Kurven in $L^q(\mathbb{R}_N)$, d.h. w_1 , $w_2 \in C([0,\vartheta],L^q(\mathbb{R}_N))$. Dann

1º existieren die Fundamentallösungen G_{w_1} und G_{w_2} der paparabolischen Differentialgleichungen zweiter Ordnung

$$c(x,t)\partial_t T - \partial_h(a_{lh}(x,t)\partial_l T) + w_1(x,t)T = 0$$

und

$$c(x,t)\partial_t T - \partial_h(a_{lh}(x,t)\partial_l T) + w_2(x,t)T = 0$$

mit beschränkten und glatten Koeffizienten a_{lh} , $c > 0$ im $\mathbb{R}_N \times [0,\vartheta]$,

und

2º gilt folgende Abschätzung

(i) $$\|G_{w_1}(x,.,t) - G_{w_2}(x,.,t)\|_{L^1} \leq C\|w_1(.,t) - w_2(.,t)\|_{L^q}$$

für $q > N/2$.

Mit Hilfe der Energieabschätzungen für symmetrische,hyperbolischen Systeme (siehe z.B. [2],Kap.VI,§8,Ab.2) und der Abschätzung (i) können wir zeigen (siehe[3],Beweis des Satzes 1) die Existenz eines positiven ϑ derart,dass S die Menge X in sich abbildet und diese Abbildung S kontrahierend (=LIPSCHITZstetig mit LIPSCHITZkonstante <1) ist. Daraus, nach BANACHschem Fixpunktprinzip, existiert Fixpunkt U der Abbildung S in X , d.h. SU=U. Es folgt hieraus,dass die Anfangswertaufgaben für (13),(14) mit (15) danach für (10),(11) und für (6),(7) mit (5) lösbar in

$C([0,\vartheta],H^s)$ sind.

Damit ist gezeigt:

SATZ 1. Seien u^o von $H^s(\mathbb{R}_3,\mathbb{R}_3)$, u^1 von $H^{s-1}(\mathbb{R}_3,\mathbb{R}_3)$, T_o von $H^s(\mathbb{R}_3,\mathbb{R})$ für $s > r + 3/2$, $r \geq 4$ und $F=0=Q$. Dann gibt es positive ϑ derart, dass die Anfangswertaufgabe für (6),(7) mit (5) eindeutig lösbar im BANACHraum $C([0,\vartheta],H^s)$ ist. Die Lösung u, T hängt stetig in der H^s-Topologie von den Anfangswerten u^o, T_o ab.

3. INHOMOGENES UND ANISOTROPES MEDIUM. In diesem Falle reduzieren wir unmittelbar die Anfangswertaufgabe für (1),(2) mit (5) auf die äquivalente (s. [2], Kap.VI, §3, Ab.9) Anfangswertaufgabe in der Gestalt

$$(17)\qquad A_o(x,t)\partial_t U = A_j(x,t)\partial_j U + B(x,t)U + \Theta$$

$$(18)\qquad c(x,t)\partial_t T = \partial_h(a_{lh}(x,t)\partial_l T) + Tb_{lk}(x,t)(\partial_l U_{4(k-1)}+\partial_k U_{4(l-1)})$$

$$(19)\qquad U(x,+0) = U^o(x)\quad,\quad T(x,+0) = T_o(x)\quad,$$

worin (17) das hyperbolische Differentialgleichungssystem erster Ordnung (siehe [6], Kap.3) ist,

$$U = \begin{bmatrix} U_0\\ U_1\\ U_2\\ U_3\\ U_4\\ U_5\\ U_6\\ U_7\\ U_8\\ U_9\\ U_{10}\\ U_{11}\end{bmatrix},\qquad \Theta = \begin{bmatrix} T\partial_h b_{1h} + b_{1h}\partial_h T\\ 0\\ 0\\ 0\\ T\partial_h b_{2h} + b_{2h}\partial_h T\\ 0\\ 0\\ 0\\ T\partial_h b_{3h} + b_{3h}\partial_h T\\ 0\\ 0\\ 0\end{bmatrix},\qquad U^o = \begin{bmatrix} u_1^1\\ \partial_1 u_1^o\\ \partial_2 u_1^o\\ \partial_3 u_1^o\\ u_2^1\\ \partial_1 u_2^o\\ \partial_2 u_2^o\\ \partial_3 u_2^o\\ u_3^1\\ \partial_1 u_3^o\\ \partial_2 u_3^o\\ \partial_3 u_3^o\end{bmatrix}.$$

Die Matrix $A_o(x,t)$ ist im Anhang 1 dargestellt, sie ist quadratisch, 12-reihig, symmetrisch und nichtsingulär für jedes x von $\mathbb{R}_3$ und $t > 0$. Die Matrizen $A_j(x,t)$ für $j=1,2,3$ sind im ANHANG 2 dargestellt, sie sind auch quadratisch, 12-reihig aber nicht symmetrisch. Daraus folgt, dass das System (17) nicht symmetrisch ist. Für die Vektorfunktion Θ vereinbaren wir folgende Abkürzung $\Theta = T\partial b + b\partial T$.

Um die nichtlineare Anfangswertaufgabe für (17), (18) mit (19) auch mit Hilfe des BANACHschen Fixpunktprinzips zu lösen, setzen wir voraus, dass das System (17) bezüglich t-Koordinate <u>stark hyperbolisch</u> (siehe [6], Kap.3, Ab.1 und Kap.2, Ab.4) ist, d.h. das Polynom $\det(A_0(x,t)\tau - A_j(x,t)z_j)$ bezüglich τ nur einfache reelle Nullstellen für $z \neq 0$, x von $\mathbb{R}_3$ und $t > 0$ hat, und die Anfangswerte U^0 zu $H^{s-1}(\mathbb{R}_3, \mathbb{R}_{12})$ und T_0 zu $H^s(\mathbb{R}_3, \mathbb{R})$ für $s > r + 3/2$, $r \geq 2$ gehören.

Unter den oben gemachten Voraussetzungen sind wir in der Lage, ebenso wie im Abschnitt 2 die Existenz der Lösung der Anfangswertaufgabe für (17), (18) mit (19) zu beweisen, nur statt der Energieabschätzungen für die symmetrischen hyperbolischen Systeme werden wir die GÅRDINGschen Energieabschätzungen für die stark hyperbolischen Systeme erster Ordnung (siehe [6], Kap. Satz 1.3) benutzen.

Wir betrachten die Menge X aller Kurven W vom BANACHraum $C([0,\vartheta], H^{s-1}(\mathbb{R}_3, \mathbb{R}_{12}))$ derart, dass $W(0) = U^0$ und für t von $[0,\vartheta]$ $\|W(t) - U^0\|_{s-1} \leq M$. Die Abbildung S der Menge X in $C([0,\vartheta], H^{s-1}(\mathbb{R}_3, \mathbb{R}_{12}))$ ist durch

$$(20) \qquad X \ni W \longrightarrow (SW)(t) = U^0 + \int_0^t \Big[A_0^{-1} A_j \partial_j (SW) + A_0^{-1} B(SW) + \\ + \big(T_0 *_3 G_{b_{lk}} (\partial_l W_{4(k-1)} + \partial_k W_{4(l-1)})\big) A_0^{-1} \partial b \quad + \\ + A_0^{-1} b_h \big(\partial_h T_0 *_3 G_{b_{lk}} (\partial_l W_{4(k-1)} + \partial_k W_{4(l-1)})\big) \Big](\sigma)\, d\sigma$$

definiert. Diese Abbildung ist wohldefiniert aufgrund von Lemma 1 und von der GÅRDINGschen Theorie der Anfangswertaufgabe für hyperbolische Gleichungen (siehe [6]).

Mit Hilfe der Energieabschätzungen (siehe [6], Satz 1.3) und die Abschätzung (i) von Lemma 1 können wir zeigen die Existenz eines positiven ϑ derart, dass S die Menge X in sich abbildet und in X die Abbild. S kontrahierend ist. Daraus, nach BANACHschem Fixpunktprinzip, existiert Fixpunkt der Abbildung S in X, d.h. $SU = U$. Es folgt hieraus, dass die Anfangswertaufgaben für (17), (18) mit (19) und danach für (1), (2) mit (5) lösbar im BANACH-raum $C([0,\vartheta],H^s)$ sind. Damit haben wir:

SATZ 2. Seien u^0 von $H^s(\mathbb{R}_3,\mathbb{R}_3)$, u^1 von $H^{s-1}(\mathbb{R}_3,\mathbb{R}_3)$, T_0 von $H^s(\mathbb{R}_3,\mathbb{R})$ für $s > r + 3/2$, $r \geqq 2$ und $F=0=Q$. Für $z \neq 0$, x von $\mathbb{R}_3$ und $t > 0$ habe das Polynom $\det(A_0(x,t)\tau - A_j(x,t)z_j)$ bezüglich τ nur einfache reelle Nullstellen. Dann gibt es positive ϑ derart, dass die Anfangswertaufgabe für (1), (2) mit (5) eindeutig lösbar im BANACH-raum $C([0,\vartheta],H^s)$ ist. Die Lösung u, T hängt stetig in der H^s-Topologie von den Anfangswerten u^0, T_0 ab.

ANHANG 1

$$
A_o = \begin{bmatrix}
1 & 0 & 0 & 0 & 0 & 0 & 0 & 0 & 0 & 0 & 0 & 0 \\
0 & a_{1111} & a_{1112} & a_{1113} & 0 & 0 & 0 & 0 & 0 & 0 & 0 & 0 \\
0 & a_{1211} & a_{1212} & a_{1213} & 0 & 0 & 0 & 0 & 0 & 0 & 0 & 0 \\
0 & a_{1311} & a_{1312} & a_{1313} & 0 & 0 & 0 & 0 & 0 & 0 & 0 & 0 \\
0 & 0 & 0 & 0 & 1 & 0 & 0 & 0 & 0 & 0 & 0 & 0 \\
0 & 0 & 0 & 0 & 0 & a_{2111} & a_{2112} & a_{2113} & 0 & 0 & 0 & 0 \\
0 & 0 & 0 & 0 & 0 & a_{2211} & a_{2212} & a_{2213} & 0 & 0 & 0 & 0 \\
0 & 0 & 0 & 0 & 0 & a_{2311} & a_{2312} & a_{2313} & 0 & 0 & 0 & 0 \\
0 & 0 & 0 & 0 & 0 & 0 & 0 & 0 & 1 & 0 & 0 & 0 \\
0 & 0 & 0 & 0 & 0 & 0 & 0 & 0 & 0 & a_{3111} & a_{3112} & a_{3113} \\
0 & 0 & 0 & 0 & 0 & 0 & 0 & 0 & 0 & a_{3211} & a_{3212} & a_{3213} \\
0 & 0 & 0 & 0 & 0 & 0 & 0 & 0 & 0 & a_{3311} & a_{3312} & a_{3313}
\end{bmatrix}
$$

ANHANG 2

$$A_j = \begin{bmatrix}
0 & a_{1j11} & a_{1j12} & a_{1j13} & 0 & a_{1j21} & a_{1j22} & a_{1j23} & 0 & a_{1j31} & a_{1j32} & a_{1j33} \\
a_{111j} & 0 & 0 & 0 & 0 & 0 & 0 & 0 & 0 & 0 & 0 & 0 \\
a_{121j} & 0 & 0 & 0 & 0 & 0 & 0 & 0 & 0 & 0 & 0 & 0 \\
a_{131j} & 0 & 0 & 0 & 0 & 0 & 0 & 0 & 0 & 0 & 0 & 0 \\
0 & a_{2j11} & a_{2j12} & a_{2j13} & 0 & a_{2j21} & a_{2j22} & a_{2j23} & 0 & a_{2j31} & a_{2j32} & a_{2j33} \\
0 & 0 & 0 & 0 & a_{211j} & 0 & 0 & 0 & 0 & 0 & 0 & 0 \\
0 & 0 & 0 & 0 & a_{221j} & 0 & 0 & 0 & 0 & 0 & 0 & 0 \\
0 & 0 & 0 & 0 & a_{231j} & 0 & 0 & 0 & 0 & 0 & 0 & 0 \\
0 & a_{3j11} & a_{3j12} & a_{3j13} & 0 & a_{3j21} & a_{3j22} & a_{3j23} & 0 & a_{3j31} & a_{3j32} & a_{3j33} \\
0 & 0 & 0 & 0 & 0 & 0 & 0 & 0 & a_{311j} & 0 & 0 & 0 \\
0 & 0 & 0 & 0 & 0 & 0 & 0 & 0 & a_{321j} & 0 & 0 & 0 \\
0 & 0 & 0 & 0 & 0 & 0 & 0 & 0 & a_{331j} & 0 & 0 & 0
\end{bmatrix}$$

LITERATUR

[1] ARSENIEV A.A., Singuläre Potentiale und Resonanz /in russischer Sprache/, Moskau 1974

[2] COURANT R., Partial Differential Equations, New York-London 1962

[3] DOMANSKI Z., PISKOREK A., On the initial value problem in nonlinear thermoelasticity, Archives of Mechanics - Archiwum Mechaniki Stosowanej, №4(1976)

[4] DUVAUT G., LIONS J.L., Les inéquations en mécanique et en physique, Dunod, Paris 1972

[5] EBIN D.G., MARSDEN J., Groups of diffeomorphisms and the motion of an incompressible fluid, Annals of Math., 92, №1(1971)

[6] GÅRDING L., Cauchy's Problem for Hyperbolic Equations, Chicago 1957

[7] HÖRMANDER L., Linear Partial Differential Operators, Springer 1963

[8] SCHALOV V.M., Gleichungen der Kontinuumsmechanik /in russischer Sprache/, Diff.Equations, IX, №5(1973)

[9] ŞUHUBI E.S., Thermoelastic Solids, Chap.2 in ERINGEN A.C., Continuum Physics, Vol.II, Acad.Press 1975, S.174-261.

Adam PISKOREK
DEPARTMENT OF MATHEMATICAL METHODS OF PHYSICS, WARSAW UNIVERSITY
00-862, WARSZAWA, HOŻA 74

ON THE UNIQUENESS AND REGULARITY OF THE SOLUTIONS OF NAVIER-STOKES PROBLEMS.

R. Rautmann

Summary: (I.) Uniqueness and stability of weak solutions holds for the initial-boundary value problem of Navier and Stokes with a mollification[1]) E. Hopf's existence proof can be applied and leads to a convergent approximation method for this problem. In addition any essentially bounded Hopf solution is the limit of the solutions of equations containing a mollification, the radius of which is approaching zero. (II.) By means of the fundamental solutions of the heat equation and the potential equation, the initial value problem of Navier and Stokes is transformed into a fixedpoint equation, which makes evident the regularity of (local) solutions constructed by application of the contracting mapping principle. (III.) In the case of the initial value problem of Navier and Stokes with a mollification, by this means we get a global classical solution. (IV.) These results enable us to construct non-solenoidal Hopf-Galerkin approximations, which are more convenient for the numerical realization than the solenoidal ones constructed in [14,15].

I. Unique Weak Solutions of the Navier-Stokes Initial-Boundary Value Problem with a Mollification.

In his famous paper [5] E. Hopf proved the existence of (global) weak solutions for the initial-boundary value problem of Navier and Stokes. What he stated on the Galerkin method ("dieses einfachsten und nächstliegenden Annäherungsverfahrens" [loc. cit. p. 226]) anticipated the general direction of modern developements in numerical methods for related problems [2, 16]. However, uniqueness being basic for the stability of a Galerkin method, the numerical applicability of the Hopf-Galerkin method depends essentially on the uniqueness of the Hopf solutions, which cannot be guaranteed in the physically most interesting case of 3 (and more) dimensions. As Ladyzenskaja [8] has shown, the solutions of the Hopf type might be not unique for certain problems in 3-dimensional regions. On the other hand, for

1) In section (I.) we consider weak solutions in the Hopf class only. Concerning uniqueness of generalized and classical solutions (including the pressure-function, too), Heywood has proved new results in [4] .

dimensions $n \geq 3$ the uniqueness has been proved only in certain subclasses of the Hopf class of weak solutions, but then the global existence remains an open question. As Serrin [19] points out, the trouble stems from the special feature of the nonlinear term in the weak form of the Navier-Stokes problem. This term contains the weak solution u and a test function φ and has (generally speaking) no limit when φ approximates u. We can overcome this difficulty by a device, which Leray in [10] has used in the case of the Navier-Stokes initial value problem on the unbounded 3-space: We take the (directional) derivative contained in the nonlinear term in the direction of an mollified function u_h rather than in that of u. In this way we obtain a "stabilized" equation which can be solved by constructive means. The unique solution is stable on each compact time interval, i.e. it depends continuously on the initial values.

I.1. The Problem with a Mollification.

We denote by Ω an open, not necessarily bounded set of the Euclidean n-space R^n with points $x = (x^1,\ldots,x^n)$ and write $\Omega_T=[0,T) \times \Omega$ for any (time-) value $T \in (0,\infty]$. For reasons of economy, normally we will omit the differentials of the variables of integration. On the Sobolev space of all vector functions $u(t,x) = (u^1,\ldots,u^n)$, which together with their first weak spatial derivatives belong to the class $L^2(\Omega_T)$, we use the norm

$$(I.1) \qquad |u|_H = \{\int_{\Omega_T} (u\cdot u + \nabla u\cdot\nabla u)\}^{\frac{1}{2}}.$$

We will formulate a variant of Hopf's weak form of the Navier-Stokes initial-boundary value problem within Hopf's class V of all measurable vector function $v(t,x) = (v^1,\ldots,v^n)$ on Ω_∞, which

a) for any $T \in (0,\infty)$ belong to the closure with respect to the norm $|\cdot|_H$ of class D of all divergence-free vector functions having partial derivatives of any order and a compact support in Ω_∞ and

b) have restrictions $v(t,\cdot)$ bounded in $L^2(\Omega)$ uniformly in $t \geq 0$.

For a vector function $u \in V$ (representing the velocity field of a viscous incompressible flow in Ω for all time-values $t \geq 0$) we consider the equation

$$\text{(I.2)} \qquad \int_{\Omega} u\cdot\varphi \Big|_{o}^{T} = \int_{\Omega_T} u\cdot\{\varphi_t + (u_h\cdot\nabla)\varphi + \Delta\varphi\},$$

being the weak form of the Navier-Stokes initial-boundary value problem with the mollification

$$\text{(I.3)} \qquad u_h(t,x) = b_h \cdot \int_{R^n} \omega_h(x-y)\tilde{u}(t,y)dy.$$

We get u_h from the continuation $\tilde{u}(t,x) = u(t,x)$ for $x \in \Omega$, $\tilde{u}(t,x)=0$ for $x \in R^n - \Omega$ of u with the help of the mollifier

$$\text{(I.4)} \qquad \omega_h(x) = \begin{cases} 0 & \text{for } |x| \geq h \\ \exp\{|x|^2 (|x|^2 - h^2)^{-1}\} & \text{for } |x| < h \end{cases}$$

with any radius $h > 0$ of the mollification and the constant $b_h = \{\int_{R^n} \omega_h(x)\, dx\}^{-1}$. In (I.2), we have $(u_h\cdot\nabla)\varphi = \sum_{j=1}^{n} u_h^j \cdot \frac{\delta}{\delta x^j}\varphi$. We start with the

Definition: Any $u \in V$ solving (I.2) for all $T \in (0,\infty)$ and $\varphi \in D$ will be called a weak solution of the stabilized Navier-Stokes initial-boundary value problem.

Evidently, any such u is weakly continuous with respect to the variable $t \geq 0$ in the L^2-closure V_o of the class D_o of all divergence-free vector functions having partial derivatives of any order and a compact support in Ω.

The definition of the class V takes into account (in the L^2-sense) the initial condition $u = u_o$ for $t = 0$ and at the boundary $\delta\Omega$ the condition of adherence $u = 0$, which is physically plausible in case of a viscous flow on Ω_∞.

I.2. A Note on the Physical Meaning of the Stabilized Equation.

The classical Navier-Stokes equation

$$\text{(I.5)} \qquad \frac{\delta}{\delta t} u + (u\cdot\nabla)\, u = \Delta u - \nabla p$$

relating to the velocity field $u(t,x)$ and pressure function $p(t,x)$ of a flow in Ω_∞ expresses Newton's law of motion for any test particle moving along the integral curve $x(t)$ of the velocity field $\frac{d}{dt}x = u(t,x)$. The stabilized equation

$$\text{(I.6)} \qquad \frac{\delta}{\delta t}u + (u_h \cdot \nabla)u = \Delta u - \nabla p$$

says exactly the same, if we let the test particles move along the integral curves $x(t)$ of the mollified velocity field $\frac{d}{dt}x = u_h(t,x)$. Correspondingly, equation (I.6) (the weak form of which is (I.2)) suggests an interpretation in the context of statistical fluid mechanics.

I.3. The Energy Equality.

In addition to the spatial mollifications u_h of a function $u(t,x)$ we consider the mollifications

$$\text{(I.7)} \qquad u_\tau = b_\tau \int_0^T \omega_\tau(t-t')\, u(t',x)dt'$$

for $\tau>0$, ω_τ and b_τ from (I.4) with τ instead of h. By standard argumentation we are led to the statements

(A) $\int_{\Omega_T} \{u_{\tau,t} \cdot v + u \cdot v_{\tau,t}\} = 0$, (B) $\int_{\Omega_T} \{u \cdot (w_h \cdot \nabla)v + v \cdot (w_h \cdot \nabla)u\} = 0$,

(C) $|w_h(t,\cdot)|_\infty^2 \leq c_h \cdot \int_\Omega (w \cdot w)(t,\cdot)$

being valid for any $u,v,w \in V$ and

(D) $\int_\Omega (u \cdot v_\tau)(t,\cdot) \to \frac{1}{2}\int_\Omega (u \cdot v)(t,\cdot)$ with $\tau \to 0$,

for $t = 0,T$, if additionally $v(t,\cdot)$ is weakly continuous in the Hilbert space V_0 with respect to the variable $t \geq 0$.

Let us assume u is a weak solution of the stabilized Navier-Stokes initial-boundary value problem.[2] We take in D a sequence

[2] Due to the mollification in (I.2), a simplified version of the method in [19, 79-88] applies to our problem without the restriction on solutions belonging to subclasses of V.

(u_k) approximating u within the norm $|\cdot|_H$, $k=1,2,\ldots$, and set in (I.2) $\varphi = u_{k\tau}$ with $\tau > 0$. In the limit[3] $k \to \infty$, the time derivative drops out because of (A) (with u = v).

Now in the limit[3] $\tau \to 0$ the directional derivative is cancelled by (B) (with u=v). Using (D) we get the energy equality

$$\text{(I.8)} \qquad \frac{1}{2} \int_\Omega u\cdot u \Big|_0^T + \int_{\Omega_T} \nabla u\cdot\nabla u = 0$$

from (I.2). Because $u(t,\cdot)$ (u being a solution of (I.2)) is weakly continuous in V_o, we deduce from (I.8) the

Corollary: Any solution $u \in V$ of (I.2) is strongly continuous in t with respect to $L^2(\Omega)$.

I.4 Uniqueness and Stability

Assume u and v are weak solutions of the stabilized Navier-Stokes initial-boundary value problem.[4] We take in D two sequences (u_k), (v_k) approximating u or v respectively within the norm $|\cdot|_H$. We set $\varphi = v_{k\tau}$ in the equation (I.2) for u and $\varphi = u_{k\tau}$ in (I.2) for v. In the limit[3] $k \to \infty$, after adding up the two equations, the time derivatives drop out by (A). We let τ approach 0[3]. Using (B), (D) we get

$$\int_\Omega u\cdot v \Big|_0^T = - \int_{\Omega_T} \{u\cdot((v_h - u_h)\cdot\nabla)v + 2\nabla u\cdot\nabla v\}.$$

The addition of this equation multiplied by (-2) to the energy equations of u and v leads to the equation

$$\int_\Omega w\cdot w\Big|_0^T = 2 \int_{\Omega_T} \{u\cdot(w_h\cdot\nabla)w - \nabla w\cdot\nabla w\}$$

for the difference $w = v-u$. We estimate the nonlinear term by means of Cauchy's inequality, thus getting a linear integral inequality for the square of the norm $|w(t,\cdot)|_{L^2(\Omega)}$ being continuous due to the corollary above. From this, the inequality

---

[3] The continuity properties used here for the mollifying operator are proved as usual.

[4] See footnote on the page above.

$$(\text{I.9}) \qquad \int_{\Omega} (w\cdot w)(t,\cdot) \le |w_o|^2_{L^2(\Omega)} \cdot e^{c_h \cdot |u_o|^2_{L^2(\Omega)} \cdot t}$$

follows by Gronwall's Lemma [20,p.15], with $u_o = u(0,\cdot)$, $w_o = v(0,\cdot) - u(0,\cdot) \in V_o$. This inequality shows the uniqueness in V of the solutions of (I.2) and their continuous dependence of the initial values in the sense of $L^2(\Omega)$.

I.5. On the Construction of the Weak Solution.

As formulated in [15], the weak solution of the stabilized Navier-Stokes initial-boundary value problem can be constructed by means of the Galerkin-method in [5]. Due to the uniqueness (I.9) of solutions of (I.2) in V, the whole sequence of all Galerkin approximations converges to the desired solution. Therefore, we have the following:

Theorem 1: To any initial value $u_o \in V_o$ there is exactly on weak solution $u \in V$ of (I.2). u is the limit in $L^2([0,T]\times(\Omega \cap Q))$ of the whole sequence of all Hopf-Galerkin approximations for any bounded open cube $Q \subset R^n$, and u is stable on any compact time-interval [0,T].

I.6. On the Approximation of Bounded Weak Solutions by Means of Solutions of Equations with Mollification.

Let $u \in V$ be a weak Hopf solution of the Navier-Stokes initial-boundary value problem, i.e. a solution of (I.2) with the vector function u itself in place of its mollification u_h in the nonlinear term. If u is bounded almost everywhere, the conclusions of section I.4. performed with the difference $v_h - u = v_h - u_h + u_h - u$ instead of w, result in the

Theorem 2: Any Hopf solution being bounded almost everywhere on Ω_∞ is the limit (for any $t \ge 0$ in the sense of $L^2(\Omega)$) of the solutions of equations with a mollification , the radius of which is going to zero.

II. A Fixedpoint Equation for the Construction of Local Classical Solutions of the Navier-Stokes Initial Value Problem.

We consider the initial value problem

(II.1) $$u_t-\Delta u = -(u\cdot\nabla)u-\nabla q,\ \nabla\cdot u=0,\ t\in(0,T],$$

$$u=u_o \quad \text{for } t=0,\ |u(t,x)|\to 0 \text{ with } |x|\to\infty$$

for the vector function $u(t,x) = (u^1,\ldots,u^n)$ and the real function $q(t,x)$, (u providing a model of the velocity-field of a viscous incompressible flow on the n-space R^n during the time interval $[0,T)$, q being the sum of the pressure function of the flow and of the potential of the density of the given forces acting on the fluid,) u_o being prescribed on R^n. With the help of the fundamental solution

(II.2) $$\Gamma(t,x) = \begin{cases} (4\pi t)^{-\frac{n}{2}} e^{-\frac{x\cdot x}{4t}} & \text{for } t>0 \\ 0 & \text{for } t<0 \end{cases}$$

of the heat equation in R^n, we get from (II.1) a fixedpoint equation for u. Using the additional condition u being divergence-free, we eliminate the unknown function q by means of the fundamental solution

(II.3) $$\gamma(x) = \begin{cases} \dfrac{|x|^{2-n}}{(n-2)\omega_n}, & n=3,4,\ldots \\ -\dfrac{1}{2\pi}\ln|x|, & n=2 \end{cases}$$

of the potential equation in R^n, ω_n denoting the area of the unit sphere in R^n. Then we will see, that the operator in the fixedpoint equation is in fact the product of an operator S with the projection F of a space of vector functions onto its subspace of all divergence-free ones. The operator S establishes the fixed-point formulation of a nonlinear parabolic initial value problem for which the maximum-minimum principle holds.

In order to formulate this and the application of the contracting mapping principle in detail, we need the following definitions: We denote by J a given interval, by C_i (or $C_{0;i}$) the class of vector functions being bounded and continuous on $J'\times R^n$ for any compact interval $J'\subset J$ together with all their partial derivatives (or together with their spatial derivatives, respectively) up to the order $i = 0,1,\ldots$. In the class C_o we define the (semi-) norms

$$[u(t,\cdot)]_o^p = \sup_{x\in R^n} |u(t,x)p(x)| \text{ with } p(x)=(1+xx)^\lambda,\ \lambda\ge 0,$$

$$[u(t,\cdot)]_\alpha^p = \sup_{x\neq y} \{|u(t,x)-u(t,y)|\ |x-y|^{-\alpha}\cdot\min\{p(x),\,p(y)\}\},$$

and in the class $C_{0;i}$

$$|u(t,\cdot)|^p_{i,\alpha} = \max_{\substack{\iota=0,\alpha;0\le j_1,\dots,j_n,\\ j_1+\dots+j_n=j\le i}} \left\{ \left[\frac{\delta^j}{(\delta x^1)^{j_1}\dots(\delta x^n)^{j_n}} u(t,\cdot)\right]^p_\iota \right\}$$

for any $\alpha \in (0,1)$. Let $C^p_{0;i,\alpha}$ denote the subspace containing all $u \in C_{0;i}$ with $\sup_{t\in J} |u(t,\cdot)|^p_{i,\alpha} < \infty$ and, finally, let $C^p_i(J \times R^n) \subset C_i$ be the subspace formed by all vector functions, the values of which together with the values of their derivatives up to the order i behave like $1/p(x)$ with $|x| \to \infty$. We use the weight-functions $p(x) = 1+xx$ and $p(x) \equiv 1$ only. In the latter case we will omit the exponent p. By u_0 we always denote a vector function being not dependend on the variable $t \in J = [0,T]$.

For (vector-) functions $u_0 \in C_0$, $f \in C_0$, $g \in C^p_0$ the convolution products

$$(\Gamma * u_0)(t,x) = \int_{R^n} \Gamma(t,x-y)\, u_0\, (y)dy,$$

$$(\Gamma *_2 f)(t,x) = \int_o^t \int_{R^n} \Gamma(t-s,\, x-y) f(s,y)dy\, ds$$

and

$$(\nabla\gamma)*g = \int_{R^n} (\nabla\gamma)(x-y)g(t,y)dy$$

are well defined. Moreover, the projection F of the space $C^p_{0;1,\alpha}$ on its subspace of divergence-free elements is given by the formula

$$\text{(II.4)} \qquad F\, v = v + \nabla\, (\nabla\cdot\gamma * v).^{5)}$$

The map F is defined on $C^p_{0;o\alpha}$ and F^2 equals F.

Due to a theorem of Calderon and Zygmund [13], F can be

---

5) More explicitly we have

$$\nabla\cdot\gamma*v = \sum_{j=1}^n \left(\frac{\delta}{\delta x^j}\gamma\right)*v^j.$$

defined on the space $L^2(R^n)$, too.

With a function $N(t)$ being continuous on $J = [0,T]$ we define the closed bounded set

$$A = \{u | u \in C^p_o,\ [u(t,\cdot)]^p_o \le N(t),\ t \in J\}$$

in the Banach space C^p_o with the norm $|\cdot|^p_o$. Concerning the map $\Gamma' : f \to \Gamma *_2 f$ we have

Lemma 1: Γ' is a bounded linear map of C^p_o in $C^p_{o;1,\alpha}$ and of $C^p_{o;i,\alpha}$ in $C^p_{o;i+2,\alpha'}$, $i = 0,1,.,\alpha' < \alpha$. For any $f \in A$ the estimate

$$|(\Gamma *_2 f)(t,\cdot)|^p_{1,\alpha} \le \sum_{j=0,\iota=0}^{j=1,\iota=\alpha} c_{j\iota} \int_o^t (t-t')^{-\frac{j+\iota}{2}} \cdot N(t')dt'$$

holds on $[0,T]$.

On the map $F' : g \to F\nabla \cdot g$, defined on the space $C^p_{o,1,\alpha}$ of vector functions having n^2 components, we state

Lemma 2: Assume $n \ge 3$ for R^n. Then F' is a bounded linear map of $C^p_{o;i+1,\alpha}$ in $C^p_{o;i,\alpha}$ with

$$|(F'g)(t,\cdot)|^p_{i,\alpha} \le c'_i |g|^p_{i+1,\alpha}.$$

The proof of Lemma 1 uses the methods of [20,§28] and the estimate (13.1) in [6,p376]. Lemma 2 follows by straightforward calculations using essentially the assumption $\nabla \cdot g \in C^p_{o;\alpha}$ and $g \in C^p_{o;\alpha}$, cp. the note [13,p57].

Now let u be a classical (local) solution of (II.1) for a given vector function $u_o \in C^p_1$, $\nabla \cdot u_o = 0$, $n \ge 3$. Some integrations by parts lead us from (II.1) to the fixedpoint equation

$$\text{(II.5)} \qquad u = F\{\Gamma * u_o - \nabla \cdot \Gamma *_2 u \otimes u\} \equiv Gu,$$

containing the n x n-matrix $u \otimes u = (u^i \cdot u^j)_{i,j=1,\dots,n}$. As Lemma 1 and 2 show, equation (II.5) has the size of a Volterra integral equation and there holds $GA \subset A$, if the function N (defining the class A) solves the integral inequality

(II.6) $$|u_o|_o^p \cdot e^{c^*t} + \int_o^t c^* \sum_{i=0,\iota=0}^{i=1,\iota=\alpha} (t-t')^{-\frac{i+\iota}{2}} N^2(t')dt' \leq N(t),$$

c^* denoting a constant.

Conclusions analog to [20,p.61] ensure us that the contracting mapping principle applies on (II.5) locally, i.e. on a sufficiently small time interval $J=[0,T^*]$. By this means we get the

Theorem 3 : To any divergence-free initial value $u_o \in C_1^p$ there exists exactly one local solution $u \in C_\infty^p((0,T^*] \times R^n) \cap C_{o;1}^p$ of the classical initial value problem with $\nabla q = (\frac{\partial}{\partial t} - \Delta)(F-1)\nabla \cdot \Gamma^*_2 u \otimes u$.

Corollary: In case $u_o \in C_\infty^p$, we have $u \in C_\infty^p([0,T^*] \times R^n)$.

The regularity properties of a fixedpoint $u \in A$ of G are established by Lemma 1 and 2 and by the wellknown properties of the heat potentials.

Note: The equation

(II.7) $$u = \Gamma^* u_o - \Gamma^*_2 (u \cdot \nabla) u,$$

which we get from (II.5) by cancelling the projection F and using $\nabla \cdot u = 0$, represents just the fixedpoint formulation of the parabolic initial value problem

(II.8) $$u_t - \Delta u = -(u \cdot \nabla) u \quad , \quad t > 0$$
$$u = u_o,\ t = 0,\ u(t,x) \to 0 \text{ with } |x| \to \infty.$$

Because the maximum-minimum-principle holds for any component u^j, $j=1,\ldots,n$ of any classical solution of (II.8), the proof of the existence o f a global classical solution of (II.7) or (II.8) might be considered as an exercise in the methods of [20 ,p. 61 , 215, 268].

III. Global Classical Solutions of the Navier-Stokes Initial Value Problem Containing a Mollification

Structurally, as the note above shows, the limitation of theorem 3 , stating local existence only, is due to the projection F additionally contained in (II.5). Technically this limitation stems from the estimation of the term $\nabla\cdot\Gamma*_2\, u\otimes u$ by the square of the norm $|u(t,\cdot)|^p_o$. Using mollification (as in (I.2)), we get the estimate

$$\text{(III.1)} \qquad [(u_h\otimes u)(t,\cdot)]^p_o \leq c_h\cdot|u(t,\cdot)|_{L^2(R^n)}\cdot[u(t,\cdot)]^p_o$$

which is linear in $[u(t,\cdot)]^p_o$ with the constant c_h from I.(C). Therefore, the initial value problem

$$\text{(III.2)} \qquad \begin{aligned} u_t-\Delta u &= -(u_h\cdot\nabla)u-\nabla q,\ \nabla\cdot u=0,\ t>0 \\ u &= u_o,\ t=0, \end{aligned}$$

considered in [10], is more convenient than the original (II.1). Corresponding to (III.2), the fixedpoint formulation established in the same way as (II.5) is given by

$$\text{(III.3)} \qquad u = F\{\Gamma*u_o-\nabla\cdot\Gamma*_2u_h\otimes u\}.$$

For short, we consider the 3-dimensional case only, physically the most interesting one. Due to the special form of the weight-function p we have $C^p_o(R^3) \subset L^2(R^3)$, and the energy equation (I.8) holds for any classical solution $u \in C^p_1(R^3_T) \cap C^p_{o;2}$ of (II.1) or (III.2) respectively. The detailed inquiry of the relation between the length of the time interval, on which the iteration scheme for (III.3) is converging, and of the bounds for the solution leads to the

Theorem 4: To any divergence-free initial value $u_o \in C^p_1(R^3)$ (or $u_o \in C^p_\infty(R^3)$) there is exactly one global classical solution

$$u \in C^p_\infty((0,\infty) \times R^3) \cap C^p_{0;1}(R^3_\infty)$$

(or $u \in C^p_\infty([0,\infty) \times R^3)$, respectively) of the initial-value problem (III.2) with mollification, c.p. [10].

IV. Application: Non-solenoidal Hopf-Galerkin Approximations for the Navier-Stokes Initial Value Problem with a Mollification.

The approximation of weak solutions in [15] is based on the use of a complete system of solenoidal (i.e. divergence-free) elements, the numerical construction of which is by no means an easy task in concrete cases. In order to simplify numerical calculations (i.e. to make them possible for more complicated 3-dimensional problems) in the following we propose non-solenoidal approximations. To handle them we start with a weak formulation of the Navier-Stokes initial-value problem within the class V* of all measurable vector functions u on $[0,\infty) \times R^n = R^n_\infty$, which

a) for any $T \in (0,\infty)$ belong to the closure with respect to the Hopf norm $|\cdot|_H$ of the class D* of all C_∞-vector functions having compact support in R^n_∞ and

b) have restrictions $u(t,\cdot)$ bounded in $L^2(R^n)$ uniformly in $t \geq 0$.

Because the (generally speaking non-solenoidal) test functions $\varphi \in D^*$ are not necessarily orthogonal in $L^2(R^n_T)$ to the gradients, the weak formulation of (III.2) results in the equation

$$\text{(IV.1)} \qquad \int_{R^n} u\cdot\varphi \Big|_0^T = \int_{R^n_T} (u\cdot\{\varphi_t + (u_h\cdot\nabla)\varphi + \Delta\varphi\} - (F_1 u)\cdot(\nabla\cdot\varphi) - (F_2 u)\cdot\varphi)$$

with

$$F_1 u = \sum_{i,j=1}^{n} \left(\frac{\partial}{\partial x^i}\gamma\right) * \left(\frac{\partial}{\partial x^j} u_h^i\right) u^j \quad ,$$

$$F_2 u = \sum_{i,j=1}^{n} (\nabla\gamma) * \left(\frac{\partial^2}{\partial x^i \partial x^j} u_h^i\right) u^j \quad ,$$

γ from (II.3).

<u>Definition (IV.1)</u> We call weak solution of (III.2) any $u \in V^*$, which solves (IV.1) for any value $T > 0$ and any $\varphi \in D^*$.
Using the methods from section I, for such weak solutions we get an energy inequality and an estimate establishing uniqueness as in (I. 9); but now both estimates hold only locally i.e., on small time intervals, because on V* there is no equivalent to $\nabla\cdot u = 0$. With the conclusions from [5, p.224] we see, that it suffices to require (IV.1) for all elements $e_i(x)$ of a complete orthonormal system suitably chosen in $D^* \subset L^2(R^n)$. Naturally for the numerical realization of the Hopf-Galerkin method below it will be more convenient to dispense with the (e_i) being orthogonal and belonging to D*. Here we make these assumptions on the (e_i) for short only.
We set for the kth approximations in (IV.1)

(IV.2) $$u_k(t,x) = \sum_{i=1}^{k} \lambda_{ki}(t) e_j(x)$$

with the k unknown functions $\lambda_{ki} \in C_1[0,\infty)$ and require that u_k is a solution of (IV.1) for $\varphi = e_1, \ldots, e_k$. Differentiating these k equations with respect to t leads us to the system

(IV.3) $$\frac{d}{dt} \int_{R^n} u_k \cdot e_i = \int_{R^n} (u_k \cdot \{\varphi_t + (u_{kh} \cdot \nabla)\varphi + \Delta\varphi - (F_1 u_k)(\nabla \cdot \varphi) - (F_2 u_k) \cdot \varphi)$$

of k ordinary differential equations for the $\lambda_{ki}(t) = \int_{R^n} u_k \cdot e_i$. The initial condition is

(IV.3a) $$\lambda_{ki}(0) = \int_{R^n} u_o \cdot e_i$$

with the given $u_o \in V_o$. Now, Hopf's existence proof from [5 ,p.226-231] applies locally, i.e. on a time interval, the length δ of which depends on the norm $|u_o|_{L^2(R^n)}$ only. Therefore, all approximate solutions u_k exist on $[0,\delta]$. Due to the uniqueness of (IV.1) in V* mentioned above, the whole sequence (u_k) is converging to the solution $u \in V^*$ of (IV.1).

We state this in the

Theorem 5: Locally, i.e. on a time interval of length $\delta=\delta(|u_o|_{L^2(R^n)})$, for any given $u_o \in V_o$ there is exactly one solution $u \in V^*$ of (IV.1), u being the limit in $L^2([0,\delta] \times Q)$ of the whole sequence of all approximations u_k from (IV.2) for any bounded open cube $Q \subset R^n$.

Especially in the case n=3, $u_o \in C_1^p$ being divergence-free, any classical solution u* of (III.2) is a weak solution in the sense of definition (IV.1) Therefore it holds u* = u and, due to (I.8), the inequality

$$|u(\delta,\cdot)|_{L^2(R^3)} \leq |u_o|_{L^2(R^3)}$$

follows. From this, we get the

Corollary: In the case $u_o \in C_1^p$ being divergence free, $\Omega = R^3$, the weak solution of (I.2) can be approximated by the non-solenoidal solutions u_k of (IV.2.,3.) step by step on successive time intervals of a length $\geq \delta(|u_o|_{L^2(R^3)})$.

Bibliography

[1] ADAMS, Robert A.: Sobolev Spaces, Academic Press, New York (1975).

[2] AUBIN, J.-P.: Approximation of Elliptic Boundary-Value Problems, John Wiley & Sons, Inc., New York, (1972).

[3] FABES, E.B., JONES, B.F., RIVIERE, N.M.: The Initial Value Problem for the Navier-Stokes Equations with Data in L^p, Arch. Rat. Mech. An. 45 (1972).

[4] HEYWOOD, John G.: On Uniqueness Questions in the Theory of Viscous Flow, preprint from Department of Mathematics, The University of British Columbia, Vancouver (1975).[6]

[5] HOPF, E.: Über die Anfangswertaufgabe für die hydrodynamischen Grundgleichungen, Math. Nachr. 4 (1951), 213-231.

[6] LADYZENSKAJA, O.A., SOLONNIKOV, V.A., URAL'CEVA, N.N.: Linear and Quasilinear Equations of Parabolic Type, American Mathematical Society, Providence, Rhode Island (1968).

[7] LADYZENSKAJA, O.A.: The Mathematical Theory of Viscous Incompressible Flow. 2.Ed., New York (1969).

[8] LADYZENSKAJA. O.A.: Example of Nonuniqueness in the Hopf Class of Weak Solutions for the Navier-Stokes-Equations, Math, USSR-Izvestija 3 (1969), 229-236.

[9] LADYZENSKAJA, O.A.: Mathematical Analysis of Navier-Stokes-Equations for Incompressible Liquids, Annual Rev. Fluid Mech. 7 (1975), 249-272.

[10] LERAY, J.: Sur le mouvement d'un liquide visqueux emplissant l'espace. Acta math. 63 (1934), 193-248.

[11] LIONS, J.L. ; PRODI, G.: Un théorème d' existence et unicité dans les équations de Navier-Stokes en dimension 2, C.R. Acad. Sci. Paris 248 (1959), 3519-3521

---

[6] For the knowlegde of [4], the author is indebted to Professor Velte.

[12] PRODI, G.: Un theorem di unicita per le equazioni di Navier-Stokes, Annali di Mat. 48 (1959) 173-182.

[13] MIKHLIN, S.G.: Multidimensional Singular Integrals and Integral Equations, Pergamon Press, Oxford (1965).

[14] RAUTMANN, R.: Bemerkungen zur Anfangswertaufgabe einer stabilisierten Navier-Stokesschen Gleichung, ZAMM 55, T 217-221 (1975).

[15] RAUTMANN, R.: On the Convergence of a Galerkin Method to Solve the Initial Value Problem of a Stabilized Navier - Stokes Equation , ISNM 27 Birkhäuser Verlag, Basel, Stuttgart, 255 - 264 (1975).

[16] RAVIART, P.-A.: Finite Element Methods for Solving the Stationary Stokes and Navier-Stokes Equations. Paper presented at the 3rd Conference on Basic Problems of Numerical Analysis, PRAGUE, August 27-31 (1973) preprint No. 73011, Université Paris VI Laboratoire Analyse Numérique.

[17] SCHMEIDLER, W.: Zur mathematischen Theorie der Wärmeleitung und Strömungslehre, J. reine angew. Math. 242, 115-133 (1970).

[18] SERRIN, J.: On the Interior Regularity of Weak Solutions of the Navier-Stokes Equations, Arch. Rat. Mech. Anal. 9 (1962), 187-195.

[19] SERRIN, J.: The Initial Value Problem for the Navier-Stokes Equations, in: Nonlinear Problems (ed. R.E. Langer) MRC Madison (1963) 69-98.

[20] WALTER, W.: Differential and Integral Inequalities, Springer Berlin (1970).

Generalized Multiparameter Spectral Theory

by

G. F. Roach
Department of Mathematics
University of Strathclyde
Scotland

and

B. D. Sleeman
Department of Mathematics
University of Dundee
Scotland

1. Introduction

Abstract multiparameter spectral theory arises as a generalisation of the well known spectral theory associated with the operator equation $Ax = \lambda x$ and is usually related to linear operator systems of the form

$$A_i u^i = \sum_{j=1}^{n} \lambda_j S_{ij} u^i, \quad i = 1,\ldots,n. \tag{1.1}$$

Here $\lambda_j \in C$, $j = 1,\ldots,n$ are the spectral parameters, $S_{ij} : H_i \to H_i$ is usually an Hermitian operator defined on the whole of some separable Hilbert space H_i and $A_i : H_i \supset D(A_i) \to H_i$ is self-adjoint. A bibliography of the more recent contributions to the study of (1.1) is given at the end of this paper.

The system (1.1) is an example of a "weakly coupled" system in that coupling is affected only through the parameters λ_j, $j = 1,2,\ldots,n$. Had coupling been only through the unknowns then the system would have been called "strongly coupled". When coupling is by means of spectral parameters and unknowns we shall say that the system is "completely coupled".

In this paper we shall be concerned with completely coupled systems, regarding them as a generalisation of the system (1.1). Typically these systems will take the form of sets of operator matrix equations of the form

$$A_k x_k = \Lambda B_k x_k, \quad k = 1,\ldots,n \tag{1.2}$$

where A_k, B_k are $n \times n$ matrices with operator entries, Λ is an $n \times n$ matrix with complex scalar entries and x_k is an $n \times 1$ column vector. We introduce the notation $A_k = [A_{ij}^k]$, $B_k = [B_{ij}^k]$, $x = (x_1^k,\ldots,x_n^k)^T$, $k = 1,2,\ldots,n$, $\Lambda = [\lambda_{ij}]$, $\lambda_{ij} \in \mathbb{C}$, $i,j = 1,2,\ldots,n$, and denote by H_i^k, $i = 1,2,\ldots,n$ a collection of separable Hilbert spaces. Here the entries A_{ij}^k, B_{ij}^k are linear operators which can be regarded in one of the following ways

$$A_{ij}^k,\ B_{ij}^k : H_j^k \to H_j^k \tag{1.3a}$$

$$A_{ij}^k,\ B_{ij}^k : H_j^k \to H_i^k \tag{1.3b}$$

In both cases the system (1.2) has an interpretation as an operator equation in a certain form of direct sum of the spaces H_i^k, $i = 1,2,\ldots,n$.

We would remark that systems of the form (1.2) are not entirely new to the literature; for instance in the particular case when $k = 1$ and $H_i^1 \equiv H_i$, $i = 1,2,\ldots,n$ there are contributions by Anselone [10] and Dunford [12]. However, to the authors' knowledge, the general system (1.2) has not been considered before. Consequently, it is the main aim of this paper to lay the foundations for a spectral theory associated with (1.2) and so form a basis from which the theory of such a system can be developed and applied to significant physical problems.

In order to fix ideas we formally consider the case $n = 2$ and consider only one member of the system (1.2). Thus we have an operator equation of the form:

$$\begin{pmatrix} A_{11} & A_{12} \\ A_{21} & A_{22} \end{pmatrix} \begin{pmatrix} x_1 \\ x_2 \end{pmatrix} = \begin{pmatrix} \lambda_{11} & \lambda_{12} \\ \lambda_{21} & \lambda_{22} \end{pmatrix} \begin{pmatrix} B_{11} & B_{12} \\ B_{21} & B_{22} \end{pmatrix} \begin{pmatrix} x_1 \\ x_2 \end{pmatrix} \tag{1.4}$$

We shall rewrite (1.4) as an element of a "weakly coupled system" of the form (1.2). Once this is done we reformulate each member of (1.2) in a similar manner and then employ the spectral theory of "weakly coupled" systems as a means of developing a spectral theory for "completely coupled systems".

Applying the usual matrix operations we may formally write (1.4) in the form

$$\begin{pmatrix} A_{11}x_1 + A_{12}x_2 \\ A_{21}x_1 + A_{22}x_2 \end{pmatrix} = \begin{pmatrix} (\lambda_{11}B_{11} + \lambda_{12}B_{21})x_1 + (\lambda_{11}B_{12} + \lambda_{12}B_{22})x_2 \\ (\lambda_{21}B_{11} + \lambda_{22}B_{21})x_1 + (\lambda_{21}B_{12} + \lambda_{22}B_{22})x_2 \end{pmatrix} \tag{1.5}$$

Now we make the transformation

$$\lambda_{ij} = \lambda\alpha_{ij} + \mu\beta_{ij}, \quad i,j = 1,2, \tag{1.6}$$

where λ and μ are certain complex quantities and α_{ij}, β_{ij} are as yet arbitrary real constants, then (1.5) takes the form

$$\begin{aligned} A_1x &= \lambda C_{11}x + \mu C_{12}x \\ A_2x &= \lambda C_{21}x + \mu C_{22}x . \end{aligned} \tag{1.7}$$

Here $\qquad x = (x_1,x_2)^T$ and

$$A_1x = A_{11}x_1 + A_{12}x_2,$$
$$A_2x = A_{21}x_1 + A_{22}x_2,$$
$$C_{11}x = (\alpha_{11}B_{11} + \alpha_{12}B_{21})x_1 + (\alpha_{11}B_{12} + \alpha_{12}B_{22})x_2$$
$$C_{21}x = (\alpha_{21}B_{11} + \alpha_{22}B_{21})x_1 + (\alpha_{21}B_{12} + \alpha_{22}B_{22})x_2$$
$$C_{12}x = (\beta_{11}B_{11} + \beta_{12}B_{21})x_1 + (\beta_{11}B_{12} + \beta_{12}B_{22})x_2$$
$$C_{22}x = (\beta_{21}B_{11} + \beta_{22}B_{21})x_1 + (\beta_{21}B_{12} + \beta_{22}B_{22})x_2$$

Although it may appear that some generality has been lost in assuming the parameters λ_{ij} should take the form (1.6) nevertheless it turns out that in most cases of interest λ_{ij} has the form (1.6). Finally writing (1.7) in matrix form we have

$$Ax = \lambda C_1x + \mu C_2x \tag{1.8}$$

where A is the 2×2 matrix with entries A_{ij}, $i,j = 1,2$,

$$C_1 = \begin{pmatrix} (\alpha_{11}B_{11} + \alpha_{12}B_{21}) & (\alpha_{11}B_{12} + \alpha_{12}B_{22}) \\ (\alpha_{21}B_{11} + \alpha_{22}B_{21}) & (\alpha_{21}B_{12} + \alpha_{22}B_{22}) \end{pmatrix}$$

$$C_2 = \begin{pmatrix} (\beta_{11}B_{11} + \beta_{12}B_{21}) & (\beta_{11}B_{12} + \beta_{12}B_{22}) \\ (\beta_{21}B_{11} + \beta_{22}B_{21}) & (\beta_{11}B_{12} + \beta_{22}B_{22}) \end{pmatrix}$$

Thus the matrix equation (1.4) has, formally at least, been rewritten as a member of weakly coupled system (1.1).

2. Direct Sums of Hilbert Spaces

Let H_i, $i = 1,2,\ldots,n$ denote separable Hilbert spaces with inner product (norm) denoted by $(\cdot,\cdot)_i$ $(||\cdot||_i)$. Algebraically, these Hilbert spaces, considered simply as linear spaces, can be combined to form a new linear space, H_e, called the external direct-sum of the H_i, $i = 1,2,\ldots,n$ and we write

$$H_e := H_1 \oplus \ldots \oplus H_n \equiv \bigoplus_{i=1}^{n} H_i \tag{2.1}$$

The underlying set of H_e is the cartesian product $H_1 \times \ldots\ldots \times H_n$ of the underlying sets H_i, $i = 1,2,\ldots,n$. Thus an element $g \in H_e$ is an ordered n-tuple

$$g = \{g_1, g_2, \ldots, g_n\} \ , \quad g_i \in H_i, \quad i = 1,2,\ldots,n \tag{2.2}$$

On H_e addition and multiplication by a scalar are defined componentwise in the form

$$\{g_1, g_2, \ldots, g_n\} + \{h_1, h_2, \ldots, h_n\} = \{g_1+h_1, g_2+h_2, \ldots, g_n+h_n\}$$

$$c\{g_1, g_2, \ldots, g_n\} = \{cg_1, cg_2, \ldots, cg_n\} \tag{2.3}$$

Moreover if θ_i and I_i denote the zero and identity elements respectively in H_i, $i = 1,2,\ldots,n$ then the corresponding quantities in H_e are $\{\theta_1, \theta_2, \ldots, \theta_n\}$ and $\{I_1, I_2, \ldots, I_n\}$. With this structure H_e is clearly a linear space.

Frequently it is the case that $H_i \subset H$, $i = 1,2,\ldots,n$ where H is a containing Hilbert space. In this case the H_i can be combined algebraically in two ways to form new linear spaces; the external direct sum as before and the internal direct sum. The internal direct sum, H_I, of the linear spaces $H_i \subset H$, $i = 1,2,\ldots,n$ is written

$$H_I := H_1 + H_2 + \ldots H_n \equiv \sum_{i=1}^{n} H_i \tag{2.4}$$

and is a linear subspace of H comprising elements $g \in H_I$ of the form

$$g = g_1+g_2 \ldots +g_n, \quad g_i \in H_i, \quad i = 1,2,\ldots,n \tag{2.5}$$

As we have indicated for $H_i \subset H$, $i = 1,2,\ldots,n$ we can also form the sum H_e. However, for $n > 1$ we notice that H_e is a subspace of the cartesian product of n copies of H. For example if $H \equiv R$, $H_1 := [0,1]$, $H_2 := [0,2]$, $H_3 := [0,3]$ then H_I is clearly a subspace of R whilst H_e is a subspace of R^3. Although H_I and H_e are different linear spaces, nevertheless, it is possible to compare them. In fact there is a natural mapping $\phi = H_e \to H_I$. This mapping is clearly linear and onto but it need not be (1-1). For instance in the case $n = 2$ with $H_1 = H_2$ then it is easily seen that the kernel of ϕ is the set of all elements of the form $\{h,-h\}$. $h \in H_1 = H_2$. Consequently, unless H_1 and H_2 are trivial then ϕ has a nontrivial kernel. However, ϕ is an isomorphism if and only if the H_i, $i, = 1,2,\ldots,n$ are adjoint and in this case we say that H_e and H_I are algebraically equivalent.

If now we wish to consider the H_i, $i = 1,2,\ldots,n$ as Hilbert spaces and not simply as linear spaces then as before we can form the external and internal direct sums. However, in this case the structure of the H_i, $i = 1,2,\dots,n$ induces an inner product and norm in both H_e and H_I according to

$$(f,g) = \sum_{i=1}^{n} (f,g)_i \tag{2.6}$$

$$||f|| = \sum_{i=1}^{n} ||f_i||_i \tag{2.7}$$

where f and g are elements of either H_e or H_I. Since we are not concerned here with more than a finite number of Hilbert spaces there are no immediate convergence problems and it is clear that H_e and H_I equipped with structure of the form (2.6) (2.7) are Hilbert spaces. Moreover, in the finite case there is no distinction between algebraic and topological sums. Finally, although H_e and H_I are quite different spaces, nevertheless, as before, they can be compared. Indeed, in the case when the H_i, $i = 1,2,\dots,n$ are mutually orthogonal then H_e and H_I are unitarily equivalent.

3. Operators on Direct sums of Hilbert spaces.

In this section we give operator realizations of the formal matrix operations introduced in (1.2). For the sake of simplicity we shall consider the case when $n = 2$.

I. A matrix operator $A : H_I \to H_I$. In the case $n = 2$ we write $H_I = H_1 + H_2$, and notice that each $x \in H_I$ has the decomposition:

$$x = x_1+x_2, \quad x_i \in H_i, \quad i = 1,2. \tag{3.1}$$

Moreover, since H_I is a linear space there exist elements z_1, $z_2 \in H_I$ such that

$$x = z_1+z_2, \quad z_1,z_2 \in H_I \tag{3.2}$$

The two statements (3.1) and (3.2) enable elements in H_I and H_k, $k = 1,2$ to be identified and we can rewrite (3.1) in the form

$$x = x_1+x_2 = (x_1+\theta_2) + (\theta_1+x_2) \tag{3.3}$$

The action of a linear operator $A : H_I \to H_I$ can be described by

$$Ax = A(x_1+x_2) = A(x_1+\theta_2) + A(\theta_1+x_2) \tag{3.4}$$

Since both elements on the right are elements of H_I we have the decomposition

$$A(x_1+\theta_2) = y_{11} + y_{21}$$
$$A(\theta_1+x_2) = y_{12} + y_{22} \qquad (3.5)$$

where $y_{kj} \varepsilon H_k$, $j,k = 1,2$. Now the elements $y_{kj} \varepsilon H_k$, $j,k = 1,2$ depend upon the elements $x_j \varepsilon H_j$, $j = 1,2$ and we can write

$$y_{ij} = A_{ij}x_j \quad i,j = 1,2$$
$$A_{ij} : H_j \to H_i \; i,j = 1,2 \qquad (3.6)$$

Therefore corresponding to each linear operator $A : H_I \to H_I$ there is a matrix whose entries are the linear operators $A_{ij} : H_j \to H_i$, $i,j = 1,2$.

Combining (3.4) to (3.6) we obtain

$$Ax = (A_{11}x_1 + A_{12}x_2) + (A_{21}x_1 + A_{22}x_2) \qquad (3.7)$$

Consequently, an operator equation of the form $Ax = f \varepsilon H_I$ can be considered in the following component form

$$A_{11}x_1 + A_{12}x_2 = f_1 \varepsilon H_1$$
$$A_{21}x_1 + A_{22}x_2 = f_2 \varepsilon H_2 \qquad (3.8)$$

It is useful to notice that the system (3.8) can be expressed in terms of certain operators $H_I \to H_I$. To see this recall that the elements $f_i \varepsilon H_i$, $i = 1,2$ induce in H_I the elements

$$g_1 = f_1 + \theta_2, \quad g_2 = \theta_1 + f_2 \qquad (3.9)$$

Therefore, for linear operators $\hat{A}_i : H_I \to H_i$, $i = 1,2$ defined by $\hat{A}_i x := A_{i1}x_1 + A_{i2}x_2$, $x_i \varepsilon H_i$, $i = 1,2$, we see that images under $\hat{A}_i$, will induce elements in H_I according to (3.9). Consequently, we can define $A_i : H_I \to H_I$, $i = 1,2$, by

$$A_i x := \hat{A}_i x + \theta_j, \quad i \neq j, \quad i = 1,2, \quad x \varepsilon H_I \qquad (3.10)$$

and rewrite (3.8) in the form: $\begin{pmatrix} A_1x \\ A_2x \end{pmatrix} = \begin{pmatrix} g_1 \\ g_2 \end{pmatrix}$

II. <u>A matrix operator</u> $A : H_I \to H_I \oplus H_I$: Let $x \in H_I$ with $x = x_1 + x_2$, $x_k \in H_k$, $k = 1,2$. We define a linear operator $A : H_I \to H_I \oplus H_I$ by $Ax := \{A_1x, A_2x\}$ where $A_k : H_I \to H_I$, $k = 1,2$ are linear operators defined by $A_1x := A_{11}x_1 + A_{12}x_2$, $A_2x := A_{21}x_1 + A_{22}x_2$, with, $A_{ij} : H_j \to H_i$, $i,j = 1,2$. Consequently an operator equation of the form $Ax = \theta \in H_I \oplus H_I$ can be considered in the following component form

$$A_kx = \theta \in H_I \quad k = 1,2 \tag{3.11}$$

In the sequel we will be mainly concerned with operators $A : H_I \to H_I$ and the associated matrices with entries $A_{ij} : H_j \to H_i$, $i,j = 1,2,\ldots,n$. Such arrays appear to have more interesting properties and lead to a more widely applicable theory than the operators $A : H_I \to H_I \oplus H_I$. To illustrate the restrictive nature problems associate with this latter operator notice that the equations (3.11) can be decomposed in the form

$$A_{11}x_1 = \theta_1, \quad A_{12}x_2 = \theta_2$$

$$A_{21}x_1 = \theta_1, \quad A_{22}x_2 = \theta_2$$

In general it is clear that this system has no non-trivial solution.

We are now in a position to effect the reduction of (1.2) to a weakly coupled system.

4. <u>Reduction of Completely Coupled Systems.</u>

As in the previous section we consider matrix operators $A = \{A_{ij}\}$, $B = \{B_{ij}\} : H_I \to H_I$. Where $A_{ij} : H_j \to H_i$, $B_{ij} : H_j \to H_i$, $i,j = 1,\ldots,n$. We consider operator equations of the form

$$Ax = \underset{\sim}{\Lambda} Bx \tag{4.1}$$

where $x \in H_I$ and $\underset{\sim}{\Lambda} = \{\lambda_{ij}\}$ is an $n \times n$ matrix with scalar entries $\lambda_{ij} \in \mathbb{C}$.

Again, for ease of presentation, consider the case $n = 2$, i.e.,

$$H_I = H_1 + H_2 \tag{4.2}$$

Thus in the notation of the previous section we may write (4.1) as

$$\begin{pmatrix} A_1x \\ A_2x \end{pmatrix} \begin{matrix} = \\ = \end{matrix} \begin{pmatrix} \lambda_{11}\lambda_{12} \\ \lambda_{21}\lambda_{22} \end{pmatrix} \begin{pmatrix} B_1x \\ B_2x \end{pmatrix} \tag{4.3}$$

If we set

$$\lambda_{ij} = \lambda\alpha_{ij} + \mu\beta_{ij}, \quad i,j = 1,2 \tag{4.4}$$

where λ and μ are complex quantities and α_{ij}, β_{ij}, $i,j = 1,2$ are arbitrary real constants then (4.3) may be expressed in the form:

$$\begin{aligned} A_1x &= \lambda C_1x+\mu D_1x = \hat{A}_1x+\theta_2, \quad x \in H_I \\ A_2x &= \lambda C_2x+\mu D_2x = \theta_1+\hat{A}_2x, \quad x \in H_I \end{aligned} \tag{4.5}$$

where C_i, $D_i : H_I \to H_I$ are defined, for $i = 1,2$, by

$$\begin{aligned} C_ix &:= \alpha_{i1}B_1x + \alpha_{i2}B_2x = \alpha_{i1}(\hat{B}_1x+\theta_2) + \alpha_{i2}(\theta_1+\hat{B}_2x) \\ D_2x &:= \beta_{i1}B_1x + \beta_{i2}B_2x = \beta_{i1}(\hat{B}_1x+\theta_2) + \beta_{i2}(\theta_1+\hat{B}_2x) \end{aligned} \tag{4.6}$$

Since $B_i : H_I \to H_i$, $i = 1,2$ it is clear from (4.5) and (4.6) that we must have the compatibility condition:

$$\begin{aligned} \alpha_{12}\hat{B}_2x &= \theta_2, \quad \beta_{12}\hat{B}_2x = \theta_2 \\ \alpha_{21}\hat{B}_1x &= \theta_1, \quad \beta_{21}\hat{B}_1x = \theta_1 \end{aligned} \tag{4.7}$$

Since in general this system is only solvable for trivial x it is necessary to choose $\alpha_{ij},\beta_{ij} = 0$, $i \neq j$. Thus we are naturally led to consider problems in which the matrix $\underset{\sim}{\Lambda}$ is diagonal. With this condition the system (4.5) is compatible and we may rewrite (4.5) as the two parameter operator equation

$$Ax = \lambda Cx + \mu Dx \tag{4.5}$$

where $A \equiv \begin{bmatrix} A_{11} & A_{12} \\ A_{21} & A_{22} \end{bmatrix}$, $C \equiv \begin{bmatrix} \alpha_{11}B_{11} & \alpha_{11}B_{12} \\ \alpha_{22}B_{21} & \alpha_{22}B_{22} \end{bmatrix}$ and $D \equiv \begin{bmatrix} \beta_{11}B_{11} & \beta_{11}B_{12} \\ \beta_{22}B_{21} & \beta_{22}B_{22} \end{bmatrix}$

with $A, C, D : H_I \to H_I$. The procedure for the reduction of the general system (1.2) is now clear. Here $\underset{\sim}{\Lambda}$ is a diagonal matrix with entries λ_{ii}, $i = 1,\ldots,n$. and for each fixed k we have the internal direct sum Hilbert Space.

$$H_I^k = H_1^k + \ldots + H_n^k \tag{4.9}$$

with elements $x_k \in H_I^k$.

If we let

$$\lambda_{ii} = \sum_{j=1}^{n} \alpha_{ij}\lambda_j, \tag{4.10}$$

then we arrive at the "weakly coupled" system

$$A_k x_k = \sum_{j=1}^{n} \lambda_j C_{kj} x_k, \quad k = 1,2,\ldots,n \tag{4.11}$$

where

$$A_k := \left\{A_{ij}^k\right\} : H_I^k \to H_I^k \qquad k = 1,2,\ldots,n \tag{4.12}$$

$$C_{kj} := \begin{bmatrix} \alpha_{ij}B_{11}^k & \alpha_{ij}B_{1n}^k \\ \alpha_{2j}B_{21}^k & \alpha_{2j}B_{2n}^k \\ \cdots & \cdots \\ \alpha_{nj}B_{n1}^k & \alpha_{nj}B_{nn}^k \end{bmatrix} : H_I^k \to H_I^k, \quad k,j = 1,2\ldots n \tag{4.13}$$

In the subsequent sections we describe the multiparameter spectral theory related to systems of the form (4.11) or (1.1) and interpret the results for the completely coupled system (1.2).

5. Spectral theory for Weakly Coupled Systems.

Let H_I^k, $k = 1,2,\ldots,n$, defined by (4.9), be separable Hilbert spaces and let $\mathcal{H} = \bigotimes_{k=1}^{n} H_I^k$ be their tensor product. Suppose that in each space H_I^k operators A_k, C_{kj}, $j = 1,2,\ldots n$ are defined as in (4.12), (4.13). We now make the following assumptions

Al: A_k, $C_{kj} : H_I^k \to H_I^k$, $k,j = 1,2,\ldots n$ are Hermitian and continuous.

In a further development of the theory we shall consider the situation when A_k is allowed to be self-adjoint and not necessarily bounded.

We shall also require a certain "definiteness" condition which may be introduced as follows. Let $f = f_1 \otimes \ldots \times, f_n$ be a decomposed element of $\mathcal{H}$ with $f_k \in H_I^k$, $k = 1,\ldots,n$ and let $\alpha_0, \alpha_1, \ldots, \alpha_n$

be a given set of real numbers not all zero. Then an operator $\Delta : \mathcal{H} \to \mathcal{H}$, $k = 0,1,\dots,n$ may be defined by the equation

$$\Delta f := \sum_{k=0}^{n} \alpha_k \Delta_k f = \det \begin{vmatrix} \alpha_0 & \alpha_1 & \cdots\cdots & \alpha_n \\ -A_1 f_1 & C_{11} f_1 & \cdots & C_{1n} f_1 \\ \cdot & \cdot & \cdot & \cdot \\ -A_n f_n & C_{n1} f_n & \cdots & C_{nn} f_n \end{vmatrix} \tag{5.1}$$

where the determinant is to be expanded formally using the tensor product. This defines Δf for decomposable $f \in \mathcal{H}$ and we can extend the definition to arbitrary $f \in \mathcal{H}$ by linearity and continuity. The required definiteness condition can now be stated as

<u>A2</u>: $\Delta : \mathcal{H} \to \mathcal{H}$ is positive definite in the sense that if $f = f_1 \otimes \dots \otimes f_n$ is a decomposed element of $\mathcal{H}$ then

$$\langle \Delta f, f \rangle = \det \begin{vmatrix} \alpha_0 \cdots\cdots\cdots & \alpha_1 \cdots\cdots\cdots & \alpha_n \\ \langle -A_1 f_1, f_1 \rangle_1 \cdots & \langle C_{11} f_1, f_1 \rangle_1 \cdots & \langle C_{1n} f_1, f_1 \rangle_1 \\ \cdot & \cdot & \cdot \\ \langle -A_n f_n, f_n \rangle_n \cdots & \langle C_{n1} g_n, f_n \rangle_n \cdots & \langle C_{nn} f_n, f_n \rangle_n \end{vmatrix}$$

$$\geq C \, |||f_1|||_1^2, \dots, |||f_n|||_n^2 ,$$

where C is a positive constant $\langle \cdot , \cdot \rangle$ denotes the inner product in $\mathcal{H}$ and $\langle \cdot , \cdot \rangle_k$ $(|||\cdot|||_k)$, denotes the inner product (norm) in H_I^k, $k = 1,2,\dots,n$.

On the basis of A1 and A2 Källstrom and Sleeman [9] have developed, under quite general conditions, a multiparameter spectral theory associated with a system of operators such as A_k, C_{kj}, $k,j = 1,2,\dots,n$. In this paper we shall restrict attention to the particular case when $\alpha_0 = 1$, $\alpha_k = 0$, $k = 1,2,\dots,n$, the more general situation can be discussed as in [9]. For the case $\alpha_0 = 1$, $\alpha_k = 0$, $k = 1,2,\dots,n$ we notice that A2 reduces to

A3 $\qquad \Delta_0 : \mathcal{H} \to \mathcal{H}$ is positive definite $\qquad$ (5.3)

This condition implies that $\Delta_0^{-1} : \mathcal{H} \to \mathcal{H}$ exists as a bounded linear operator.

Rather than use the inner product $\langle\cdot,\cdot\rangle$ in $\mathcal{H}$ generated by the inner products $\langle\cdot,\cdot\rangle_k$ in H_I^k, we use the inner product $\langle\Delta_0\cdot,\cdot\rangle$ which will be denoted by $[\cdot,\cdot]$. The norms induced by these inner products are equivalent and so topological concepts such as continuity and convergence may be discussed unambiguously. Certain algebraic concepts however may depend on the particular inner product.

For $L : \mathcal{H} \to \mathcal{H}$ we denote by $L^{\#}$ the adjoint of L with respect to $[\cdot,\cdot]$, i.e. for all f, $g \in \mathcal{H}$ we have

$$[Lf,g] = [f,L^{\#}g]. \tag{5.4}$$

The notation L^* will denote the adjoint of L with respect to $\langle\cdot,\cdot\rangle$. Further for $i = 1,\ldots,n$, Γ_i will denote the operator $\Delta_0^{-1}\Delta_i : \mathcal{H} \to \mathcal{H}$.

<u>Theorem</u> 1: $\Gamma_i^{\#} = \Gamma_i$, $i = 1,\ldots,n$.

<u>Proof</u>: Since the operators $\Delta_i ; \mathcal{H} \to \mathcal{H}$ are formed from the Hermitian operators A_k, C_k, $k, = 1,\ldots,n$, it is clear that $\Delta_i = \Delta_i^*$, $i = 0,1,\ldots,n$. Thus if f, $g \in \mathcal{H}$ we have for $i = 1,\ldots,n$

$$\begin{aligned}[\Gamma_i f,g] &= \langle\Delta_0\Delta_0^{-1}\Delta_i f,g\rangle = \langle f,\Delta_i g\rangle \\ &= \langle f,\Delta_0\Delta_0^{-1}\Delta_i g\rangle = \langle\Delta_0 f,\Gamma_i g\rangle \\ &= [f,\Gamma_i g],\end{aligned}$$

and the result is proved.

For a continuous operator $L : H_I^k \to H_I^k$, we denote by L^+ the induced continuous operator $L^+ : \mathcal{H} \to \mathcal{H}$ defined on decomposable elements $f = f_1 \otimes\ldots\otimes f_n$ by

$$L^+f = f_1 \otimes\ldots\otimes f_{k-1} \otimes Lf_k \otimes f_{k+1} \otimes\ldots\otimes f_n, \tag{5.5}$$

and extended to arbitrary $f \in \mathcal{H}$ by linearity and continuity.

<u>Lemma</u> 1 [4]. Let $A_k : H_I^k \to H_I^k$, $k = 1,\ldots,n$ be a continuous linear operator. Then

$$\bigcap_{k=1}^{n} \mathrm{Ker}(A_k^+) = \bigotimes_{k=1}^{n} \mathrm{Ker}\, A_k.$$

We now establish a fundamental property enjoyed by the operators Γ_i.

Theorem 2: The operators Γ_i, $i = 1,\ldots,n$ are pairwise commutative.

Proof: Let $f \in \mathcal{H}$ and suppose that the elements $g^k \in \mathcal{H}$, $k = 1,2,\ldots,n$ uniquely solve the system.

$$A_k^+ f = \sum_{j=1}^{n} C_{kj}^+ g^j \tag{5.6}$$

That such a solution exists has been established by Källström and Sleeman in [7]. If for the moment we let $f = g^0$ then it follows from (5.6) that

$$\Delta_p g^q = \Delta_q g^p, \quad p,q = 0,1,\ldots,n \tag{5.7}$$

In particular $\Delta_0 g^q = \Delta_q f$, i.e. $g^q = \Gamma_q f$. Thus for $p,q = 1,\ldots,n$ we have

$$\Delta_p \Gamma_q f = \Delta_q \Gamma_p f, \tag{5.8}$$

for all $f \in \mathcal{H}$. An application of Δ_0^{-1} to both sides of (5.8) proves the result.

An immediate consequence of this result is

Corollary 1: $A_k^+ f = \sum_{j=1}^{n} C_{kj}^+ \Gamma_j f, \quad k = 1,\ldots,n.$

6. Spectral theory for several commuting Hermitian operators

Working with the inner product $[\cdot,\cdot]$ in $\mathcal{H}$ the operators Γ_i, $i = 1,\ldots,n$ form a family of n commuting Hermitian operators. Let $\sigma(\Gamma_i)$ denote the spectrum of Γ_i and σ_0 the Cartesian product of the $\sigma(\Gamma_i)$, $i = 1,\ldots,n$. Then since $\sigma(\Gamma_i)$ is a non-empty compact subset of R it follows that σ_0 is a non-empty compact-subset of R^n.

We now give a slight generalisation of the usual concept of "numerical range". By the numerical range of the system $\{A_k, C_{kj}\}$ we shall mean the subset Ω of R^n defined by

$$\Omega = \{(\langle\Delta_1 f^1, f^1\rangle,\ldots,\langle\Delta_n f^n, f^n\rangle) \mid \langle\Delta_0 f^i, f^i\rangle = 1, f^k \in \mathcal{H}, i = 1,\ldots,n\}$$

or

$$\Omega = \{([\Gamma_1 f^1, f^1],\ldots[\Gamma_n f^n, f^n]) \mid [f^i, f^i] = 1,\; f^i \in \mathcal{H},\; i = 1,\ldots,n\}$$

Theorem 3: (i) Ω is a convex set
(ii) $\sigma_0 \subset \overline{\Omega}$

Proof: (i) Ω is the cartesian product of the sets $\{[\Gamma_i f,f] \mid [f,f] = 1,\ i = 1,\dots,n\}$ each of which is convex [11, p.388] consequently Ω is convex.

(ii) Suppose $\lambda = (\lambda_1,\dots,\lambda_n) \notin \overline{\Omega}$. Then there exists i, $1 \leq i \leq n$ such that $\lambda_i \notin \{[\Gamma_i f,f] \mid [f,f] = 1\}$. Thus $\lambda_i \notin \sigma(\Gamma_i)$ and so $\lambda \notin \sigma_0$. Thus we have $\sigma_0 \in \overline{\Omega}$.

Let $E_i(\cdot)$ denote the resolution of the identity for the operator Γ_i and let $M_i \in \mathbb{R}$ be a Borel set, $i = 1,\dots,n$. We then define

$$E(M_1 x \dots x M_n) = \prod_{i=1}^{n} E_i(M_i).$$

Notice that the projections $E_i(\cdot)$ will commute since the operators Γ_i commute. Thus in this way we obtain a spectral measure $E(\cdot)$ on the Borel subsets of $\mathbb{R}^n$ which vanishes outside σ_0. Thus for each $f, g \in \mathcal{H}$, $[E(\cdot)f,g]$ is a complex valued Borel measure vanishing outside σ_0. Measures of the form $[E(\cdot)f,f]$ will be non-negative finite Borel measures vanishing outside σ_0.

The spectrum σ of the system $\{A_k,C_{kj}\}$ is defined to be the support of the operator valued measure $E(\cdot)$,. Alternately σ is the smallest closed set with the property $E(M) = E(M \cap \sigma)$ for all Borel sets $M \subset \mathbb{R}^n$. Thus σ is a compact subset of $\mathbb{R}^n$ and if $\lambda \in \sigma$, then $E(M) \neq 0$ for all non-degenerate closed rectangles M with $\lambda \in M$. Thus the measures $[E(M)f,g]$, $f, g \in \mathcal{H}$ actually vanish outside σ. We are now in a position to state our main result namely the Parseval equality and expansion theorem.

Theorem 4. Let $f \in \mathcal{H} = \bigotimes_{k=1}^{n} H_I^k$. Then

(i) $\langle \Delta_0 f,f\rangle = \int_\sigma [E(d\lambda)f,f] = \int_\sigma \langle E(d\lambda)f,\Delta_0 f\rangle.$

(ii) $f = \int_\sigma E(d\lambda)f,$

where this integral converges in the norm of $\mathcal{H}$.

The proof of this theorem is an easy consequence of the theory of functions of several commuting Hermitian operators and is omitted (See for example Prugovečki [13, pp. 270-285].

7. Eigenvalues

An eigenvalue for the system $\{A_k, C_{kj}\}$ is defined as the n-tuple of complex numbers $\lambda = (\lambda_1, \ldots, \lambda_n)$ for which there exists a non zero decomposable element $f = f_1 \otimes \ldots \otimes f_n \in \mathcal{H}$, such that

$$A_k f_k = \sum_{j=1}^{n} \lambda_j C_{kj} f_k, \quad k = 1, \ldots, n \tag{7.1}$$

Theorem 5. Let $\lambda = (\lambda_1, \ldots, \lambda_n)$ be an eigenvalue for the system $\{A_k, C_{kj}\}$. Then each λ_i, $i = 1, \ldots, n$ is real.

Furthermore, if f, g are distinct eigenvectors corresponding to the eigenvalues λ and μ then $[f, g] = 0$

Proof: If $f = f_1 \otimes \ldots \otimes f_n$ is an eigenvector corresponding to λ, then

$$\langle A_k f_k, f_k \rangle_k = \sum_{j=1}^{n} \lambda_j \langle C_{kj} f_k, f_k \rangle_k, \quad k = 1, \ldots, n$$

and

$$\langle f_k, A_k f_k \rangle_k = \sum_{j=1}^{n} \overline{\lambda}_j \langle f_k, C_{kj} f_k \rangle_k,$$

Since A_k is self-adjoint and each C_{kj} Hermitian we have

$$\sum_{j=1}^{n} (\lambda_j - \overline{\lambda}_i) \langle C_{kj} f_k, f_k \rangle_k = 0, \quad k = 1, \ldots, n .$$

It now follows from the definiteness condition (5.3) that $\lambda_j = \overline{\lambda}_j$, $j = 1, \ldots, n$, thus proving the reality of the eigenvalue λ. That f, g are orthogonal in the sense described in the theorem follows by a similar argument and is ommitted.

Theorem 6. If $\lambda \in \sigma$ is such that $E(\{\lambda\}) \neq 0$, then λ is an eigenvalue. Conversely if λ is an eigenvalue then $\lambda \in \sigma$ $E(\{\lambda\}) \neq 0$.

Proof: Suppose $\lambda = (\lambda_1, \ldots, \lambda_n) \in \sigma$ is such that $E(\{\lambda\}) \neq 0$ and let $g \in E(\{\lambda\})\mathcal{H}$, $g \neq 0$. Then since the operators $E_i(\{\lambda_i\})$, $i = 1, \ldots, n$, commute $E_i(\{\lambda_i\})g = g$ for each i. From standard spectral theory we deduce that $\Gamma_i g = \lambda_i g$ and so

$$A_k^{+} g - \sum_{j=1}^{n} \lambda_j C_{kj}^{+} g = \left(A_k^{+} - \sum_{j=1}^{n} C_{kj}^{+} \Gamma_j \right) g = 0,$$

$k = 1,\dots,n$ by Corollary 1 to theorem 2. Thus

$$0 \neq g \in \bigcap_{k=1}^{n} \operatorname{Ker}\left(A_k - \sum_{j=1}^{n} \lambda_j C_{kj}\right)^{+} = \bigotimes_{k=1}^{n} \operatorname{Ker}\left(A_k - \sum_{j=1}^{n} \lambda_j C_{kj}\right),$$

and so there must exist a non-zero decomposable element $f = f_1 \otimes \dots \otimes f_n \in \mathcal{H}$ such that

$$A_k f_k - \sum_{j=1}^{n} \lambda_j C_{kj} f_k = 0, \quad k = 1,\dots,n.$$

This shows that λ is an eigenvalue.

Conversely, if $\lambda = (\lambda_1,\dots,\lambda_n)$ is an eigenvalue with non zero decomposable eigenvector $f = f_1 \otimes \dots \otimes f_n$, we have

$$A_k^{+} f = \sum_{j=1}^{n} \lambda_j C_{kj}^{+} f, \quad k = 1,\dots,n.$$

Hence it follows from the proof of theorem 2 that for $k = 1,\dots,n$, $\Gamma_k f = \lambda_k f$. Thus $\lambda_k \in \sigma(\Gamma_k)$ and $E_k(\{\lambda_k\}) f = f$. This shows that $E(\{\lambda\}) f = f$ and the result follows.

8. Compact operators.

In this section we consider the nature of spectrum, σ, under the additional assumption that each of the operators $A_k : H_I^k \to H_I^k$, $k = 1,\dots,n$ is compact. For each $\lambda \in \mathbb{R}^n$ we define the operators $S_k(\lambda) : H_I^k \to H_I^k$ by

$$S_k(\lambda) = A_k - \sum_{j=1}^{n} \lambda_j C_{kj}, \quad k = 1,\dots,n \tag{8.1}$$

Then as in [4] we may prove

<u>Theorem</u> 7. (i) $\lambda \in \sigma$ if and only if 0 is in the spectrum of each $S_k(\lambda)$, $k = 1,\dots,n$.

(ii) λ is an eigenvalue if and only if 0 is an eigenvalue of each $S_k(\lambda)$, $k = 1,\dots,n$.

(iii) if each of the operators $A_k : H_I^k \to H_I^k$ is compact and $\lambda \neq 0$, then it is impossible for 0 to be in the continuous spectrum of each $S_k(\lambda)$.

Immediate consequences of this result are

Corollary 1. Suppose each of the operators $A_k : H_I^k \to H_I^k$, $k = 1,\dots,n$ is compact. If $\lambda = (\lambda_1,\dots,\lambda_n)$ is a non-zero point in σ then at least one of the equations

$$A_k f_k - \sum_{j=1}^{n} \lambda_j C_{kj} f_k = 0, \quad k = 1,\dots,n$$

has a non-trivial solution.

Corollary 2. Suppose each of the operators $A_k : H_I^k \to H_I^k$, $k = 1,\dots,n$ is compact. Then $0 \in \sigma$.

9. Concluding Remarks

In this paper we have shown how certain systems of matrix operator equations which are completely coupled through general spectral parameters and unknowns may be reduced to what we call weakly coupled systems. This reduction is achieved through a direct sum Hilbert space structure and a simple linear transformation (4.10). The abstract theory for weakly coupled systems of Hermitian operators as developed by Browne [4] and Källström and Sleeman [9] is then invoked in order to provide a foundation for a spectral theory relating to completely coupled systems. All the results we give for weakly coupled systems have a ready interpretation for completely coupled systems via the transformation (4.10) and the decomposition of the direct sum spaces involved.

It is hoped to develop the theory outlined in this paper to cover the case when some or all of the operators A_k are unbounded and then to investigate spectral and eigenvalue problems for systems of strongly coupled differential operators.

Bibliography to Abstract Weakly Coupled Systems

[1] Atkinson, F. V., Multiparameter Spectral Theory, Bull. Amer. Math. Soc., 74, 1-27 (1968).

[2] Atkinson, F. V., Multiparameter Eigenvalue Problems, Vol. 1, Matrices and Compact Operators, Academic Press, New York (1972).

[3] Browne, P.J., A multiparameter eigenvalue problem, J. Math. Anal. Appl., 38, 553-568 (1972).

[4] Browne, P.J., Multiparameter spectral theory, Indiana Univ. Math. J., 24, 249-257, (1974).

[5] Källström, A., Sleeman, B. D., An abstract multiparameter eigenvalue problem, Uppsala University Mathematics Report, No. 1975:2.

[6] Källström, A., Sleeman, B. D., An abstract multiparameter spectral theory, Dundee University Mathematics Report, No. DE 75:2.

[7] Källström, A., Sleeman, B. D., Solvability of a linear operator system., J. Math. Anal. Appl. (to appear).

[8] Källström, A., Sleeman, B.D., An abstract relation for multiparameter eigenvalue problems., Proc. Roy. Soc. Edinburgh, (to appear).

[9] Källström, A., Sleeman, B. D., Multiparameter spectral theory, Dundee University Mathematics Report, No. DE 75:4.

[10] Anselone, P.M., Matrices of Linear Operators, Enseignement Math 9, 191-197 (1963).

[11] Bachman, G., Narici, L.. Functional Analysis, Academic Press, New York (1966).

[12] Dunford, N., A spectral theory for certain operators on a direct sum of Hilbert spaces, Math. Ana. 162, 294-330 (1965/1966)

[13] Prugovečki, E., Quantum Mechanics in Hilbert Space, Academic Press, New York (1971).

On a control of systems with distributed parameters.

S.Rolewicz (Warszawa)

In engineering problems we often meet systems described by partial differential equations. Such systems are called by engineers "systems with distributed parameters".

The subject of my talk will concern a problem of controls of linear systems with distributed parameters. At the beginning I want to present several systems of that type.

A) Control of a string with fixed position of the first and changing position of the second end.

B) Control of the temperature in a slab before entering into a rolling mill.

C) Control of pulsation in a gas tube.

D) Control of diffusion.
and many, many others (see for example, Butkovskiy [1]).

Generally we have following situation. Two Banach spaces over reals X,Y are given. The space X is called the space of entrances, the space Y is called the space of exits. Now there are different types of models. The first one is called a fixed time model, the second type is called a time depending model.

In the first type [4], [8] we consider one linear continuous operator C mapping X into Y and we write our linear system in the form

(1) $$(X \xrightarrow[C]{} Y)$$

In the second type [5], [6], [8] we consider a family of linear continuous operators C_t mapping X into Y, where t is a real parameter, $0 \leq t \leq T \leq +\infty$ called by tradition a time. In that case we write our linear system in the form

$(2)_t$ $$(X \xrightarrow[C_t]{} Y)$$

There are many problems concerning systems of type (1) as well as systems of type $(2)_t$. In the talk, because of the limit of time, only a small number of such problems will be considered.

At the beginning problems concerning systems of type (1) will be discussed. We say that a system (1) is controllable if CX=Y (i.e. when the operator C is surjective).

Suppose now that a system

(1) $$(X \xrightarrow{C} Y)$$

is controllable. Thus by classical perturbation theory there is a positive number δ such that for all operators C_0 such that $\|C - C_0\| < \delta$ the system

$(1)_0$ $$(X \xrightarrow{C_0} Y)$$

is controllable.

Unfortunately the result presented above is not satisfactory for many engineering applications.

Example 1. Let $X = Y = C([0, 2\pi])$ be the space of continuous 2π-periodic functions. Let $C_t x(s) = x(s-t)$. Observe that $\|C_t - C_{t_0}\| = 2$ for $t \neq t_0$ and on the other hand if t is near to t_0 it is natural to require that C_t be "near" t_0.

For that reason we shall introduce another notion of continuity, namely image continuity.

Let E be a linear topological space. We say that a family of subsets $E(t) \subset E$, where the index t belongs to a metric space T, is continuous at a point t_0, if for each neighbourhood of zero U in E there is a positive number δ such that for t satisfying $\rho(t,t_0) < \delta$

(3) $$E(t) \subset E(t_0) + U$$

and simultaneously

(4) $$E(t_0) \subset E(t) + U$$

If only (3) holds we say that the family E(t) is upper semicontinuous at the point t_0. If E (t) is continuous (upper semicontinuous) for all $t_0 \in T$ we say that the family E (t) is continuous (upper semicontinuous).

We say that a family of continuous linear operators C_t mapping a Banach space X into a Banach space Y is image continuous, if there is a bounded convex closed set $U \subset X$ with non-void interior such that $C_t(U)$ is a continuous family of subsets in Y.

Observe that the family of operators in Example.1 is image continuous. It is enough to take as U the closed unit ball in X.

Theorem 1. Let X,Y be two Banach spaces over reals. Let C_t, t belonging to a metric space T, be an image continuous family of linear operators. If the system

$$(2)\ t_0 \qquad (X \xrightarrow[C_{t_0}]{} Y)$$

is controllable, then there is a positive number δ such that for all t such that $\rho(t,t_0)<\delta$ the system

$$(2)_t \qquad (X \xrightarrow[C_t]{} Y)$$

is controllable.

The proof is based on Rådström's cancellation lemma

Lemma 1. (4). Let A,B be closed convex sets in a normed space X. Let C be a bounded set. If A+C=B+C, then A=B.

Lemma 1 trivially implies

Lemma 2. Let X be a normed space. Let K_r be the ball of radius r in X. Then for all $r'<r$ and arbitrary closed convex set Γ without interior K_r is not contained in $\Gamma+K_{r'}$.

Theorem 1 follows from Lemma 2 and Banach's theorem on open mappings.

The notion of image continuity is independent of the notion of continuity of families of operators in strong operator topology.

Example 2. Let X = Y be one dimensional complex plane. Let T=[0,1]. We define C_t by formula

$$(5) \qquad C_t z = \begin{cases} z & \text{for } t=0 \\ e^{i/t} z & \text{for } 0<t\leq 1 \end{cases}$$

It is easy to verify that C_t is an image continuous family of linear operators, and on the other hand it is not continuous in the strong operator topology.

Example 3. Let $X = Y = L^1 [0,1]$. Let $T = [0,1]$. Let

$$C_t x(\cdot) = \chi_{[t,1]} x(\cdot) \tag{6}$$

of course C_t is continuous in the strong operator topology and it is not image continuous.

Observe that in Example 3, the conclusion of theorem 1 does not hold.

The image continuity has also a serious disadvantage. Namely, there are examples showing that the sum, the superposition, the conjugate of two image continuous families of operators may not be image continuous.

Problem 1 Is it possible to introduce a topology τ in the space of linear continuous operators mapping a Banach space X into a Banach space Y in such a way, that

1) Theorem 1 holds for continuity in topology τ,
2) the sum, the superposition, the conjugate of two τ-continuous families of operators are τ-continuous,
3) family of operators in Example 1 is τ-continuous ?

Now we shall come to the problem of minimalization of a convex continuous functional $F(x)$ defined on X.

The number

$$a = \inf \{F(x) : Cx=y_o\}$$

is rather easy to calculate in the case when Y is one-dimensional. If $F(x)$ is a norm in the space X and Y is one-dimensional, the problem of calculation of a can be easily reduced to the calculation of the norm of linear continuous functional C. For that reason an important role is played by the following formula called a theorem of moments [4]

$$\inf \{F(x) : Cx = y_o\} = \sup_{\phi\in Y^*} \inf \{F(x) : \phi(C(x)) = \phi(y_o)\} \tag{7}$$

which holds when the sets $\Gamma_r = C\{x : F(x) \leq r\}$ are closed and $\inf\{F(x) : x \in X\} < a$.

The hypothesis that Γ_r is closed is essential. Counter-examples that without this hypothesis formula (7) may not hold were given by I.Singer

[12] and the author [8] . Recently S. Dolecki has shown that formula (7) is valid for all $Y_o \in CX$ if and only if

$$\bigcap_{\varepsilon>0} C\{x : F(x) < a+\varepsilon\} = \bigcap_{\varepsilon>0} \overline{C\{x : F(x) \leq a+\varepsilon\}}$$

Formula (7) has one serious disadvantage. Namely on the right hand side it is necessary to take supremum of an infinite family (sometimes even infinite-dimensional) of linear continuous functionals. It would be much more convenient if we could replace supremum on the right hand side of formula (7) by one fixed functional ϕ_0, namely when there is a continuous linear functional $\phi_o \in Y^*$ such that

(8) $\inf\{F(x) : Cx = Y_o\} = \inf\{F(x) : \phi_o(Cx) = \phi_o(Y_o)\}$

When (8) holds for all $Y_o \in CX$ we say that the Maximum Principle of Pontriagin (briefly MPP) holds.

Using the Hahn-Banach theorem we trivially see that the MPP holds if $CX = Y$. On the other hand we have

Theorem 2. If the sets Γ_r are closed and the MPP holds, then $CX = Y$.

The proof is based on Wojtaszczyk's lemma [13].

Lemma 3 [13]. Let X be a Banach space. Let Γ be a closed convex set in X such that

1) $\operatorname{lin} \Gamma = X$
2) 0 belongs to the algebraic interior of Γ
3) If each algebraic boundary point is a supporting point then Γ has the topological interior.

S.Kurcyusz has shown that the hypothesis that Γ_r are closed is essential, but an example given by him has a slightly artificial character.

Problem 2. May we replace in Theorem 2, the hypothesis that Γ_r are closed by the hypothesis that F(x) is a norm in X?

Now we shall pass to the problems concerning systems depending on time.

The problem considered first will be a problem of controllability. We consider now the system

$$(2)_t \qquad (X \xrightarrow{C_t} Y)$$

where X is of the form of a Cartesian product

$$(9) \qquad X = X_1 \times X_2$$

In models X_1 generally is a space of initial states and X_2 is a space of controls.

We say that system $(2)_t$ is controllable, (controllable to zero), if for all $x_1 \in X_1$, $y \in Y$ (respectively $x_1 \in X_1$) there is t and $x_2 \in X_2$ such that

(10) $$C_t(x_1,x_2) = Y$$

(respectively

(10') $$C_t(x_1,x_2) = 0)$$

Let us write

(11) $$X_t = \{x_1 \in X_1 : 0 \in C_t(x_1, x_2)\}$$

Theorem 3 [11], If

(12) $$X_t \subset X_{t_1} \quad \text{for } t \leq t_1$$

and system $(2)_t$ is controllable to zero, then there is a universal time t_u such that for all $x_1 \in X_1$ there is $x_2 \in X_2$ such that

(13) $$C_{t_u}(x_1,x_2) = 0$$

The proof of the theorem is based on the Baire cathegory method.
Let

(14) $$B_t = C_t(0,X_2)$$

As a consequence of theorem 3 we get

Theorem 4 [11]. If system $(2)_t$ is controllable and

(15) $$B_t \subset B_{t'}, \quad \text{for } t \leq t'$$

then there is a universal time t_u such that for all $x_1 \in X_1$, $y \in Y$, there is $x_2 \in X_2$ such that

(16) $$C_{t_u}(x_1,x_2) = y$$

Theorem 3 can be applied to systems described by differential equations in Banach spaces. Theorem 4, at the moment, can be applied to systems described by differential equations in Banach spaces with operator coefficients, which are not depending on time.

Theorems 3 and 4 are extended by Dolecki [2] to systems with approximative controls. We say that systems $(2)_t$ is approximative controllable (controllable to zero), if for all $x_1 \in X_1$, $y \in y$ (respectively $x_1 \in X_1$) there is t such that for arbitrary $\epsilon > 0$, there is $x_2 \in X_2$ such that

(17) $$\| C_t (x_1, x_2) - y \| < \epsilon$$

(respectively

(17') $$\| C_t (x_1, x_2) \| < \epsilon)$$

S. Dolecki in [2] proved the existence of universal time for approximative controllability.

The last problem considered is a problem of reduction of a minimum time control problem to a minimum norm control problem.

We assume now that in the space X a closed convex set U is given. Let Y(t) be a family of subsets of the space Y. Let

(18) $$T = \inf{}' \{ t > 0 : C_t (U) \cap Y(t) \neq 0 \}$$

We say that a minimum time control problem has a solution if

(19) $$C_T(U) \cap Y(T) \neq 0$$

There are several theorems garrantueeing the existence of the minimum time control problem based on investigations of compactness in different topologies. [6], [8]. Unfortunately those theorems are not effective. For that reason the following procedure is used. We assume additionally that U has an interior and that 0 belongs to this interior.

Let F (x) be the Minkowski functional generated by U

(21) $$F(x) = \inf \{ t > 0 : x/t \in U \}$$

Let

(22) $$\Lambda(t) = \inf{}' \{ F(x) : C_t x \in Y(t) \}$$

For the investigation of the minimum time control problem it is important to know properties of the function $\Lambda(t)$. Unfortunately even in simple examples $\Lambda(t)$ may not be continuous. [8].

However the following theorem holds

Theorem 5. Let $C_t(U)$ and $Y(t)$ be continuous families of sets. If $C_{t_o}(U)$ has an interior, then $\Lambda(t)$ is continuous at t_o.

In a previous version of theorem 5 [8] it was required that $C_t(U)$ has an interior for t to belong to a certain neighbourhood of t_o. Now using Lemma 2 we can get theorem 2 under a weaker hypothesis [9].

References

[1] A.G.Butkovskiy, Methods of control of systems with distributed parameters (Russian), Moscow 1975

[2] S. Dolecki, A classification of controllability concepts for infinite dimensional linear systems (to appear)

[3] H.Rådström, An embedding theorem for spaces of convex sets, Proc. Amer. Math. Soc. 3 (1952), pp 165-169.

[4] S.Rolewicz, On a problem of moments, Stud.Math. 30(1968), pp. 183-191.

[5] S.Rolewicz, On minimum time control problem for linear systems, Ann.Acad. Bras.Ci. 42(1970) pp 653-654.

[6] S.Rolewicz, On minimum time control problem, IRIA Seminar Reports, Analyse et Contrôle des Systemes, 1975, pp 253-259.

[7] S.Rolewicz, On general theory of linear systems, Beiträge zur Analysis 8 (1976) pp 119-127.

[8] S.Rolewicz, Functional analysis and theory of control (Polish) PWN, Warszawa 1974, will be published in German by Springer-Verlag.

[9] S.Rolewicz, On minimum time control problem and continuous families of convex sets, Stud. Math. (in print).

[10] S.Rolewicz, Linear systems in Banach spaces, Proc.of Symposium on Variational Calculus and Control Theory, MRC, University-Wisconsin-Madison, September 22 - 24, 1975 (in print).

[11] S.Rolewicz, On universal time for controllability of time - depending linear control systems, Stud. Math. (in print).

[12] I.Singer, On a problem of moments of Rolewicz, Stud. Math. 48(1973) pp. 95-98.

[13] P.Wojtaszczyk, A theorem on convex sets related to the Abstract Pontriagin Maximum Principle, Bull.Acad. Pol.Sc. 21(1973) pp. 93-95.

Instytut Matematyczny
Polskiej Akademii Nauk
Institute of Mathematics
of the Polish Academy of Sciences

ON THE MAPPING PROBLEM FOR SECOND ORDER ELLIPTIC EQUATIONS IN THE PLANE

Stephan Ruscheweyh

University of Dortmund

In the present paper we shall study univalent 'holomorphic' solutions of the equations

$$w_{z\bar{z}} = A(|z|^2)|z|w_{|z|} \tag{1}$$

with an 'admissible' function $A(t^2)$, i. e. $A(t^2)$ analytic in the closed unit disc $\bar{\Delta}$ ($\Delta = \{t \,|\, |t| < 1\}$) and realvalued for real t.

It is well known that the mapping problem for generalized CAUCHY-RIEMANN equations has a beautiful and complete solution analogous to RIEMANN's mapping theorem. On the other hand, analytic univalent functions in Δ, represented by their powerseries, are under permanent consideration, in particular in view of the famous BIEBERBACH conjecture. We shall extend this consideration of analytic functions as harmonic functions (i. e. solutions of $w_{z\bar{z}} = 0$) with one-sided expansions to the equations (1). Our work is rather incomplete and some open problems will be mentioned in a concluding paragraph.

It follows from VEKUA's approach for solving equations of this type (see [4]) that there exists a RIEMANN function $R(\zeta,\zeta^*,z,z^*)$ analytic in $\bar{\Delta}^4$ which generates a complete system of solutions

$$w_k(z) = e^{ik\varphi}\, d_{|k|}(r), \quad k \in \mathbb{Z}, \quad z = re^{i\varphi}, \tag{2}$$

where $d_0(r) \equiv 1$, $d_k(r) = c_k(r)/c_k(1)$ and

$$c_k(r) = r^k \{\exp[2 \int_0^r tA(t^2)dt] - \int_0^1 t^k \, dR(rt,o,r,r)\}. \tag{3}$$

It should be noted that $d_k(r)$ are the uniquely determined solutions of

$$d_k''(r) + \left(\frac{1}{r} - 4r\,A(r^2)\right)d_k'(r) - \frac{k^2}{r^2}d_k(r) = o \tag{4}$$

which are regular at $r = o$ and satisfy $d_k(1) = 1$.

Our standard example for an equation (1) will be

$$\begin{aligned} w_{z\bar{z}} &= A_s(|z|^2)\,|z|\,w_{|z|} \\ A_s(|z|^2) &:= \frac{-2s^2}{1-(s^2|z|^2)^2}, \quad o \le s < 1, \end{aligned} \tag{5}$$

where we obtain

$$d_k(r) = r^k \frac{1 + k\dfrac{1-(sr)^2}{1+(sr)^2}}{1 + k\dfrac{1-s^2}{1+s^2}}, \quad k \in \mathbb{N} \cup \{o\}.$$

A complexvalued function $w \in C^2(\Delta)$ is a solution of (1) in Δ if and only if there exist sequences $\{a_k\}$, $\{b_k\}$ with

$$\limsup_{k\to\infty} |a_k|^{1/k} \le 1, \quad \limsup_{k\to\infty} |b_k|^{1/k} \le 1$$

such that

$$w(z) = \sum_{k=o}^{\infty} [a_k e^{ik\varphi} + b_k e^{-ik\varphi}]d_k(r), \quad z = re^{i\varphi}, \tag{6}$$

and these series are absolutely and locally uniformly convergent in Δ. We shall consider only those solutions of (1) which have an expansion (6) with $b_k = o$, $k \ge o$, and denote their collection by $H(A)$. Obviously the operator

$$\tau_A : f = \sum_{k=0}^{\infty} a_k z^k \longmapsto w = \sum_{k=0}^{\infty} a_k d_k(r) e^{ik\varphi}$$

generates a one-to-one mapping of $H := H(o)$ (the set of analytic functions in Δ) onto $H(A)$.

We wish to point out the following important property of this operator: if $f \in H$ extends continuously to $\bar{\Delta}$ then the same is true for $w = \tau_A f$ and we have $w(e^{i\varphi}) = f(e^{i\varphi})$, $\varphi \in \mathbb{R}$. This result (including a solution of DIRICHLET's problem for (1)) follows from a combination of HOPF's maximum principle, which shows that the functions

$$P_\rho(z) := \sum_{k=0}^{\infty} d_k(\rho) z^k \tag{7}$$

satisfy $\operatorname{Re} P_\rho(z) > \frac{1}{2}$, $z \in \Delta$, $o < \rho \leq 1$, and KOROVKIN's famous theorem on positive operators.

UNIVALENT SOLUTIONS

Let $S(A)$ denote the set of solutions in $H(A)$ which are univalent in Δ ($S(o) := S$). Furthermore let $\tau_A^{-1}(S(A)) := S(A)^{-1}$. Our first theorem reflects the distinguished position of the analytic case $A \equiv o$.

THEOREM 1. i) Let $A \not\equiv o$ be admissible. Then $S(A)^{-1}$ is a proper subset of S.
ii) Let $w = \tau_A f \in S(A)$. Then $w(\Delta) = f(\Delta)$.

We give a sketch of the proof. Let $w = \tau_A f \in S(A)$ and put

$$f_\rho(z) := P_\rho(z) * f(z), \quad o < \rho < 1, \tag{8}$$

where $*$ denotes the HADAMARD product in H. Obviously $f_\rho(e^{i\varphi}) = w(\rho e^{i\varphi})$

and hence $f_\rho \in S$, $o < \rho < 1$. But $f_\rho \to f$ locally uniformly in Δ for $\rho \to 1$ and hence $f \in S$ (HURWITZ' theorem). An application of CARATHÉODORY's 'kernel theorem' and a continuity argument completes the proof of ii).

Now assume $S(A)^{-1} = S$ for an admissible A. Hence for each $f \in S$ the functions f_ρ (see (8)) form a univalent and differentiable subordination chain in the sense of POMMERENKE [1]. Hence

$$\mathrm{Re}\,\frac{\frac{\partial}{\partial\rho} f_\rho(z)}{z f_\rho'(z)} > o, \quad z \in \Delta, \quad o < \rho < 1. \tag{9}$$

Using the notation

$$P(z) := \lim_{\rho\to 1} \frac{\partial}{\partial \rho} P_\rho(z) \tag{10}$$

one easily deduces that the limiting case $\rho \to 1$ of (8) is equivalent to

$$\left(P(z) + a\frac{z}{(1-z)^2}\right) * f \neq o, \quad o < |z| < 1$$

for arbitrary $a = \alpha + i\beta$, $\alpha > o$. This implies that $P(z) + a\frac{z}{(1-z)^2}$ is a starlike univalent function and the known properties of coefficient growth of starlike functions lead to the conclusion

$$P(z) = \sum_{k=1}^{\infty} d_k'(1) z^k = d_1'(1)\frac{z}{(1-z)^2}. \tag{11}$$

Now let $\psi_k(r) := r^{-k} d_k(r)$. From (3) with $Q(r) := -2\int_r^1 tA(t^2)dt$ we obtain

$$\psi_k^{(n)}(r) = (\exp Q(r))^{(n)} + o(1), \quad k \to \infty, \tag{12}$$

for fixed r,n. On the other hand, from (4),

$$\psi_k''(r) + \{\frac{2k+1}{r} - 4rA(r^2)\}\,\psi_k'(r) - 4kA(r^2)\,\psi_k(r) = o. \tag{13}$$

(11) and (12) exclude the possibility $d_1'(1) = o$ and hence give $\psi_k'(1) = k\psi_1'(1) = Q'(1) + o(1)$ which implies $\psi_k'(1) = Q'(1) = o$, $k \in \mathbb{N}$. Since $Q'(1) = 2A(1)$ we obtain from (13) $\psi_k''(1) = o$, $k \in \mathbb{N}$. Hence from (12): $A'(1) = o$ and by differentiating (13): $\psi_k'''(1) = o$. By induction we get $A^{(k)}(1) = o$, $k \in \mathbb{N} \cup \{o\}$, and thus $A \equiv o$.

Although we are not able to present a description of the classes $S(A)^{-1}$ we can state that their union over all admissible $A \not\equiv o$ is dense in S. This, in fact, is already true for the set $\bigcup_{o<s<1} S(A_s)^{-1}$ with A_s from (5).

CONVEX UNIVALENT SOLUTIONS

It is a consequence of HOPF's maximum principle that $w = \tau_A f$ is in $S(A)$ if the level curves $w(re^{i\varphi})$, $o \leq \varphi < 2\pi$, are JORDAN curves surrounding convex domains for $o < r < 1$. Using the common description of this fact (monotonic turning tangent vector) and the relation

$$-i\frac{\partial}{\partial\varphi}\tau_A f = \tau_A(zf') \qquad (\varphi = \arg z)$$

one obtains the following sufficient condition for univalence in $H(A)$.

THEOREM 2: $f \in S(A)^{-1}$ _if_

$$\text{i)}\ \frac{1}{z}\,\tau_A zf' \neq o,\ z \in \Delta,$$

$$\text{ii)}\ \operatorname{Re}\frac{\tau_A z^2 f''}{\tau_A zf'} + 1 \geq o,\ z \in \Delta. \tag{14}$$

If f satisfies (14) we call $w = \tau_A f$ _convex univalent_, denoting the class of these functions $K(A)$ $(K(o) := K)$. Obviously $K(A)^{-1} \in K$ for each admissible A. The convex univalent solutions have a number of pleasing properties which may be deduced by applying the theory of HADAMARD products in K (see [2]).

THEOREM 3. i) Let $f,g \; K(A)^{-1}$. Then $f * g \in K(A)^{-1}$.

ii) (Principle of subordination). Let $v \in H(A)$, $w \in K(A)$ satisfy $v(o) = w(o$ $v(\Delta) \subset w(\Delta)$. Then for $o < r < 1$ $v(\{|z| < r\}) \subset w(\{|z| < r\})$.

iii) (ROGOSINSKI's Lemma). Let v, w be as in ii). Then for each $p > o$, $o < r < 1$

$$\int_o^{2\pi} |v(re^{i\theta})|^p d\theta \le \int_o^{2\pi} |w(re^{i\theta})|^p d\theta.$$

THEOREM 4. Let $\sum_{k=2}^{\infty} k^2 |a_k| \le 1$. Then for each admissible A

$$w(z) = d_1(r)e^{i\varphi} + \sum_{k=2}^{\infty} a_k d_k(r) e^{ik\varphi} \in K(A), \qquad (z = re^{i\varphi}).$$

Proof. A look into the second order differential equation satisfied by $d_k(r)/d_1(r)$, $k \in \mathbb{N}$, shows $o < d_k(r) < d_1(r) < 1$, $k \in \mathbb{N}$, $o < r < 1$. Hence with $f := z + \sum_{k=2}^{\infty} a_k z^k$

$$|\tau_A z^2 f''| \le \sum_{k=2}^{\infty} |a_k| k(k-1) d_k(|z|)$$

$$< d_1(|z|) - \sum_{k=2}^{\infty} |a_k| k d_k(|z|)$$

$$\le |\tau_A z f'|.$$

Since $w = \tau_A f$ the conclusion follows from Theorem 2.

Remarks: 1) It is well known that in the analytic case the condition

$$\sum_{k=2}^{\infty} k^{\delta} |a_k| \le 1 \tag{15}$$

with $\delta = 1$ leads to an univalent function. Our standard example (5) with $s \to 1$ can be used to show that there is no exponent $\delta < 2$ in (15) such tha the corresponding class of solutions is in S(A) for every admissible A.

2) Theorem 4 implies the interesting estimate (see (2))

$$\left| 1 - \frac{w_k(z_2)}{w_k(z_1)} \right| \le k^2 \left| 1 - \frac{w_1(z_2)}{w_1(z_1)} \right|$$

for $k \in \mathbb{N}$, $|z_2| \le |z_1| \le 1$. Our standard example (5) shows that the factor k^2 is best possible. Note that in the analytic case we have k instead of k^2.

THE SINGULAR CASE

If we let $s \to 1$ in (5) we obtain an equation which is singular on the unit circle:

$$w_{z\bar{z}} = \frac{-2}{1-|z|^4} |z| w_{|z|} \tag{16}$$

The system

$$w_k(z) = e^{ik\varphi}\left(1 + |k| \frac{1-|z|^2}{1+|z|^2}\right), \quad k \in \mathbb{Z},$$

is still complete. In [3] it has been shown that in this particular case every univalent 'holomorphic' solution is convex univalent in the sense of the previous paragraph. A somewhat weaker result can be proved for a class of singular equations similar to (16).

THEOREM 5. <u>Let</u> A(t) <u>be analytic in</u> $t \in \Delta$, <u>realvalued for real</u> t, <u>and let</u> $t = 1$ <u>be a simple pole with residuum</u> $\alpha > 1/4$. <u>Then every 'holomorphic' univalent solution of</u> (1) <u>in</u> Δ <u>maps</u> Δ <u>onto a convex domain</u>.

<u>Remarks:</u> 1) Theorem 4 remains valid also in this singular case.
2) Our standard example (16) belongs to the case $\alpha = \frac{1}{2}$.

An outline of the proof is as follows: From the FUCHSian theory of regular-singular points of the second order equation (4) one deduces the behaviour of $d_k(r)$ close to $r = 1$:

$$d_k(r) = 1 + \frac{k^2}{2 - 8\alpha}(r-1)^2 + o((r-1)^2).$$

Hence $d_k'(1)/d_1'(1) = k^2$ and the limiting case of (9) (necessary for f to be in $S(A)^{-1}$) becomes $\mathrm{Re}\left(\frac{zf''(z)}{f'(z)} + 1\right) \geq 0$, $z \in \Delta$. Hence $f \in K$ and the conclusion follows from Theorem 1,ii).

OPEN PROBLEMS

The present paper is only a first step into a possible theory of S(A). We finally mention a few problems which probably can be treated by variations of the common methods.

1) It is clear that, for $w \in S(A)$, $|w|$ is bounded by certain upper and lower bounds depending on A: What are the distortion theorems for S(A)?

2) This leads immediately to the question if S(A) is compact for every A This property is urgently needed for the treatment of a lot of other questions.

3) In connection with 2) the problem of estimating the JACOBIan of $w \in S(A)$ arises. Is $w \in S(A)$ always a diffeomorphism? Or is it even locally uniformly quasiconformal with nontrivial bounds for the dilatation depending only on A? Something in this direction has been done for (16).

4) Is it true that for every admissible $A \neq o$ there are simply connected domains with more than one boundary point which are not images of Δ under a function $w \in S(A)$? Note that this is not contained in Theorem 1.

5) According to the distinguished role of convexity within H(A) we conjecture $K(A)^{-1} = K$ for every admissible A. This has been established for several values of s in our standard example. From the known convolution properties in K one easily deduces that this conjecture holds (for a certain A) if and only if $(1-z)^{-1} \in K(A)^{-1}$.

6) It is very likely that the Principle of Subordination and ROGOSINGSKI's Lemma (as mentioned in Theorem 3,ii) for K(A)) holds in S(A).

7) What are the coefficient bounds for functions in $S(A)^{-1}$. For which A does the BIEBERBACH conjecture hold? Subordination chains combined with the CARATHÉODORY-TOEPLITZ theorem, however, may help but will not do the full job.

REFERENCES

1. Chr. Pommerenke, Über die Subordination analytischer Funktionen, J. reine angew. Math. 218 (1965), 159 - 173.
2. St. Ruscheweyh and T. Sheil-Small, Hadamard products of schlicht functions and the Pólya-Schoenberg conjecture, Comment. Math. Helv. 48 (1973), 119 - 135.
3. St. Ruscheweyh and K.-J. Wirths, Riemann's mapping theorem for n-analytic functions, to appear.
4. I. N. Vekua, New Methods for Solving Elliptic Equations, North-Holland: 1968.

Regularisierung singulärer Integralgleichungen vom nicht normalen Typ mit stückweise stetigen Koeffizienten

Baldur M. Schüppel

(Darmstadt)

In [1] und [2] untersuchte S. Prößdorf singuläre Integralgleichungen nicht normalen Typs auf geschlossenen Kurven und mit stetigen Koeffizienten. Er stellte eine Theorie für solche Gleichungen auf und gab unter gewissen einschränkenden Bedingungen Regularisatoren an. Chvedelidse [5] und Gochberg [6] untersuchten Gleichungen vom Normaltyp mit stetigen Koeffizienten auch auf nicht geschlossenen Kurven. Gachov ([7] und [8]) und Gochberg [4] erforschten Gleichungen mit stückweise stetigen Koeffizienten. Dabei gelang es Gochberg, das Symbol solcher Gleichungen in den $L_p(\Gamma,\rho)$-Räumen anzugeben. Er fand, daß das Symbol eine Funktionenmatrix zweiter Ordnung ist und, daß es sowohl von p als auch von der Gewichtsfunktion ρ des Raumes abhängt. Aufbauend auf diesen beiden Forschungen wird nun mit dieser Arbeit ein Regularisator für Gleichungen nicht normalen Typs mit stückweise stetigen Koeffizienten angegeben.

Im Unterschied zu [9] und [10] wird hier der wirkliche Wert des Symbols berücksichtigt. Deshalb muß zusätzlich zur in [9] und in [10] geforderten Annahme $a^2(c_k) - b^2(c_k) \neq 0$, $k = 1,2,\dots,s$ und $a^2(c_k+0) - b^2(c_k+0) \neq 0$ für $k = 2r+1, 2r+2,\dots,s$ auch noch $\det A(c_k,x) \neq 0$ für $x \in [0,1]$ und $k = 1,2,\dots,s$ verlangt werden. Dies bedeutet, daß bei einer als normalauflösbar angenommenen singulären Integralgleichung nach der Wahl von $p \in (1,\infty)$ nicht mehr alle Werte aus dem Intervall $(-1, p-1)$ für die Exponenten γ_k, $k = 1,2,\dots,s$ der Gewichtsfunktion frei wählbar sind.

Wir betrachten die eindimensionale singuläre Integralgleichung

$$(\mathfrak{A}\varphi)(t) := a(t)\,\varphi(t) + \frac{b(t)}{\pi i}\int_\Gamma \frac{\varphi(\tau)}{\tau - t}\,d\tau + (\mathfrak{T}\varphi)(t) = f(t),\quad t\in\Gamma \qquad (1)$$

unter folgenden Voraussetzungen:

1. Γ sei die Vereinigung von endlich vielen ebenen, disjunkt liegenden und orientierten Kurven aus der Klasse $C^{m+1,\mu}, m\in\mathbb{N}_0, \mu\in(0,1)$. Γ kann sowohl geschlossene als auch nicht geschlossene Kurven enthalten. Mit $c_1,\dots,c_r$ bzw. $c_{r+1},\dots,c_{2r}$ werden die Anfangs- bzw. Endpunkte aller nicht geschlossenen Kurven von Γ bezeichnet. Außerdem seien $c_{2r+1}, c_{2r+2},\dots,c_s$ festgewählte Punkte auf Γ, von denen keiner mit einem der Punkte $c_1,\dots,c_{2r}$ zusammenfalle.

2. Sei $\nu := (p,\gamma_1,\gamma_2,\dots,\gamma_s)$ ein Vektor mit $(s+1)$ reellen Komponenten, die die Bedingungen

$$1 < p < \infty \quad , \quad 0 < \gamma_k < p-1 \qquad (k = 1,\dots,s)$$

erfüllen. Damit definieren wir die Gewichtsfunktion ρ

$$t \mapsto \rho(t) := \prod_{k=1}^{s} |t-c_k|^{\gamma_k} \; , \; t\in\Gamma \tag{2}$$

und die Hilfsfunktion f_k, $k = 1,2,\dots,s$

$$x \mapsto f_k(x) := \left\{ \begin{array}{ll} \dfrac{\sin(\vartheta_k x)}{\sin\vartheta_k}\, e^{i\vartheta_k(x-1)} & \text{falls } \vartheta_k \neq 0 \\ x & \text{falls } \vartheta_k = 0 \end{array} \right\} x\in[0,1],$$

wobei $\vartheta_k := \pi - \dfrac{2\pi(1+\gamma_k)}{p}$, $k = 1,2\dots,s$ gesetzt wurde.

3. Sei $a, b\in \hat{C}^{m,\mu}(\Gamma)$. Die Determinante des Symbols hat dann nach [4] den Wert $\det A(t,x) =$

$$= \left\{ \begin{array}{ll} W(t,x)\,\beta^2(t) & \text{für } t\in\Gamma\setminus\{c_{2r+1},c_{2r+2},\dots c_s\} \\ W(c_k,x)\beta(c_k)\beta(c_k+0) & \text{für } t=c_k,\ k=2r+1,2r+2,\dots,s \end{array} \right\}, \; x\in[0,1],$$

wobei gesetzt wurde:

$(t,x) \mapsto W(t,x) :=$

$$:= \left\{\begin{array}{ll} \dfrac{\alpha(t)}{\beta(t)} & \text{für } t \in \Gamma \setminus \{c_1 \ldots, c_s\} \\ f_k(x)\beta^{-1}(c_k)\alpha(c_k)+1-f_k(x) & \text{für } t=c_k; k=1,\ldots,r \\ (1-f_k(x)\beta^{-1}(c_k)\alpha(c_k)+f_k(x) & \text{für } t=c_k; k=r+1,\ldots,2r \\ f_k(x)\beta^{-1}(c_k+0)\alpha(c_k+0)+(1-f_k(x)\beta^{-1}(c_k)\alpha(c_k) & \\ & \text{für } t=c_k; k=2r+1,\ldots,s \end{array}\right\}, x \in [0,1]$$

und $\alpha(t) := a(t) + b(t)$, $\beta(t) := a(t) - b(t)$, $t \in \Gamma$.

4. α besitze auf Γ nur endlich viele Nullstellen in den Punkten $t_{11}, t_{12}, \ldots, t_{1r_1}$. Die natürlichen Zahlen $m_{11}, m_{12}, \ldots, m_{1r_1}$ seien die Ordnungen dieser Nullstellen. β besitze auf Γ nur endlich viele Nullstellen in den Punkten $t_{21}, t_{22}, \ldots, t_{2r_2}$. Die natürlichen Zahlen $m_{21}, m_{22}, \ldots, m_{2r_2}$ seien die Ordnungen dieser Nullstellen. Es gelte $m \geq \max\{m_{ij} : j = 1,2,\ldots r_i, i=1,2\}$. Ferner sei $\det A(c_k,x) \neq 0$ für $x \in [0,1]$ und $k = 1,2,\ldots,s$.

5. $\mathcal{T}$ sei ein kompakter Operator von $L_p(\Gamma,\rho)$ in $W_p^m(\Gamma,\rho)$.

6. $f \in W_p^m(\Gamma,\rho)$ ist gegeben.

7. Die Lösung φ von (1) soll im Raum $L_p(\Gamma,\rho)$ gesucht werden.

Bezeichnungen: Es sei $m \in \mathbb{N}$ und $\mu \in (0,1)$. Wir sagen, daß die rektifizierbare Kurve Γ zur Klasse $C^{m,\mu}$ gehört, wenn eine Parameterdarstellung

$$t(s) = x(s) + iy(s), \quad s \in [0,\gamma], \quad i^2 = -1$$

(s = Bogenabszisse von $t(s) \in \Gamma$, γ = Gesamtlänge von Γ, $x(s)$ und $y(s)$ die beiden kartesichen Koordinaten des Punktes $t(s) \in \mathbb{C}$) existiert, in der x und y stetige Funktionen der Bogenlänge s sind, die stetige Ableitungen bis einschließlich m-ter Ordnung besitzen und darüber hinaus die Ableitungen m-ter Ordnung je eine Hölderbedingung mit dem Exponenten μ auf $[0,\gamma]$ erfüllen.

Es sei nun Γ eine Kurve der Klasse $C^{m+1,\mu}$ oder auch die Vereinigung von endlich vielen, einfachen und paarweise disjunkten Kurven dieser Klasse. Mit $C^{m,\mu}(\Gamma)$ bezeichnen wir den Banachraum aller komplexwertigen m-mal stetig differenzierbaren Funktionen , deren m-te Ableitungen $\varphi^{(m)}$ je eine Hölderbedingung auf Γ mit dem Exponenten μ erfüllen. Mit $\hat{C}^{m,\mu}(\Gamma)$ bezeichnen wir die Menge aller komplexwertigen auf $\Gamma \setminus \{c_k : k=2r+1,\ldots,s\}$ m-mal stetig differenzierbaren Funktionen φ, deren m-te Ableitungen $\varphi^{(m)}$ auf $\Gamma \setminus \{c_k : k=2r+1,\ldots,s\}$ je eine Hölderbedingung gleichmäßig mit dem Exponenten μ erfüllen, die bis zur einschließlich m-ten Ableitung $\varphi^{(m)}$ auf Γ linksseitig stetig sind (d. h. es gilt $\varphi^{(\nu)}(c_k-0) = \varphi^{(\nu)}(c_k)$ für $\nu = 0,1,\ldots,m$ und $k = 2r+1,\ldots,s$) und die in c_k, $k = 2r+1,\ldots,s$, einschließlich ihrer m-ten Ableitung $\varphi^{(m)}$ endliche Grenzwerte $\varphi^{(\nu)}(c_k+0)$, $\nu = 0,1,\ldots,m$, von rechts besitzen.

Mit $L_p(\Gamma,\rho)$ bezeichnen wir den Banachraum aller auf Γ definierten komplexwertigen Funktionen φ, für die $|\varphi|^p\rho$ auf Γ summierbar ist. Dabei sei ρ die in der Formel (2) definierte Funktion. Schließlich bezeichnen wir mit $W_p^m(\Gamma,\rho)$ den Raum aller Funktionen $\varphi \in L_p(\Gamma,\rho)$, die verallgemeinerte Ableitungen $\varphi^{(k)} \in L_p(\Gamma,\rho)$, $k = 0,1,\ldots,m$ besitzen.

Ein abgeschlossener Operator A von einem lokalkonvexem Raum X in einen anderen lokalkonvexen Raum Y heißt normalauflösbar, wenn sein Wertebereich abgeschlossen ist. A heißt dann Plus-Semifredholmoperator, wenn außerdem noch $\dim \operatorname{Ker} A < \infty$ ist. Wenn A normalauflösbar ist und $\dim \operatorname{Coker} A < \infty$ gilt, dann heißt A ein Minus-Semifredholmoperator. Ein normalauflösbarer Operator A , für den sowohl $\dim \ker A < \infty$ als auch $\dim \operatorname{coker} A < \infty$ ist, wird Fredholm-Operator genannt.

Der Operator $B : Y \to X$ heißt linker Regularisator des Operators $A : X \to Y$, wenn die Beziehungen

$$D(B) \supset R(A) \quad \text{und} \quad BA = I + K$$

gelten, wobei K ein kompakter Operator in X ist und I bezeichnet den identischen Operator X .

Satz von Rakovščik (siehe [3] Th. 3, oder [2] S. 29 Satz 4.1): Die Existenz eines beschränkten linken Regularisators des abgeschlossenen Operators $A : X \to Y$ ist notwendig und hinreichend dafür, daß A ein Plus-Semifredholmoperator ist.

I. C. Gochberg (siehe [4] und [6]) bewies, daß der Operator $\mathfrak{A}$ in Formel (1) mit $D(\mathfrak{A}) = L_p(\Gamma,\rho)$ keinen beschränkten Regularisator auf dem Raume $L_p(\Gamma,\rho)$ besitzt, wenn $m > 0$ ist. Der Operator $\mathfrak{A}$ ist dann notwendig kein Fredholmoperator (die Normalauflösbarkeit gilt nicht mehr).

Zur Vereinfachung schreiben wir den Operator $\mathfrak{A}$ in (1) in der Form

$$\mathfrak{A} = a\,I + b\,S + \mathfrak{T}.$$

S ist dabei der singuläre Integraloperator mit Cauchyschem Kern (d. h. das Integral (1) ist im Sinne des Cauchyschen Hauptwertes zu verstehen).

Definition: Durch Einschränkung des Operators $\mathfrak{A}$ auf einen gewissen Teilraum von $L_p(\Gamma,\rho)$, der in $L_p(\Gamma,\rho)$ dicht liegt, gewinnen wir den neuen Operator $\overline{\mathfrak{A}}$,

$$\overline{\mathfrak{A}} := \mathfrak{A}|D(\overline{\mathfrak{A}}) \;, \quad D(\overline{\mathfrak{A}}) := \{\varphi \in L_p(\Gamma,\rho) : \mathfrak{A}\varphi \in W_p^m(\Gamma,\rho)\}.$$

Bemerkung 1: $\overline{\mathfrak{A}}$ ist ein linearer abgeschlossener Operator mit dichtem Difinitionsgebiet $D(\overline{\mathfrak{A}}) \subset L_p(\Gamma,\rho)$, denn $\dot{C}^\infty(\Gamma)$ (Menge aller unendlich oft differenzierbaren Funktionen φ auf Γ, für deren Träger Γ_φ gilt: $\overline{\Gamma}_\varphi \subset \overset{\circ}{\Gamma}$) ist eine Teilmenge von $D(\overline{\mathfrak{A}})$ und $C^\infty(\Gamma)$ liegt dicht im Raume $L_p(\Gamma,\rho)$. Wir zeigen nun die Abgeschlossenheit des Operators $\overline{\mathfrak{A}}$:

Sei $\varphi_k \in D(\overline{\mathfrak{A}})$, $\varphi_k \to \varphi \in L_p(\Gamma,\rho)$ und $\overline{\mathfrak{A}}\varphi_k \to \Psi$ in der Norm des Raumes $W_p^m(\Gamma,\rho)$. Dann konvergiert insbesondere $\mathfrak{A}\varphi_k \to \Psi$ in der Norm von $W_p(\Gamma,\rho)$ und aufgrund der Stetigkeit des Operators $\mathfrak{A}$ im Raume $L_p(\Gamma,\rho)$ ist somit $\Psi = \mathfrak{A}\varphi$. Also ist $\varphi \in D(\overline{\mathfrak{A}})$ und $\overline{\mathfrak{A}}\varphi = \mathfrak{A}\varphi = \Psi$, d. h. $\overline{\mathfrak{A}}$ ist abgeschlossen.

Satz 1: Der Operator $\overline{\mathfrak{A}} : L_p(\Gamma,\rho) \to W_p^m(\Gamma,\rho)$ besitzt den beschränkten linken Regularisator $\mathcal{L}_\ell$

$$\mathcal{B}_\ell : = \frac{1}{2\Pi_\lambda}\{\frac{1}{\alpha}(I-\mathcal{H}_1)(I+S) + \frac{1}{\beta}(I-\mathcal{H}_2)(I-S)\}\Pi_\lambda I \; . \tag{3}$$

Dabei bezeichne $\mathcal{H}_i\varphi$, $i = 1,2$, das Hermitesche Interpolationspolynom, von φ mit den Eigenschaften

$$(\mathcal{H}_i\varphi)^{(\nu)}(c_k) = \varphi^{(\nu)}(c_k) \quad \text{für} \quad k = 1,2,\ldots,s \;,\; \nu = 0,1,\ldots,m-1, i=1,2$$

und

$$(\mathcal{H}_i\varphi)^{(\nu)}(t_{ij}) = \varphi^{(\nu)}(t_{ij}) \quad \text{für} \quad \nu = 0,1,\ldots,m_{ij}-1, j = 1,2\ldots,r_i, i=1,2$$

Π_λ ist die Funktion

$$\Pi_\lambda(t) : = \rho^{\frac{\lambda}{p}}(t)\prod_{i=1}^{s}(t-c_i)^m \;,\; t\in\Gamma \;,\; \lambda\in(0, \tfrac{p}{p-1}] \; .$$

<u>Beweis:</u> Der durch (3) gegebene Operator $\mathcal{B}$ bildet nach den Hilfssätzen 5.5, 2.1 und 4.5 aus [9] den Raum $W_p^m(\Gamma,\rho)$ stetig in den Raum $L_p(\Gamma,\rho)$ ab. Wir zeigen nun, daß die Gleichung

$$\mathcal{B}_\ell\overline{\mathcal{O}\!\mathcal{L}} = I + K \tag{4}$$

gilt, wobei K einen kompakten Operator im Raume $L_p(\Gamma,\rho)$ bezeichnet. Nach der Folgerung 6.6 aus [9] ergibt sich nämlich sofort:

$$\begin{aligned}(I+S)\Pi_\lambda\overline{\mathcal{O}\!\mathcal{L}} &= \Pi_\lambda a + \Pi_\lambda bS + S\Pi_\lambda a + S\Pi_\lambda bS + (I+S)\Pi_\lambda\mathcal{T}\\ &= \Pi_\lambda a + \Pi_\lambda bS + \Pi_\lambda aS + [S\Pi_\lambda aI-\Pi_\lambda aS] + \Pi_\lambda b+K_b+(I+S)\Pi_\lambda\mathcal{T}\\ &= \Pi_\lambda(a+b)(I+S) + \mathcal{T}_1\\ &= \Pi_\lambda\alpha(I+S) + \mathcal{T}_1 \quad ,\end{aligned}\tag{5}$$

wobei $\mathcal{T}_1 : L_p(\Gamma,\rho) \to W_p^m(\Gamma,\rho^{1-\lambda})$ kompakt.
Ebenso gilt

$$\begin{aligned}(I-S)\Pi_\lambda\overline{\mathcal{O}\!\mathcal{L}} &= \Pi_\lambda a + \Pi_\lambda bS - S\Pi_\lambda a - S\Pi_\lambda bS + (I-S)\Pi_\lambda\mathcal{T}\\ &= \Pi_\lambda a + \Pi_\lambda bS - \Pi_\lambda aS + [\Pi_\lambda aS-S\Pi_\lambda aI]-\Pi_\lambda b-K_b+(I-S)\Pi_\lambda\mathcal{T}\\ &= \Pi_\lambda(a-b)(I-S) + \mathcal{T}_2\\ &= \Pi_\lambda\beta(I-S) + \mathcal{T}_2 \quad ,\end{aligned}\tag{6}$$

wobei $\mathcal{T}_2 : L_p(\Gamma,\rho) \to W_p^m(\Gamma,\rho^{1-\lambda})$ kompakt.

Da die Operatoren $\frac{1}{2\Pi_\lambda \alpha}(I-\mathfrak{h}_1)$ und $\frac{1}{2\Pi_\lambda \beta}(I-\mathfrak{h}_2)$ den Raum $W_p^m(\Gamma,\rho^{1-\lambda})$ stetig in den Raum $L_p^\lambda(\Gamma,\rho)$ abbilden, ist

$$\mathfrak{T}_\ell := \frac{1}{2\Pi_\lambda \alpha}(I-\mathfrak{h}_1)\,\mathfrak{T}_1 + \frac{1}{2\Pi_\lambda \beta}(I-\mathfrak{h}_2)\,\mathfrak{T}_2 \qquad (7)$$

ein kompakter Operator im Raume $L_p(\Gamma,\rho)$. Aus $\overline{\mathfrak{A}}\varphi \in W_p^m(\Gamma,\rho)$ folgt nach Hilfssatz 5.6 von [9] sofort, daß

$$(I+S)\Pi_\lambda \overline{\mathfrak{A}} \in W_p^m(\Gamma, \rho^{1-\lambda}) \quad \text{und} \quad (I-S)\Pi_\lambda \overline{\mathfrak{A}} \in W_p^m(\Gamma,\rho^{1-\lambda})$$

für beliebiges $\varphi \in D(\overline{\mathfrak{A}})$ gilt. Nun erfüllt die Gewichtsfunktion $\rho^{1-\lambda}$ die Voraussetzungen des Hilfssatzes 2.1 aus [9] (wie man leicht nachprüft). Somit sind wegen (5) und (6) die beiden Funktionen

$$\varphi_1 := \Pi_\lambda \alpha(I+S)\varphi \quad \text{und} \quad \varphi_2 := \Pi_\lambda \beta(I-S)\varphi$$

Elemente des Raumes $C^{m-1}(\Gamma)$ für $\varphi \in D(\overline{\mathfrak{A}})$ und es gilt offenbar

$$\mathfrak{h}_1\varphi_1 = \mathfrak{h}_2\varphi_2 = 0 .$$

Hieraus folgt unter Berücksichtigung von (5) bis (7) die Gleichung (4).

Aus Satz 1, aus dem Satz von Rakovščik und aus der Bemerkung 1 ergibt sich die

Folgerung 1: Der Operator $\overline{\mathfrak{A}}$ ist ein Plus-Semifredholmoperator.

Bemerkung 2: Im Gegensatz zu [9] und [10] wird in dieser Arbeit $\beta(c_k) \neq 0$ für $k = 1,2,\ldots,s$, $\beta(c_k+0) \neq 0$ für $k = 2r+1, 2r+2,\ldots,s$ und $W(c_k,X) = 0$ für $X \in [0,1]$ und $k = 1,2,\ldots,s$ statt nur $\alpha(c_k)\beta(c_k) \neq 0, k = 1,2,\ldots,s$ und $\alpha(c_k+0)\beta(c_k+0) \neq 0$ für $k=2r+1,2r+2,\ldots,s$ gefordert. Diese Korrektur sollte noch in den Arbeiten [9] und [10] erfolgen.

Bemerkung 3: Ebenso wie in [9] läßt sich der Satz 1 auch für Systeme von singulären Integralgleichungen aussprechen. In der Gleichung (1) seien dann die Funktionen φ bzw. f nicht skalar und aus $L_p(\Gamma,\rho)$ bzw. aus $W_p^m(\Gamma,\rho)$, sondern sie seien Vektorfunktionen aus $L_{p,n}(\Gamma,\rho) := [L_p(\Gamma,\rho)]^n$ bzw. $W_{p,n}^m(\Gamma,\rho) := [W_p^m(\Gamma,\rho)]^n$, $n \in \mathbb{N}$. Entsprechend sind dann a und b quadratische Funktionenmatrizen der Ordnung n, deren Elemente a_{ij} und $b_{ij} \in \hat{C}^{m,n}(\Gamma)$, $i,j = 1,\ldots,n$

sind, während $W(t,x)$ folgende Gestalt annimmt

$$(t,x) \mapsto W(t,x) :=$$

$$:= \begin{cases} \beta^{-1}(t)\alpha(t) & \text{für } t \in \Gamma \setminus \{c_k, k=1,\dots,2r\} \\ f_k(x)\beta^{-1}(c_k)\alpha(c_k) + (1-f_k(x))E_n & \text{für } t=c_k, k=1,2,\dots,r \\ (1-f_k(x)^{\beta-1}(c_k)\alpha(c_k) + f_k(x)E_n & \text{für } t=c_k, k=r+1,r+2,\dots,2r \\ f_k(x)^{\beta-1}(c_k+0)\alpha(c_k+0)+(1-f_k(x)^{\beta-1}(c_k)\alpha(c_k) & \\ & \text{für } t=c_k, k=2r+1,2r+2,\dots,s. \end{cases}$$

Dabei bezeichnet E_n die Einheitsmatrix der Ordnung n (Wie oben ist auch hier $\alpha(t) := a(t) + b(t)$, $\beta(t) := a(t) - b(t)$, $t \in \Gamma$).
Die Determinante des Symbols hat dann nach [4] den Wert

$$\det A(t,x) = \left.\begin{cases} \det W(t,x) \det \beta^2(t) & \text{für } t \in \Gamma \setminus \{c_{2r+1},\dots,c_s\} \\ \det W(t,x) \det \beta(c_k) \det \beta(c_k+0) & \text{für } t=c_k, k=2r+1,2r+2,\dots,s \end{cases}\right\} x \in [0,1].$$

Wenn man diese Verallgemeinerungen beachtet, dann kann man die ersten drei Voraussetzungen von Satz 1 einfach übernehmen.
Die 4. Voraussetzung von Satz 1 für Systeme lautet dann:
Die Determinante $\Delta_1 := \det \alpha$ besitze auf Γ nur endlich viele Nullstellen in den Punkten $t_{11},\dots,t_{1r_1}$.
Die Determinante $\Delta_2 := \det \beta$ besitze auf Γ nur endlich viele Nullstellen in den Punkten $t_{21},\dots,t_{2r_2}$.
Es gelte $m \geqq \max\{m_{ij} : j = 1,2,\dots,r_i, \ i = 1,2\}$.
Ferner sei $\det A(c_k,x) \neq 0$ für $x \in [0,1]$ und $k = 1,2,\dots,s$.

5. $\mathcal{T}$ sei ein kompakter Operator von $L_{p,n}(\Gamma,\rho)$ in $W^m_{p,n}(\Gamma,\rho)$
6. $f \in W^m_{p,n}(\Gamma,\rho)$ ist gegeben
7. Die Lösung φ von (1) soll im Raum $L_{p,n}(\Gamma,\rho)$ gesucht werden. Ebenso wie oben schränkt man den Operator $\mathfrak{A}$ auf $D(\bar{\mathfrak{A}}) := \{\varphi \in L_{p,n}(\Gamma,\rho) : \mathfrak{A}\varphi \in W^m_{p,n}(\Gamma,\rho)\}$ ein und erhält so den Operator $\bar{\mathfrak{A}}$. Für den gilt der

<u>Satz 2:</u> Der Operator $\bar{\mathfrak{A}} := aE_n + bS_n + \mathcal{T}$ besitzt unter diesen Voraussetzungen den beschränkten linken Regularisator

$$\mathcal{L}_\ell := \frac{1}{2\Pi_\lambda}\{\frac{1}{\Delta_1}(I-\mathcal{h}_1)\tilde{\alpha}(E_n+S_n)+\frac{1}{\Delta_2}(I-\mathcal{h}_2)\tilde{\beta}(E_n-S_n)\}\Pi_\lambda I. \qquad (8)$$

Dabei wurde $S_n = E_n S$ gesetzt und $\tilde{\alpha}$ bzw. $\tilde{\beta}$ bezeichne diejenige Matrix, deren Elemente $\tilde{\alpha}_{ij}$ bzw. $\tilde{\beta}_{ij}$ die Adjunkte zu α_{ji} bzw. β_{ji} aus der Matrix α bzw. aus der Matrix β sind, d. h. es gilt $\tilde{\alpha}\alpha = \alpha\tilde{\alpha} = \Delta_1 E_n$ und $\tilde{\beta}\beta = \beta\tilde{\beta} = \Delta_2 E_n$. Der Beweis von Satz 2 ist analog zu dem von Satz 1.

Folgerung 1: Der Operator $\overline{\mathcal{A}}$ (von Satz 1 und von Satz 2) ist ein Plus-Semifredholmoperator.

Definition: Der Operator $B : Y \to X$ (X,Y lokalkonvexe topologische Vektorräume) heißt rechter Regularisator des Operators $A : X \to Y$, wenn die Beziehungen

$$R(B) \subset D(A) \quad \text{und} \quad AB = I + K$$

gelten, wobei K ein kompakter Operator in Y ist.

Satz von Prößdorf (siehe [11]):
Aus der Existenz eines beschränkten rechten Regularisators eines abgeschlossenen linearen Operators A folgt, daß A ein Minus-Semifredholmoperator ist.
Den Satz für die rechte Regularisierung formulieren wir gleich für Systeme von singulären Integralgleichungen. Den skalaren Fall ($n=1$) erhält man dann durch die Spezialisierung $E_n = 1$, $\tilde{\alpha} = \tilde{\beta} = 1$, $S_n = S$, $L_{p,1}(\Gamma,\rho^{1-\lambda}) = L_p(\Gamma,\rho^{1-\lambda})$ und $W^m_{p,1}(\Gamma,\rho^{1-\lambda}) = W^m_p(\Gamma,\rho^{1-\lambda})$.

Satz 3: Sei λ aus dem offenen Intervall $(0, \frac{p}{p-1})$. Wir betrachten die Gleichung (1) unter den Voraussetzungen
1. bis 4. aus Satz 2 und außerdem sei

5. $\mathcal{T}$ ein linearer kompakter Operator von $L_{p,n}(\Gamma,\rho^{1-\lambda})$ in $W^m_{p,n}(\Gamma,\rho^{1-\lambda})$,
6. $f \in W^m_{p,n}(\Gamma,\rho^{1-\lambda})$ und
7. die Lösung φ von (1) soll in diesem Fall im Raume $L_{p,n}(\Gamma,\rho^{1-\lambda})$ gesucht werden. Ferner sei der Operator $\mathcal{A}$ zu $\overline{\overline{\mathcal{A}}}$ eingeschränkt:

$$\overline{\overline{\mathcal{A}}} := \mathcal{A}|D(\overline{\overline{\mathcal{A}}}), \text{ wobei } D(\overline{\overline{\mathcal{A}}}) := \{\varphi \in L_{p,n}(\Gamma,\rho^{1-\lambda}) : \mathcal{A}\varphi \in W^m_{p,n}(\Gamma,\rho^{1-\lambda})\}$$

Der Operator $\overline{\overline{\mathcal{A}}}$ besitzt dann den beschränkten rechten Regularisator $\mathcal{R}_r$,

$$\mathcal{R}_r := \frac{1}{2}\Pi_\lambda\{\tilde{\alpha}(E_N+S_n)\frac{1}{\Delta_1\Pi_\lambda}(I-\mathcal{T}_1)+\tilde{\beta}(E_n-S_n)\frac{1}{\Delta_2\Pi_\lambda}(I-\mathcal{T}_2)\}I. \tag{9}$$

Beweis: $\mathcal{L}_r$ bildet nach den Hilfssätzen 5.5, 2.1 und 4.5 aus [9] den Raum $W^m_{p,n}(\Gamma,\rho^{1-\lambda})$ stetig in den Raum $L_{p,n}(\Gamma,\rho^{1-\lambda})$ ab. Wir zeigen nun, daß die Beziehung

$$\bar{\bar{\mathfrak{A}}}\mathcal{L}_r = I + K \tag{10}$$

gilt, wobei K einen kompakten Operator im Raume $W^m_{p,n}(\Gamma,\rho^{1-\lambda})$ bezeichnet. Nach Folgerung 6.6 aus [9] ergibt sich nämlich:

$$\begin{aligned}
\bar{\bar{\mathfrak{A}}}\Pi_\lambda\tilde{\alpha}(E_n+S) &= a\Pi_\lambda\tilde{\alpha}+bS_n\Pi_\lambda\tilde{\alpha}I+a\Pi_\lambda\tilde{\alpha}S_n+bS_n\Pi_\lambda\tilde{\alpha}S_n+\mathcal{T}\Pi_\lambda\tilde{\alpha}(E_n+S_n)\\
&= \Pi_\lambda a\tilde{\alpha}E_n+\Pi_\lambda b\tilde{\alpha}S_n+b\left[S_n\Pi_\lambda\tilde{\alpha}I-\Pi_\lambda\tilde{\alpha}S_n\right]+\Pi_\lambda a\tilde{\alpha}S_n\\
&\qquad +\Pi_\lambda b\tilde{\alpha}E_n+bK_\alpha+\mathcal{T}\Pi_\lambda\tilde{\alpha}(E_n+S_n)\\
&= \Pi_\lambda a\tilde{\alpha}E_n+\Pi_\lambda b\tilde{\alpha}E_n+\Pi_\lambda a\tilde{\alpha}S_n+\Pi_\lambda b\tilde{\alpha}S_n\\
&\qquad +b\left[S_n\Pi_\lambda\tilde{\alpha}I-\Pi_\lambda\tilde{\alpha}S_n\right]+bK_{\tilde{\alpha}}+\mathcal{T}\Pi_\lambda\tilde{\alpha}(E_n+S_n)\\
&= \Pi_\lambda(a+b)\tilde{\alpha}(E_n+S_n)+b\left[S_n\Pi_\lambda\tilde{\alpha}I-\Pi_\lambda\tilde{\alpha}S_n\right]+bK_\alpha+\mathcal{T}\Pi_\lambda\tilde{\alpha}(E_n+S_n)\\
&= \Pi_\lambda\Delta_1(E_n+S_n)+\mathcal{T}_3
\end{aligned}\tag{11}$$

und ebenso

$$\begin{aligned}
\bar{\bar{\mathfrak{A}}}\Pi_\lambda\tilde{\beta}(E_n-S_n) &= a\Pi_\lambda\tilde{\beta}E_n+bS_n\Pi_\lambda\tilde{\beta}I-a\Pi_\lambda\tilde{\beta}S_n-bS_b\Pi_\lambda\tilde{\beta}S_n+\mathcal{T}\Pi_\lambda\tilde{\beta}(E_n-S_n)\\
&= \Pi_\lambda(a-b)\tilde{\beta}(E_n-S_n)+b\left[S_n\Pi_\lambda\tilde{\beta}I-\tilde{\beta}\Pi_\lambda S_n\right]+bK_{\tilde{\beta}}+\mathcal{T}\Pi_\lambda\tilde{\beta}(E_n-S_n)\\
&= \Pi_\lambda\Delta_2(E_n-S_n)+\mathcal{T}_4.
\end{aligned}\tag{12}$$

Dabei sind die Operatoren $\mathcal{T}_3$, $\mathcal{T}_4$: $L_{p,n}(\Gamma,\rho) \to W^m_{p,n}(\Gamma,\rho^{1-\lambda})$ kompakt. Wir rechnen nun mit den anderen beiden (den charakteristischen) Summanden in (11) und (12) weiter:

$$\begin{aligned}
\Pi_\lambda\Delta_1(E_n+S_n)\frac{1}{\Delta_1\Pi_\lambda}(I-\mathfrak{h}_1) &= E_n(I-\mathfrak{h}_1)+\left[\Pi_\lambda\Delta_1S_n-S_n\Pi_\lambda\Delta_1\right]\frac{1}{\Pi_\lambda\Delta_1}(I-\mathfrak{h}_1)+S_n(I-\mathfrak{h}_1)\\
&= E_n(I-\mathfrak{h}_1)+\mathcal{T}_5+S_n(I-\mathfrak{h}_1)\\
\Pi_\lambda\Delta_2(E_n-S_n)\frac{1}{\Pi_\lambda\Delta_2}(I-\mathfrak{h}_2) &= E_n(I-\mathfrak{h}_2)-\left|\Pi_\lambda\Delta_2S_n-S_n\Pi_\lambda\Delta_2\right|\frac{1}{\Pi_\lambda\Delta_2}(I-\mathfrak{h}_2)-S_n(I-\mathfrak{h}_2)\\
&= E_n(I-\mathfrak{h}_2)-\mathcal{T}_6-S_n(I-\mathfrak{h}_2).
\end{aligned}$$

Die beiden Operatoren $\mathcal{T}_5$ und $\mathcal{T}_6$ sind nach Folgerung 6.6 und Satz 4.3 aus [9] im Raume $W^m_{p,n}(\Gamma,\rho^{1-\lambda})$ kompakt. Somit erhalten wir aus [11] und [12] sofort:

$$2\overline{\mathfrak{A}}\mathfrak{B}_r = \Pi_\lambda \Delta_1 (E_n + S_n)\frac{1}{\Pi_\lambda \Delta_1}(I-\mathfrak{h}_1) + \mathfrak{T}_3 \frac{1}{\Pi_\lambda \Delta_1}(I-\mathfrak{h}_1)$$

$$+ \Pi_\lambda \Delta_2 (E_n - S_n)\frac{1}{\Pi_\lambda \Delta_2}(I-\mathfrak{h}_2) + \mathfrak{T}_4 \frac{1}{\Pi_\lambda \Delta_2}(I-\mathfrak{h}_2)$$

$$= E_n(2I-\mathfrak{h}_1-\mathfrak{h}_2) + \mathfrak{T}_5 + \mathfrak{T}_6 + S_n(\mathfrak{h}_1-\mathfrak{h}_2) + \mathfrak{T}_3 \frac{1}{\Pi_\lambda \Delta_1}(I-\mathfrak{h}_1)$$

$$+ \mathfrak{T}_4 \frac{1}{\Pi_\lambda \Delta_2}(I-\mathfrak{h}_2) .$$

Daraus ergibt sich unmittelbar die Gleichung (10), wenn man dabei beachtet, daß:

1. der Operator $S_n(\mathfrak{h}_1-\mathfrak{h}_2)$ wegen

$$\{(\mathfrak{h}_1-\mathfrak{h}_2)\varphi\}^{(\nu)}(c_k) = 0, \quad k = 1,\ldots,s \quad , \quad \nu = 0,1,\ldots, m-1$$

   stetig ist im Raume $W^m_{p,n}(\Gamma,\rho^{1-\lambda})$ und daß

2. $\mathfrak{h}_1$ und $\mathfrak{h}_2$ endlichdimensionale (d. h. Operaotren, deren Wertebereich endlichdimensional ist) und damit kompakte Operatoren sind. Q.e.d.

Satz 4: Sei nun $\overline{\mathfrak{A}}$ wie in Satz 2 vorausgesetzt. Dann besitzt $\overline{\mathfrak{A}}$ einen beschränkten rechten Regularisator $\hat{\mathfrak{B}}_r$.

Beweis: Wir setzen $\lambda_o := 1-\frac{1}{p-1}\max\limits_k \gamma_k$ und es sei $\lambda\in(0,\lambda_o)$. Ferner führen wir die Größen $\hat{\gamma}_k := \frac{\gamma_k}{1-\lambda}$, $k=1,2,\ldots,s$, ein, die offensichtlich die Forderungen $\hat{\gamma}_k \in (0, p-1)$ erfüllen. Damit läßt sich die Gewichtsfunktion $\hat{\rho}$,

$$\hat{\rho}(\tau) := \prod_{k=1}^{s} |\tau-c_k|^{\hat{\gamma}_k}, \quad \tau\in\Gamma,$$ definieren. Schließlich brauchen wir noch den Faktor $\hat{\Pi}_\lambda$,

$$\hat{\Pi}_\lambda(\tau) := \prod_{k=1}^{s} |\tau-c_k|^{m+\frac{\lambda}{p}\hat{\gamma}_k}, \quad \tau\in\Gamma .$$

Der Operator $\overline{\mathfrak{A}}: D(\overline{\mathfrak{A}}) \to W^m_{p,n}(\Gamma,\rho) = W^m_{p,n}(\Gamma,\hat{\rho}^{1-\lambda})$ erfüllt wegen $L_{p,n}(\Gamma,\hat{\rho}^{1-\lambda}) = L_{p,n}(\Gamma,\rho)$ damit die Voraussetzungen unseres Satzes 3 und besitzt damit den beschränkten Regularisator $\hat{\mathfrak{B}}_r$,

$$\hat{\mathfrak{B}}_r := \tfrac{1}{2}\hat{\Pi}_\lambda\{\tilde{\alpha}(E_n+S_n)\frac{1}{\hat{\Pi}_\lambda\Delta_1}(I-\mathfrak{h}_1) + \tilde{\beta}(E_n-S_n)\frac{1}{\hat{\Pi}_\lambda\Delta_2}(I-\mathfrak{h}_2)\} .$$ Q.e.d.

Aus dem Satz von Prößdorf und aus dem Satz 4 ergibt sich

Folgerung 2: Der Operator $\bar{\mathfrak{A}}$ ist ein Minus-Semifredholmoperator.

Folgerung 1 und Folgerung 2 zusammen ergeben

Folgerung 3: Der Operator $\bar{\mathfrak{A}}$ ist ein Fredholmoperator.

Folgerung 4: Der Operator $(\mathcal{B}_{\ell} - \mathcal{B}_{r})$ ist kompakt vom Raume $W^{m}_{p,n}(\Gamma,\rho)$ in den Raum $L_{p,n}(\Gamma,\rho)$ (Dies folgt sofort aus [3] Satz 1).

Literatur

[1] Prößdorf, S.
Eindimensionale singuläre Integralgleichungen und Faltungsgleichungen nicht normalen Typs in lokalkonvexen Räumen, Habilitationsschrift K.-M.-Stadt (Chemnitz) 1967.

[2] Prößdorf, S.
Einige Klassen singulärer Gleichungen, Birkhäuser Verlag Basel und Stuttgart 1974.

[3] Rakovščik, L. S.
Einige Sätze über Regularisatoren linearer Operatoren in Banachräumen (russ), Doklady akad. nauk SSSR 140, Nr. 5, 1023-1025 (1961).

[4] Gochberg, I. C. and N. Ja. Krupnik
Singular integral operators with piecewise continuous coefficients and their symbols, Math. USSR Izvestija, Vol. 5 (1971), No. 4.
Übers. aus Izv. Akad. Nauk SSSR, ser. Mat. Tom 35(1971)Nr.4

[5] Chvedelidse, B. V.
Lineare unstetige Randwertaufgaben der Funktionentheorie, singuläre Integralgleichungen und einige ihrer Anwendungen (russ.), Trudy Tbil. mat. inst. 23, (1956), 3 - 158.

[6] Gochberg, I. C. and N. Ja. Krupnik
Singular integral equations with continuous coefficients on a compound contour (russ.), Mat. Issled. 5 (1970), no. 2, 89 - 103.

[7] Gachov, F. D.
Randwertaufgaben der Theorie analytischer Funktionen und singuläre Integralgleichungen" (russ.) Habilitationsschrift, Tiflis 1941.

[8] Gachov, F. D.
Randwertaufgaben für analytische Funktionen und singuläre Integralgleichungen (russ.) Isv. Kazansk, fiz.-matem. O.-va 14, ser. 3, 1949, 75 - 160.

[9] Schüppel, B. M.
Regularisierung singulärer nicht elliptischer Integralgleichungen mit unstetigen Koeffizienten, Inaugur. Diss., Universität München 1973.

[10] Schüppel, B. M.
Regularisierung singulärer Integralgleichungen vom nicht normalen Typ mit unstetigen Koeffizienten, ZAMM 56, T 265 - T 266 (1976).

[11] Prößdorf, S.
Ein Satz über die äquivalente Regularisierung abgeschlossener Operatoren, Wissensch. Zeitschr. d. TH KMSt., Jahrg. 10, Heft 5 (1968), 555 - 556.

Towards the Validity of the Geometrical Theory of Diffraction

by B D Sleeman

Department of Mathematics, University of Dundee

Dundee DD1 4HN, Scotland

§1 Introduction

The following classic problem is of fundamental interest in mathematical physics.

Determine a scalar valued function $U(\underset{\sim}{r};k)$ such that

(i) $(\Delta + k^2)U = f(\underset{\sim}{r})$ in an infinite domain D with boundary $\partial\Omega$.

(ii) $\alpha U + \beta \frac{\partial U}{\partial n} = g(\underset{\sim}{r}), \quad \underset{\sim}{r} \in \partial\Omega$

where α, β are given constants, $g(\underset{\sim}{r})$ is prescribed on $\partial\Omega$ and $\frac{\partial}{\partial n}$ is the outward directed (in to D) normal derivative to $\partial\Omega$. (1.1)

(iii) $U(\underset{\sim}{r};k) = U^i(\underset{\sim}{r};k) + U^s(\underset{\sim}{r};k)$ where $U^i(\underset{\sim}{r};k)$ is the prescribed incident field $U^s(\underset{\sim}{r};k)$ is the scattered field which satisfies $(\Delta + k^2)U^s = 0$

(iv) $U^s(\underset{\sim}{r};k)$ satisfies the Sommerfeld radiation condition, e.g.

$$\lim_{|\underset{\sim}{r}|\to\infty} |\underset{\sim}{r}|\left\{\frac{\partial U^s}{\partial n} \pm ikU^s\right\} = 0$$

where $\pm$ is chosen depending on the assumed harmonic time dependence which has been suppressed.

Exact solutions to such problems are known only in certain special cases. For example the so-called canonical problems wherein $\partial\Omega$ is a normal surface in a particular orthogonal curvilinear coordinate system. In this case the problem may be solved using the classic separation of variables technique.

In general exact solutions cannot be obtained and even for canonical problems the solutions are often too intractible for useful application. Thus there is considerable interest in approximate methods of solution. If the wave number k is large ($k = 2\pi/\lambda$, λ = wave length) the modern geometrical theories of optics and diffraction provide what are conjectured to be asymptotic approximations of exact solutions in the illuminated and shadow portions of D.

For the problem of scattering by an arbitrary convex cylinder of finite cross section it has been established that the field predicted by the geometrical theory of optics is an asymptotic solution in the illuminated region. This was established for the Neumann problem (set $\alpha = 0$ in (1.1) ii) by Ursell [12] in his now classic paper of 1957. Morawetz and Ludwig [9] proved the corresponding result for the Dirichlet problem (set $\beta = 0$ in (1.1) ii). In addition they established the validity of the geometrical theory of optics for the case of a point source radiating outside a closed convex surface on which the Dirichlet condition is imposed. As far as the geometrical theory of optics is concerned these appear to be the most general results known.

The geometrical theory of diffraction however has been verified in only a few special situations. To cite the most complete confirmation we mention Ursell [12] who verified the theory for the case of scattering by a circular cylinder and Leppington [6], Ursell [13] who verified the theory for the case of scattering by an elliptic cylinder.

The first significant attack on establishing the validity of the geometrical theory of diffraction for non canonical problems, that is where the method of separation of variables cannot be applied, is that of Bloom and Matkowsky [1]. In that work they considered the scattering of a wave from an infinite line source by an infinitely long cylinder C say. The line source is parallel to the axis of C and the cross section Ω of the cylinder is smooth, closed and convex. Furthermore Ω is formed by joining a pair of smooth convex arcs to a circle Ω_0, one on the illuminated side, and one on the dark side, so that Ω is circular near the points of diffraction. By a rigorous argument Bloom and Matkowsky derived the asymptotic behaviour of the field at high frequencies in the "deep" shadow of Ω. The leading term of the asymptotic expansion was shown to be the field predicted by the geometrical theory of diffraction.

In this paper we describe a genuine extension to the results of Bloom and Matkowsky. The extension is based on the fundamental observation that given an arbitrary closed convex curve, then for any two points on this curve we can construct a one-parameter family of ellipses passing through the two points and having two-point contact with the given curve at these two points. This observation suggests that we consider Ω to be formed by joining a pair of smooth convex arcs to an ellipse Ω_0, one on the illuminated side and one on the dark side, so that Ω is elliptic near the points of diffraction.

The analysis then falls naturally into two parts. Firstly we consider the scattering of a circular wave by a smooth convex curve Ω_1 which is elliptic on its dark side and also near the points of "diffraction". Thus Ω_1 may be considered as being formed by pasting a convex curve B_1 to the illuminated side of an ellipse Ω_0. (See Figure 1).

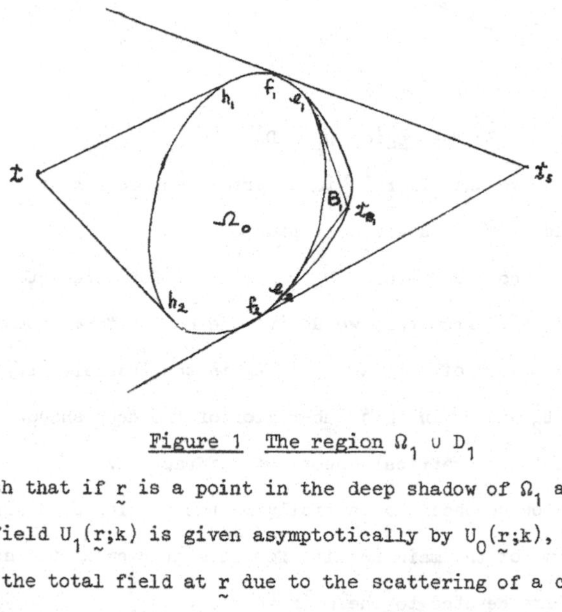

Figure 1 The region $\Omega_1 \cup D_1$

We establish that if $\underset{\sim}{r}$ is a point in the deep shadow of Ω_1 and k is large then the total field $U_1(\underset{\sim}{r};k)$ is given asymptotically by $U_0(\underset{\sim}{r};k)$, where $U_0(\underset{\sim}{r};k)$ represents the total field at $\underset{\sim}{r}$ due to the scattering of a circular wave by Ω_0. We obtain a uniform asymptotic expansion of the field $U_1(\underset{\sim}{r};k)$ in the form predicted by the theory of Lewis, Bleistein and Ludwig [8]. From this the non-uniform series expansions of the extended geometrical theory of diffraction [15] is obtained. The insensitivity of $U_1(\underset{\sim}{r};k)$ to the geometry of B_1 as $k \to \infty$ is that predicted by the geometrical theory of diffraction.

Next we consider the more general case where a circular wave is scattered by a smooth convex curve Ω_2 which coincides with Ω_0 only near the points of diffraction. Thus Ω_2 may be thought of as formed by pasting a convex bump B_2 to the dark side of Ω_1 (see Figure 2).

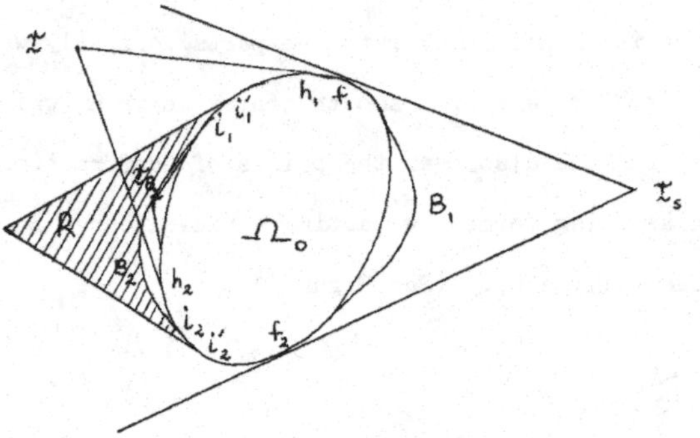

Figure 2 The region $\Omega_2 \cup D_2$

From the above results we prove that if $\underset{\sim}{r}$ is in a certain sub-region of the deep shadow of Ω_2 the total field $U_2(\underset{\sim}{r};k)$ is also asymptotic to the field $U_0(\underset{\sim}{r};k)$ as $k \to \infty$. As in the above we obtain $U_2(\underset{\sim}{r};k)$ in the form predicted by Lewis, Bleistein and Ludwig [8]. From this we derive the non-uniform expansions of the extended geometrical theory of diffraction. Again the insensitivity of $U_2(\underset{\sim}{r};k)$ to the geometry of B_2 and B_1 in this sub-region of the deep shadow confirms the predictions of the geometrical theory of diffraction.

As the analysis to follow depends substantially on the problem of diffraction by an ellipse we outline some of the main results for this problem in the next section. Sections 3 and 4 are devoted to the main results.

§2 Diffraction by an ellipse

Introduce elliptic coordinates (ξ,η) related to Cartesian coordinates (x,y) by the equations

$$x = c \cosh \xi \cos \eta, \quad y = c \sinh \xi \sin \eta, \tag{2.1}$$

$$\xi > 0, \quad 0 < \eta < 2\pi.$$

The curve $\xi = \xi_0$ describes an ellipse whose semi-major and minor axes are $c \cosh \xi_0$ and $c \sinh \xi_0$ respectively. Any ellipse can be represented in this form by an appropriate choice of c and ξ_0. Suppose we consider the ellipse $\xi = \xi_0$ and that a source of unit strength is situated at the point $\underset{\sim}{r}_s = (\xi_s, \eta_s)$, $\xi_s > \xi_0$. The total field $U_0(\underset{\sim}{r};\underset{\sim}{r}_s;k)$ due to $\xi = \xi_0$ (Ω_0) is the unique solution to the problem

$$\text{(i)} \quad \frac{\partial^2 U_0}{\partial \xi^2} + \frac{\partial^2 U_0}{\partial \eta^2} + k^2c^2(\cosh^2 \xi - \cos^2 \eta)\, U_0 = \delta(\xi - \xi_s)\delta(\eta - \eta_s) \tag{2.2}$$

$$\text{(ii)} \quad \frac{\partial U_0}{\partial \xi} = 0 \quad \text{on} \quad \xi = \xi_0 \tag{2.3}$$

$$\text{(iii)} \quad \rho^{\frac{1}{2}} \left(\frac{\partial U_0}{\partial \rho} - ik\, U_0\right) \to 0 \text{ as } \rho \to \infty \tag{2.4}$$

where

$$\rho^2 = c^2(\cosh^2 \xi \cos^2 \eta + \sinh^2 \xi \sin^2 \eta).$$

Separable solutions of the wave equation (2.2) may be obtained in the form $u(\eta)v(\xi)$ provided u and v satisfy the respective Mathieu equations

$$u''(\eta) + N^2(b^2 - \cos^2 \eta)u(\eta) = 0, \tag{2.5}$$

$$v''(\xi) - N^2(b^2 - \cosh^2 \xi)v(\xi) = 0, \tag{2.6}$$

where $N = kc$ and b is an arbitrary complex separation constant.

Solutions of (2.5) may be expressed in terms of the fundamental pair $c(\eta;b;N)$ and $s(\eta;b;N)$ which satisfy (2.5) and are normalised so that

$$\begin{aligned} c(0;b;N) &= 1, \quad c'(0;b;N) = 0 \\ s(0;b;N) &= 0, \quad s'(0;b;N) = 1. \end{aligned} \tag{2.7}$$

The two fundamental solutions of (2.6) are chosen to be those which correspond to outgoing and incoming waves at infinity and are denoted by $v^{(1)}(\xi;b;N)$ and $v^{(2)}(\xi;b;N)$ respectively. That is

$$v^{(i)}(\xi;b;N) \sim H_\alpha^{(i)}(N \cosh \xi) \text{ as } \xi \to \infty, \quad (i = 1,2), \tag{2.8}$$

where $H_\alpha^{(i)}$ is a Hankel function and α is a function of b and N.

The solution $U_0(\underset{\sim}{r};\underset{\sim}{r}_s;k)$ may now be constructed in terms of the Mathieu functions $c(\eta)$, $s(\eta)$ and $v^{(1)}(\xi), v^{(2)}(\xi)$ by use of the Watson transformation [14] as

$$U_0(\underset{\sim}{r};\underset{\sim}{r}_s;k) = \frac{1}{2\pi i}\int_B N^2 b\, F(\eta;\eta_s;b)G(\xi;\xi_0;\xi_s;b)db \tag{2.9}$$

where

$$F(\eta,\eta_s;b) = \frac{s(\eta-\pi;b)s(\eta_s;b)}{s(\pi,b)} - \frac{c(\eta-\pi;b)c(\eta_s;b)}{c'(\pi,b)}\;, \quad \text{for } \eta \geq \eta_s, \tag{2.10}$$

$$F(\eta,\eta_s;b) = F(\eta_s;\eta;b), \quad \text{for } \eta \leq \eta_s,$$

and

$$G(\xi;\xi_0;\xi_s;b) = \frac{\pi}{4i}\,\frac{v^{(2)}(\xi_s)v^{(1)'}(\xi_0) - v^{(1)}(\xi_s)v^{(2)'}(\xi_0)}{v^{(1)'}(\xi_0)}\cdot v^{(1)}(\xi)$$

$$\text{for } \xi \geq \xi_s$$

$$G(\xi;\xi_0;\xi_s;b) = G(\xi_s;\xi_0;\xi,b), \quad \text{for } \xi \leq \xi_s\;. \tag{2.11}$$

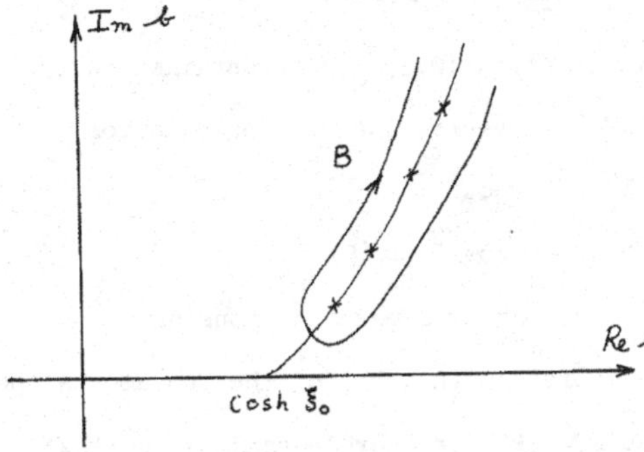

Figure 3 The contour B

The contour B in (2.9) is that shown in Figure 3, and is seen to enclose the poles of the integrand due to the zeros of the Mathieu function $v^{(1)'}(\xi_0)$. Thus U_0 can be expressed as an infinite sum of the residues at these points. Implicit in this procedure is the assumption that the contour B can be suitably closed at infinity. This is verified in [3]. It is known that for points in the deep shadow of the ellipse the first few terms of this series are small and decrease exponentially as $N = kc \to \infty$. Later terms may in fact be large [6, 13] but their sum is exponentially small. Thus the leading terms of the residue series give a good estimate for U_0 in the deep shadow.

By employing the analysis of Leppington [6] and Levy [7] it follows that the leading term $T(\underset{\sim}{r};\underset{\sim}{r}_s;k)$ of the radial expansion is an asymptotic representation of $U_0(\underset{\sim}{r};\underset{\sim}{r}_s;k)$ as $k \to \infty$ uniformly in $\underset{\sim}{r}$ for the observation point $\underset{\sim}{r}$ in the shadow of the ellipse.

That is

$$U_0(\underset{\sim}{r};\underset{\sim}{r}_s;k) = T(\underset{\sim}{r};\underset{\sim}{r}_s;k)[1 + O(\exp\{-k^{1/3}\gamma d(\underset{\sim}{r};\underset{\sim}{r}_s)\})] \tag{2.12}$$

where $\gamma > 0$ depends linearly on the first two zeros of $Ai'(q)$ (the derivative of the Airy function $Ai(q)$) and $d(\underset{\sim}{r},\underset{\sim}{r}_s)$ is the minimum distance travelled by a ray from $\underset{\sim}{r}_s$ to the observation point $\underset{\sim}{r}$.

In order to develop the form of the leading term $T(\underset{\sim}{r};\underset{\sim}{r}_s;k)$ in (2.12) we require uniform asymptotic expansions of the various Mathieu functions appearing in the representation (2.9). Many of these expansions are known (see [6, appendix 1]) or may be developed afresh using the recently developed theory of Olver [10] which includes an account of the errors involved in neglecting higher order terms. To prevent this paper becoming of inordinate length we outline the main steps and give sufficient information which will enable the reader to develop the full expansions.

If b_1 denotes the first zero of $v^{(1)'}(\xi_0)$ in (2.9) (2.11) then on noting the direction of the contour B in figure 3 we obtain the result

$$U_0(\underset{\sim}{r},\underset{\sim}{r}_s) \sim \frac{\pi}{4i} N^2 b_1 \, F(\eta,\eta_s;b) \, \frac{v^{(1)}(\xi)v^{(1)}(\xi_s) \, v^{(2)'}(\xi_0)}{\frac{\partial}{\partial b} v^{(1)'}(\xi_0) \, |_{b=b_1}} \, , \tag{2.13}$$

for $\underset{\sim}{r} = \underset{\sim}{r}(\xi,\eta)$ in the shadow of the ellipse $\xi = \xi_0$. Now the Wronksian of $v^{(1)}(\xi)\, v^{(2)}(\xi)$ is constant and is easily shown to have the value $-\frac{4i}{\pi}$. Further since $v^{(1)'}(\xi_0) = 0$ when $b = b_1$ we see that $v^{(2)'}(\xi_0) = -\frac{4i}{\pi} v^{(1)}(\xi_0)^{-1}$. Consequently we may write (2.13) in the form

$$U_0(\underset{\sim}{r},\underset{\sim}{r}_s) \sim -N^2 b_1 F(\eta,\eta_s;b_1) \, \frac{v^{(1)}(\xi)\, v^{(1)}(\xi_s)}{v^{(1)}(\xi_0) \, \frac{\partial}{\partial b} \, v^{(1)'}(\xi_0)|_{b=b_1}} \tag{2.14}$$

$$= T(\underset{\sim}{r},\underset{\sim}{r}_s;k).$$

Next we estimate each of the terms appearing in (2.14) as $N \to \infty$. The first few zeros b_n $(n = 1,2,\ldots)$ of the function $v^{(1)'}(\xi_0)$ are given by

$$b_n \sim \cosh \xi_0 + \frac{|q_n'|}{N^{2/3}} \, 2^{-1/3} e^{\pi i/3} \, (\cosh \xi_0)^{-1/3} (\sinh \xi_0)^{2/3} + \ldots . \tag{2.15}$$

To estimate the function $F(\eta,\eta_s;b_1)$ we employ the following uniform asymptotic expansions of the Mathieu functions $c(\eta)$ $s(\eta)$, i.e.

$$\begin{aligned} c(\eta) &\sim (b^2 - 1)^{1/4} \cos(N\zeta)\,(b^2 - \cos^2\eta)^{-1/4} \\ s(\eta) &\sim N^{-1}(b^2 - 1)^{-1/4} \sin(N\zeta)\,(b^2 - \cos^2\eta)^{-1/4} \end{aligned} \tag{2.16a,b}$$

where

$$\zeta(\eta) = \int_0^{\eta} (b^2 - \cos^2 t)^{1/2} dt.$$

Thus for example if $\eta \geq \eta_s$ we may express $T(\underset{\sim}{r};\underset{\sim}{r}_s;k)$ defined by (2.14) as the sum of two terms namely

$$T(\underset{\sim}{r};\underset{\sim}{r}_s;k) = \frac{N}{2} \sum_{m=1}^{2} \exp i N\phi_m \, L(\eta,\eta_s;\xi,\xi_s) \tag{2.17}$$

where

$$\phi_m = (-1)^{m+1} [\zeta(\eta - \pi) + \zeta(\eta_s)], \quad m = 1,2 \tag{2.18}$$

and

$$L(\eta,\eta_s;\xi,\xi_s) = b_1 \frac{(b_1^2 - \cos^2\eta)^{-1/4}(b_1^2 - \cos^2\eta_s)^{-1/4}}{\sin(N\zeta(\pi))} \times \frac{v^{(1)}(\xi)\, v^{(1)}(\xi_s)}{v^{(1)}(\xi_0) \frac{\partial}{\partial b_1} v^{(1)'}(\xi_0)} . \tag{2.19}$$

The uniform asymptotic expansion for the radial function $v^{(1)}(\xi)$ is given by

$$v^{(1)}(\xi) \sim 2e^{-i\pi/3}(Nb_1)^{-1/3} f(b_1,\xi) \mathrm{Ai}(N^{2/3} b^{2/3} \chi e^{2\pi i/3}) \tag{2.20}$$

where

$$f(b_1,\xi) = \left(\frac{4b_1^2 \chi}{b_1^2 - \cosh^2 \xi} \right)^{1/4}$$

and χ is defined by

$$2/3\, \chi^{3/2} = -\frac{1}{b_1} \int_{\cosh^{-1} b_1}^{\xi} (b_1^2 - \cosh^2 x)^{1/2} dx. \tag{2.21}$$

From (2.20) and (2.21) we find that

$$[v^{(1)}(\xi_0)\frac{\partial}{\partial b_1}v^{(1)'}(\xi_0)] \sim 6\, 2^{-2/3} e^{-i\pi/3} N^{2/3} \cosh^{2/3}\xi_0 \sinh \xi_0^{-2/3} |a_1|^{-2} \times$$

$$Ai^2(N^{2/3} b_1^{2/3} \chi(\xi_0) e^{2\pi i/3}) \qquad (2.22)$$

Using these formulae in (2.17) and employing, where appropriate, the known uniform asymptotic expansions of the Airy function Ai for large argument it may be shown that $T_1(\underset{\sim}{r},\underset{\sim}{r}_s)$ can be identified with the geometrical diffraction form predicted by Lewis, Bleisten and Ludwig [8]. If $\underset{\sim}{r}$ is in the shadow of the ellipse but is not on $\xi = \xi_0$ then we may expand the Airy function $Ai(N^{2/3} b_1^{2/3} \chi(\xi)\, e^{2\pi i/3})$ for large N to write T_1 in terms of the "diffraction modes" of Keller. This expansion fails at the portions of $\xi = \xi_0$ in the shadow since these points lie on a "caustic". Likewise if ξ lies in the shadow of the ellipse and is allowed to take value on $\xi = \xi_0$ then we obtain the "creeping wave modes" discussed by Franz and Deppermann [2] and Keller [5]. Again $\xi = \xi_0$ is a caustic for this expansion.

The upshot of this analysis is that since $T_1(\underset{\sim}{r},\underset{\sim}{r}_s)$ verifies the geometrical theory of diffraction for an ellipse then, if we can show that for the more general geometries described in section 1 the field in the shadow has $U_0(\underset{\sim}{r},\underset{\sim}{r}_s; k)$ as its leading term as $k \to \infty$ it follows that the geometrical theory of diffraction is valid for these geometries also. This is essentially the content of the following sections.

§3 The First Extension

Let $U_1(\underset{\sim}{r};\underset{\sim}{r}_s;k)$ be the solution to the scattering problem

$$\begin{aligned} &\text{(i)} && \Delta U + k^2 U = \delta(\underset{\sim}{r},\underset{\sim}{r}_s), \quad \underset{\sim}{r},\underset{\sim}{r}_s \in D_1 \\ &\text{(ii)} && \partial_n^{(1)} U = 0, \quad \underset{\sim}{r} \in \partial\Omega_1 \\ &\text{(iii)} && \lim_{|\underset{\sim}{r}|\to\infty} \rho^{1/2}|U_\rho - i k U| = 0 \quad \rho = |\underset{\sim}{r} - \underset{\sim}{r}'|, \quad \underset{\sim}{r}' \in D_1. \end{aligned} \qquad (3.1)$$

Here D_1 is the exterior of a smooth convex curve Ω_1 formed by "pasting" the ends e_1 and e_2 of a convex arc B_1 to that part of the ellipse Ω_0 illuminated by a point source at $\underset{\sim}{r}_s$; as shown in figure 1. The symbol $\partial_n^{(1)}$ denotes differentiation in the direction of the outward normal to Ω_1 and $\underset{\sim}{r}, \underset{\sim}{r}_s$ are the "source" and "observation" points respectively.

We prove the following theorem

<u>Theorem 1</u>

As $k \to \infty$

$$U_1(\underset{\sim}{r};\underset{\sim}{r}_s;k) = U_0(\underset{\sim}{r};\underset{\sim}{r}_s;k)[1 + O(\exp\{-k^{1/3}\sigma\})], \tag{3.2}$$

uniformly in $\underset{\sim}{r}$, $\underset{\sim}{r} \in S_1^{<}(\underset{\sim}{r}_s)$.

In this theorem σ is a positive constant independent of $\underset{\sim}{r}$ and k, $U_0(\underset{\sim}{r};\underset{\sim}{r}_s;k)$ is the solution to the scattering problem (2.2, 2.3, 2.4) for the ellipse Ω_0 $(\xi = \xi_0)$ and $S_1^{<}(r_s)$ is the "deep shadow" of Ω_1 defined as follows: $\underset{\sim}{r} \in S_1(\underset{\sim}{r}_s)$, the shadow of Ω_1, if and only if $\underset{\sim}{r} \in D_1 \cup \Omega_1$ and the straight line through $\underset{\sim}{r}$ and $\underset{\sim}{r}_s$ cuts Ω_1 at two distinct points. $S_1^{<}(\underset{\sim}{r}_s)$ is any closed bounded subset of $S_1(\underset{\sim}{r}_s)$. If $\underset{\sim}{r} \in S_1^{<}(\underset{\sim}{r}_s)$ we say that $\underset{\sim}{r}$ is in the "deep shadow" of Ω_1. Note: the shadow $S_0(\underset{\sim}{r}_s)$ of the ellipse Ω_0 is identical to $S_1(\underset{\sim}{r}_s)$ and so $S_1^{<}(\underset{\sim}{r}_s) = S_0^{<}(\underset{\sim}{r}_s)$.

<u>Proof of Theorem 1</u>

We begin by expressing $U_1(\underset{\sim}{r};\underset{\sim}{r}_s;k)$ as a perturbation of $U_0(\underset{\sim}{r},\underset{\sim}{r}_s;k)$. That is we write

$$U_1(\underset{\sim}{r};\underset{\sim}{r}_s;k) = U_0(\underset{\sim}{r};\underset{\sim}{r}_s;k) + I_1(\underset{\sim}{r};\underset{\sim}{r}_s;k). \tag{3.3}$$

An application of Green's theorm to the region $D_1 \cup \Omega_1$ shows that the perturbation $I_1(\underset{\sim}{r};\underset{\sim}{r}_s;k)$ is given by

$$I_1(\underset{\sim}{r};\underset{\sim}{r}_s;k) = \int_{\partial B_1} \partial_n^{(1)} U_0(\underset{\sim}{r};\underset{\sim}{r}_{B_1};k)U_1(\underset{\sim}{r}_{B_1},\underset{\sim}{r}_s;k)ds_1. \tag{3.4}$$

Here ∂B_1 is the boundary of B_1 extending from e_1 to e_2 and ds_1 is the element of arc length along ∂B_1.

To estimate I_1 for $\underset{\sim}{r} \in S_1^{<}(\underset{\sim}{r}_s)$ we make use of the following results:

$$U_1(\underset{\sim}{r}_{B_1};\underset{\sim}{r}_s;k) = O(\frac{1}{k^{1/6}|\underset{\sim}{r}_{B_1} - \underset{\sim}{r}_s|}) = o(1), \tag{3.5}$$

as $k \to \infty$. This is the bound derived by Grimshaw [4]. Next, Leppington [6], we have

$$|\partial_n^{(1)} U_0(\underset{\sim}{r};\underset{\sim}{r}_{B_1};k)| \leq Ak^{1/3} \exp(-Ak^{1/3}\sigma_1) \tag{3.6}$$

where A is a positive constant and σ_1 is that part of the minimum optical path from r_{B_1} to h_1 (or h_2) as shown in figure 1 which coincides with Ω_0. Furthermore an analysis of the problem for the ellipse (Leppington [6], Bloom and Matkowsky [1] are helpful here) shows that

$$U_0^{-1}(\underset{\sim}{r};\underset{\sim}{r}_s;k) \sim O(k^{2/3}\exp(Ak^{1/3}\sigma_2)) \tag{3.7}$$

where σ_2 is the minimum of the arcs f_1h_1, f_2h_2, as shown in figure 1.

From (3.4) (3.5) and (3.6) we obtain

$$I_1(\underset{\sim}{r};\underset{\sim}{r}_s;k) \sim O(k^{1/6}\exp(-Ak^{1/3}\sigma_3)), \tag{3.8}$$

where σ_3 is the minimum of the arcs e_1h_1, e_2h_2 shown. we now rewrite (3.8) in the form

$$I_1(\underset{\sim}{r};\underset{\sim}{r}_s;k) \sim O\{\frac{k^{1/6}\exp(-Ak^{1/3}\sigma_3)}{U_0(\underset{\sim}{r};\underset{\sim}{r}_s;k)}\} U_0(r;r_s;k)$$

$$\sim O\{k^{5/6}\exp(-Ak^{1/3}[\sigma_3 - \sigma_2])\}U_0(\underset{\sim}{r};\underset{\sim}{r}_s;k). \tag{3.9}$$

By construction we see that $\sigma_3 - \sigma_2 > 0$; setting $\sigma = A[\sigma_3 - \sigma_2]$ and combining (3.9) with (3.3) we obtain the desired result (3.2).

§4 The Second Extension

Let $U_2(r;r_s;k)$ be the solution to the scattering problem

$$\begin{aligned}
&\text{(i)} \quad \Delta U + k^2U = \delta(\underset{\sim}{r},\underset{\sim}{r}_s), \quad \underset{\sim}{r},\underset{\sim}{r}_s \in D_2 \\
&\text{(ii)} \quad \partial_n^{(2)}U = 0 \qquad \underset{\sim}{r} \in \partial\Omega_2 \\
&\text{(iii)} \quad \lim_{|\underset{\sim}{r}|\to\infty} \rho^{1/2}|U_\rho - ikU| = 0, \quad \rho = |\underset{\sim}{r} - \underset{\sim}{r}'|, \quad \underset{\sim}{r}' \in D_2
\end{aligned} \tag{4.1}$$

Here D_2 is the exterior of the smooth convex curve Ω_2 formed by "pasting" the ends i_1 and i_2 of a smooth convex arc B_2 to the "dark" side of the convex curve Ω_1 as shown in figure 2. $\partial_n^{(2)}$ denotes differentiation in the direction of the outward drawn normal to Ω_2 and $\underset{\sim}{r}$ and $\underset{\sim}{r}_s$ again denote the "source" and "observation" points respectively.

We now establish

<u>Theorem 2</u>

As $k \to \infty$

$$U_2(\underset{\sim}{r};\underset{\sim}{r}_s;k) = U_0(\underset{\sim}{r};\underset{\sim}{r}_s;k)[1 + O(\exp\{-k^{1/3}\mu\})], \tag{4.2}$$

uniformly in $\underset{\sim}{r}$, $\underset{\sim}{r} \in S_2^-(r_s) = S_2^<(\underset{\sim}{r}_s) - R$, where μ is positive and independent of k and $\underset{\sim}{r}$.

$S_2(\underset{\sim}{r}_s)$ is the "shadow" of Ω_2. That is $\underset{\sim}{r} \in S_2(\underset{\sim}{r}_s)$ if and only if $\underset{\sim}{r} \in \Omega_2 \cup D_2$ and the straight line joining $\underset{\sim}{r}$ and $\underset{\sim}{r}_s$ cuts Ω_2 at two distinct points. $S_2^<(\underset{\sim}{r}_s)$ is any closed bounded subset of $S_2(\underset{\sim}{r}_s)$. If $\underset{\sim}{r} \in S_2^<(\underset{\sim}{r}_s)$ we say that $\underset{\sim}{r}$ lies in the "deep shadow" of Ω_2.

R is the region of influence of B_2 and is constructed as follows. Consider the smaller of the two elliptic arcs f_1i_1 and f_2i_2 of $\Omega_2 \cap S_2(\underset{\sim}{r}_s)$. Suppose it is f_1i_1. Let i_1' be a point on f_1i_1 arbitrarily close to i_1. Now take i_2' to be the point on the elliptic arc of $\Omega_2 \cap S_2(\underset{\sim}{r}_s)$ for which $|f_1i_1'| = |f_2i_2'|$. R is then the region bounded by the tangents to Ω_2 at i_1', i_2' and the arc $i_1'i_1i_2i_2'$ of $\Omega_2 \cap S_2(\underset{\sim}{r}_s)$.

We remark that the region $S_2^-(\underset{\sim}{r}_s)$ is not in general the maximal region in which we expect theorem 2 to hold on the basis of the geometrical theory of diffraction.

The maximal region in fact consists of $S_2^-(\underset{\sim}{r}_s)$ and those points $\underset{\sim}{r} \in R$ for which;

(i) there is an optical path P in $D_2 \cup \Omega_2$ from $\underset{\sim}{r}_s$ to $\underset{\sim}{r}$ that does not intersect B_2.

(ii) $\int_{\Omega_2 \cap P} \kappa^{2/3}(\underset{\sim}{r})|d\underset{\sim}{r}| < \int_{\Omega_2 \cap P'} \kappa^{2/3}(\underset{\sim}{r})|d\underset{\sim}{r}|$

where $\kappa(\underset{\sim}{r})$ is the curvature of Ω_2 at $\underset{\sim}{r}$ and P' is the shortest of the other optical paths in $D_2 \cup \Omega_2$ from $\underset{\sim}{r}_s$ to $\underset{\sim}{r}$.

Proof of Theorem 2

In order to establish (4.2) we begin by showing that as $k \to \infty$ then

$$U_2(\underset{\sim}{r};\underset{\sim}{r}_s;k) = U_1(\underset{\sim}{r};\underset{\sim}{r}_s;k)[1 + O(\exp\{-k^{1/3}\alpha\})]. \qquad (4.3)$$

uniformly in $\underset{\sim}{r}$, $\underset{\sim}{r} \in S_2^-(\underset{\sim}{r}_s)$, where α is positive and independent of k and $\underset{\sim}{r}$. Theorem 2 will then follow from this result and theorem 1 since $S_2^-(\underset{\sim}{r}_s) \subseteq S_1^<(\underset{\sim}{r}_s)$.

In order to establish (4.3) we proceed as we did in the proof of theorem 1. That is we consider $U_2(\underset{\sim}{r};\underset{\sim}{r}_s;k)$ to be a perturbation of $U_1(\underset{\sim}{r};\underset{\sim}{r}_s;k)$ and write

$$U_2(\underset{\sim}{r};\underset{\sim}{r}_s;k) = U_1(\underset{\sim}{r};\underset{\sim}{r}_s;k) + I_2(\underset{\sim}{r};\underset{\sim}{r}_s;k). \qquad (4.4)$$

An application of Green's theorem in $D_2 \cup \Omega_2$ applied to U_1 and U_2 leads to the representation

$$I_2(\underset{\sim}{r};\underset{\sim}{r}_s;k) = \int_{\partial B_2} \partial_n^{(2)} U_1(\underset{\sim}{r}_{B_2};\underset{\sim}{r}_s;k) \cdot U_2(\underset{\sim}{r};\underset{\sim}{r}_{B_2};k)ds_2. \qquad (4.5)$$

Here ∂B_2 is the boundary of B_2 extending from i_1 to i_2 and ds_2 is the element of arc length along ∂B_2. To estimate I_2 for $\underset{\sim}{r} \in S_2^-(\underset{\sim}{r}_s)$ we make use of the following results which hold as $k \to \infty$:

$$U_2(\underset{\sim}{r};\underset{\sim}{r}_{B_2};k) = O\left(\frac{1}{k^{1/6}|\underset{\sim}{r} - \underset{\sim}{r}_{B_2}|}\right) = o(1). \qquad (4.6)$$

This again follows from the work of Grimshaw [4]. Next Leppington [6], we have

$$|\partial_n^{(2)} U_1(\underset{\sim}{r}_{B_2};\underset{\sim}{r}_s;k)| \leq Ak^{1/3}\exp(-Ak^{1/3}\mu_1), \qquad (4.7)$$

where A is a positive constant and μ_1 is that part of the minimum optical path from $\underset{\sim}{r}_{B_2}$ to f_1 (or f_2) as shown in figure 2 which coincides with Ω_0.

Finally from the results given in section 2 and following the arguments of Bloom and Matkowsky [1] we find that

$$U_1^{-1}(\underset{\sim}{r};\underset{\sim}{r}_s;k) = O(U_0^{-1}(\underset{\sim}{r};\underset{\sim}{r}_s;k))$$

$$= O(k^{2/3}\exp(Ak^{1/3}\mu_2)) \qquad (4.8)$$

where μ_2 is the minimum of the arcs f_1h_1, f_2h_2 shown in figure 2.

From (4.5, (4.6) and (4.7) we obtain

$$I_2(\underset{\sim}{r};\underset{\sim}{r}_s;k) \sim O(k^{1/6}\exp(-Ak^{1/3}\mu_3)) \qquad (4.9)$$

where μ_3 is the minimum of the arcs i_1f_1, i_2f_2, as shown. Now rewrite (4.9) in the form

$$I_2(r;r_s;k) \sim O\left\{\frac{k^{1/6}\exp(-Ak^{1/3}\mu_3)}{U_1(\underset{\sim}{r};\underset{\sim}{r}_s;k)}\right\} U_1(\underset{\sim}{r};\underset{\sim}{r}_s;k)$$

$$\sim O\,\{k^{5/6}\exp(-Ak^{1/3}[\mu_3 - \mu_2])\}U_1(\underset{\sim}{r};\underset{\sim}{r}_s;k). \qquad (4.10)$$

By construction we see that $\mu_3 - \mu_2 > 0$ and setting $\alpha = A[\mu_3 - \mu_2]$ and combining (4.10) with (4.4) we obtain (4.3) as required.

Theorem 2 then follows from (4.3) and theorem 1.

§5 Concluding Remarks

The results of theorems 1 and 2 represent, we believe, the most general verifications to date of the validity of the geometrical theory of diffraction. The idea depends essentially on a knowledge of a general canonical problem, namely diffraction by an ellipse. In order to extend the analysis to convex bodies rather than cylinders one would require a complete knowledge of diffraction by a spheroid and this in turn would depend on an extension of the Watson transformation as treated for example in [11].

Perhaps a complete verification of the geometrical theory of diffraction for convex cylinders could be obtained by basing the argument on a more general region Ω_0 rather than the ellipse as we have done. For example one might consider Ω_0 to be formed by a pair of ellipses of four point contact in the regions of diffraction and joined by suitable convex curves. However such a region leads to a problem which cannot be treated by the separation of variables technique, i.e. it is not a canonical problem. Thus the central part of the analysis would give rise to added difficulties of its own. These remarks suggest that our approach to the problem of verifying the validity of the geometrical theory of diffraction cannot be substantially improved upon and recourse must be made to other techniques.

References

[1] C. A. Bloom, and B. J. Matkowsky, On the validity of the geometrical theory of diffraction by convex cylinders. Arch. Rat. Mech. Anals. **33** 71-90 (1969).

[2] W. Franz and K. Deppermann, Theorie der Beugung am Zylinder unter Berücksichtigung der Kriechwelle, Ann der Phys. **10** 361-373 (1952).

[3] R. F. Goodrich and N. D. Kazarinoff, Scalar diffraction by prolate spheroids whose eccentricities are almost one. Proc. Camb. Phil. Soc. **59** 167-183 (1963).

[4] R. Grimshaw, High frequency scattering by finite convex regions. Comm. Pure Appl. Math. **19** 167-198 (1966).

[5] J. B. Keller, Diffraction by a convex cylinder. Trans. I.R.E. AP-**4**, 312-321 (1956).

[6] F. G. Leppington, On the short-wave asymptotic solution of a problem in acoustic radiation. Proc. Camb. Phil. Soc. **64** 1131-1150 (1968).

[7] B. R. Levy, Diffraction by an elliptic cylinder. J. Math. Mech. **9** 147-166 (1960).

[8] R. M. Lewis, N. Bleistein and D. Ludwig, Uniform asymptotic theory of creeping waves. Comm. Pure Appl. Math. **20** 295-327 (1967).

[9] C. S. Morawetz and D. Ludwig, An inequality for the reduced wave operator and the justification of geometrical optics. Comm. Pure Appl. Math. **21** 187-203 (1968).

[10] F. W. J. Olver, Asymptotics and special functions. Academic Press, London (1974) (see also, second-order linear differential equations with two turning points. Phil. Trans. Roy. Soc. 278 137-185, (1974)).

[11] B. D. Sleeman, Integral representations associated with high-frequency non-symmetric scattering by prolate spheroids. Quart. J. Mech. and Appl. Math. 22 405-426 (1969).

[12] F. Ursell, On the short wave asymptotic theory of the wave equation $(\nabla^2 + k^2)\phi = 0$. Proc. Camb. Phil Soc. 53 115-133 (1957).

[13] F. Ursell, Creeping modes in a shadow. Proc. Camb. Phil. Soc. 64 171-191 (1968).

[14] G. N. Watson, The diffraction of electric waves by the earth. Proc. Roy. Soc. London Ser. A 95 83-99 (1918).

[15] T. T. Wu and S. R. Seshadri, The electromagnetic theory of light II. Harvard University, Cruft Laboratory, Scientific Report No 22 (1958).

Department of Mathematics
University of Dundee
DUNDEE DD1 4HN
Scotland UK.

Über verallgemeinerte Faltungsoperatoren und ihre Symbole

F.-O. Speck

(Darmstadt)

0. Einleitung. Einige Klassen $\mathfrak{A}$ linearer stetiger Operatoren von einem Sobolevraum $L^p_s(E_m)$ in einen anderen $L^p_t(E_m)$ $(1 < p < \infty,\ s,t \in E_1,\ m \geq 2)$ sind homomorph abbildbar in eine Funktionenalgebra $\mathfrak{B}$, so daß die folgenden Bedingungen äquivalent sind:

(i) $A \in \mathfrak{A}$ ist Fredholm-Operator

(ii) $\Phi_A \in \mathfrak{B}$ ist regulär in $\mathfrak{B}$

(iii) ess inf $|\Phi_A| > 0$.

Dies wird für einige Fälle nachgewiesen, die Summen und Produkte von Operatoren der folgenden Gestalt einschließen: L^1-Faltungen, singuläre Integraloperatoren vom CALDERON-MIKHLIN-Typ, CAUCHY-Operatoren bezüglich verschiedener Richtungen, Multiplikationen mit passenden Funktionen Differentialoperatoren, kompakte Operatoren und andere. Dabei werden Charakterisierungen der Symbol-Algebren $\mathfrak{B}$ angegeben.

1. Faltungsoperatoren auf Sobolevräumen

1.0 Bezeichnungen. E_m bezeichne den m-dimensionalen Euklidischen Raum $(m \geq 2)$, $\mathcal{S}$, $\mathcal{S}'$ die Schwartz'schen Funktionen bzw. Distributionen, F die Fouriertransformation, definiert durch

$$F\,\phi(\xi) := \frac{1}{\sqrt{2\pi}^m} \int_{E_m} e^{-ix\xi}\,\phi(x)\,dx \quad , \quad \phi \in \mathcal{S}$$

$$< F\,\Phi, \phi > := < \Phi, F^{-1}\phi >, \quad \Phi \in \mathcal{S}', \quad \phi \in \mathcal{S} .$$

Den Sobolevraum L^p_s $(1 < p < \infty,\ s \in E_1)$ verstehe man als isometrisches Bild von L^p unter dem Besselpotentialoperator

$$J^s = j_s * \ , \quad d^{-s}(\xi) := F j_s(\xi) = (1+|\xi|^2)^{-s/2} .$$

In den linearen Räumen $\mathcal{S} \subset L^p_s \subset \mathcal{S}'$ (dicht) benutzen wird die Identifizierung

$$\langle u, \phi \rangle = \int \overline{u(x)}\, \phi(x) dx$$

für $u, \phi \in \mathcal{S}$ bzw. durch Approximation in L^p_s und $\mathcal{S}'$. Es bezeichne $\mathcal{L} = \mathcal{L}(L^p_s, L^p_t)$ die linearen stetigen Operatoren, $\mathcal{C}$ die kompakten, $\mathcal{F}$ die Fredholm- und $\mathfrak{T}$ die translationsinvarianten Operatoren, d. h. diejenigen $A \in \mathcal{L}$ mit

$$T_h A = A T_h \quad \text{für } h \in E_m, \text{ wobei}$$

$$T_h u(x) = u(x-h) \quad \text{für } u \in \mathcal{S}', \quad x, h \in E_m .$$

Φ_A sei das Symbol von $A \in \mathfrak{T}$, falls $\Phi_A \in \mathcal{S}'$ und

$$A \phi = F^{-1} \Phi_A \cdot F \phi \quad \text{für } \phi \in \mathcal{S} \text{ (und } \Phi_A \text{ eindeutig)},$$

$\mathfrak{S}(L^p_s, L^p_t)$ der Symbolraum zu $\mathfrak{T}(L^p_s, L^p_t)$, falls Φ_A existiert für alle $A \in \mathfrak{T}$.

Es seien einige bekannte Ergebnisse nebst offensichtlichen Folgerungen erwähnt:

1.1 Satz (HÖRMANDER 1960):
Zu jedem $A \in \mathfrak{T}(L^p_s, L^p_t)$ existiert das Symbol $\Phi_A \in L^\infty_{loc}$.

$$\mathfrak{T} \to \mathfrak{S}, \quad A \mapsto \Phi_A \text{ ist algebraischer Isomorphismus,}$$

$$\mathfrak{T}(L^p_s, L^p_t) \to \mathfrak{T}(L^p, L^p), \quad A \mapsto \tilde{A} := J^{-t} A J^s \text{ Isometrie.}$$

$$\mathfrak{S}(L^p_s, L^p_t) = d^{s-t} \cdot \mathfrak{S}(L^p, L^p) \subset d^{s-t} L^\infty$$

$$\mathfrak{S}(L^2_s, L^2_t) = d^{s-t} \cdot L^\infty$$

$$\| \tilde{\Phi}_A \| := \| d^{t-s} \Phi_A \|_{L^\infty} \leq \| A \|$$

$$\| \tilde{\Phi}_A \| = \| A \| \quad \text{für } p = 2 .$$

1.2 Satz: Für den dualen Operator A' zu $A \in \mathfrak{T}(L^p_s, L^p_t)$ gilt

$$A' \in \mathfrak{T}(L^{p'}_{-t}, L^{p'}_{-s}) \quad \text{und} \quad \Phi_{A'} = \overline{\Phi_A},$$

$$A' \in \mathfrak{T}(L^p_s, L^p_t) \quad \text{mit} \quad A' = S Q A S Q$$

$$(SQu(x) = \overline{u(-x)}) .$$

1.3 Satz : $\mathcal{Y}(L^p_s, L^p_t)$ ist abgeschlossen in $\mathcal{L}$.
Produkte und Inverse von $\mathcal{Y}$-Operatoren sind wieder $\mathcal{Y}$-Operatoren.
Für $s = t$ bildet $\mathcal{Y}$ eine involutorische, kommutative Banachalgebra, für $p = 2$ sogar eine B^*-Algebra.

1.4 Satz : Die Defektzahlen von $A \in \mathcal{Y}(L^p_s, L^p_t)$ sind entweder Null oder unendlich. Daraus folgt

$$A \in \mathcal{F} \quad \Leftrightarrow \quad A \text{ bijektiv.}$$

(das gilt für weit allgemeinere Teilräume von $\mathcal{S}'$ anstelle L^p_s, L^p_t, s. SPECK 1974).

1.5 Beispiel : Die in der Einleitung genannten $\mathcal{Y}$-Operatoren sind von der Gestalt

$$A = \sum_{|\nu| \leq n} A_\nu D^\nu ,$$

wobei für die Symbole $\Phi = \Phi_{A_\nu}$ von $A_\nu \in \mathcal{Y}(L^p, L^p)$ $(t = s + n)$ die Stabilitätsbedingung gilt

$$\Phi = \overset{0}{\Phi} + \Phi_0$$

(S) mit $\overset{0}{\Phi}(\rho\xi) = \overset{0}{\Phi}(\xi)$ für $\rho > 0$, d. h. homogen vom Grade Null,

und $\Phi_0 \in L^\infty_0$, d. h. $\| \chi_\rho \Phi_0 \|_{L^\infty} \underset{\rho \to \infty}{\longrightarrow} 0$, $\chi_\rho(\xi) := \begin{cases} 1 & |\xi| \geq \rho \\ 0 & \text{sonst} \end{cases}$.

Die beiden Anteile von A_ν nennt man "singulär" bzw. "nichtsingulär" und A_ν vom "zusammengesetzten Typ", falls keiner der beiden Anteile verschwindet (und $\overset{0}{\Phi} \not\equiv$ const.).

Es gilt

$$A \in \mathcal{F} \Leftrightarrow A \text{ bijektiv} \quad \Rightarrow \quad \operatorname{essinf} |\tilde{\Phi}_A| > 0 .$$

Die Umkehrung gilt für $p = 2$ allgemein, für $p \neq 2$ nur für bestimmte Operatorenklassen und sei deshalb hier hervorgehoben $(s = t = 0)$:

(B1) $\mathfrak{A} \subset \mathfrak{L}(L^p, L^p)$

für alle $A \in \mathfrak{A}$, $\operatorname{essinf} |\Phi_A| > 0$ gilt $F^{-1} \Phi_A^{-1} \cdot F \in \mathfrak{A}$.

Beispiele sind:

a) L^1-Faltungen 2. Art $(\Phi = a + \Phi_0)$,

b) singuläre (polunabhängige) CALDERON-MIKHLIN-Operatoren $(\Phi = \overset{0}{\Phi})$,

c) als zusammengesetzte Operatoren: solche, deren Symbol die MIKHLIN-Bedingung erfüllen:

$D^{\nu}\Phi$ existiere für $|\nu| \leq m$, $\nu_j \leq 1$,
und sei stetig für $|\nu| < m$; für alle diese Ableitungen gelte

$$|\xi|^{|\nu|} |D^{\nu}\Phi(\xi)| \leq M \quad \text{für} \quad \xi \in E_m .$$

Es lassen sich viele weitere Beispiele konstruieren, etwa auch mit stückweise stetigen Symbolen.

2. Lokale Eigenschaften

2.0 Bezeichnungen. $\overline{E_m}$ sei eine der folgenden Kompaktifizierungen des E_m:

$$\dot{E}_m = E_m \cup \{\infty\}$$

$$\ddot{E}_m = [-\infty, +\infty]^m$$

$$\tilde{E}_m = E_m \cup \mathfrak{N}, \qquad \mathfrak{N} \text{ homöomorph } \partial K_1(0) = \{\xi : |\xi| = 1\},$$

mit den üblichen Topologien, vgl. [12], $\operatorname{mes}(\overline{E}_m \setminus E_m) := 0$.
$U(x)$ sei Umgebung von $x \in \overline{E_m}$, χ_E charakteristische Funktion einer meßbaren Menge E und

$$W_{x_o} := \{w \in C^{\infty}(\overline{E_m}) : w(x) \equiv 1 \text{ in } U(x_o)\}, \quad x_o \in \overline{E_m} .$$

Für $A, B \in \mathcal{L}(L^p_s, L^p_t)$ sei, vgl. [12] ,

$A \in \mathcal{L}_o$ "Operator vom Lokaltyp" <=>
$w\cdot A - Aw\cdot \in \mathcal{C}$ für alle $w \in C^\infty(\overline{E_m})$

$A \in \mathcal{F}_x$ "lokal Fredholmsch im Punkt $x \in \overline{E_m}$" <=>
es existieren $w \in W_x$, $L_x, R_x \in \mathcal{L}(L^p_t, L^p_s)$
mit $(L_x A - I)w\cdot$, $w\cdot(AR_x - I)$ kompakt.

$A \overset{x}{\sim} B$ "lokal äquivalent in $x \in \overline{E_m}$" <=>
zu jedem $\varepsilon > 0$ existiert ein $w \in W_x$ mit
$\| (A - B)w\cdot \|_{\mathcal{L}/\mathcal{C}} < \varepsilon$, $\| w\cdot(A - B)\|_{\mathcal{L}/\mathcal{C}} < \varepsilon$,
wobei $\| A\|_{\mathcal{L}/\mathcal{C}} : = \inf_{K\in\mathcal{C}} \| A + K\|$.

2.1 Folgerung : $\mathcal{L}_o(L^p_s, L^p_t)$ ist abgeschlossen in $\mathcal{L}$. Summen und Produkte von $\mathcal{L}_o$-Operatoren sind wieder vom Lokaltyp. Also bildet $\mathcal{L}_o$ für $s = t$ eine Banachalgebra und entsprechendes gilt für $\mathcal{F} \cap \mathcal{L}_o$.

2.2 Satz (SIMONENKO 1964 - RABINOVIČ 1972) :

I : Sei $A \in \mathcal{L}_o(L^p_s, L^p_t)$ bezüglich $\dot{E}_m$. Dann gilt

$A \in \mathcal{F}$ <=> $A \in \mathcal{F}_x$ für alle $x \in \dot{E}_m$

$A \in \mathcal{C}$ <=> $A \overset{x}{\sim} 0$ für alle $x \in \dot{E}_m$.

II : Sind $A, B \in \mathcal{L}_o$, $x \in \dot{E}_m$, $A \overset{x}{\sim} B$, so gilt

$A \in \mathcal{F}_x$ <=> $B \in \mathcal{F}_x$

$A \overset{x}{\sim} 0$ <=> $B \overset{x}{\sim} 0$.

Alle Bedingungen sind äquivalent den entsprechenden Bedingungen für

$$\tilde{A} : = J^{-t} A J^s , \tilde{B} : = J^{-t} B J^s \in \mathcal{L}_o(L^p, L^p) .$$

Beweis: SIMONENKO zeigte den Satz für $s = t = 0$ mit äquivalenten Definitionen; aus der Arbeit von RABINOVIČ folgt: $J^r \in \mathcal{L}_o(L^p_s, L^p_{s+r})$ für alle $s, r \in E_1$ bezüglich $\dot{E}_m$, s. [10; Prop. 1.3, S. 48] .

2.3 Beispiele : Vom Lokaltyp (bezüglich $\dot{E}_m$) sind folgende (L^p_s, L^p_s)-Operatoren:

$a \cdot I$, das ist trivial,

$B_k = k *$, $k \in L^1$, s. RAKOVČIK 1963 ,

$B = F^{-1} \Phi \cdot F$ mit $\Phi = \Phi_0 \in L^\infty_0$ für $p = 2$, da

$\|(I - T_h)B\|_{\mathcal{L}} = \|(1-e^{-ih\xi})\Phi(\xi)\|_{L^\infty} \to 0$

für $|h| \to 0$, aufgrund des Lemmas von RIESZ & WEIL,

$C_{f/\cdot}$: CALDERON-MIKHLIN-Operatoren mit der Charakteristik f , s. [2] , und damit für $p = 2$ alle Operatoren mit $\tilde{E}_m$-stabilem Symbol (d. h. Gültigkeit von (S)) und stetigem $\overset{0}{\Phi}$.

C_j : CAUCHY-Operatoren bezüglich der x_j-Koordinate (analoger Beweis).

Bei linearen Transformationen im Symbol bleibt die $\mathcal{L}_0$-Eigenschaft erhalten (aus $T : u(x) \mapsto u(M x - h)$ folgt $F^{-1}(T \Phi) \cdot F = \hat{T} F^{-1} \Phi \cdot F \hat{T}^{-1}$ mit $\hat{T} = e^{ih\cdot} T_{M^{*-1}}$). So erhält man aus $\frac{1}{2}(I + C_j)$ mit $\Phi = \text{sgn}\, \xi_j$ singuläre Operatoren mit stückweise stetigen und $\tilde{E}_m$-instabilen Symbolen. Weitere Möglichkeiten erschließt die Betrachtung glatter Transformationen ("quasistabile" Symbole) mit Anwendungen z. B. in der Geodäsie (Schwerepotential der Erde ohne Idealisierung als Kugel).

Die ersten drei Typen sind vom Lokal-Typ auch bezüglich $\ddot{E}_m$ und $\tilde{E}_m$, vgl. SIMONENKO 1967, nicht jedoch singuläre Operatoren $C \neq a \cdot I$, s. SPECK 1974. Deshalb wird fernerhin nur der Fall $\overline{E_m} = \dot{E}_m$ diskutiert. Abschließend sei erwähnt: $a(.)\cdot$ und $D^\nu \in \mathcal{L}_0$ auf passenden Sobolevräumen.

2.4 Satz : Sei $A \in \mathcal{Y}(L^p_s, L^p_t)$. Dann gilt:

$A \overset{\infty}{\sim} 0 \iff A = 0$

$A \in \mathcal{F}_\infty \iff A$ bijektiv.

2.5 Satz : Sei $A \in \mathcal{Y}(L^p_s, L^p_t)$, $\tilde{A} = \overset{0}{\tilde{A}}$, $x \in E_m$. Dann ist

$A \overset{x}{\sim} 0 \iff A = 0$

$A \in \mathcal{F}_x \iff A$ bijektiv.

Beweise : Für $s = t = 0$ s. SPECK 1974, S. 43 - 48.
Der allgemeine Fall ergibt sich direkt aus dem Satz von RABINOVIČ :
$J^r \in \mathcal{Y} \cap \mathcal{L}_0(L^p_s, L^p_{s+r})$ für $r, s \in E_1$.

2.6 Satz : Sei $A \in \mathcal{Y}(L^p_s, L^p_t)$, $\widetilde{A} = \overset{0}{\widetilde{A}} + \widetilde{A}_0$, $x \in E_m$.
Dann ist für $p = 2$

$$A \overset{x}{\sim} 0 \quad \Longleftrightarrow \quad \overset{0}{\widetilde{A}} = 0$$

$$A \in \mathcal{F}_x \quad \Longleftrightarrow \quad \overset{0}{\widetilde{A}} \text{ bijektiv.}$$

Für $p \neq 2$ gilt dies unter den zusätzlichen Voraussetzungen

$\widetilde{A}_0 \in \mathcal{L}_0$ und es existiert eine Folge

(B2) $\{\Phi_j\} \subset L^\infty_{00}$(komp. Träger) : $\overset{-1}{F}\ \Phi_j \cdot F \to \widetilde{A}_0$ in $\mathcal{L}$.

Beweis : Ohne Einschränkung sei $s = t = 0$, also $A = \widetilde{A}$.

1. Ist $p = 2$, $w \in W_x$, supp w kompakt, so gilt $\chi_\rho\ w \cdot A_0 \to 0$ für $\rho \to \infty$ und $(I - T_h) w \cdot A_0 \to 0$ für $h \to 0$ in $\mathcal{L}$, also wA_0 und analog $A_0 w \cdot \in \mathcal{L}$ und damit $A_0 \overset{x}{\sim} 0$.

2. Für $p \neq 2$, w wie oben, ist $B_j := w \cdot \overset{-1}{F}\ \Phi_j \cdot Fw \cdot$ ein HILLE-TAMARKIN-Operator, da der Faltungskern $w(x) \cdot (\overset{-1}{F}\ \Phi_j)(x-y) w(y) \in L^\infty_{00}$, und damit ist B_j kompakt, s. [7].
Wegen $A_0 \in \mathcal{L}_0$ ist $w^2 \cdot A_0 - w \cdot A_0 w \cdot \in \mathcal{L}$ und wegen (B2) ist $A_0 \overset{x}{\sim} 0$.
In beiden Fällen folgen nun die Aussagen aus Satz 2.2 und 2.5.

2.7 Folgerung : Hat $\widetilde{A}$ ein $\widetilde{E}_m$-stabiles Symbol und gilt für $p \neq 2$ (B1) für $\overset{0}{\widetilde{A}}$ und (B2) für $\widetilde{A}_0$, so ist für $x \in E_m$

$$A \in \mathcal{F}_x \quad \Longleftrightarrow \quad \text{essinf } |\overset{0}{\widetilde{\Phi}}| > 0 .$$

Für $x = \infty$ gilt dagegen

$$A \in \mathcal{F}_x \quad \Longleftrightarrow \quad \text{essinf } |\widetilde{\Phi}| > 0 .$$

im Falle $p = 2$ und für $p \neq 2$, falls $\widetilde{A}$ (B1) erfüllt.

3. Verallgemeinerte Faltungsoperatoren. Stabile Symbole.

3.0 Bezeichnungen. $A \in \mathcal{L}(L_s^p, L_t^p)$ heiße "einhüllender Operator" der Familie $\{A_x\}_{x \in E_m}$, wenn

$$A \overset{x}{\sim} A_x \ , \ A_x \in \mathcal{L}_o \quad \text{für alle} \quad x \in \dot{E}_m$$

und

$$(x \mapsto A_x) \in [\dot{E}_m \ ; \mathcal{L}]_{\mathcal{L},\mathrm{loc}} \quad ,$$

d. h. für jedes $\varepsilon > 0$ und $x_o \in \dot{E}_m$ existiert eine Funktion $w \in W_{x_o}$, so daß für alle $x \in \operatorname{supp} w$

$$\| w\cdot(A_{x_o} - A_x)\|_{\mathcal{L}/\mathcal{L}} < \varepsilon \text{ und } \| (A_{x_o} - A_x)w\cdot \|_{\mathcal{L}/\mathcal{L}} < \varepsilon \ .$$

Ist $\{A_x\} \subset \mathcal{F}$, so heißt A "verallgemeinerter Faltungsoperator", kurz $A \in \mathcal{F}_{(\cdot)}$.

3.1 Satz : Sei $A \in \mathcal{L}(L_s^p, L_t^p)$, $\widetilde{A} := J^{-t} A J^s$.

A ist Einhüllender von $\{A_x\}$

$\Longleftrightarrow$ $\widetilde{A}$ ist Einhüllender von $\{\widetilde{A}_x\} = \{J^{-t} A_x J^s\}$.

Beweis : folgt aus 3.0 wegen $J^r \in \mathcal{L}_o$.

3.2 Folgerung aus den Sätzen von SIMONENKO [12] : Dann gilt:

I : A ist durch $\{A_x\}$ modulo $\mathcal{L}$ eindeutig bestimmt.

II : $A \in \mathcal{F}_{(\cdot)}(L_s^p, L_t^p) \Longleftrightarrow \widetilde{A} \in \mathcal{F}_{(\cdot)}(L^p, L^p)$.

III: $A \in \mathfrak{F}(L_s^p, L_t^p) \Longleftrightarrow \widetilde{A} \in \mathfrak{F}(L^p, L^p)$

$\Longleftrightarrow \widetilde{A}_{x_o} \in \mathfrak{F}_{x_o}(L^p, L^p)$ für alle $x_o \in \dot{E}_m$.

3.3 Definition :

1. Sei

$$\sigma := \{\Phi : \dot{E}_m \times \widetilde{E}_m \to \mathbb{C} \text{ meßbar}, \ \overset{0}{\Phi}(x,\cdot) \text{ existiert für } x \in \dot{E}_m ,$$
$$\Phi(x,\cdot) = \overset{0}{\Phi}(x,\cdot) \text{ für } x \in E_m \ , \ (x \mapsto \overset{0}{\Phi}(x,\cdot)) \in [\dot{E}_m ; L^\infty]\}$$

mit der Norm

$$\|\Phi\|_{\sigma} := \sup_{x \in \dot{E}_m} \|\Phi(x,\cdot)\|_{L^\infty} .$$

2. Es sei $A \in \mathcal{A}_{(\cdot)}$ Einhüllender von $\{A_x\} \subset \mathcal{A}$,

$\overset{0}{\widetilde{\Phi}}_x$ existiere für alle $x \in \dot{E}_m$,

für $p \neq 2$ gelte (B2) für $\widetilde{A}_{x_o}$ $(x_o \in E_m)$.
Dann heiße

$$(x,\xi) \mapsto \widetilde{\Phi}(x,\xi) := \begin{cases} \overset{0}{\widetilde{\Phi}}_x(\xi) & x \in E_m \\ \widetilde{\Phi}_x(\xi) & x = \infty \end{cases} \qquad \xi \in \widetilde{E}_m$$

Symbol von $\widetilde{A}$ und $\Phi_A(x,\xi) := d^{s-t}(\xi)\cdot\widetilde{\Phi}(x,\xi)$
Symbol von A . Kurz : $\widetilde{\Phi}$ ist "$\widetilde{E}_m$-stabil" und "$A \in \mathcal{A}_{(\cdot,\sim)}$" .

<u>3.4</u> <u>Satz :</u>

I : Zu $A \in \mathcal{A}_{(\cdot,\sim)}$ ist Φ_A eindeutig definiert.

II : $\tau : \mathcal{A}_{(\cdot,\sim)} \to \sigma$, $A \mapsto \widetilde{\Phi}$

ist stetiger Homomorphismus mit $\ker \tau = \mathcal{L}$.

III: σ ist Banachalgebra, isomorph zu

$L_o^\infty \times [\dot{E}_m ; L^\infty(\mathfrak{M})]$.

<u>Beweis:</u>

I : folgt aus den Sätzen 2.4 - 2.6 .

II : Zu $\Phi_A \in \sigma$ ist nur die Stetigkeit von $x \mapsto \overset{0}{\widetilde{\Phi}}(x,\cdot)$ zu prüfen
Als nichtsingulären Anteil von $\widetilde{A}_x$ $(x \in E_m)$ wählt man o.E. den von $\widetilde{A}_\infty$, da er zum Symbol nichts beiträgt. Damit folgt aus

$$\|w\cdot(\widetilde{A}_{x_o}-\widetilde{A}_x)\|_{\mathcal{L}/\mathcal{L}} < \varepsilon \Rightarrow \|w\cdot(\overset{0}{\widetilde{A}}_{x_o}-\overset{0}{\widetilde{A}}_x)\|_{\mathcal{L}/\mathcal{L}} < \varepsilon$$

$$\Rightarrow \|\overset{0}{\widetilde{A}}_{x_o}-\overset{0}{\widetilde{A}}_x\|_{\mathcal{L}} < \varepsilon \quad , \text{ s. } [14 ; \text{S. } 47]$$

$$\Rightarrow \|\overset{0}{\widetilde{\Phi}}_{x_o}-\overset{0}{\widetilde{\Phi}}_x\|_{L^\infty} < \varepsilon .$$

Per Definitionem ist τ Homomorphismus und aus $\Phi = 0$ folgt $\tilde{A}_x \overset{x}{\sim} 0$ für $x \in \dot{E}_m$ mit Hilfe von 2.4 und 2.6; damit ist A kompakt nach 2.2.

Ähnlich erkennt man die Stetigkeit von τ mittels Satz 1.1.

III: ist offensichtlich.

<u>3.5 Folgerung :</u>

$\mathfrak{A}_{(\cdot,\sim)}(L_s^p, L_t^p)$ und $\mathfrak{A}_{(\cdot,\sim)}/\mathcal{L}$ sind abgeschlossen. Für $s = t$ ist $\mathfrak{A}_{(\cdot,\sim)}$ Banachalgebra und $\mathfrak{A}_{(\cdot,\sim)}/\mathcal{L}$ kommutative involutorische Banachalgebra $(\Phi^* := \overline{\Phi})$. Der Index von $A \in \mathfrak{F} \cap \mathfrak{A}_{(\cdot,\sim)}$ ist Null, vgl. 1.2.

<u>3.6 Satz :</u> Sei $A \in \mathfrak{A}_{(\cdot,\sim)}(L_s^p, L_t^p)$.

Für $p = 2$ sind folgende Aussagen äquivalent:

1. $A \in \mathfrak{F}$
2. $\tilde{\Phi}$ ist regulär in σ, d. h.

$$\operatorname*{essinf}_{\xi} |\tilde{\Phi}(\infty, \xi)| > 0 \quad \text{und} \quad \min_{x \in \dot{E}_m} \operatorname*{essinf}_{\xi} |\overset{0}{\tilde{\Phi}}(x, \xi)| > 0.$$

Für $p \neq 2$ gilt die notwendige Bedingung; die Umkehrung ist gültig, wenn

$$\tilde{A}_\infty, \ \overset{0}{\tilde{A}}_x \in \mathfrak{A} \ (x \in E_m) \ \text{mit (B1) und (B2) für } \tilde{A}_{x_0} \ (x \in E_m) \text{ gilt.}$$

<u>Beweis:</u> Zusammenfassung vorangegangener Aussagen.

<u>3.7 Beispiel :</u> Sei $p = 2$, $\Phi \in \sigma$ und $\overset{0}{\tilde{\Phi}}_x$ stückweise stetig (auf $\mathfrak{N}$) für alle $x \in \dot{E}_m$ in folgendem Sinne:

$$\overset{0}{\tilde{\Phi}}_x = \sum_{j=0}^{\infty} \overset{0}{\tilde{\Phi}}_{x,j} \chi_{K_j}$$

mit stetigem $\overset{0}{\tilde{\Phi}}_{x,j}$ und Konen K_j mit ebenen Rändern. Dann ist $A_x \in \mathcal{L}_0$ (vgl. 2.3), $A \in \mathfrak{A}_{(\cdot,\sim)}$ existiert mit Symbol Φ und es gelten die obigen Aussagen.

<u>3.8 Bemerkung :</u> Definiert man $\mathfrak{A}_{(\sim,\cdot)}$ analog durch Vertauschung der beiden Kompaktifizierungen $\dot{E}_m$ und $\tilde{E}_m$, so erhält man ähnliche Resultate für ganz andere Operatorklassen. Beispiele findet man bei SIMONENKO 1967 [13] und PRÖSSDORF 1972 [9]. Entsprechend kann man $\ddot{E}_m$ statt $\tilde{E}_m$ diskutieren mit CAUCHY-Operatoren mit verschobenen

Symbolen anstelle von CALDERON-MIKHLIN-Operatoren bzw. mit anderen Multiplikatoren.

4. Operatoren mit unstabilen Symbolen

Wir beschränken uns auf den Fall $p = 2$ und (ohne Einschränkung) $s = t = 0$. Es wird ein allgemeines Konzept dargestellt, ohne auf die Beschreibung der Symbolräume einzugehen. Deren Charakterisierung soll im Zusammenhang mit Kompaktheitskriterien für Faltungsoperatoren an anderer Stelle vorgenommen werden. Man vergleiche z. B. die Arbeit von BREUER & CORDES [1].

4.1 Definition : Sei $A \in \mathfrak{A}_{(\cdot)}$. Eine meßbare Funktion

$$\Phi \quad : \quad \dot{E}_m \times E_m \to \mathbb{C}$$

heiße "Präsymbol" von A, wenn A Einhüllender von $\{A_x\}$ und $\Phi_{A_x} = \Phi(x,\cdot)$ für alle $x \in \dot{E}_m$.

σ_1 sei die Menge aller Präsymbole aller $A \in \mathfrak{A}_{(\cdot)}$,

$$\sigma_0 \; := \{\Phi \in \sigma_1 \; : \; F^{-1}\, \Phi(x_0,\cdot) \cdot F \overset{x_0}{\sim} 0 \quad \text{für jedes} \quad x_0 \in \dot{E}_m\} .$$

Sei

$$\sigma_1\text{-inf}\, \Phi := \inf_{x \in \dot{E}_m} \operatorname*{ess\,inf}_{\xi \in E_m} \Phi(x, \xi) ,$$

$$\sigma\text{-inf}\, \Phi := \sup_{\Phi_0 \in \sigma_0} \sigma_1\text{-inf}\, (\Phi - \Phi_0).$$

4.2 Folgerung : Zu jedem $A \in \mathfrak{A}_{(\cdot)}$ ist ein Präsymbol modulo σ_0-Funktionen eindeutig bestimmt. σ_0 ist abgeschlossenes Ideal in σ_1.

4.3 Satz :

I : $\mathfrak{A}_{(\cdot)}/\mathcal{L}$ und σ_1/σ_0 sind isomorph.

II : Ist $A \in \mathfrak{A}_{(\cdot)}$ mit Präsymbol Φ und $\Phi + \sigma_0$ regulär in σ_1/σ_0, d. h. $\sigma\text{-inf}\, |\Phi| > 0$, so ist A Fredholm-Operator.

Beweis :

I : Die Homomorphie ist klar. Ist $A \in \mathcal{L}$, so ist $A \overset{x_o}{\sim} 0$ für alle $x_o \in \dot{E}_m$ und damit auch $F^{-1}\Phi(x_o,\cdot)\cdot F \overset{x_o}{\sim} 0$, d. h. $\Phi \in \sigma_0$. Dasselbe gilt umgekehrt, s. 2.2.

II : Aus der Regularität folgt: für alle $x_o \in \dot{E}_m$ existiert ein $B_{x_o} \in \mathcal{Y} \cap \mathcal{L}_o$ mit $B_{x_o} \overset{x_o}{\sim} 0$ und

$$\operatorname*{essinf}_{\xi \in E_m} |\Phi(x_o,\xi) - \Phi_{B_{x_o}}(\xi)| > 0 \text{ ; da } A \overset{x_o}{\sim} F^{-1}\Phi(x_o,\cdot)\cdot F - B_{x_o}$$

gilt, ist $A \in \mathcal{F}_{x_o}$ für alle $x_o \in \dot{E}_m$, also $A \in \mathcal{F}$.

Literatur

[1] Breuer, M. und Cordes, H. O.:
On Banach Algebras with σ-Symbol I.
Jour. Math. Mech. 13 (1964), 313 - 323.

[2] Calderón, A. P. und Zygmund, A.:
On singular integrals.
Amer. Jour. Math. 78 (1956), 289 - 300.

[3] Calderón, A. P. und Zygmund, A.:
Algebras of certain singular operators.
Amer. Jour. Math. 78 (1956), 310 - 320.

[4] Cordes, H. O.:
On compactness of commutators of multiplications and convolutions, and boundedness of pseudodifferential operators.
Jour. Funct. Anal. 18 (1975), 115 - 131.

[5] Herman, H.:
The symbol of the algebra of singular integral operators.
Jour. Math. Mech. 15 (1966), 147 - 155.

[6] Hörmander, L.:
Estimates for translation invariant operators in L^p spaces.
Acta Math. 104 (1960), 93 - 140

[7] Jörgens, K.:
Lineare Integraloperatoren.
Teubner, Stuttgart 1970.

[8] Mikhlin, S. G.:
Multidimensional singular integrals and integral equations.
Pergamon Press, Oxford 1965.

[9] Prössdorf, S.:
Über eine Algebra von Pseudodifferentialoperatoren im Halbraum.
Math. Nachr. 52 (1972), 113 - 149.

[10] Rabinovič, V. S.:
Pseudodifferential operators on a class of noncompact manifolds.
Mat . Sbornik 89 (131) (1972), No. 1.
Math. USSR Sbornik, Vol. 18 (1972), 45 - 59.

[11] Rakovščik, L. S.:
Zur Theorie der Integralgleichungen vom Faltungstyp.
Usp. Mat. Nauk 18 (1963), 171 - 178.

[12] Simonenko, I. B.:
A new general method of investigating linear operator equations of the type of singular integral equations.
Izv. Akad. Nauk SSSR, Ser. Mat. 29 (1965), 567 - 586.
Soviet Mat. Dokl. 5 (1964), 1323 - 1326.

[13] Simonenko, I. B.:
Operators of convolution type in cones.
Mat. Sbornik 74 (1967), 298 - 313.
Soviet Mat. Dokl. 8 (1967), 1320 - 1323.
Math. USSR-Sbornik 3 (1967), 279 - 293.

[14] Speck, F.-O.:
Über verallgemeinerte Faltungsoperatoren und eine Klasse von Integrodifferentialgleichungen.
Dissertation, TH Darmstadt, 1974.

INTEGRAL REPRESENTATIONS FOR LINEAR ANALYTIC ELLIPTIC SYSTEMS AND ITS APPLICATIONS

CHUNG-LING YU

I. INTRODUCTION

In this paper we shall study the integral representations for solutions $W(z) = (W_1(z), \ldots\ldots, W_n(z))$ of the elliptic systems of the following type:

$$(1) \qquad W_{\bar{z}} = A(z, \bar{z})W + B(z, \bar{z})\bar{W} + C(z, \bar{z})$$

in a simply connected domain G of the $z = x + iy$ plan, where $A(z, \zeta) = (a_{ij}(z, \zeta))$, $B(z, \zeta) = (b_{ij}(z, \zeta))$ are two $n \times n$ holomorphic matrices for $z, \zeta \in C$, and $C(z, \zeta) = (C_i(z, \zeta))$ is a $n \times 1$ holomorphic vector for $z, \zeta \in C$. We shall also study their applications to the boundary behavior of systems (1) and to the analytic continuation properties.

For the case $n = 1$, the integral representation of the solutions of the above system in G has been established by Vekua [8] and later extended to the boundary ∂G of G by Yu [9]. The boundary value problems in this case have been studied by many authors; the references can be found in Haack and Wendland [6], Vekua [7].

From now on, we will assume D to be a simply connected domain on the $z = x + iy$ plane whose boundary D is supposed to contain a segment σ, $\sigma = \{x \mid a < x < b\}$. We assume σ contains the origin as an interior point. We call $f^*(z_1, \ldots\ldots, z_n)$ the * conjugate function to a function $f(z_1, \ldots\ldots, z_n)$ and is defined according to the formula

$$f^*(z_1, \ldots\ldots, z_n) = \overline{f(\bar{z}_1, \ldots\ldots \bar{z}_n)}$$

Some other conventional notation will be used without specifically defined such as

$$|A(z, \zeta)| = \sum_{i,j=1}^{n} |a_{ij}(z, \zeta)|$$

II. INTEGRAL REPRESENTATIONS OF THE SOLUTIONS OF SYSTEMS

(1) $W_{\bar z} = A(z,\bar z)\, W + B(z, \bar z)\, \bar W + C(z, \bar z)$

We shall first establish the local representation of the solution $W(z)$, i.e. the representation of $W(z)$ in a neighborhood of any point z_0 belonging to G. We then study the analytic extension property of $W(z)$ and of its representation. Finally, by utilizing the above arguments, we are able to establish the global integral representation of $W(z)$ in the entire domain G, and its closure $G \cup \partial G$. The global representation of $W(z)$ in G obtained by Vekua for the case $n = 1$ was based upon the global fundamental solution; our approach here without reference to the fundamental solution is therefore different from Vekua. The main theorem of this section is the following:

Theorem 1

Let $W(z)$ be a solution of the elliptic system (1) in G and continuous in $G \cup \partial G$, then its analytic continuation $W(z,\zeta)$ is a holomorphic function of $(z,\zeta) \in G \times \bar G$ and has the representation

$$(2)\quad w(z) = \mathcal{E}(z,\bar z)\left\{ \phi(z) + \int_{z_0}^{z} K_1(z, \bar z, t, \bar z_0)\, \phi(t)\, dt + \int_{\bar z_0}^{\bar z} K_2(z, \bar z, z_0, t)\, \phi^*(t)\, dt + U_0(z, \bar z) \right\} = W(z,\bar z),$$

where $\mathcal{E}(z,\zeta)$ is an entire function with $\det \mathcal{E}(z,\zeta) \neq 0$,

$(z_0, \bar{z}_0)$ is a fixed point in $G \times \bar{G}$, $\phi(z)$ is a holomorphic function in G and continuous in $G \cup \partial G$, $U_0(z, \zeta)$, $K_1(z, \zeta, t, s)$, $K_2(z, \zeta, t, s)$ are three holomorphic functions of the four variables $z, \zeta, t, s \in C^4$. Furthermore, $\mathcal{E}, U_0, K_1, K_2$ depend upon the coefficient matrices $A(z, \zeta)$, $B(z, \zeta)$, $C(z, \zeta)$ only, and can be constructed explicitly. Conversely, the function W(z) which is given by the formula (2) is a solution of the system (1) in G, and is continuous in $G \cup \partial G$.

The proof will be based upon the following lemmas.

Lemma 1 Let z_0 be any point in G, then W(z) has the representation (2) in a neighborhood G_0 of z_0 where the function $\phi(z)$ is only known to be holomorphic in G_0, the functions U_0, K_1, K_2 have the same properties as stated in Theorem 1.

Proof It is well known that the solution $W(z) = W(x, y)$ are analytic functions of $(x, y) \in G$. By continuing x, y into complex field, we obtain a holomorphic function $W(z, \zeta) = w\left(\frac{z+\zeta}{2}, \frac{z-\zeta}{2i}\right)$ of two complex variables $z = x + iy$, $\zeta = x - iy$, and $W(z, \bar{z}) = w(x, y)$ when x, y assume real values. The function $W(z, \zeta)$ satisfies the following equations

$$(3) \quad W_{\zeta}(z, \zeta) = A(z, \zeta)\, W(z, \zeta) + B(z, \zeta)\, W^*(\zeta, z)$$

These equations is equivalent to the system of n integral equations

$$(4) \quad W(z, \zeta) = \phi(z) + \int_{\bar{z}_0}^{\zeta} \left\{ A(z, t)\, W(z, t) + B(z, t)\, W^*(t, z) \right\} dt$$

Introducing its conjugate integral equations, we obtain

$$(5)\quad W^*(\zeta, z) = \phi^*(\zeta) + \int_{z_0}^{z} \left\{ A^*(\zeta, t)\, W^*(\zeta, t) + B^*(\zeta, t)\, W(t,\zeta) \right\} dt$$

The above two systems (4) and (5) can be considered as a system of 2n Volterra integral equations with 2n unknown functions $W_1(z,\zeta),\dots W_n(z,\zeta)$, $W_1^*(\zeta,z),\dots\dots, W_n^*(\zeta, z)$. Solving this system by method of successive approximation, we get the desired result.

The next lemma is concerned with the analytic extension property of the local representation of the solution W(z).

Lemma 2 Let D_1, D_2 be two simply connected domains such that $D_1 \cap D_2 \neq 0$, and (without loss of generality) $z = 0$ and a portion of real axis $\sigma \in D_1 \cap D_2$, $D_1 \cup D_2$ symmetric with respect to x axis. Assuming that $w(z)$ has integral representations (We may assume that $\mathcal{E}^i = I$)

$$(6)\quad w(z) = W^i(z, \bar{z}) = \phi^i(z) + \int_{z_i}^{z} K_1(z, \bar{z}, t, \bar{z}_i)\, \phi^i(t)$$

$$+ \int_{\bar{z}_i}^{\bar{z}} K_2(z, \bar{z}, z_i, t)\, \phi^{i*}(t)\, dt + U_0^i(z, \bar{z})$$

in the domains D_i, $i = 1, 2$, respectively. Then $\phi^i(z)$, $W^i(z,\zeta)$ can be analytic continued for z, ζ into entire domain $D_1 \cup D_2$. Further $W^1(z,\zeta) = W^2(z,\zeta)$ for $z,\zeta \in D_1 \cup D_2$.

<u>Proof</u> The functions W^1 and W^2 agree the same function $g(x)$ on σ, i.e.

$$(7) \qquad W^1(x, x) = W^2(x, x) = w(x) = g(x)$$

then $g(z)$ is a holomorphic function for $z \in D_1 \cap D_2$, and

$$(8) \qquad g(z) = \phi^i(z) + \int_{z^i}^{z} K_1^i(z, z, t, \bar{z}_i)\, \phi^i(t)\, dt + \int_{\bar{z}_i}^{\bar{z}} K_2^i(z, z, z_i, t)\phi^{i*}(t)\,dt$$

$$+ U_\circ^i(z, z)$$

$$= \phi^i(z) + \int_0^{z} K_z^i(z, z, t, \bar{z}_i)\, \phi^i(t)\, dt + \int_0^{\bar{z}} K_2^i(z, z, z_i, t)\phi^{i*}(t)\,dt$$

$$+ V_\circ^i(z, z)$$

where,

$$(9) \quad V_\circ^i(z,\zeta) = \int_{z_i}^{0} K_1^i(z,\zeta, t, \bar{z}_i)\, \phi^i(t)\, dt + \int_{\bar{z}_i}^{0} K_2^i(z,\zeta, z_i, t)\phi^{i*}(t)\, dt$$

$$+ U_\circ^i(z, \zeta)$$

Recall that $V_\circ^i(z, z)$, $i = 1,2$, are entire functions.

The above relations suggest the following integral equations:

$$(10) \quad \phi^1(z) + \int_0^z K_1^1(z, z, t, \bar{z}_1)\, \phi^1(t)\, dt - \int_0^z K_2^2(z, z, z_2, t)\, \phi^{2*}(t)\, dt = a(z)$$

$$(11) \quad \phi^{2*}(z) + \int_0^z K_1^{2*}(z, z, t, z_2)\, \phi^{2*}(t)\, dt - \int_0^z K_2^{1*}(z, z, \bar{z}_1, t)\, \phi^1(t)\, dt = b(z)$$

where $a(z)$, $b(z)$ are holomorphic functions for $z \in D_2$. Therefore the integral equations (10) and (11) have the unique solution ($\phi^1(z)$, $\phi^{2*}(z)$) which is holomorphic in D_2 and coincides with the given ($\phi^1(z)$, $\phi^{2*}(z)$) on $D_1 \cap D_2$. It therefore follows $W^1(z,\zeta) \equiv W^2(z,\zeta)$ for $z, \zeta \in D_1 \cup D_2$, and our lemma is proven.

Proof of Theorem 1

To any point $(x_i, y_i) \in D$, there exists a disc D_i with center (x_i, y_i) such that the analytic continuation $W^i(z,\zeta)$ of $w(z)$ is defined there. Since any bounded domain $D \subset G$ can be covered by finite number of such disc, we only need apply lemma 2 finite number of times and obtain the representation (2) in a neighborhood of D. Since we can choose D as close to G as possible, we get the representation for $w(z)$ in G.

We now extend this representation to the boundary of G, it will no loss of generality, we assume that G is a unit disc. Let z_1, z_2 be any two points in G. Also let $\overline{z_\circ z_1}$, $\overline{z_1 z_2}$ denote segments from $z_\circ$ to z_1 and z_1 to z_2 respectively. Let (we assume that $\mathcal{E} = I$)

$$(12) \qquad k(z) = w(z) - U_\circ(z, \bar{z})$$

therefore, (2) becomes

$$(13)\ k(z) = \phi(z) + \int_{z_\circ}^{z} K_1(z, \bar{z}, t, \bar{z}_\circ)\, \phi(t)\, dt + \int_{\bar{z}_\circ}^{\bar{z}} K_2(z, \bar{z}, z_\circ, t)\, \phi^*(t)\, dt$$

$$\text{for } z \in \overline{z_1 z_2} \cup \overline{z_\circ z_1}$$

In view of the method of successive approximation, we have

$$(14)\ \phi(z) = k(z) + \int_{z_\circ}^{z} H_1(z, t)\, k(t)\, dt + \int_{\bar{z}_\circ}^{\bar{z}} H_2(z, t)\, k^*(t)\, dt$$

for $z \in \overline{z_1 z_2} \cup \overline{z_\circ z_1}$. Furthermore, $|H_1|$, $|H_2| < M$ for $z, t \in \overline{z_1 z_2}$ where M is independent of k, z_1, z_2.

Since

$$(15)\ \phi(z_2) - \phi(z_1) = k(z_2) - k(z_1) + \int_{z_1}^{z_2} H_1(z,t)\, k(t)\, dt$$

$$+ \int_{\bar{z}_1}^{\bar{z}_2} H_2(z, t)\, k^*(t)\, dt$$

We have

$$(16)\quad |\phi(z_2) - \phi(z_1)| \leqslant |k(z_2) - k(z_1)| + 2 M_1 \, p \, |z_2 - z_1|$$

where $p = \max_{z \in G} |k(z)|$

Since k(z) is uniform continuous in G, by (13), $\phi(z)$ is uniform continuous there; therefore, $\phi(z)$ has continuous extension to $G \cup \partial G$ (See Graves [4] p. 117).

<u>Corollary 1</u> (Maximal Principle). Let w(z) be a solution of the homogeneous system (1) in G. If $|w(z)| < k$ on the boundary ∂G of G, then there exists a constant N which is independent of w such that

$$|w(z)| \leqslant NK \quad ; \quad z \in G \cup \partial G$$

<u>Proof</u> Choosing z_0 on ∂G, and solving the equation (2) for $\phi(z)$ on ∂G, we then obtain the estimate of the bound for $|\phi(z)|$ on $G \cup \partial G$ in terms of K. The desired inequality then follows from (2).

<u>Remark 1</u> We remark that the Monodromy theorem for the solutions of (1) is an easy consequence of Lemma 2.

III. UNIQUE CONTINATUION THEOREMS

Theorem 3

Let $w(z) = (w_1(z), \ldots\ldots, w_n(z))$ be a solution of the homogeneous system (1) in G and assume that w does not vanish identically. Then the zeros of w are isolated, and at least one of the functions $\{w_i(z)\}$ have the property that the zeros are isolated and their orders of zeros are positive integers.

For the case n = 1, Carleman has established the above theorem for the system (1) with Hölder continuous coefficients. Bers and Nirenberg [2], Vekua [8] have extended this result to the case of bounded measurable coefficients. Recently, Gilbert and Wendland [5] have proved this theorem for some large elliptic system with analytic coefficients in the plane. Our method here can be considered as the continuation of the one given by Vekua [8]. We further remark that Theorem 3 also holds even for the case of bounded measurable coefficients (my forthcoming paper).

Proof of Theorem 3

Without loss of generality, we can assume that w(o) = 0. The function w(z) has the representation

$$(17)\quad w_m(z) = \phi_m(z) + \sum_{j=1}^{n} \left\{ \int_o^z K_1^{mj}(z, \bar{z}, t, o)\, \phi_j(t)\, dt + \int_0^{\bar{z}} K_2^{mj}(z, \bar{z}, o, t)\, \phi_j^*(t)\, dt \right\},\ m = 1, \ldots, n$$

(We may assume that $\mathcal{E} = I$).

It is clear that $\phi_m(z) = z^{k_m} f_m(z)$, where k_m is a positive integer and $f_m(z)$ is a holomorphic function in G with $f_m(o) \neq 0$. Following Vekua [8], we can write (17) as

$$(18)\quad w_m(z) = r^{k_m} e^{ik_m\theta} + \sum_{j=1}^{n} r^{k_j+1} \left\{ e^{i(k_j+1)\theta} \int_0^1 G_1^{mj}(z,zt)t^{k_j} \phi_j(zt)\,dt \right.$$

$$\left. + e^{-i(k_j+1)\theta} \int_0^1 G_2^{mj}(z,\bar{z}t)\,t^{k_j}\, \overset{*}{\phi}_j(\bar{z}t)\,dt \right\}$$

Suppose $k_1 = \min \{ k_m ; \quad m = 1, \ldots\ldots, n \}$, then

$$(19)\quad w_1(z) = r^{k_1} U_1(z)$$

It is clear that $U_1(o) \neq 0$ and nonvanishing in a neighborhood of 0. This shows that $z = 0$ is an isolated zero of $w_1(z)$ and has order k_1. This completes the proof.

Corollary 2

(The extreme value theorem) Let $U(z) = (U_1, \ldots\ldots, U_n)$ be a solution of analytic system

$$\Delta U + a U_x + bU_y = 0$$

in G, where a, b are n x n analytic matrices in G, then the local extreme values for functions $U_1(z), \ldots, U_n(z)$ will not occur at a same interior point z_0 of G unless U reduces to a constant vector $(C_1, \ldots\ldots, C_n)$.

Proof Assuming that $U_1(z_0)$, ... $U_n(z_0)$ are local extreme values for the functions U_i, respectively, and $z_0 \in G$. The function $w(z) = U_x - i\,U_y$ is a solution of the system (1). Hence, unless $U = (C_1, \ldots, C_n)$, z_0 is an isolated zero for a function $w_i(z) = U_{ix} - i\,U_{iy}$. Since (by Theorem 3) the order of zero for $w_i(z)$ at z_0 is positive, its index at z_0 is again positive, but since $U_i(z_0)$ is an extreme value, by a simple geometric argument, the index of w at z_0 is -1. This contradication proves our asseration.

The above geometric interpretation was observed in the book by Bers, John and Schechter [1].

By means of the representation (2) many analytic continuation theorems for the case n = 1, Yu [8], may be extended to the case $n > 1$. For instance, we have

Theorem 4

Let $A_i(z,\zeta)$, $B_i(z,\zeta)$, $C_i(z,\zeta)$, i = 1,2, be holomorphic functions for $z, \zeta \in D \cup \sigma \cup \bar{D}$. Consider the equations

$$(20)_i \qquad w_{i\bar{z}} = A_i(z, \bar{z})\, w_i + B(z, \bar{z})\, \bar{w}_i + C_i(z, \bar{z}), \quad i = 1,2$$

If $w_1(z)$ is a solution of $(20)_1$ in D, continuous in $D \cup \sigma$, $w_2(z)$ is a solution of $(20)_2$ in $\bar{D}$, continuous in $\bar{D} \cup \sigma$, such that $w_1(x) = w_2(x) = g(x)$ on σ, then $g(x)$ is analytic on σ, and its analytic continuation $g(z)$ is holomorphic in $D \cup \sigma \cup \bar{D}$. Furthermore, $w_1(z)$ and $w_2(z)$ can be continued analytically into whole $D \cup \sigma \cup \bar{D}$ as solutions of $(20)_1$ and $(20)_2$ respectively.

The proof is very similar to the proof of lemma 2.

Theorem 5

Let $w(z)$ be a solution of (1) in D. If $\lim_{y \to 0} w(z) = 0$ for $x \in \sigma$, $(x,y) \in \Lambda_x$, where Λ_x is an arc with endpoint (x,o), then $w(z) \equiv 0$.

Proof By the same argument in the proof of Theorem 1, we can show that the functions $\phi_i(z)$ in the representation (2) has the limit $\lim_{y \to 0} \phi_i(z)$ for $x \in \sigma$. Therefore the integrals in the representation (2) is defined on the paths Λ_x from (x, o) to (x, y) for all $x \in \sigma$, and

$$(21) \quad \phi(z) + \int_{z_o}^{z} K_1(z, \bar{z}, t, \bar{z}_o)\, \phi(t)\, dt + \int_{\bar{z}_o}^{\bar{z}} K_2(z, \bar{z}, z_o, \tau)\, \phi^*(\tau)\, d\tau$$

$$+ U_o(z, \bar{z}) \longrightarrow 0 \quad \text{as } y \to 0 \quad \text{for } x \in \sigma$$

Since $K_1(z,\zeta, t, \bar{z}_o)$ and $K_2(z, \zeta, z_o, \tau)$ are holomorphic functions for $z, \zeta \in C^2$, therefore

$$(22) \quad \lim_{y \to 0} \int_{z_o}^{z} K_1(z, z, t, \bar{z}_o)\, \phi(t)\, dt = \lim_{y \to 0} \int_{z_o}^{z} K_1(z, \bar{z}, z_o, t)\, \phi(t)\, dt$$

$$\lim_{y \to 0} \int_{\bar{z}_o}^{\bar{z}} K_2(\bar{z}, \bar{z}, z_o, \tau)\, \phi^*(\tau)\, d\tau = \lim_{y \to 0} \int_{\bar{z}_o}^{\bar{z}} K_2(z, \bar{z}, z_o, \tau)\, \phi^*(\tau)\, d\tau$$

Set

(23) $$f(z) = \emptyset(z) + \int_{z_\circ}^{z} K_1(z, z, t, \bar{z}_\circ)\, \emptyset(t)\, dt$$

(24) $$g(z) = -\int_{\bar{z}_\circ}^{\bar{z}} K_2(\bar{z}, \bar{z}, z_\circ, \tau)\, \emptyset^*(\tau)\, d\tau - U_\circ(\bar{z}, \bar{z})$$

It is clear

$$\lim_{y \to 0} f(z) = \lim_{y \to 0} g(\bar{z})$$

By a theorem of analytic continuation (See Bagemihl [3]), f(z) is holomorphic in $D \cup \sigma \cup \bar{D}$. Solving (23) we get $\emptyset(z)$ is holomorphis in $D \cup \sigma \cup \bar{D}$.

Q.E.D.

Theorem 6

Let w(z) be a solution of (1) in $|z| < 1$, and

(25) $$\lim_{r \to 1} w(re^{i\theta}) = 0$$

for $a < \theta < b$, then $w(z) \equiv 0$.

Proof By conformal transform and Theorem 5, we get the desired result.

R E F E R E N C E S

1. Bers, L., John, F., and Schechter, M.
Partial Differential Equations, Interscience, New York, 1964.

2. Bers, L. and Nirenberg, L.
On a representation theorem for linear elliptic systems with discontinuous coefficients and its applications, Atti del Convegno internazionale sulle Equazioni alle derivate parzialli, Trieste, 1954, pp. 111 - 140.

3. Bagemihl, F.
Analytic continuation and the Schwarz reflection principle. Proc. N.A.S. Vol. 51, 1964 pp. 378 - 380.

4. Graves, L.
The theory of functions of real variables, McGraw-Hill, New York, 1965.

5. Gilbert, R. P. and Wendland, W.
Analytic, generalized, hyperanalytic function theory and an application to elasticity, Proc. A Royal Soc. Edinburgh 73, 1975, 317-331.

6. Haack, W. and Wendland, W.
Lectures on partial and pfaffian differential equations. Pergamon, Oxford, 1972.

7. Vekua, I.N.
Generalized analytic functions, Pergamon, London 1962.

8. Vekua, I.N.
New methods for solving elliptic equations. John Wiley, New York, 1967.

9. Yu, C.L.
Cauchy problem and analytic continuation for systems of first order elliptic equations with analytic coefficients. Trans. Amer. Math. Soc. 185. 1973, 429-443.

Engineering Faculty, Benghazi University.

Über eine Klasse von Problemen mit freiem (unbekanntem) Rand für elliptische Gleichungen

I. I. DANILJUK

(Donezk)

1. Einleitung: Eine Reihe nichtlinearer Probleme der Mathematischen Physik enthält den unbekannten Rand - oder einen Teil davon - des betrachteten Bereichs. Dazu gehören die meisten Probleme der theoretischen Hydrodynamik: Theorie der Gleichgewichtsfiguren gravitierender flüssiger Medien; freistrahlende, kavitierende oder wellenförmige Strömungen flüssiger Kontinua etc.. Ein weiterer derartiger Problemkreis kommt in der Theorie der thermodynamischen Prozesse vor, die den Übergang eines Stoffes von einem Aggregatzustand in einen anderen (Stefan-Problem) beschreiben: Als unbekannt erscheint dabei die Fläche, die die verschiedenen Phasen der wärmeleitenden Medien trennt. Die hauptsächliche mathematische Schwierigkeit bei der Untersuchung solcher Probleme besteht darin, daß ihre Randbedingungen einerseits gesuchte hydrodynamische oder thermodynamische Charakteristika der Medien in nichtlinearer Form enthalten, aber andererseits auf zunächst unbekanntem Rand erfüllt werden müssen. In allgemeiner Problemstellung sind diese Schwierigkeiten - trotz der großen Bedeutung des Problems für die Theorie und für die Anwendungen - bis heute noch nicht überwunden.

Wir werden weiter unten diese Probleme in einem spezifischen (statischen oder quasistationären) Fall betrachten, wenn ihre Lösungen in einem beweglichen Koordinatensystem, das mit dem unbekannten Rand verbunden ist, nicht von der Zeit abhängen. Es zeigt sich dabei, daß solche Probleme vom Variationstyp sind: das unbekannte Paar, die Funktion und das Gebiet, ist die Lösung des nichtlinearen Randwertproblems genau dann, wenn es "kritischer Punkt" eines bestimmten Funktionals mit veränderlichem Integrationsgebiet ist. In der hydrodynamischen Interpretation entspricht dieses Funktional - bis auf einen konstanten Faktor - der totalen Wellenenergie. Ein solches Integral aus der Theorie der freien Randwertprobleme wurde schon bei klassischen Untersuchungen von Oberflächen schwerer rotierender Flüssigkeiten benutzt. Den ebenen Kavitationsströmungen entsprechend beruht das Variationsprinzip von Riabouchinskij (s. z. B. [1] , Kap. 4, § § 10, 11) auf analogen Überlegungen. In der Theorie der Strahlströmungen - sowohl der ebenen als auch der achsensymmetrischen - wurde dieses Prinzip von K.O. Friedrichs [2] zum Beweis der Eindeutigkeitssätze benutzt. In den Jahren nach dem Krieg lösten Garabedian, Spencer, Lewy und Schiffer mit Hilfe der Variationsmethode Existenz- und Eindeutigkeitsprobleme von achsensymmetrischen Kavitationsströmungen (siehe [2], [4]). Später wurde in mehreren Arbeiten des Autors (siehe [5] - [10], eine vollständige Darstellung enthält [9]) die Variationsmethode auf die Theorie der Wellenströmung in einem beliebigen Potentialfeld der äußeren Kräfte angewendet. Auf thermodynamische Probleme wurde die Methode in den Arbeiten [11] - [15] angewendet.

Die Methode der Integralfunktionale mit variablem Integrationsgebiet gestattet, Existenzprobleme zu lösen und die Voraussetzungen zu finden, unter denen die Probleme eindeutig lösbar sind. Darüber hinaus liefert dieses Vorgehen als Ergebnis ein Problem in der Variationsrechnung, dessen Analyse gestattet, einige Sätze über globale Eigenschaften der Menge aller möglichen Lösungen des betrachteten nichtlinearen Problems aufzufinden. Schließlich eröffnet diese Methode die Gewinnung effektiver Verfahren zur approximativen und numerischen Lösung.

2. Hydrodynamische Probleme und ihre Formulierung in der Variationsrechnung

Sei Γ eine einfache hinreichend glatte Kurve in der Ebene der Veränderlichen $z = x + iy$. Wir führen auf Γ eine Orientierung ein, derart, daß das von Γ begrenzte endliche einfach zusammenhängende Gebiet G zur Linken liegt. Wir bezeichnen mit G_γ das zweifach zusammenhängende Gebiet zur Linken von Γ, das als zweite Randkurve eine hinreichend glatte Kurve γ besitzt, die eventuell auf einer nicht einblätterigen Riemannschen Fläche liegt. Die Kurve γ heißt freier Rand und ist im voraus unbekannt. Bei ihrer Bestimmung müssen die folgenden Bedingungen erfüllt werden: 1) im Gebiet G_γ muß eine Lösung $\Psi(x,y)$ der Laplace'schen Gleichung $\Psi_{xx} + \Psi_{yy} = 0$ existieren, die auf der Menge $G_\gamma \cup \Gamma \cup \gamma$ stetig ist; 2) das Potential $\Psi(x,y)$ muß die Eigenschaften der "Stromfunktion": $\Psi = 0$ auf Γ und $\Psi = 1$ auf γ besitzen; 3) die ersten Ableitungen Ψ_x, Ψ_y müssen auf der Menge $G_\gamma \cup \gamma$ existieren und die Bedingung

$$|\operatorname{grad} \Psi(x,y)| = Q(x,y) \quad , \quad (x,y) \in \gamma \tag{1}$$

erfüllen, wobei $Q(x,y)$ eine vorgegebene positive Funktion in der z-Ebene ist. Die Beziehung (1) ist die "verallgemeinerte Bernoulli-Gleichung". Die Funktion $Q(x,y)$ kann man als Potential des äußeren Kraftfeldes interpretieren. Der freie Rand beschreibt das Wellenprofil.

Dieses formulierte Problem wurde das erste Mal betrachtet im Artikel [16] von A. Beurling. Für konkrete Funktionen $Q(x,y)$ gestattet das Problem eine physikalische Interpretation. Wir betrachten zum Beispiel die Theorie der eingeschwungenen (permanenten) periodischen Wellen einer schweren inkompressiblen idealen Flüssigkeit endlicher Tiefe über ebenem Grunde im Schwerkraftfeld der Erde. Wir beschreiben die pyhsikalische Ebene durch die Veränderlichen ξ, η in einem Koordiantensystem, das mit dem Wellenprofil fest verbunden ist. Wir bezeichnen mit $\ell \in (0,\infty)$ die Wellenperiode und mit c_1 ihren "Durchsatz". Dann betrachten wir die Abbildung $z = \exp\{2\pi i\zeta / \ell\}$, $\zeta = \xi + i\eta$, bei der die Strecke des Bodens $\eta = 0$, $0 \leq \xi \leq \ell$ dem Einheitskreis $\Gamma : |z| = 1$ und das Wellenprofil einer Kurve von der Gestalt γ entsprechen. Die Eulerschen Gleichungen, die eine stationäre Bewegung ohne Quellen und Wirbel beschreiben, führen zum sogenannten Bernoulli-Integral, das eine Bedingung von der Gestalt (1) mit der Funktion

$$Q(x,y) = \frac{\ell}{2\pi c_1 r}\sqrt{c + \frac{g\ell}{\pi}\ln r} = \frac{\ell}{2\pi c_1 r}\sqrt{U^2 + \frac{g\ell}{\pi}\ln \frac{r}{a}}, \quad r^2 = x^2 + y^2 \tag{2}$$

erzeugt, wobei c die Bernoulli'sche Konstante ist; U ist der absolute Betrag der Strömungsgeschwindigkeit im fixierten Punkt $(0,\eta_0)$ auf dem Wellenprofil, $0 < \eta_0 < \infty$; $a = \exp\{-2\pi\eta_0 / \ell\}$.

Wir betrachten jetzt das Funktional

$$J(\Psi,\gamma) = \iint_{G_\gamma} \{\Psi_x^2 + \Psi_y^2 + Q^2(x,y)\}\, dx\, dy , \tag{3}$$

das von dem Paar (Ψ,γ) abhängt. Wir werden annehmen, daß das Gebiet G_γ veränderlich ist und die oben beschriebene Gebietsklasse durchläuft, während $\Psi(x,y)$ im Gebiet G_γ definiert, hinreichend glatt ist und die Bedingung 2) erfüllt. Die erste Variation des Funktionals (3) (siehe zum Beispiel [17], Kapitel IV, § 11) in dem von uns betrachteten Fall kann dargestellt werden in der Gestalt

$$\delta J(\Psi,\gamma\ ;\ \overline{\delta\Psi},\ \overrightarrow{\delta z}) = -2 \iint_{G_\gamma} \overline{\delta\Psi}\{\Psi_{xx} + \Psi_{yy}\}\, dx\, dy \tag{4}$$

$$+ \int_\gamma \{Q^2(x,y) - |\mathrm{grad}\ \Psi|^2\}\vec{n}\cdot\overrightarrow{\delta z}\, ds\ ,$$

wobei $\overline{\delta\Psi}$ die Variation der Funktion $\Psi(x,y)$ bei fixiertem Gebiet G_γ ist ; $\overrightarrow{\delta z} = (\delta x, \delta y)$ ist das Paar von solchen Funktionen, die den Übergang von G_γ zu einem hinreichend benachbarten Gebiet beschreiben; $\vec{n}$ ist die äußere Normale an $\Gamma \cup \gamma$. Auf der Formel (4) fußt der Beweis der folgenden Behauptung (siehe [5] , [9] , § 1) :

Lemma 1. Dafür, daß das Paar (Ψ, γ), das die klassischen Differenzierbarkeitseigenschaften besitzt – γ habe keine mehrfachen Punkte – , Lösung des nichtlinearen hydrodynamischen Problems ist, ist notwendig und hinreichend, daß dieses Paar kritisch ist für das Funktional (3) auf der beschriebenen Menge der Paare. Dabei ist die verallgemeinerte Bernoulli-Bedingung (1) die natürliche Randbedingung.

Eine analoge Behauptung gilt auch für den Fall kompressibler Medien (siehe [9] , § 1, Punkt 1.4); ebenfalls für das Problem der Strömung zweier nichtvermischter Flüssigkeiten [18].

3. Das thermodynamische Problem und seine Formulierung in der Variationsrechnung

Im physikalischen Raum der Veränderlichen (x,y,z) betrachten wir das zylindrische Gebiet $\mathcal{D}$, dessen Erzeugende parallel zur z-Achse sind. Wir nehmen an, daß $\mathcal{D}$ von einem Stoff eingenommen wird, der sich in zwei Aggregatzuständen befinden kann, im flüssigen (im oberen Teil) und im festen (im unteren Teil). Wir bezeichnen mit S eine im voraus unbekannte Fläche, die diese zwei Phasen trennt und mit $\mathcal{D}_s$ die untere (feste) Phase. Der Einfachheit halber nehmen wir an, daß die thermodynamischen Parameter beider Phasen nicht von der Temperatur abhängen. Auf der Oberfläche S findet ein Wärmeübergang gemäß der Bedingung (in dimensionsloser Form)

$$\frac{\partial u}{\partial u} + \omega_o u = 0\ ,\quad \omega_o > 0\ , \text{ auf } \partial\mathcal{D}_s \setminus S \tag{5}$$

statt, wobei n die äußere Normale ist. Außerdem gilt

(6) $u = 0$ für $z = -\infty$, $u = 1$ auf S .

Wir werden den quasistationären Fall betrachten, wenn gemäß Definition der ganze Vorgang in einem sich mit konstanter Geschwindigkeit von unten nach oben bewegenden Koordinatensystem nicht von der Zeit abhängt. Dann erfüllt die unbekannte Funktion u die elliptische Gleichung

$$(7) \qquad \frac{\partial}{\partial x}\left(e^{\omega z}\frac{\partial u}{\partial x}\right) + \frac{\partial}{\partial y}\left(e^{\omega z}\frac{\partial u}{\partial y}\right) + \frac{\partial}{\partial z}\left(e^{\omega z}\frac{\partial u}{\partial z}\right) = 0, \quad \omega > 0 .$$

Die thermodynamische STEFAN-Bedingung kann man in diesem Fall in der Gestalt

$$(8) \qquad |\mathrm{grad}\ u|_-^2 = \kappa^2 |\mathrm{grad}\ u|_+^2 + \lambda_1 \frac{\partial u_-}{\partial z} + \lambda_2 \frac{\partial u_+}{\partial z} \equiv Q^2 \quad \text{auf} \quad S$$

schreiben, wobei κ, λ_1, λ_2 positive Parameter sind; das Minuszeichen bezieht sich auf die feste, das Pluszeichen auf die flüssige Phase. Wir beschränken uns auf das vereinfachte Einphasenproblem von STEFAN im quasistationären Fall, wenn die Funktion Q auf der rechten Seite von (8) als bekannt und als im gesamten Gebiet $\mathcal{D}$ definiert angenommen wird. Dann fügen wir zu den Bedingungen (5) und (6) die Gleichung

$$(9) \qquad |\mathrm{grad}\ u|_- = Q(x,y,z) \quad \text{auf} \quad S$$

hinzu. Man kann ebenfalls das vereinfachte Zweiphasenproblem von STEFAN im quasistationären Fall betrachten, wenn man als bekannt nur die Summe zweier aufeinanderfolgender Summanden auf der rechten Seite von (8) annimmt, während die Temperatur $u_+(x,y,z)$ der flüssigen Phase ebenso wie u_- gesucht ist. In diesem letzten Fall kann man das Gebiet $\mathcal{D}$ als halbunendlich nach unten annehmen, während auf seinem oberen Querschnitt entweder der Wert $\nu \equiv u_+$ oder der Wärmestrom gegeben sind. Die Bedingung (9) ist analog zur Bedingung (1).

Die vereinfachten quasistationären STEFAN-Probleme sind "selbstadjungiert" und gestatten eine Formulierung in der Variationsrechnung. Wir betrachten das Funktional

$$(10) \qquad J(u,S) = \iiint\limits_{\mathcal{D}_s} e^{\omega z}\{u_x^2+u_y^2+u_z^2+Q^2\}\,dx\,dy\,dz + \omega_0 \iint\limits_{\partial\mathcal{D}_s \setminus S} e^{\omega z}u^2\,dS ,$$

das von dem veränderlichen Integrationsgebiet $\mathcal{D}_s$ und der in ihm definierten Funktion u abhängt. Wir nehmen an, daß S eine hinreichend glatte zweidimensionale Mannigfaltigkeit ist, deren Rand auf $\partial\mathcal{D}$ liegt und durch einen festen Punkt auf $\partial\mathcal{D}$ geht. Die Funktion u sei auf $\mathcal{D}_s$ definiert, hinreichend glatt und erfülle die Bedingungen (6).

In vollständiger Analogie zu Lemma 1 gilt die Behauptung, daß das Randwertproblem (5), (6), (7) und (9) äquivalent ist zu dem Variationsproblem für das Funktional (10), wobei die Bedingungen (5) und (9) natürlich sind (siehe [11]). Diese Tatsache läßt sich auch auf das Zweiphasenproblem (siehe [19]) verallgemeinern.

4. Eindeutigkeitsprobleme

Zum Beispiel beschränken wir uns auf das thermodynamische Problem und betrachten dabei den achsensymmetrischen Fall. Wir setzen voraus, daß die auf $\mathcal{D}_s \cup \partial\mathcal{D}_s$ stetige Funktion $u(r,z)$ (im Zylinderkoordinatensystem) nicht nur den Randbedingungen (6) genügt, sondern stetige erste Ableitungen besitzt, wobei

$$u_z(r,z) > 0 \quad \text{auf} \quad \mathcal{D}_s \cup \partial\mathcal{D}_s . \tag{11}$$

Dann erzeugt die Zuordnung $(r,z) \mapsto (r,u)$ eine eineindeutige Abbildung von $\mathcal{D}_s \cup \partial\mathcal{D}_s$ auf die Menge $\Delta \cup \partial\Delta$, wobei Δ das Quadrat $0 < r,\ u < 1$ ist. Die Linien u-const. sind <u>Isothermen</u>, während die Kristallisationsfront S der Isotherme $u = 1$ entspricht und man kann S in expliziter Form darstellen

$$S : z = z(r,1) \;,\; 0 \le r \le 1 \;;\; u[r, z(u,r)] \equiv u . \tag{12}$$

Man überzeugt sich leicht davon, daß auch die Umkehrung gilt: falls die Kristallisationsfront S die explizite Darstellung (12) gestattet, erfüllt das Temperaturfeld $u(r,z)$ die Bedingung (11).

Das Funktional (10) nimmt im achsensymmetrischen Fall für die neue unbekannte Funktion $w = \exp \omega z(r,u)$ die Gestalt

$$\mathfrak{F}(w) = \frac{1}{\omega} \iint_\Delta \{w_r^2 + \omega^2 w^2 + w_u^2 Q^2(r, \tfrac{1}{\omega}\ln w)\} \frac{r\,dr\,du}{w_u} + \frac{\omega_0}{\omega}\int_0^1 u^2 w_u\,du \tag{13}$$

an, während jedoch die Randbedingungen sich in der folgenden Weise formulieren lassen: 1. $w(r,u)$ ist definiert und nichtnegativ auf $\Delta \cup \partial\Delta$ und nimmt den Wert null nur auf der Basis $u = 0,\ 0 \leq r \leq 1$ an; 2. $w_u(r,u) > 0$ auf $\Delta \cup \partial\Delta$; 3. $w(1,1) = 1$ (Diese Bedingung bedeutet, daß S durch den vorgegebenen Punkt $(r,z)=(1,0)$ geht. Die Menge aller zulässigen Funktionen $w(r,u)$ bildet eine konvexe Menge Ω. Sei $w^0,\ w^1 \in \Omega,\ \delta w = w^1 - w^0$. Dann gehört die einparametrige Familie $w^\varepsilon = w^0 + \varepsilon\delta w,\ 0 \leq \varepsilon \leq 1$, zu Ω. Das Funktional (13) - im Unterschied zu (10)- besitzt einen festen Integrationsbereich. Durch unmittelbares Nachprüfen zeigt sich, daß, falls

(14) $\qquad Q_z(r,z) \geq 0\ ,\quad 0 \leq r \leq 1\ ,\quad -\infty < z < +\infty\ ,$

die zweite Ableitung $d^2J(w^\varepsilon)/d\varepsilon^2$ positiv definit bezüglich δw ist. Falls man die Methodik aus [2] benutzt, erhält man hieraus die Behauptung [11] :

Theorem 1. $w^o(r,z)$ sei klassische Lösung des Problems (5), (6), (7) und (9) und die vorgegebene Funktion $Q(r,z)$ erfülle (14). Dann macht $w^o(r,z)$ das Funktional (13) zu einem absoluten Minimum auf Ω: $\tilde{J}(w^1) > \hat{J}(w^0)\ ,\ w^1 \not\equiv w^0$.

Eine analoge Eigenschaft besitzt ein geeignetes Paar (u^0, S^0) in bezug auf das Funktional (10). Aus dem Theorem 1 folgt, daß in der Klasse (11) (oder (12)) das vereinfachte Einphasenproblem von STEFAN nicht mehr als eine Lösung haben kann. Eine ähnliche Behauptung gilt im Zweiphasen-Fall [19] und ebenfalls bei hydrodynamischen Problemen [20], [18] . Wir bemerken jedoch, daß die Funktion (2) eine Bedingung von der Form (14) nicht erfüllt.

5. Über globale Eigenschaften der Menge aller Lösungen eines hydrodynamischen Problems

Das Randwertproblem aus Abschnitt 2 ist eine konforme Invariante, folglich kann man immer annehmen, daß Γ der Einheitskreis $|z| = 1$ ist. Sei $\rho \in (0,1)$ der Konformitätsmodul des Gebietes G_γ und $z = z_\gamma(\tau)$ eine holomorphe Funktion, die das Ringgebiet $G_\rho : \rho < |\tau| < 1$ auf G_γ eineindeutig abbildet und die die Bedingung $z_\gamma(1) = 1$ erfüllt (eine solche Funktion existiert und ist eindeutig bestimmt). Falls $\tau = \tau_\gamma(z)$ die Umkehrfunktion ist, dann ist $\Psi(x,y) = \ln|\tau_\gamma(z)|/\ln\rho$ offensichtlich eine harmonische Stromfunktion des Gebietes G_γ. Für ein solches Paar (Ψ,γ) nimmt das Funktional (3) die Gestalt

$$(15) \qquad J(\Psi,\gamma) \equiv J_1(z,\rho) = -\frac{2\pi}{\ln\rho} + \iint_{G_\rho} Q^2(x,y)\left|\frac{dz(\tau)}{d\tau}\right|^2 d\xi d\eta$$

an und der erste Summand $\nu = -2\pi/\ln\rho$ gibt die "Zirkulation" der Strömung an.

Wir setzen voraus, daß die Funktion $z_\gamma(\tau)$ die Bedingung

$$(16) \qquad \sup_{\rho<t<1} \int_0^{2\pi} |z'_\gamma(te^{i\sigma})|^2 d\sigma < \infty, \quad z'_\gamma(\tau) = \frac{dz_\gamma(\tau)}{d\tau}$$

erfüllt. Bei fixiertem $\rho\in(0,1)$ kann die Menge aller in G_ρ holomorphen Funktionen, die (16) erfüllen, zu einem vollständigen Hilbertraum $H_2^{(1)}(G_\rho)$ über dem Körper der reellen Zahlen gemacht werden, während die Menge aller Funktionen von der Form $z_\gamma(\tau)$ eine unendlichdimensionale Mannigfaltigkeit $M_\rho \subset H_2^{(1)}(G_\rho)$ ohne Rand ([6], Punkt 3; [9], §3) ist. Die Parametrisierung $z = z_\gamma(\rho e^{i\sigma})$, $0 \leq \sigma \leq 2\pi$, des freien Randes γ heißt <u>holomorph.</u> Die Kurve γ (oder die ihr entsprechende Funktion $z_\gamma(\tau)\in M_\rho$) heißt <u>kritisch</u> für das Funktional (15) auf der Mannigfaltigkeit M_ρ, falls die erste Variation des im Punkte z_γ genommenen Funktionals (15) verschwindet für alle Vektoren aus dem Tangentialraum $T_{z_\gamma}M_\rho$. Die hydrodynamische Bedeutung der kritischen Punkte $z_\gamma\in M_\rho$ besteht darin, daß sie ein und dieselbe Zirkulation $\nu = -2\pi/\ln\rho$ besitzen, während der Durchsatz der Strömung c_1 (in der dimensionslosen Aufgabenstellung aus Abschnitt 2 ist er gleich eins) eine von ν abhängende Größe (siehe [9], § 10) ist.

Dafür, daß γ kritisch ist, ist notwendig und hinreichend, daß ihre holomorphe Parametrisierung $z_\gamma(\rho e^{i\sigma}) \equiv x(\sigma) + iy(\sigma)$ ein System von nichtlinearen singulären Integrodifferentialgleichungen ([6]; siehe auch [9], § 6) erfüllt. Wir schreiben hier nur die dritte (aber wichtigste) Gleichung hin:

$$(17) \qquad W_3(z,\rho)(\sigma) \equiv$$

$$\equiv Q^2[x(\sigma),y(\sigma)]y'(\sigma) + \frac{1}{2\pi}\int_0^{2\pi} x'(s)Q^2[x(s),y(s)]\{\operatorname{ctg}\frac{s-\sigma}{2} - L(\sigma,s;z,\rho)\}ds = 0,$$

in der $L(\sigma,s;z,\rho)$ ein effektiv konstruierbarer Kern ([9] § 4) ist. Weil andererseits $x(\sigma)$ bzw. $y(\sigma)$, der Real- bzw. der Imaginärteil der Randwerte der im Ringgebiet G_ρ analytischen Funktion $z_\gamma(\tau)$ ist, erfüllen sie auch die Gleichung ([6], aber auch [9], § 2):

$$(18)\quad \begin{aligned} &x'(\sigma) + \frac{1}{2\pi}\int_0^{2\pi} y'(s)\,\mathrm{ctg}\,\frac{s-\sigma}{2}\,ds - \frac{1}{\pi}\int_0^{2\pi} y'(s)\,\zeta_0(s-\sigma;\rho)ds \\ &+\frac{1}{2\pi}\int_0^{2\pi} \mathrm{Re}\left[\frac{1}{i}\frac{d}{d\sigma} z_\gamma(e^{i\sigma})\right] \mathrm{Re}\,\zeta(\sigma-s-i\ln\rho)ds - \frac{\eta(\rho)}{\pi}\int_0^{2\pi} y(s)ds = 0 , \end{aligned}$$

in der ζ die klassische ζ-Funktion von Weierstraß mit den Halbperioden π, $i\ln\rho$; $\eta(\rho)=\zeta(\pi)$; $\zeta_0(\sigma-s;\rho) = \zeta(\sigma-s) - \frac{1}{2}\mathrm{ctg}\,\frac{\sigma-s}{2} - \eta(\rho)(\sigma-s)/\pi$ ist.

Wir setzen nun voraus, daß die Funktion $Q(x,y)$ hinreichend glatt ist und die Bedingungen

$$(19)\qquad 0 < Q_0 \leq Q(x,y) \leq Q_1 < +\infty$$

erfüllt, wobei Q_0, Q_1 feste Zahlen sind. Dann ist das System (17), (18) linear in x' und y' und vom Normaltyp (siehe [21], § 44). Auf dieser Tatsache basiert der Beweis der Glattheit jeder kritischen Kurve γ in Abhängigkeit von den Glattheitseigenschaften der Funktion $Q(x,y)$ ([5] und ebenso [9], § 6) und ebenso auch die Aufstellung von Kriterien darüber, wann diese kritische Kurve γ lokal eindeutig bestimmt ist (isoliert ist) ([15], und ebenfalls [9], § 7). Bezüglich des Kerns $L(\sigma,s;z,\rho)$ folgt aus der Gleichung (17) das

<u>Theorem 2 [22].</u> Zu jedem $\varepsilon > 0$ existiert ein solches $\rho_0(\varepsilon)\in(0,1)$, so daß für alle $z_\gamma \in M_\rho$, $s, \sigma\in[0,2\pi]$ und für beliebiges $\rho\in(0,\rho_0]$ die a priori-Abschätzung

$$(20)\qquad |L(\sigma,s;z,\rho)| < \varepsilon$$

gilt.

Der Beweis dieses Theorems steht in der im Druck befindlichen englischsprachigen Ausgabe der Arbeit [9]. Es spielt eine wesentliche Rolle bei der Untersuchung globaler Eigenschaften der Menge N_ρ aller kritischen Punkte des Funktionals (15) auf M_ρ. Für jedes $a\notin\Gamma$ und für beliebiges $b \in (0, +\infty)$ betrachten wir die Menge

$$N_\rho^b(a) = N_\rho \cap \{z_\gamma : z_\gamma \in M_\rho,\ J_1(z,\rho) \leq b,\ \gamma \ni a\}. \tag{21}$$

Theorem 3 [22]. Die Funktion $Q(x,y)$ sei glatt und sie erfülle die Bedingungen (19). Dann existiert eine solche nur von Q_0 und Q_1 abhängende Zahl $\rho_0 \in (0,1)$, so daß die Menge (21) im Raum $H_2^{(1)}(G_\rho)$ kompakt ist für beliebiges $b \in (0, +\infty)$ und für jedes $\rho \in (0, \rho_0]$. Falls jedoch alle kritischen Punkte des Funktionals (15) ausgeartetet sind (siehe [7], Punkt 5, und auch [9], § 12), ist unter denselben Bedingungen die Menge (21) endlich.

Der Beweis dieses Theorems beruht auf der Ungleichung (20) und dem System (17), (18) und steht in der englischen Ausgabe der Arbeit [9]. Das Theorem 3 dient auch als Grundlage für den Aufbau der verallgemeinerten Theorie von MORSE für Funktionale von der Gestalt (15) (siehe [8] und ebenfalls [9], § 13).

6. Eine Anwendung auf die Theorie der Wellen

Wir veranschaulichen Theorem 3 am Beispiel der Theorie der periodischen Wellen einer idealen Flüssigkeit über ebenem Grunde, wenn die Funktion $Q(x,y)$ die Gestalt (2) hat. Dafür, daß diese Funktion die Bedingungen von Theorem 3 erfüllt - insbesondere die Ungleichungen (19) -, genügt es vorauszusetzen, daß: 1. alle betrachteten Wellen eine beliebige aber feste Periode $\ell \in (0, +\infty)$ und beliebigen aber festen Durchsatz $c_1 \in (0, +\infty)$ haben; 2. jede Welle durch einen vorgegebenen Punkt $(0,\eta_0)$, $\eta_0 \in (0, +\infty)$ geht und in ihm den Absolutbetrag der Geschwindigkeit $U \in (0, +\infty)$ hat; 3. alle Wellen sich in dem Streifen

$$0 < \eta \leq \eta_0 + \frac{U^2-\delta^2}{2g}, \qquad 0 < \delta < U^2 \tag{22}$$

befinden; 4. die Totalenergie jeder Welle nicht einen vorgegebenen Wert $E \in (0, +\infty)$ übersteigt.

Theorem 4. Sei N eine Menge von Wellen der betrachteten Gestalt, die die Bedingungen 1) bis 4) erfüllen. Dann existiert eine solche Zahl $\nu_0 = \nu_0(\ell, c_1, \eta_0, U, \delta, E) > 0$, daß für die Zirkulation $\nu \in (0,\nu_0]$ jeder Welle die Menge N kompakt bzw. endlich ist in der

Metrik, die von dem Raum $H_2^{(1)}(G_\rho)$ erzeugt wird.

7. Das Existenzproblem.

Wir setzen voraus, daß die Funktion $Q(x,y)$ aus der verallgemeinerten Bernoulli-Gleichung (1) die Bedingung

$$(23) \qquad d = \inf_{(\Psi,\gamma)} \iint_{G_\gamma} \{\Psi_x^2 + \Psi_y^2 + Q^2(x,y)\}\, dxdy < \iint_G Q^2(x,y)dxdy$$

erfüllt. In ihrer Bedeutung ist diese Ungleichung verwandt mit der Bedingung von DOUGLAS in dem Problem von PLATEAU (siehe z. B. [23], Kap. IV) und sie kann bei geeigneter Wahl des Durchsatzes c_1 erfüllt werden. Die Methode der minimisierenden Folgen zusammen mit der Methode der "inneren" Variationen von SCHIFFER gestatten die folgende Behauptung (siehe [9], § 11) zu beweisen:

Theorem 5. Wir setzen voraus, daß $Q(x,y) \in C_\alpha^{(n)}$, $0 < \alpha < 1$, $n \geq 1$, ist und die Bedingungen (19) und (23) erfüllt. Es existiert eine Kurve $\gamma_0 : z = z_0(\rho_0 e^{i\sigma})$, $\sigma \in [0,2\pi]$, $z_0(\tau) \in M_{\rho_0}$, die zusammen mit der harmonischen Stromfunktion Ψ_0 des Gebietes G_{γ_0} das Funktional (3) zum kleinsten Wert d macht und fast überall auf γ_0 läßt sich die BERNOULLI-Bedingung (1) erfüllen. Die Funktion $z_0(\rho_0 e^{i\sigma})$ ist absolut-stetig, 2π-periodisch und erfüllt eine Hölderbedingung mindestens mit dem Exponenten 1/2. Falls γ_0 eine SMIRNOW-Kurve (siehe z. B. [24], Kap. III) ist, dann ist $z_0(\rho_0 e^{i\sigma}) \in C_\alpha^{(n+1)}[0,2\pi]$ und die BERNOULLI-Bedingung ist dann auch im klassischen Sinne erfüllt.

In jedem Fall, in dem, wie im Beispiel (2), die Funktion $Q(x,y)$ nur von τ abhängt, kann Theorem 5 zu "geometrisch trivialen" Lösungen $\gamma_0 : |z| = \rho_0$, $0 < \rho_0 < 1$ führen. Zur Ermittlung der "geometrisch nichttrivialen" Lösungen wurde die Methode der Minimierung des Funktionals (15) ausgearbeitet, die auf der klassischen Verzweigungsmethode von SCHMIDT-LJAPUNOW (siehe [25], ein vollständiger Beweis siehe [26] und [9], § 8) begründet wurde. Wir formulieren nun ein Existenztheorem in bezug auf die Funktion (2).

Theorem 6. Das Problem der ebenen periodischen permanenten Wellen einer idealen inkompressiblen Flüssigkeit über ebenem Grund im Schwerkraftfeld der Erde besitzt eine Familie "geometrisch nichttrivialer" Lösungen, die durch den folgenden Satz von Parametern definiert ist: 1. durch die Zahl $\rho_0 = \exp\{-2\pi/\nu\}$, wobei ν die Zirkulation ist; im allgemeinen durch hinreichend kleines ρ_0; 2. durch die ganze Zahl m der Minima des Wellenprofils in einem Periodenintervall; 3. durch die Wahl des Vorzeichens $(\pm)$, das eine Konvexität oder eine Konkavität in einem vorgegebenen Punkt charakterisiert; 4. durch den dreidimensionalen Vektor $h = (\ell-\ell_o, c-c_o, c_1-c_1^o)$ in einer Umgebung des Punktes $(0,\mu_2,0)$, $\mu_2 < 0$.

8. Über die RITZ'sche Methode bei einem thermodynamischen Problem

Um das Minimum-Problem des Funktionals (13) näherungsweise zu lösen, setzen wir

$$(24) \qquad w(\tau,u;a_{kj}) = \sum_{k=1}^{K} u^k \sum_{j=0}^{L_k} a_{kj}\,\tau^{2j}, \quad K \geq 1,\ L_k \geq 0,\ k = 1,2,\ldots,K,$$

wobei entsprechend der Zulässigkeitsbedingung 3) für w (siehe Abschnitt 4) sein muß:

$$(25) \qquad E^o_{n-1} : \sum_{k=1}^{K} \sum_{j=0}^{L_k} a_{kj} - 1 = 0, \qquad n = \sum_{k=1}^{K} (L_k+1).$$

Die Zulässigkeitsbedingung 2) nimmt jedoch die Gestalt

$$(26) \qquad f(a_{kj}) \equiv \min_{0\leq\tau,u\leq 1} \sum_{k=1}^{K} k\,u^{k-1} \sum_{j=0}^{L_k} a_{kj}\,\tau^{2j} > 0$$

an und gewährleistet auch die Zulässigkeitsbedingung 1). Das entsprechende Ritzsche System ist nichtlinear und seine Auflösung ist verhältnismäßig schwierig. In dieser Richtung bemerken wir das Ergebnis (vergleiche [12]):

<u>Theorem 7.</u> Wir setzen voraus, daß $a_{ko} = 0$, $k = 2,3,\ldots,K$. Falls ω_0, ω und $\max Q(\tau,z)$ hinreichend klein sind, dann besitzt das RITZ'sche System (13), (24), (25) auf der Menge (26) eine einzige Lösung.

Diese Behauptung läßt sich ausdehnen auch auf das vereinfachte Zweiphasenproblem von STEFAN im quasistationären Fall [19]. Bei der numerischen Berechnung der Parameter a_{kj} wurde die Methode des steilsten Abstieges angewandt.

In letzter Zeit wurde das Existenzproblem für das quasistationäre STEFAN-Problem mit zwei geometrischen Veränderlichen (sowohl in den vereinfachten Varianten, als auch in der exakten Problemstellung) gelöst. Einige dieser Ergebnisse sind dargestellt in den Artikeln [13], [14], [15] und [27]. Es wurde auch ein "Optimierungsproblem" betrachtet, in dem die "Steuerung", die durch das Q ausgeübt werden kann, so bestimmt wird, daß der freie Rand (im Sinnes des quadratischen Mittels) minimal von einer vorgegebenen Kurve abweicht und es wurde die Lösbarkeit des Problems mit endlichdimensionalen RITZ'schen Näherungen [28] bewiesen.

L i t e r a t u r

[1] Birkhoff, G. & Zarantanello, E.
Jets and Cavities, Academic Press.

[2] Friedrich, K. O.
Über ein Minimumproblem für Potentialtrömungen mit freiem Rande, Math. Ann. 109 (1933).

[3] Garabedian, P. R., H. Lewy, M. Schiffer
Axially symmetric cavitational flow,
Ann. Math. 56 (1952), 560 - 602.

[4] Garabedian, P. R., D. C. Spencer
Extremal methods in cavitational flow,
J. Rat. Mech. Anal., 1 (1952), 359 - 409.

[5] Daniljuk, I. I.
Existenztheoreme bei einem nichtlinearen Problem mit freiem Rand (russ.), Uspechi matem. nauk 20, Nr. 1, (1968), 25.

[6] Daniljuk, I. I.
Untersuchung einer Klasse von Funktionalen, deren Werte Integral sind mit einem variablen Integrationsgebiet (Ukraïnisch). Doklady akademii nauk Ukrainsk. SSR, ser. A, Nr. 9, (1969), 783.

[7] Daniljuk, I. I.
Über die Methode des Antigradienten in der Theorie einer Klasse von Funktionalen (Ukraïnisch). Dokl. akad. nauk Ukr. SSR, ser. A, Nr. 10, (1970), 876.

[8] Daniljuk, I. I.
Eine Verallgemeinerung der Theorie von Morse für eine Klasse von Funktionalen (Ukraïnisch), Dokl. akad. nauk Ukr. SSR, ser. A, Nr. 1 (1971), 16.

[9] Daniljuk, I. I.
Über Integral-Funktionale mit veränderlichem Integrationsgebiet (russ.),
Trudy matem. instituta imeni V. A. Steklow, tom 118, (1972).

[10] Daniljuk, I. I.
Sur une classe de fonctionnelles integrales a domaine variable d'integration,
Actes du Congr. Intern., vol. 2 (1970), 703 - 715.

[11] Kaschkacha, W. Ju., I. I. Daniljuk
Über eine nichtlineare Problemstellung mit unbekanntem Rand. (Ukraïnisch), Dokl. akad. nauk Ukr. SSR, ser. A, (1973),119.

[12] Daniljuk, I. I. und W. Ju. Kaschkacha
Über ein nichtlineares Ritz-System (Ukraïnisch),
Dokl. akad. nauk. Ukr. SSR, ser. A., Nr. 10 (1973), 870.

[13] Basalij, B. W. und W. Ju. Schelepow
Über ein gemischtes Problem mit freiem Rand für die Laplacegleichung (russ.),
Doklady akad. nauk SSSR, Nr. 2 (1973), 209.

[14] Basalij, B. W. und W. Ju. Schelepow
Über ein stationäres STEFAN-Problem (Ukraïnisch),
Dokl. akad. nauk Ukr. SSR, ser. A, Nr. 1 (1974), 5.

[15] Basalij, B. W. und W. Ju. Schelepow
Über ein verallgemeinertes stationäres Stefanproblem (russ.), Sbornik "Matem. Fisika", Kiew, tom 18 (1975).

[16] Beurling, A.
On free-boundary problems for the Laplace equation,
Semin. Analyt. Func., vol. 1, N.-J., Inst. Adv. Study, (1958), 248 - 263.

[17] Courant, R. und D. Hilbert
Methoden der mathematischen Physik,
Band 1, Springer-Verlag Berlin 1968.

[18] Bogatyrew, W. A.
Über die Eindeutigkeit der Lösung eines nichtlinearen Problems (russ.),
Uspechi matem. Nauk 25, Nr. 3 (1973), 347.

[19] Kaschkacha, W. E.
Über die näherungsweise Berechnung der Kristallisationsfront von Barren mit rechteckigem Querschnitt (russ.),
Sbornik "Matem. Fisika", tom 16, Kiew (1974).

[20] Daniljuk, I. I. und M. W. Olejnik
Über die Eindeutigkeit der Lösungen eines nichtlinearen Problems mit freiem Rand (Ukraïnisch),
Doklady akad. nauk Ukr. USSR, ser. A, Nr. 3 (1972), 202.

[21] Muschelišwili, N. I.
Singuläre Integralgleichungen.

[22] Daniljuk, I. I.
Über globale Eigenschaften der Lösungsmengen nichtlinearer Probleme (Ukraïnisch),
Dokl. akad. nauk Ukr. SSR, ser. A, Nr. 7 (1975), 586.

[23] Courant, R.
Dirichlet's Principle, Conformal Mapping, and Minimal Surfaces, Interscience Publishers, New York

[24] Priwalow, I. I.
Randeigenschaften analytischer Funktionen,
Deutscher Verlag d. Wissenschaften, Berlin 1956

[25] Basalij, B. W. und I. I. Daniljuk
Über die Bestimmung der kritischen Punkte von Funktionalen, deren Werte Integral mit variablem Integrationsgebiet sind (Ukraïnisch),
Dokl. akad. nauk Ukr. SSR, ser. A, Nr. 1 (1970), 3.

[26] Basalij, B. W. und I. I. Daniljuk
Über stationäre Punkte des Funktionals, das einem Randwertproblem mit freiem Rand zugeordnet ist.
Sbornik "Matem. Fisika", vypusk 8, Kiew, "Naukowa Dumka", 1970, 3.

[27] Basalij, B. W.
Über ein quasistationäres Stefanproblem (Ukraïnisch),
Dokl. akad. nauk Ukr. SSR, ser. A, Nr. 1 (1976), 3.

[28] Daniljuk, I. I. und O. S. Minenko
Über ein Optimierungsproblem mit freiem Rand.
Dokl. akad. nauk Ukr. SSR, ser. A, (1976), Nr. 5, 391

Eine Bemerkung zur Funktionentheorie in Algebren

Klaus Habetha

Es soll eine kurze Entwicklung einer Funktionentheorie in Algebren bis zum Integralsatz von Cauchy gegeben werden, die die meisten in der Literatur behandelten Fälle umfaßt. Daran schließt sich der Beweis einer "einfachen" Integralformel von Cauchy an, was zu einer Charakterisierung der Funktionentheorien in den Quaternionen (und in $\mathbb{C}$) führt. Auch die Funktionentheorie in Cliffordschen Algebren erweist sich als besonders einfach.

Gegeben sei eine reelle, assoziative Algebra $\mathfrak{A}$ der Dimension m mit Einzelelement e, das mit $1 \in \mathbb{R}$ identifiziert wird. $\{e_1,\ldots,e_m\}$ sei eine Basis von $\mathfrak{A}$ und für $n \leq m$ sei $V_z := \{h_z | z = \sum_{i=1}^{n} x_i e_i, x_i \in R\}$ ein Untervektorraum von $\mathfrak{A}$. Gebiete im $\mathbb{R}^n$ bzw. $\mathbb{R}^m$ werden mit den entsprechenden Mengen in V_z bzw. $\mathfrak{A}$ identifiziert, $|z| = \sqrt{\Sigma x_i^2}$ sei der absolute Betrag.

Betrachtet werden in einem Gebiet $G \subset V_z$ Abbildungen $w : G \to \mathfrak{A}$, die durch m reelle Koordinatenfunktionen $w_i(z)$ beschrieben werden. Solche Abbildungen werden auf verschiedene Weise als regulär (analytisch o. ä.) definiert. Recht umfassend ist die folgende Definition, dabei sei $\Omega = \sum_{i=1}^{m} \omega_i e_i$ eine Differentialform in G, wenn die ω_i Differentialform auf $G \subset \mathbb{R}^n$ sind (Ω sei von der Stufe p, wenn alle ω_i das sind).

1. Definition:

$w : G \to \mathfrak{A}$ sei stetig differenzierbar und Ω eine stetig differenzierbare, geschlossene Differentialform der Stufe p, also $d\Omega = o$ (Differenzierbarkeit wird koordinatenweise bezüglich $x_1,\ldots, x_n$ erklärt).

Dann heißt w rechtsregulär bezüglich Ω, wenn

$$dw \wedge \Omega = o\,,$$

linksregulär bezüglich Ω, wenn

$$\Omega \wedge dw = o\,.$$

Unter " $\wedge$ " wird natürlich die äußere Multiplikation von Differentialformen verstanden, die unter Beachtung der Multiplikationsregeln in der Algebra auszuführen ist, ebenso

$$d\Omega = d\sum_{i=1}^{m} \omega_i e_i = \sum_{i=1}^{m} (d\omega_i) e_i .$$

2. Hilfssatz:

Zu $\Omega := \sum_{i_1 < \ldots < i_p} a_{i_1 \ldots i_p} dx_{i_1} \wedge \ldots \wedge dx_{i_p}$

mit $d\Omega = o$ existieren Differentialoperatoren

$$D_{j_1 \ldots j_{p+1}} := \sum_{k=1}^{p+1} (-1)^{k-1} a_{j_1 \ldots j_{k-1} j_{k+1} \ldots j_{p+1}} \frac{\partial}{\partial x_{j_k}}$$

für alle $1 \leq j_1 \ldots j_{p+1} \leq n$, so daß w genau dann linksregulär (rechtsregulär) bezüglich Ω ist, wenn für alle $j_1 < \ldots < j_{p+1}$

$$D_{j_1 \ldots j_{p\beta 1}} w = o \quad (wD_{j_1 \ldots j_{p+1}} = o).$$

Speziell gehört zu

$$\Omega = dz := \sum_{i=1}^{n} e_i dx_i^* := \sum_{i=1}^{n} e_i (-1)^{i-1} dx_1 \wedge \ldots \wedge dx_{i-1} \wedge dx_{i+1} \wedge \ldots \wedge dx_n$$

$$D := \sum_{j=1}^{n} e_j \frac{\partial}{\partial x_j} .$$

Der Beweis besteht in einer einfachen Rechnung, z. B. ist die Linksregularität von w gleichbedeutend mit

$$0 = \Big(\sum_{i_1 < \ldots < i_p} a_{i_1 \ldots i_p} dx_{i_1} \wedge \ldots \wedge dx_{i_p} \Big) \Big(\sum_{k=1}^{n} \frac{\partial w}{\partial x_k} dx_k \Big)$$

$$= \sum_{i_1 < \ldots < i_p} \sum_{k = i_1, \ldots, i_p} a_{i_1 \ldots i_p} \frac{\partial w}{\partial x_k} dx_{i_1} \wedge \ldots \wedge dx_{i_p} \wedge dx_k$$

$$= (-1)^p \sum_{j_1 < \ldots < j_{p+1}} D_{j_1 \ldots j_{p+1}} w .$$

Entsprechend diesem Hilfssatz liegt für $p < n - 1$ ein überbestimmtes Differentialgleichungssystem vor, das im allgemeinen nur in Spezialfällen Lösungen besitzt. Für $p = n - 1$ liegt ein normal bestimmtes System vor.

Daß man auch von der Differentialgleichung ausgehend zu solchen regulären Funktionen gelangen kann, zeigt z. B. der

3. Hilfssatz

Existiert eine Funktion $f : G \to \mathfrak{A}$ mit

$$\sum_{k=1}^{n} (-1)^{k-1} \frac{\partial}{\partial x_k} (f a_k) = o ,$$

so daß $f(z)$ stets in $\mathfrak{A}$ invertierbar ist, so gibt es zu

$$D := \sum_{k=1}^{n} a_k \frac{\partial}{\partial x_k}$$

ein Ω der Stufe $n-1$, so daß die Lösungen von $Dw = o$ mit den linksregulären Funktionen bezüglich Ω übereinstimmen.

Der Beweis ist wieder einfach: Die Lösungsmenge von $Dw = o$ wird durch Multiplikation der Gleichung von links mit f nicht beeinflußt. Soll dann

$$\Omega := \sum_{k=1}^{n} f a_k \, dx_k^*$$

geschlossen sein, so ist dies gleichwertig mit der im Satz angegebenen Bedingung für f.

Nun einige Hinweise zur Literatur:
Bezüglich des obigen Zugangs sei auf Kriszten [5] verwiesen. Für $n > 2$ ist im allgemeinen $\Omega = dz^*$ behandelt worden, besonders weit entwickelt ist die Funktionentheorie in der Quaternionenalgebra (Fueter [2], Moisil [6]) und in der Cliffordalgebra (Delanghe [1], Iftimie [4]). Für $n = 2$ hat Douglis $V_z = \mathbb{C}$ und die komplexe Algebra $\mathfrak{A}$ benutzt mit der Basis $\{1, e, \dots, e^{r-1}\}$, $e^r = o$. Elliptische Differentialgleichungssysteme 1. Ordnung in der Ebene lassen sich in der Form
$Dw := \left(\frac{\partial}{\partial \bar z} + \left(a \frac{\partial}{\partial \bar z} + b \frac{\partial}{\partial z}\right) e\right) w = o$ schreiben. Mit einer geeigneten Lösung von $Dt = o$ wird

$$\Omega = \frac{t_z}{1+ae} (dz + (a\,dz - b\,d\bar z)e) .$$

Es sei auf Gilbert [3] verwiesen, jeweils auch bei allen Zitaten auf die dort angegebene Literatur.

Weiter ergibt sich als direkte Folgerung aus dem Satz von Stokes:

4. Satz

a) Für beliebige, stetig differenzierbare Funktionen $w, W : G \to \mathcal{O}\!\mathcal{C}$, und eine geschlossene Differentialform Ω der Stufe p und eine $(p+1)$-dimensionale Mannigfaltigkeit $M \subset G$ (einschließlich Rand ∂M hinreichend glatt) gilt

$$\int_{\partial M} W\Omega w = \int_M (dW \wedge \Omega w + (-1)^p W\, \Omega \wedge dw)$$

(Greensche Formel).

b) Sind w linksregulär und W rechtsregulär bezüglich Ω, so gilt

$$\int_{\partial M} W\,\Omega\, w = o .$$

c) Ist w linksregulär (rechtsregulär) bezüglich Ω, so gilt

$$\int_{\partial M} \Omega w = o \qquad \left(\int_{\partial M} w\Omega = o \right)$$

(Cauchyscher Integralsatz).

In a) ist der Stokessche Satz auf $W\,\Omega\, w$ anzuwenden,
in b) $dW\wedge \Omega = o = \Omega \wedge dw$ zu berücksichtigen und
in c) $W = 1$ zu setzen.

Der Satz reduziert sich im Fall der kleinsten Funktionentheorie $V_z = \mathcal{O}\!\mathcal{C} = \mathbb{C}$ mit $\Omega = dz$ auf den bekannten Integralsatz. Die Natur dieses Satzes scheint mir mehr der allgemeinen Analysis zuzuordnen zu sein, während in funktionentheoretischer Hinsicht die Integralformel von Cauchy eine wichtige Grundlage für weitere Sätze ist (Reihenentwicklungen, Satz von Liouville usw.).

Der Beweis der Cauchyformel im klassischen Fall ist wegen des einfachen Kerns leicht. Verwendet man etwas Ähnliches im allgemeinen Fall, so erhält man:

5. Satz

Ist $\Omega = dz^*$, so ist $W(z-z_o)$ mit

$$W(z) := |z|^{-n} \sum_{j=1}^{n} b_j\, x_j$$

rechtsregulär bezüglich Ω für $z \neq z_o$, wenn es Konstanten $b_j \in \mathcal{O}\!\mathcal{C}$ gibt mit

$$(*) \qquad \begin{aligned} & b_i\, e_j + b_j\, e_i = o, \quad i \neq j,\ i \text{ und } j = 1,\dots,n \\ & \frac{1}{n} \sum_{i=1}^{n} b_i e_i = b_k e_k = 1\,, \quad k = 1, \dots, n . \end{aligned}$$

<u>Ist dann</u> G <u>ein Gebiet mit hinreichend glattem Rand und</u> w <u>linksregulär bezüglich</u> Ω <u>in einer Umgebung von</u> $\bar{G}$, <u>so gilt die Cauchysche Integralformel</u> $(z_o \in G)$:

$$w(z_o) = \frac{1}{\omega_n} \int_{\partial G} W(z - z_o)\, dz^*\, w(z)$$

(ω_n <u>Oberfläche der Einheitskugel im</u> $\mathbb{R}^n$) .

Die Bedingung (*) wird im nächsten Satz zur Charakterisierung der Algebren ausgenutzt, für die solch einfacher Cauchykern vorhanden ist.

<u>Beweis:</u> Die Rechtsregularität des angegebenen W bedeutet

$$o = \sum_{i=1}^{n} W_{x_i} e_i = |z|^{-n-2} \{|z|^2 \sum_{i=1}^{n} b_i e_i - n \sum_{i,j=1}^{n} b_i x_i e_j x_j\} .$$

Die geschweifte Klammer ist genau dann Null, wenn das (*) des Satzes gilt, allerdings ist $b_k a_k = c$ für $k = 1,\dots,n$ mit invertierbarem c hinreichend. Der Ansatz $b_k = c\, \tilde{b}_k$ führt dies auf den angegebenen Fall zurück.

Der Beweis der Cauchyformel verläuft dann wie üblich:
Sei $K_\varepsilon = \{|z - z_o| \leqq \varepsilon\}$, die Anwendung von Satz 4b auf $W(z-z_o)dz^*w(z)$ ergibt

$$(\int_{\partial K_\varepsilon} W(z-z_o)dz^*)w(z_o) =$$

$$= \int_{\partial G} W(z-z_o)dz^*w(z) - \int_{\partial K_\varepsilon} W(z-z_o)dz^*[w(z)-w(z_o)].$$

Setzt man auf ∂K_ε an: $z = z_o + \varepsilon y$, so wird $dz^* = \varepsilon^{n-1}(\sum_{j=1}^{n} y_j e_j)do(y)$ mit dem Oberflächenelement $do(y)$ der Einheitskugel $\{|y| = 1\}$ im R^n. Dies wird zusammen mit dem gegebenen W in die letzte Formelreihe eingesetzt. Wegen

$$W(\varepsilon y) \sum_{j=1}^{n} y_j e_j = \varepsilon^{-n+1} \sum_{i,j=1}^{n} x_i b_i y_j e_j = \varepsilon^{-n+1}$$

folgt

$$(\int_{|y|=1} do(y))\, w(z_o) = \int_{\partial G} W(z-z_o)dz^*w(z) - \int_{|y|=1} [w(z_o+\varepsilon y)-w(z_o)]do(y)$$

und für $\varepsilon \to o$ die behauptete Integralformel.

Jetzt sei als Letztes formuliert, für welche Algebren die Bedingung (*) von Satz 5 lösbar ist. Auf jeden Fall müssen $e_1,\dots,e_n$ invertierbar sein.

6. Satz:

a) Ist $1 \in V_z$, etwa $e_1 = 1$, so besitzt (*) in Satz 5 nur die Lösung

$$b_1 = 1, \quad b_k = -e_k$$

und es gilt

$$e_i^2 = -1, \quad e_i\, e_j = -e_j\, e_i$$

$$\text{für} \quad i,j = 2,\dots,n; \quad i \neq j .$$

Ist V_z eine Algebra $\tilde{\mathfrak{A}}$, so handelt es sich entweder um $\mathbb{C}$ oder um die Quaternionenalgebra $\mathfrak{Q}$.

b) Ist $1 \notin V_z$, so folgt aus der Bedingung (*) in Satz 5:

$$e_i\, e_j = -\, e_j\, e_i\,, \quad i \text{ und } j = 1,\dots,n$$

ist gleichwertig mit der Existenz eines invertierbaren Elements $d \in \mathfrak{A}$ mit $e_i^2 = d$, $i = 1,\dots,n$. Hierunter fallen die Cliffordsche Algebra mit $d = 1$ oder der von e_2, e_3, e_4 in der Quaternionenalgebra aufgespannte Untervektorraum mit $d = -1$

Teil a) kennzeichnet $\mathbb{C}$ und $\mathfrak{Q}$ nicht durch die Eigenschaft, ein Körper zu sein. Teil b) ist immerhin ein Indiz dafür, daß die Funktionentheorie in der Cliffordalgebra besonders der Untersuchung wert ist.

Beweis: Es muß natürlich $b_i = e_i^{-1}$ und $e_i^{-1} e_j = -e_j^{-1} e_i$ sein. Nimmt man $e_i\, e_j = -e_j\, e_i$ an, so folgt $e_i^{-1} e_j^2 = -e_j^{-1}\, e_i \cdot e_j = e_j^{-1} e_j \cdot e_i = e_i$ und

$$e_j^2 = e_i^2 =: d\,, \quad i,\, j = 1,\dots,n\,.$$

d ist natürlich invertierbar. Umgekehrt folgt aus der letzten Gleichung $e_j^{-1} = d^{-1}\, e_j = b_j$, also aus $b_i\, e_j + b_j e_i = 0$ wie gewünscht $e_i e_j + e_j e_i = 0$. Damit ist b) bewiesen.

Ist nun etwa $e_1 = 1$, so auch $b_1 = 1$ und

$$e_i^{-1} e_1 + e_1 e_i = 0, \quad i \neq 1\,,$$

mithin

$$k_i = e_i^{-1} = e_i\,, \quad e_i^2 = -1 \quad , \quad i \neq 1$$

$$e_i\, e_j + e_j\, e_i = 0, \quad i \neq j;\ i,j = 1,\dots,n\,.$$

Ist nun V_z eine Algebra $\tilde{\mathfrak{A}}$, so genügen nur $\mathbb{C}$ und $\mathfrak{Q}$ diesen Forderungen. Da ich dafür keinen Beweis in der Literatur gefunden habe, sei hier einer skizziert:

Aus den Multiplikationsgesetzen

$$e_i\, e_j = \sum_{k=1}^{n} \alpha_{ijk}\, e_k = -e_j\, e_i, \quad i, j = 2,\dots,n$$

folgt für $i \neq j$ beide > 1

$$(e_i\, e_j)e_i = \sum_{k=1}^{n} \alpha_{ijk}\, e_k e_i = \sum_{k\neq 2,i} \alpha_{ijk}\, e_k e_i + \alpha_{ij1}\, e_i - \alpha_{iji}$$

$$e_i(e_j e_i) = -\sum_{k=1}^{n} \alpha_{ijk}\, e_i e_k = \sum_{k \neq 2,i} \alpha_{ijk}\, e_k e_i - \alpha_{ij1} e_i + \alpha_{iji}\,.$$

Aus der Assoziativität ergibt sich die Gleichheit dieser Ausdrücke, also

(**) $\alpha_{ij1} = o\,, \quad \alpha_{iji} = o\,.$

Sind e_i, e_j, e_k paarweise und von 1 verschieden, so erhält man weiter

$$(e_i e_j)e_k = \sum_{p=2}^{n} {}_{ijp}\, e_p e_k = \sum_{\substack{p,q=2\\ p\neq k}}^{n} \alpha_{ijp}\, \alpha_{pkq}\, e_q - \alpha_{ijk}$$

$$e_i(e_j e_k) = \sum_{p=2}^{n} \alpha_{jkp}\, e_i e_p = \sum_{\substack{p,q=2\\ p\neq i}}^{n} \alpha_{ipq}\, \alpha_{jkp}\, e_q - \alpha_{jki}\,.$$

Gleichsetzen ergibt einmal wegen $\alpha_{ijk} = -\alpha_{jik}$

(+) $\alpha_{ijk} = \alpha_{jki} = -\alpha_{jik} = -\alpha_{ikj}$

und zum anderen für $q = 2,\dots,n$

$$\sum_{p \neq k} \alpha_{ijp}\, \alpha_{pkq} = \sum_{p\neq i} \alpha_{ipq}\, \alpha_{jkp}\,.$$

Benutzt man (**) und setzt $q = i$, so ist hier die rechte Seite Null. Der Vertauschungsregel (+) liefert schließlich

(++) $\displaystyle\sum_{p=2}^{n} \alpha_{ijp}\, \alpha_{ikp} = o, \quad \sum_{p=2}^{n} \alpha_{ipj}\, \alpha_{ipk} = o\,.$

$p = i,j,k$ hat den Summanden Null zur Folge, (++) hat daher erst für $n \geqq 5$ Bedeutung. Dort schließt man zusammen mit

$$(e_i\, e_j)^2 = -1 = -\sum_{p=2}^{n} (\alpha_{ijp})^2$$

darauf, daß die Matrix $A_i := (\alpha_{ijp})_{2 \leqq j,\, p \leqq n}$ orthogonal und wegen (+) schiefsymmetrisch ist. Das geht aber nicht, so daß für $n \geqq 5$ keine solche Algebra existiert.

Für $n = 4$ muß wegen (**) $e_2 e_3 = \alpha e_y$ sein und wegen $(e_2 e_3)^2 = -1 = -\alpha^2$ noch $\alpha = \pm 1$.
Das aber ist die Quaternionenalgebra.

Für $n = 3$ muß wegen (**) $e_2 e_3 = o$ sein, was wegen $e_2(e_2 e_3) \neq o = -e_3 = (e_2 e_2) e_3$ auf einen Widerspruch führt.

Für $n = 2$ bleibt wegen $e_2^2 = -1$ nur $\mathbb{C}$.

L i t e r a t u r

[1] Delanghe, R.: On the theory of regular functions with values in a Clifford algebra. Ber. Ges. Math. Datenverarb., Bonn, 77. 43-51 (1973)

[2] Fueter, R.: Reguläre Funktionen einer Quaternionenvariablen. Vorlesungsausarbeitung Universität Zürich, 1940.

[3] Gilbert, R. P.: Constructive methods for elliptic equations. Chapter X. Lecture Notes Math. 365, Berlin-Heidelberg-New York 1974.

[4] Iftimie, V.: Fonctions hypercomplexes. Bull. math. Soc. Sci. math. Roma 9, 279-332 (1965)

[5] Krisztin, A.: Hyperkomplexe und pseudo-analytische Funktionen. Comment. math. Helv. 26, 6 - 35 (1952)

[6] Moisil, G. L.: Sur les quaternions monogênes. Bull. Sci. math. 55, 168-174 (1931).

Lehrstuhl II für Mathematik
RWTH Aachen
Templergraben 55 D-5100 A a c h e n

On the ANALYTIC FUNCTIONS' METHOD in the THEORY of PARTIAL DIFFERENTIAL EQUATIONS with SINGULAR COEFFICIENTS

L. G. Mikhailov

The analytic functions' method is the most effective for studying the theoretical questions and applied problems in mathematical physics. In this lecture we are confined to partial differential equations of elliptic type with two independent variables. If the coefficients are sufficiently smooth or analytic, in particular, we must believe the function-theoretic method to be very well suited.

We shall not give any details in this survey but refer to [2],[5].

In the monograph of I. N. Vekua [1] the function-theoretic methods for the equation

(1) $$\Delta U + A(x,y)\cdot U_x + B(x,y)\cdot U_y + C(x,y)\cdot U = f(x,y)$$

with analytic coefficients are developed.

If the functions are continued analytically to two complex variables then by integrating (1) we have a complex integral equation of Volterra type with two arbitrary analytic functions involved. Solving this equation we obtain the representation formula for the solutions. A further development of these ideas took place in the second monograph by I. N. Vekua [2]. It is devoted to the linear system of partial differential equations of the first order, called "generalized Cauchy-Riemann system", and its solutions are called "generalized analytic functions".

Starting from the complex form of the system, namely

(2) $$\partial_{\bar z} w = A(z)\cdot w + B(z)\cdot \bar w + F(z)$$

where $z = x + iy$, $\bar z = x - iy$, $2\cdot\partial_z = \partial_x - i\partial_y$, $2\partial_{\bar z} = \partial_x + i\partial_y$,
I. N. Vekua had studied it without any conditions of analyticity for the coefficients claiming only these to be continuous or summable to the power $p > 2$. This method is based on the apparatus of complex differentiation and integration by $\bar z$: $\partial_{\bar z}\phi = 0$, if and only if $\phi(z)$ is analytic; the operations

(3) $$Tf = -\frac{1}{\pi}\iint_D \frac{f(\zeta)\cdot ds_\zeta}{\zeta - z}, \quad \partial_{\bar z} Tf = f(z) ,$$

are inverse.

The representation formula for the solutions of (2) is obtained if one solves the integral equation

(4) $$w(z) = \phi(z) + T(Aw + B\bar{w} + F)$$

where $\phi(z)$ is an arbitrary analytic function.

The integral equation (4) seems to be quite analogues to Volterra type integral equations: it has no eigenfunctions and is always solvable, the solution may be constructed by the method of successive approximations. A very simple and complete investigation of a variety of solutions of (3) with different properties and the solvability of the main boundary value problems were obtained in the past.

The theory of an arbitrary second order partial differential equation has been also developed to an advanced stage.

About 20 years ago the author has begun his research on partial differential equations with singular coefficients in two or more dimensions. The equations with singular points have coefficients of the forms:

(5) $$A(z) = \frac{a(z)}{|z|}, \quad B(z) = \frac{b(z)}{|z|}, \quad C(z) = \frac{c(z)}{|z|},$$

where $a(z)$, $b(z)$, $c(z)$ are bounded.

Some equations with singular lines for the coefficients are considered too (see below 1.3 and § 2).

The integral equations' method is developed first. Representing the solutions by volume potentials or by means of Green's function of the Laplace operator we obtain a new type of singular integral operators. For systems of type (2) this will be $T(\frac{a(z)}{|z|} \cdot \bar{w})$.

In my monograph of 1963, being published in English in 1970 [3] , the first results in this direction were formulated. Let be $w_o(z) \in M$, C, and L^p, respectively, then the set of functions

$$w(z) = |z|^{-\beta} w_o(z), \quad ||w||_\beta = ||w_o||,$$

form isometrical Banach spaces M_β, C_β, and L^p_β, respectively. In these weighted spaces the operators are bounded but not completely continuous. If $a(o) = o$, $T(\frac{a(z)}{|z|} \cdot \bar{w})$ is completely continuous such that the Fredholm theorems hold if $|a(z)|$ is sufficiently small.

And in accordance with this many different results on the manifold of solutions and on the solvability of boundary value problems have been received for equations (1) and (2) with coefficients of type (5) where $|a(z)|$, $|b(z)|$, $|c(z)|$ are sufficiently small.

The author's and his colleagues' investigations after 1963 were directed firstly to release from the conditions of smallness and this was done in different directions.

The theory of a new class of singular integral equations has been developed. A method sufficiently effective to complete the theory of the large class of multidimensional equations with homogeneous kernels of order $-n$, where n is the dimension of space, has been developed [3], [4]. This method permits, for example, a full investigation of the equation with the operator $T(\frac{\lambda}{2\bar{z}} \cdot \bar{w})$ (see below § 1) and is useful for studying many other differential equations with singular coefficients. Secondly, by the method of separation of variables, simple or model equations with singular coefficients are solved. By combining with the method of integral equations we can get rid of the conditions of smallness (s. [11]). Thirdly, for the system (2) the method does not need integral equations and the conditions of smallness (see 1.2) have been obtained.

For a three-dimensional rotational body in cylindrical coordinates we arrive at the following equation with a singular line from Laplace's equation

(6) $$y \cdot \Delta u + \mu \cdot U'_y = o \, .$$

It may be called a "fundamental equation of the axial symmetric field theory", in American papers usually called a GASPT-equation ("Generalized Axially Symmetric Potential Theory") [5].

The function-theoretic methods for the equation (6) have been developed by A. Weinstein, P. Henrici (see in [5]), U. P. Krivenkov [6] and others. In § 2 we shall discuss the papers by N. Radjabov. Starting from the representations by Henrici and Krivenkov, he confines them to the curve and therefore boundary value problems of the Dirichlet and Neumann type are reduced to boundary value problems for analytic functions. In this way he gives precise theorems on the solvability and, in the case of a circular boundary, effective integral formulas for the solutions have been established.

1. Genralized Cauchy-Riemann System with a Singular Point

1.1 The first model equation is

(1.1) $$\frac{\partial w}{\partial \bar{z}} - \frac{\lambda}{2\bar{z}} \cdot \bar{w} = \frac{g(z)}{2\bar{z}}$$

being equivalent to the integral equation

$$(1.2)\qquad w(z) + \frac{\lambda}{2\pi} \cdot \iint\limits_{|\zeta| \le R} \frac{\overline{w(\zeta)} \cdot dS_\zeta}{\bar{\zeta} \cdot (\zeta - z)} = f(z), \ |z| \le R ,$$

where $f(z) = g(z) + T\,(\frac{g(z)}{2\bar{z}})$ and $\phi(z)$ is an arbitrary analytic function. We stated the following theorem [7].

<u>Theorem 1.1.</u> Let $f(z)$ and $w(z) \in M_\beta$, C_β or $L^p_{\beta - \frac{2}{p}}$, where $o < \beta < 1$ and $p \ge 1$. Then the homogenoeus equation (1.2.) has no solutions for arbitrary λ. The inhomogeneous equation is uniquely solvable for arbitrary $f(z)$ if $|\lambda| < \beta$; for $|\lambda| > \beta$ the condition

$$(1.3)\qquad \iint\limits_{|\zeta| \le R} [f(z) - \frac{\lambda}{|\lambda|} \cdot \overline{f(z)}] \cdot |z|^{|\lambda| - 2} \cdot ds_z = o$$

is necessary and sufficient for the solvability of (1.2).

Let now $f(z)$ be bounded. Then $f(z) \in M_\beta$ with an arbitrary small β and for any $\lambda \neq o$ the condition (1.3) must be satisfied. The solution $w(z)$ may be unbounded in $z = o$, however. From solvability considerations of (1.2) and condition (1.3) the following result has been obtained [7]: if the condition

$$(1.4)\qquad f_o(\tau) = \frac{1}{2\pi} \int_o^{2\pi} f(\tau, \phi) d\phi = o$$

holds the solution exists, is unique and bounded.

Furthermore, let $g(z) \equiv o$ in (1.1). Then $f(z) = \phi(z)$ is an analytic function and the condition (1.4) leads to $\phi(o) = o$. Thus a one-to-one correspondence between the family of solutions of the equation $\partial_{\bar{z}} w = \frac{\lambda}{2\bar{z}} \cdot \bar{w}$ and analytic functions $\phi(z)$, where $\phi(o) = o$ holds, is established.

The solution of (1.1) may be written in the form

$$w(z) = f(z) + \iint\limits_{|\zeta| \le R} [\lambda \cdot \Gamma_1(z,\zeta) + \bar{\lambda} \cdot \Gamma_2(z,\zeta)] \cdot \frac{ds_\zeta}{|\zeta|^2}$$

where the resolvents $\Gamma_1(z,\zeta)$, $\Gamma_2(z,\zeta)$ are given by series expansions [3].

For the more general model equations

$$(1.5)\qquad \frac{\partial w}{\partial \bar{z}} = \frac{\lambda}{2\bar{z}} \cdot e^{in\phi}\, \bar{w} , \quad \phi = \arg z ,$$

the analogous results have been formulated before [9].

Let D be an arbitrary domain containing the point $z = o$, $B(z)$ be continuous at $z = o$, and $B(z) \in M(D)$.

The equation

$$(1.6) \qquad \frac{\partial w}{\partial \bar{z}} - \frac{B(z)}{2\bar{z}} \cdot \bar{w} = g(z)$$

is equivalent to the integral equation

$$(1.7) \qquad w(z) = -\frac{1}{2\pi} \iint_D \frac{B(\zeta) \cdot \overline{w(\zeta)}}{\bar{\zeta} \cdot (\zeta - z)} \cdot ds_\zeta + f(z)$$

by subtracting $B(o)$ we separate off the model part (1.2) and leave the remaining operator $T(\frac{B(\zeta)-B(o)}{2 \cdot \bar{\zeta}} \cdot \bar{w})$ being completely continuous [3]. From general theorems in operator theory it follows that the Noether theorems hold for (1.7) with the index $\kappa = o$ in case of $|B(o)| < ß$ and with $\kappa = -1$ in case of $|B(o)| > ß$. Many conclusions may be drawn from this if $|B(\zeta) - B(o)|$ is sufficiently small.

To treat (1.5) the equation

$$(1.8) \qquad \frac{\partial w}{\partial \bar{z}} = \frac{B(z)}{2\bar{z}} \cdot \bar{w}$$

may be reduced analogously. Some results on the solvability of the boundary value problem

$$\text{Re} \quad [t^{-n} \cdot w(t)] = h(t)$$

have been obtained before [9].

1.2. The other method for the system (1.1) is based on second order equations. For general analytical systems of partial differential equations this is considered in the book [10]. For regular generalized Cauchy-Riemann systems such relations to two equations of second order are considered by I. N. Vekua [2]. For singular systems this method has been developed in our papers [11], [12], [13].

Differentiating (1.1) by z and taking into account $4 \cdot \partial^2_{z\bar{z}} w = \Delta w \; \partial_z \bar{w} = \overline{\partial_{\bar{z}} w}$, we obtain

$$(1.9) \qquad \tau^2 \cdot \Delta u - |\lambda|^2 \cdot u = f \quad , \text{ where } \quad f = \text{Re } [2z \cdot \partial_z g - \lambda \cdot \bar{g}]$$

Transforming to polar coordinates the system (1.1) may be written in the form

$$(1.10) \qquad \begin{aligned} \frac{\partial}{\partial \tau} (\tau^{-\lambda_1} \cdot e^{-\lambda_2 \theta} v) &= -\tau^{-\lambda_1 - 1} \cdot e^{-\lambda_2 \theta} \cdot [(u'_\theta + \lambda_2 u) - (d \cdot \cos\theta + c \sin\theta)] \\ \frac{\partial}{\partial \theta} (\tau^{-\lambda_1} \cdot e^{-\lambda_2 \theta} \cdot v) &= -\tau^{-\lambda_1 - 1} \cdot e^{-\lambda_2 \theta} \cdot [(\tau u'_\tau + \lambda_1 \cdot u) - \tau(c \cdot \cos\theta - d \sin\theta)], \end{aligned}$$

where $c + di = 2g$, $\lambda_1 + i\lambda_2 = \lambda$. If the solution $u(\tau,\theta)$ to (1.9) is known the solution of (1.10) will be found in terms of multi-valued functions by integrating the total differential. For example let the first boundary value problem be considered for (1.1)

$$(1.11) \qquad [\mathrm{Re}\, w]_{\tau=1} = u(1,\theta) = h(\theta), \quad o \le \theta \le 2\pi .$$

First solving (1.9) by Fourier's method we have

$$(1.12) \quad u(\tau,\theta) = \sum_{k=o}^{\infty} \tau^{\mu_k} \cdot (h_k \cdot \cos k\theta + g_k \cdot \sin k\theta) + \int_o^{2\pi} \sum_{k=o}^{\infty} Q_k(\tau,\phi) \cdot \cos k(\theta-\phi) d\phi ,$$

where $\mu_k^2 = k^2 + |\lambda|^2$, h_k, g_k being Fourier's coefficients of $h(\theta)$, $g(\theta)$, and Q_k a simple integral operator acting on $f(\tau,\theta)$. After substituting (1.12) into the right part of (1.10) we find $v(\tau,\theta)$. Taking into account the conditions of continuity and single-valuedness we have:

<u>Theorem 1.2.</u> For an arbitrary real or complex λ the homogeneous problem (1.1) with (1.11) has no nontrivial solutions, $w(z) \equiv o$, and the inhomogeneous problem is solvable for any $g(z)$ and $h(\theta)$, if λ is an arbitrary complex or real negative number; if λ is positive then the condition

$$(1.13) \qquad h_o = \int_o^{2\pi} h(\theta) \cdot d\theta = o$$

is necessary and sufficient for solvability.

The methods mentioned apply also to equations with a singular line [13].

<u>The Problem S_q:</u> It is required to find the solution for the equation

$$(1.14) \qquad \partial_{\bar{z}} w - \frac{e^{i\phi}}{2} \cdot \frac{\lambda}{\tau - q} \cdot \bar{w} = \frac{e^{i\phi}}{2} \cdot \frac{g(z)}{\tau - q} , \quad q < \tau < 1 ,$$

with the asymptotic behavior

$$(1.15) \qquad w(z) = O[(\tau-q)^{2\lambda-1}] , \text{ as } \tau \to q ,$$

on the interior boundary line and with the boundary condition (1.11) on the exterior boundary.

Expanding $w(\tau,\theta)$ into Fourier series in respect to the variable ϕ with coefficients depending on τ we may reduce the problem S_q to an infinite system of ordinary differential equations of the Fuchs class with the singular points $\tau = o$ and $\tau = q$.

<u>Theorem 1.3.</u> For any λ complex or real positive, the boundary value problem S_q is uniquely solvable. If λ is real and negative then there are eigenfunctions $i(\tau-q)^{-\lambda}$

and the condition (1.13) is necessary and sufficient for solvability.

In the equation

$$(1.16)\qquad \partial_{\bar z} w - \frac{e^{i\phi}}{2} \cdot \frac{b(z)}{\tau - q} \cdot \bar w = \frac{e^{i\phi}}{2} \cdot \frac{c(z)}{\tau - q}$$

let $b(z)$ be continuous and $b(qe^{i\phi}) \equiv \text{const} = \lambda$; after splitting into $b(z) = \lambda + [b(z)-\lambda]$ we obtain the equation of type (1.14) with $\overline{w(z)}$ on the right hand side. Under a smallness condition for $|b(z)-\lambda|$ a theorem similar to 1.3 has been established.

2. A Singular Equation of Axially Symmetric Field Theory

2.1 Considering problems of Mathematical Physics for rotationally symmetric bodies in cylindrical coordinates we arrive at the equation

$$(2.1)\qquad \Delta u + \frac{\mu}{y} U'_y = o , \quad \mu > o .$$

As was shown by P. Henrici [14] every solution of (2.1) analytical in x, y including the singular line $y = o$ may be represented by the formula

$$(2.2)\qquad U(x,y) = \alpha_\mu \cdot \int_o^1 \frac{\phi[x+iy(1-2\sigma)]}{[\sigma \cdot (1-\sigma)]^{1-\mu/2}} \cdot d\sigma \equiv \Pi_\mu(\phi) ,$$

where (z) is an analytic function and

$$(2.3)\qquad U(x,o) = \phi(x) .$$

Here and in the following by α_μ, β_μ etc. different but completely determined constants are denoted. U. P. Krivenkov [6] proved the representation formula (2.2) for solutions of (2.1) being continuous on the singular line only in case of $\mu \geq 1$ in the case of $o < \mu < 1$ the condition

$$(2.4)\qquad \lim_{y \to o} y^\mu \cdot \frac{\partial U}{\partial y} = o$$

is required to hold.

As far as we are considering real solutions to (2.1) due to (2.3) the analytical continuation of $\phi(z)$ may be performed to the lower plane by the symmetry principle, $\phi(\bar z) = \overline{\phi(z)}$.

In accordance with this formula (2.2) determines solutions to (2.1) being continuous which may be continued across the axis Ox such that $u(x,-y) = u(x,y)$.

If $o < \mu < 1$ and condition (2.4) does not hold the class of solutions to (2.1) is much wider. If

(2.5) $$\psi(x) = \lim_{y\to o} \left(\frac{y}{1-\mu}\right)^{1-\mu} \frac{\partial u}{\partial y}$$

is an analytical function, the following representation formula holds

(2.6) $$u(x,y) = \prod_{\mu}(\phi) + \beta_{\mu}\cdot y^{1-\mu} \cdot \prod_{2-\mu}(\psi) \quad .$$

2.2 The formulas (2.2) and (2.6) establishing relations between solutions to (2.1) and analytic functions in many different directions may be used particularly for the study of boundary value problems. Such problems were set up about 15 years ago and in this direction N. Radjabov has successfully studied many important boundary value problems and has given formulas for their effective solution.

Let D be the semi-circle, $x^2 + y^2 \leq 1$, $y \geq o$, with boundary Γ - the contour of the circle and the segment $[-1,1]$ being denoted by γ.

The Problem I_o To find the solutions to (2.1) $u(x,y) \in C^2(D)$, to be continuous up to the singular line γ and for $o < \mu < 1$ the condition (2.4) to be satisfied too, such that on γ

(I_o) $$u(x,y)\Big|_{\Gamma} = f(x,y) \; .$$

The problem I differs from I_o only in that (2.4) is replaced by (2.5).

The problems II_o and II differ from I_o and I only in that a condition of the Neumann-Type is given on Γ.

Let us consider the scheme for solving these problems given by N. Radjabov [15], [16]. Denote by D* a domain symmetrical to D, relative to the axis Ox, and by Γ^* a curve symmetrical to Γ. Denoting by $\tilde{f}(x,y)$ the even continuation of $f(x,y)$ we shall solve a Dirichlet problem for the circle. Let us substitute (2.2) into I_o. Putting $\zeta = x + iy\cdot(1-2\sigma)$ we shall get on $\Gamma + \Gamma^*$

(2.7) $$\alpha_{\mu} y^{1-\mu} \int_{\bar{t}}^{t} \frac{\phi(\zeta)\cdot d\zeta}{[(t-\zeta)\cdot(\bar{t}-\zeta)]^{1-\mu/2}} = \tilde{f}(x,y) \quad t = x + iy \, .$$

The integration is performed along the straight line from t to $\bar{t}$ but replaced by the curve of the circle. Let us cut the plane along the negative semi-axis then the kernel and other functions are discontinuous at the point (-1,o). Transforming to polar coordinates $\zeta = e^{i\alpha}$, $t = e^{i\theta}$, we may give (2.7) the form

(2.8) $$\int_{o}^{\theta} \omega(\alpha)\cdot[\cos\alpha - \cos]^{\mu/2-1} \cdot d\alpha = g(\theta), \quad o \leq \theta \leq \pi,$$

where $\omega(\alpha) = \mathrm{Re}\,[t^{\mu/2}\cdot\phi(t)]$, $g(\theta) = \delta_{\mu}\cdot\sin^{\mu-1}\alpha\cdot f(\cos\theta, \sin\theta)$.

Let K be an integer and λ be the fractional part of the number $\mu/2$. If we assume $\tilde{f}(x,y) \in C^K(\Gamma + \Gamma^*)$ then after differentiating (2.8) we arrive at

$$(2.9) \qquad \int_o^\theta \frac{\omega(\alpha)d\alpha}{(\cos\alpha - \cos\theta)^{1-\lambda}} = g_1(\theta), \quad o \leq \theta \leq \pi,$$

where $$g_1(\theta) = p_\mu \cdot \left(\frac{1}{\sin\theta} \cdot \frac{d}{d\theta}\right)^K g(\theta) .$$

Equation (2.9) is related to the well-known Abel equation and the inversion of it leads to

$$(2.10) \qquad \mathrm{Re}\, [t^{\mu/2} \cdot \phi(t)] = F(t) ,$$

where $$F(t) = q_{\lambda,\mu} \cdot \frac{d}{d\theta} \int_o^\theta \frac{\left(\frac{1}{\sin\alpha}\frac{d}{d\alpha}\right)^K [\sin\alpha^{\mu-1} \cdot g(\cos\alpha, \sin\alpha)] \sin\alpha \cdot d\alpha}{(\cos\alpha - \cos\theta)^\lambda}$$

We have a boundary value problem in (2.10) known from the theory of functions with the discontinuous coefficient $t^{\mu/2}$. If its index $\kappa = -K$ then K conditions on $F(t)$ are necessary and sufficient to hold. In this case all of them are available.

<u>Theorem 2.1</u> Let K be the integer part of $\mu/2$ and the continued function $\tilde{f}(x,y)$ have derivatives of the order $K + 1$ satisfying a Hoelder condition, then problem I_o is uniquely solvable and the solutions are given by the formula (2.2) where $\phi(z)$ is an integral of the Cauchy-Type with $F(t)$ as its density.

The solution of the problem I is analogous if we use formula (2.6). Its inversion leads to two boundary value problems of type (2.10).

Similar investigations of the equations of Helmholtz-type

$$(2.11) \qquad \Delta u + \mu/2 \cdot u_y + \lambda^2 \cdot u = o$$

and of other equations of second and higher order are performed by N. Radjabov.

R e f e r e n c e s

[1] Vekua, I. N.: New Methods for Solving Elliptic Equations. OGIZ, Moscow and Leningrad, 1948.

[2] Vekua, I. N.: Generalized Analytic Functions. Fizmatgiz, Moscow, 1959.

[3] Mikhailov, L. G.: A New Class of Singular Integral Equations and its Application to Differential Equations with Singular Coefficients. Wolters-Noordhoff Publishing Comp., Groningen, 1970, Netherlands, and Akademie-Verlag, Berlin 1970

[4] Mikhailov, L. G.: Multidimensional Integral Equations with Homogeneous Kernels, Proceedings of a Symposium on Continuum Mechanics and Related Problems of Analysis. Tbilisi, 1973.

[5] Gilbert, R. P.: Function Theoretic Methods in Partial Differential Equations. Academic Press, New York and London, 1969.

[6] Krivenkov, U. P.: Soviet. Math. Dokl., vol. 116, Nos. 3, 4, 1957.

[7] Mikhailov, L. G.: Soviet. Math. Dokl., vol. I, No. 1, 1970.

[8] Usmanov, Z. D.: Differential Equations. vol. VII, No. 12, Minsk, 1972.

[9] Usmanov, Z. D.: Proc. AN Tadj. SSR, No. 1 (43), 197 .

[10] Courant, R. and Hilbert, D.: Methods of Mathematical Physics. Fizmatgiz, M.-L., 1951.

[11] Mikhailov, L. G. and Usmanov, Z. D.: Soviet. Math. Dokl., vol.7 No. 4, 1966.

[12] Mikhailov, L. G.: Investigation of a System of Differential Equations with Singular Coefficients by a Method Stated in Connection with Second Order Equations. Proc. Symposium on Immersion Theorems and its Application. Alma-Ata, 1975.

[13] Kusnezov, M. I. and Mikhailov, L. G.: Dokl. AN Tadj. SSR, vol. XV , No. 8, 1972

[14] Henrici, P.: Comm. Math. Helv., 27, Nos. 3, 4, 1953.

[15] Radjabov, N.: Boundary-Value Problems of Axial Symmetric Theory of Fields in "Investigations on Boundary-Value Problems of the Theory of Functions and Differential Equations", AN Tadj. SSR-Press, Dushanbe, 1965.

[16] Radjabov, N.: Boundary Value Problems of Dirichlet- and Neumann-type for some Modell Equations of Elliptic Type with a Singular Line.
Proc. AN Tadj. SSR, No. 4 (50), 1973 and No. 4 (54), 1974.

[17] Gachov, F. D.: Boundary Value Problems.
Fizmatgiz, 1958.

Vol. 399: Functional Analysis and its Applications. Proceedings 1973. Edited by H. G. Garnir, K. R. Unni and J. H. Williamson. II, 584 pages. 1974.

Vol. 400: A Crash Course on Kleinian Groups. Proceedings 1974. Edited by L. Bers and I. Kra. VII, 130 pages. 1974.

Vol. 401: M. F. Atiyah, Elliptic Operators and Compact Groups. V, 93 pages. 1974.

Vol. 402: M. Waldschmidt, Nombres Transcendants. VIII, 277 pages. 1974.

Vol. 403: Combinatorial Mathematics. Proceedings 1972. Edited by D. A. Holton. VIII, 148 pages. 1974.

Vol. 404: Théorie du Potentiel et Analyse Harmonique. Edité par J. Faraut. V, 245 pages. 1974.

Vol. 405: K. J. Devlin and H. Johnsbråten, The Souslin Problem. VIII, 132 pages. 1974.

Vol. 406: Graphs and Combinatorics. Proceedings 1973. Edited by R. A. Bari and F. Harary. VIII, 355 pages. 1974.

Vol. 407: P. Berthelot, Cohomologie Cristalline des Schémas de Caracteristique p > o. II, 604 pages. 1974.

Vol. 408: J. Wermer, Potential Theory. VIII, 146 pages. 1974.

Vol. 409: Fonctions de Plusieurs Variables Complexes, Séminaire François Norguet 1970–1973. XIII, 612 pages. 1974.

Vol. 410: Séminaire Pierre Lelong (Analyse) Année 1972–1973. VI, 181 pages. 1974.

Vol. 411: Hypergraph Seminar. Ohio State University, 1972. Edited by C. Berge and D. Ray-Chaudhuri. IX, 287 pages. 1974.

Vol. 412: Classification of Algebraic Varieties and Compact Complex Manifolds. Proceedings 1974. Edited by H. Popp. V, 333 pages. 1974.

Vol. 413: M. Bruneau, Variation Totale d'une Fonction. XIV, 332 pages. 1974.

Vol. 414: T. Kambayashi, M. Miyanishi and M. Takeuchi, Unipotent Algebraic Groups. VI, 165 pages. 1974.

Vol. 415: Ordinary and Partial Differential Equations. Proceedings 1974. XVII, 447 pages. 1974.

Vol. 416: M. E. Taylor, Pseudo Differential Operators. IV, 155 pages. 1974.

Vol. 417: H. H. Keller, Differential Calculus in Locally Convex Spaces. XVI, 131 pages. 1974.

Vol. 418: Localization in Group Theory and Homotopy Theory and Related Topics. Battelle Seattle 1974 Seminar. Edited by P. J. Hilton. VI, 172 pages 1974.

Vol. 419: Topics in Analysis. Proceedings 1970. Edited by O. E. Lehto, I. S. Louhivaara, and R. H. Nevanlinna. XIII, 392 pages. 1974.

Vol. 420: Category Seminar. Proceedings 1972/73. Edited by G. M. Kelly. VI, 375 pages. 1974.

Vol. 421: V. Poénaru, Groupes Discrets. VI, 216 pages. 1974.

Vol. 422: J.-M. Lemaire, Algèbres Connexes et Homologie des Espaces de Lacets. XIV, 133 pages. 1974.

Vol. 423: S. S. Abhyankar and A. M. Sathaye, Geometric Theory of Algebraic Space Curves. XIV, 302 pages. 1974.

Vol. 424: L. Weiss and J. Wolfowitz, Maximum Probability Estimators and Related Topics. V, 106 pages. 1974.

Vol. 425: P. R. Chernoff and J. E. Marsden, Properties of Infinite Dimensional Hamiltonian Systems. IV, 160 pages. 1974.

Vol. 426: M. L. Silverstein, Symmetric Markov Processes. X, 287 pages. 1974.

Vol. 427: H. Omori, Infinite Dimensional Lie Transformation Groups. XII, 149 pages. 1974.

Vol. 428: Algebraic and Geometrical Methods in Topology, Proceedings 1973. Edited by L. F. McAuley. XI, 280 pages. 1974.

Vol. 429: L. Cohn, Analytic Theory of the Harish-Chandra C-Fun tion. III, 154 pages. 1974.

Vol. 430: Constructive and Computational Methods for Differ tial and Integral Equations. Proceedings 1974. Edited by D. L. Colt and R. P. Gilbert. VII, 476 pages. 1974.

Vol. 431: Séminaire Bourbaki – vol. 1973/74. Exposés 436–45 IV, 347 pages. 1975.

Vol. 432: R. P. Pflug, Holomorphiegebiete, pseudokonvexe G biete und das Levi-Problem. VI, 210 Seiten. 1975.

Vol. 433: W. G. Faris, Self-Adjoint Operators. VII, 115 page 1975.

Vol. 434: P. Brenner, V. Thomée, and L. B. Wahlbin, Bes Spaces and Applications to Difference Methods for Initial Val Problems. II, 154 pages. 1975.

Vol. 435: C. F. Dunkl and D. E. Ramirez, Representations of Co mutative Semitopological Semigroups. VI, 181 pages. 19

Vol. 436: L. Auslander and R. Tolimieri, Abelian Harmonic Analy Theta Functions and Function Algebras on a Nilmanifold. V, pages. 1975.

Vol. 437: D. W. Masser, Elliptic Functions and Transcenden XIV, 143 pages. 1975.

Vol. 438: Geometric Topology. Proceedings 1974. Edited L. C. Glaser and T. B. Rushing. X, 459 pages. 1975.

Vol. 439: K. Ueno, Classification Theory of Algebraic Variet and Compact Complex Spaces. XIX, 278 pages. 1975

Vol. 440: R. K. Getoor, Markov Processes: Ray Processes a Right Processes. V, 118 pages. 1975.

Vol. 441: N. Jacobson, PI-Algebras. An Introduction. V, 115 pag 1975.

Vol. 442: C. H. Wilcox, Scattering Theory for the d'Alemb Equation in Exterior Domains. III, 184 pages. 1975.

Vol. 443: M. Lazard, Commutative Formal Groups. II, 236 pag 1975.

Vol. 444: F. van Oystaeyen, Prime Spectra in Non-Commutat Algebra. V, 128 pages. 1975.

Vol. 445: Model Theory and Topoi. Edited by F. W. Lawve C. Maurer, and G. C. Wraith. III, 354 pages. 1975.

Vol. 446: Partial Differential Equations and Related Topi Proceedings 1974. Edited by J. A. Goldstein. IV, 389 pag 1975.

Vol. 447: S. Toledo, Tableau Systems for First Order Numb Theory and Certain Higher Order Theories. III, 339 pages. 19

Vol. 448: Spectral Theory and Differential Equations. Proceedin 1974. Edited by W. N. Everitt. XII, 321 pages. 1975.

Vol. 449: Hyperfunctions and Theoretical Physics. Proceedir 1973. Edited by F. Pham. IV, 218 pages. 1975.

Vol. 450: Algebra and Logic. Proceedings 1974. Edited by J. Crossley. VIII, 307 pages. 1975.

Vol. 451: Probabilistic Methods in Differential Equations. Proce dings 1974. Edited by M. A. Pinsky. VII, 162 pages. 1975.

Vol. 452: Combinatorial Mathematics III. Proceedings 19 Edited by Anne Penfold Street and W. D. Wallis. IX, 233 pag 1975.

Vol. 453: Logic Colloquium, Symposium on Logic Held at Bost 1972–73. Edited by R. Parikh. IV, 251 pages. 1975.

Vol. 454: J. Hirschfeld and W. H. Wheeler, Forcing, Arithme Division Rings. VII, 266 pages. 1975.

Vol. 455: H. Kraft, Kommutative algebraische Gruppen und Rin III, 163 Seiten. 1975.

Vol. 456: R. M. Fossum, P. A. Griffith, and I. Reiten, Trivial tensions of Abelian Categories. Homological Algebra of Tri Extensions of Abelian Categories with Applications to R Theory. XI, 122 pages. 1975.

l. 457: Fractional Calculus and Its Applications. Proceedings 74. Edited by B. Ross. VI, 381 pages. 1975.

l. 458: P. Walters, Ergodic Theory – Introductory Lectures. , 198 pages. 1975.

l. 459: Fourier Integral Operators and Partial Differential Equa-ns. Proceedings 1974. Edited by J. Chazarain. VI, 372 pages. 75.

l. 460: O. Loos, Jordan Pairs. XVI, 218 pages. 1975.

l. 461: Computational Mechanics. Proceedings 1974. Edited by T. Oden. VII, 328 pages. 1975.

l. 462: P. Gérardin, Construction de Séries Discrètes p-adiques. ur les séries discrètes non ramifiées des groupes réductifs ployés p-adiques«. III, 180 pages. 1975.

l. 463: H.-H. Kuo, Gaussian Measures in Banach Spaces. VI, 4 pages. 1975.

l. 464: C. Rockland, Hypoellipticity and Eigenvalue Asymptotics. 171 pages. 1975.

l. 465: Séminaire de Probabilités IX. Proceedings 1973/74. Edité r P. A. Meyer. IV, 589 pages. 1975.

l. 466: Non-Commutative Harmonic Analysis. Proceedings 1974. ited by J. Carmona, J. Dixmier and M. Vergne. VI, 231 pages. 1975.

l. 467: M. R. Essén, The Cos $\pi\lambda$ Theorem. With a paper by rister Borell. VII, 112 pages. 1975.

l. 468: Dynamical Systems – Warwick 1974: Proceedings 1973/74. ited by A. Manning. X, 405 pages. 1975.

l. 469: E. Binz, Continuous Convergence on C(X). IX, 140 pages. 75.

l. 470: R. Bowen, Equilibrium States and the Ergodic Theory of osov Diffeomorphisms. III, 108 pages. 1975.

l. 471: R. S. Hamilton, Harmonic Maps of Manifolds with Boundary. 168 pages. 1975.

l. 472: Probability-Winter School. Proceedings 1975. Edited by Ciesielski, K. Urbanik, and W. A. Woyczyński. VI, 283 pages. 75.

l. 473: D. Burghelea, R. Lashof, and M. Rothenberg, Groups of omorphisms of Manifolds. (with an appendix by E. Pedersen) 156 pages. 1975.

l. 474: Séminaire Pierre Lelong (Analyse) Année 1973/74. Edité r P. Lelong. VI, 182 pages. 1975.

l. 475: Répartition Modulo 1. Actes du Colloque de Marseille-miny, 4 au 7 Juin 1974. Edité par G. Rauzy. V, 258 pages. 1975. 75.

l. 476: Modular Functions of One Variable IV. Proceedings 1972. ited by B. J. Birch and W. Kuyk. V, 151 pages. 1975.

l. 477: Optimization and Optimal Control. Proceedings 1974. ited by R. Bulirsch, W. Oettli, and J. Stoer. VII, 294 pages. 1975.

l. 478: G. Schober, Univalent Functions – Selected Topics. V, 0 pages. 1975.

l. 479: S. D. Fisher and J. W. Jerome, Minimum Norm Extremals Function Spaces. With Applications to Classical and Modern alysis. VIII, 209 pages. 1975.

. 480: X. M. Fernique, J. P. Conze et J. Gani, Ecole d'Eté de babilités de Saint-Flour IV–1974. Edité par P.-L. Hennequin. 293 pages. 1975.

. 481: M. de Guzmán, Differentiation of Integrals in R^n. XII, 226 ges. 1975.

. 482: Fonctions de Plusieurs Variables Complexes II. Séminaire nçois Norguet 1974–1975. IX, 367 pages. 1975.

. 483: R. D. M. Accola, Riemann Surfaces, Theta Functions, and lian Automorphisms Groups. III, 105 pages. 1975.

. 484: Differential Topology and Geometry. Proceedings 1974. ted by G. P. Joubert, R. P. Moussu, and R. H. Roussarie. IX, 7 pages. 1975.

Vol. 485: J. Diestel, Geometry of Banach Spaces – Selected Topics. XI, 282 pages. 1975.

Vol. 486: S. Stratila and D. Voiculescu, Representations of AF-Algebras and of the Group U (∞). IX, 169 pages. 1975.

Vol. 487: H. M. Reimann und T. Rychener, Funktionen beschränkter mittlerer Oszillation. VI, 141 Seiten. 1975.

Vol. 488: Representations of Algebras, Ottawa 1974. Proceedings 1974. Edited by V. Dlab and P. Gabriel. XII, 378 pages. 1975.

Vol. 489: J. Bair and R. Fourneau, Etude Géométrique des Espaces Vectoriels. Une Introduction. VII, 185 pages. 1975.

Vol. 490: The Geometry of Metric and Linear Spaces. Proceedings 1974. Edited by L. M. Kelly. X, 244 pages. 1975.

Vol. 491: K. A. Broughan, Invariants for Real-Generated Uniform Topological and Algebraic Categories. X, 197 pages. 1975.

Vol. 492: Infinitary Logic: In Memoriam Carol Karp. Edited by D. W. Kueker. VI, 206 pages. 1975.

Vol. 493: F. W. Kamber and P. Tondeur, Foliated Bundles and Characteristic Classes. XIII, 208 pages. 1975.

Vol. 494: A Cornea and G. Licea. Order and Potential Resolvent Families of Kernels. IV, 154 pages. 1975.

Vol. 495: A. Kerber, Representations of Permutation Groups II. V, 175 pages. 1975.

Vol. 496: L. H. Hodgkin and V. P. Snaith, Topics in K-Theory. Two Independent Contributions. III, 294 pages. 1975.

Vol. 497: Analyse Harmonique sur les Groupes de Lie. Proceedings 1973–75. Edité par P. Eymard et al. VI, 710 pages. 1975.

Vol. 498: Model Theory and Algebra. A Memorial Tribute to Abraham Robinson. Edited by D. H. Saracino and V. B. Weispfenning. X, 463 pages. 1975.

Vol. 499: Logic Conference, Kiel 1974. Proceedings. Edited by G. H. Müller, A. Oberschelp, and K. Potthoff. V, 651 pages 1975.

Vol. 500: Proof Theory Symposion, Kiel 1974. Proceedings. Edited by J. Diller and G. H. Müller. VIII, 383 pages. 1975.

Vol. 501: Spline Functions, Karlsruhe 1975. Proceedings. Edited by K. Böhmer, G. Meinardus, and W. Schempp. VI, 421 pages. 1976.

Vol. 502: János Galambos, Representations of Real Numbers by Infinite Series. VI, 146 pages. 1976.

Vol. 503: Applications of Methods of Functional Analysis to Problems in Mechanics. Proceedings 1975. Edited by P. Germain and B. Nayroles. XIX, 531 pages. 1976.

Vol. 504: S. Lang and H. F. Trotter, Frobenius Distributions in GL_2-Extensions. III, 274 pages. 1976.

Vol. 505: Advances in Complex Function Theory. Proceedings 1973/74. Edited by W. E. Kirwan and L. Zalcman. VIII, 203 pages. 1976.

Vol. 506: Numerical Analysis, Dundee 1975. Proceedings. Edited by G. A. Watson. X, 201 pages. 1976.

Vol. 507: M. C. Reed, Abstract Non-Linear Wave Equations. VI, 128 pages. 1976.

Vol. 508: E. Seneta, Regularly Varying Functions. V, 112 pages. 1976.

Vol. 509: D. E. Blair, Contact Manifolds in Riemannian Geometry. VI, 146 pages. 1976.

Vol. 510: V. Poènaru, Singularités C^∞ en Présence de Symétrie. V, 174 pages. 1976.

Vol. 511: Séminaire de Probabilités X. Proceedings 1974/75. Edité par P. A. Meyer. VI, 593 pages. 1976.

Vol. 512: Spaces of Analytic Functions, Kristiansand, Norway 1975. Proceedings. Edited by O. B. Bekken, B. K. Øksendal, and A. Stray. VIII, 204 pages. 1976.

Vol. 513: R. B. Warfield, Jr. Nilpotent Groups. VIII, 115 pages. 1976.

Vol. 514: Séminaire Bourbaki vol. 1974/75. Exposés 453 – 470. IV, 276 pages. 1976.

Vol. 515: Bäcklund Transformations. Nashville, Tennessee 1974. Proceedings. Edited by R. M. Miura. VIII, 295 pages. 1976.